# COMPREHENSIVE BIOCHEMISTRY

ELSEVIER BIOMEDICAL PRESS

1 Molenwerf, P.O. Box 211, Amsterdam

ELSEVIER NORTH-HOLLAND INC.

52, Vanderbilt Avenue, New York, N.Y. 10017

With 165 figures and 48 tables

Library of Congress Cataloging in Publication Data

Main entry unter title:

Protein metabolism.

   (Comprehensive biochemistry, ISSN 0-444-80151-0;
v. 19B, pt. 2. Section IV, Metabolism)
   Bibliography: p.
   Includes Index.
   1. Proteins--Metabolism.  2. Metabolic regulation.
I. Neuberger, Albert.  II. Deenen, Laurens L. M. van.
III. Series: Comprehensive biochemistry; v. 19B, pt. 2.
[DNLM: 1. Proteins--Metabolism.  QU 4 C743 v. 19B pt. 1 etc.]
QD415.F54  Sect. 4, vol. 19B, pt. 2 [QP551]   81-19530
ISBN 0-444-80346-7  574.19′2s  [574.19′245]  AACR2

COPYRIGHT © 1982 BY ELSEVIER BIOMEDICAL PRESS,
AMSTERDAM
ALL RIGHTS RESERVED. NO PART OF THIS PUBLICATION MAY BE
REPRODUCED, STORED IN A RETRIEVAL SYSTEM, OR TRANSMITTED
IN ANY FORM OR BY ANY MEANS, ELECTRONIC, MECHANICAL,
PHOTOCOPYING, RECORDING, OR OTHERWISE, WITHOUT THE PRIOR
WRITTEN PERMISSION OF THE PUBLISHER,
ELSEVIER BIOMEDICAL PRESS,
MOLENWERF 1, AMSTERDAM

PRINTED IN THE NETHERLANDS

# COMPREHENSIVE BIOCHEMISTRY

### ADVISORY BOARD

Sir RUDOLPH A. PETERS, M.C., M.D., D.Sc., F.R.S.
Emeritus Professor of Biochemistry, Oxford; Department of Biochemistry,
Cambridge
*Chairman*

C.F. CORI, M.D., D.Sc.
Professor of Biochemistry, Massachusetts General Hospital, Boston, Mass.

E.F. GALE, D.Sc., F.R.S.
Professor of Chemical Microbiology, University of Cambridge

A. BAIRD HASTINGS, B.Sc., Ph.D., D.Sc.
Department of Neurosciences,
University of California, San Diego, La Jolla, Calif. 92093

E. LEDERER, Ph.D., D.Sc.
Professor of Biochemistry, Faculty of Science, University of Paris

S. OCHOA, B.A., M.D., Hon.LL.D., Hon.D.Sc.
Roche Institute of Molecular Biology, Nutley, N.J. 07110

J. ROCHE, D.Sc.
Professor of General and Comparative Biochemistry, Collège de France, Paris

KENNETH V. THIMANN, Ph.D., F.R.S. (Foreign Member)
Professor of Biology, University of California, Santa Cruz, Calif.

Sir FRANK YOUNG, M.A., D.Sc., F.R.S.
Emeritus Professor of Biochemistry, University of Cambridge

# COMPREHENSIVE BIOCHEMISTRY

SECTION I (VOLUMES 1—4)
PHYSICO-CHEMICAL AND ORGANIC ASPECTS
OF BIOCHEMISTRY

SECTION II (VOLUMES 5—11)
CHEMISTRY OF BIOLOGICAL COMPOUNDS

SECTION III (VOLUMES 12—16)
BIOCHEMICAL REACTION MECHANISMS

SECTION IV (VOLUMES 17—21)
METABOLISM

SECTION V (VOLUMES 22—29)
CHEMICAL BIOLOGY

SECTION VI (VOLUMES 30—34)
A HISTORY OF BIOCHEMISTRY

# COMPREHENSIVE BIOCHEMISTRY

ALBERT NEUBERGER
*Chairman of Governing Body, The Lister Institute
of Preventive Medicine, University of London,
London (Great Britain)*

LAURENS L.M. VAN DEENEN
*Professor of Biochemistry, Biochemical Laboratory,
Utrecht (The Netherlands)*

VOLUME 19B Part II

PROTEIN METABOLISM

ELSEVIER SCIENTIFIC PUBLISHING COMPANY

AMSTERDAM . OXFORD . NEW YORK

1982

# CONTRIBUTORS TO THIS VOLUME

R. CASEY
John Innes Institute, Colney Lane, Norwich NR4 7UH (U.K.)

P. FEIGELSON
Department of Biochemistry and the Institute of Cancer Research,
Columbia University, New York, NY 10032 (U.S.A.)

G.H. GOODWIN
Institute of Cancer Research, Royal Cancer Hospital, Chester Beatty Research
Institute, Fulham Road, London SW3 6JB (U.K.)

A.M. GOTTO Jr.
Department of Medicine, The Methodist Hospital and Baylor College
of Medicine, Houston, TX 77030 (U.S.A.)

E.W. JOHNS
Institute of Cancer Research, Royal Cancer Hospital, Chester Beatty Research
Institute, Fulham Road, London SW3 6JB (U.K.)

L.A. KILLEWICH
Department of Biochemistry and the Institute of Cancer Research,
Columbia University, New York, NY 10032 (U.S.A.)

H. LEHMANN
University of Cambridge, Department of Biochemistry, Tennis Court Road,
Cambridge CB2 1QW (U.K.)

J.P. MILLER
University Hospital of South Manchester, Manchester
M20 8LR (U.K.)

J. MONTREUIL
Laboratoire de Chimie Biologique et Laboratoire Associé au C.N.R.S.
No. 217, Université des Sciences et Techniques de Lille I,
Villeneuve d'Ascq (France)

B.J. SMITH
Institute of Cancer Research, Royal Cancer Hospital, Chester Beatty Research
Institute, Fulham Road, London SW3 6JB (U.K.)

D.I. STOTT
Department of Bacteriology and Immunology, and the Department of
Biochemistry, University of Glasgow, Glasgow, Scotland (U.K.)

J.M. WALKER
Institute of Cancer Research, Royal Cancer Hospital, Chester Beatty Research Institute, Fulham Road, London SW3 6JB (U.K.)

A.R. WILLIAMSON
Glaxo Group Research Ltd., Greenford Road, Greenford, Middlesex UB6 0HE (U.K.)

# GENERAL PREFACE

The Editors are keenly aware that the literature of Biochemistry is already very large, in fact so widespread that it is increasingly difficult to assemble the most pertinent material in a given area. Beyond the ordinary textbook the subject matter of the rapidly expanding knowledge of biochemistry is spread among innumerable journals, monographs, and series of reviews. The Editors believe that there is a real place for an advanced treatise in biochemistry which assembles the principal areas of the subject in a single set of books.

It would be ideal if an individual or a small group of biochemists could produce such an advanced treatise, and within the time to keep reasonably abreast of rapid advances, but this is least difficult if not impossible. Instead, the Editors with the advice of the Advisory Board, have assembled what they consider the best possible sequence of chapters written by competent authors; they must take the responsibility for inevitable gaps of subject matter and duplication which may result from this procedure.

Most evident to the modern biochemists, apart from the body of knowledge of the chemistry and metabolism of biological substances, is the extent to which we must draw from recent concepts of physical and organic chemistry, and in turn project into the vast field of biology. Thus in the organization of Comprehensive Biochemistry, sections II, III and IV, Chemistry of Biological Compounds, Biochemical Reaction Mechanisms, and Metabolism may be considered classical biochemistry, while the first and fifth sections provide selected material on the origins and projections of the subject.

It is hoped that sub-division of the sections into bound volumes will not only be convenient, but will find favour among students concerned with specialized areas, and will permit easier future revisions of the individual volumes. Towards the latter end particularly, the Editors will welcome all comments in their effort to produce a useful and efficient source of biochemical knowledge.

Liège/Rochester

M. Florkin[†]
E.H. Stotz

# PREFACE TO VOLUME 19B PART II

The second and final part of our treatment of protein metabolism consists of six chapters. We wish to repeat briefly the policy which we have already indicated in the preface to the first volume. The Editors have selected special areas of investigation which appear to be important and topical, and in which significant advances have been made in the last ten or fifteen years. This applies, for instance, to the chapter on lipoproteins where both structure and function and their interrelationship is much better understood now than it was twenty years ago. Particularly fast has been the development in our knowledge of glycoproteins, where again an intimate relationship between the structure and the function has been demonstrated. We also felt some emphasis should be given to the important advances which have been made in our knowledge of chromosomal proteins which comprise both histones and non-basic proteins. The topic of haemoglobin also continues to attract a great deal of interest for a variety of reasons, and this also applies to the chemistry and function of immunoglobulins. Enzyme induction is also a field of some importance, and we have chosen to concentrate on one particular enzyme, the biosynthesis of which commands widespread interest.

We are conscious that we have neglected other aspects of protein metabolism, but to aim at complete coverage would have presented an impossible task.

L.L.M. van Deenen
A. Neuberger

# CONTENTS

## VOLUME 19B PART II

## PROTEIN METABOLISM

General Preface .................................... viii
Preface to Volume 19B Part II ....................... ix

### Chapter 1. Glycoproteins
### by JEAN MONTREUIL

1. Introduction. ..................................... 1
2. Historical development of the knowledge and concepts of glycoproteins. ......................................... 3
   2.1. The early history. .............................. 4
   2.2. Middle Ages and Renaissance of glycoproteins .......... 14
   2.3. The revolution of 1968 and the birth of the molecular biology of glycoconjugates. ............................ 20
3. Definition and classification of glycans and glycoproteins ....... 23
4. Structure and conformation of glycoprotein glycans .......... 27
   4.1. Methods ...................................... 27
      4.1.1. Isolation of glycoproteins and glycans ............ 27
         4.1.1.1. Isolation of glycoproteins .............. 28
         4.1.1.2. Isolation of glycans .................. 30
      4.1.2. Determination of the primary structure of glycans. ... 30
         4.1.2.1. Sugar composition of glycans. ........... 31
         4.1.2.2. Methods for the determination of the primary structure of glycans .................. 31
   4.2. Nature of linkages between glycans and protein. .......... 40
      4.2.1. $N$-Glycosidic linkages (glycosylamine linkages) ...... 43
      4.2.2. $O$-Glycosidic linkages ......................... 44
         4.2.2.1. Linkages to L-serine and L-threonine ....... 44
         4.2.2.2. Linkages to 5-hydroxy-L-lysine .......... 46
         4.2.2.3. Linkage to hydroxy-L-proline ........... 46
         4.2.2.4. Unusual linkages .................... 47
      4.2.3. Concluding remarks. ......................... 47
   4.3. Primary structure of glycoprotein glycans .............. 49
      4.3.1. Concepts and rules. .......................... 49

|   |   |   |   |
|---|---|---|---|
| | | 4.3.1.1. $n$-Glycans and isoglycans. . . . . . . . . . . . . . . | 49 |
| | | 4.3.1.2. The concept of the common "inner-core" and of the $inv$ fraction of glycans. . . . . . . . . . . . | 50 |
| | | 4.3.1.3. The concept of the antenna and of the $var$-fraction of glycans. . . . . . . . . . . . . . . . . . | 50 |
| | | 4.3.1.4. Oligomannosidic and $N$-acetyl-lactosaminic type structures of glycans of the $N$-glycosyl-proteins. . . . . . . . . . . . . . . . . . . . . . . . | 51 |
| | | 4.3.1.5. Substitution rules . . . . . . . . . . . . . . . . . . | 52 |
| | 4.3.2. | Description of glycan structures present in $O$-glycosyl-proteins. . . . . . . . . . . . . . . . . . . . . . . . . . . . . . . . . | 53 |
| | | 4.3.2.1. Glycans conjugated through an $N$-acetyl-$\beta$-D-galactosaminyl-L-serine or L-threonine linkage. | 58 |
| | | 4.3.2.2. Glycans conjugated through a $\beta$-D-xylosyl-L-serine linkage . . . . . . . . . . . . . . . . . . . . . | 58 |
| | | 4.3.2.3. Glycans conjugated through a $\beta$-D-galactosyl-hydroxy-L-lysine linkage. . . . . . . . . . . . . . | 63 |
| | | 4.3.2.4. Glycans conjugated through a $\beta$-L-arabofuran-nosyl-hydroxy-L-proline linkage . . . . . . . . . | 63 |
| | 4.3.3. | Description of glycan structures present in $N$-glycosyl-proteins. . . . . . . . . . . . . . . . . . . . . . . . . . . . . . . . . | 63 |
| | | 4.3.3.1. Structure of glycans of the oligomannosidic type (or mannose-rich type) . . . . . . . . . . . . | 63 |
| | | 4.3.3.2. Structure of glycans of the $N$-acetyl-lactos-aminic type (or complex type). . . . . . . . . . . | 64 |
| | | 4.3.3.3. Structure of glycans of the mixed (or hybrid) type . . . . . . . . . . . . . . . . . . . . . . . . . . . . | 73 |
| | | 4.3.3.4. Unusual structures. . . . . . . . . . . . . . . . . . | 74 |
| 4.4. Microheterogeneity of glycans . . . . . . . . . . . . . . . . . . . . . . | | | 77 |
| 4.5. Spatial conformation of glycans . . . . . . . . . . . . . . . . . . . . | | | 80 |
| 5. Some selected glycoproteins . . . . . . . . . . . . . . . . . . . . . . . . | | | 85 |
| 5.1. Human serotransferrin . . . . . . . . . . . . . . . . . . . . . . . . | | | 86 |
| 5.2. $\alpha_1$-Acid glycoprotein . . . . . . . . . . . . . . . . . . . . . . . . | | | 87 |
| 5.3. Thyroglobulin. . . . . . . . . . . . . . . . . . . . . . . . . . . . . . . | | | 87 |
| 5.4. Proteoglycans . . . . . . . . . . . . . . . . . . . . . . . . . . . . . . . | | | 89 |
| 5.5. Glycophorin A . . . . . . . . . . . . . . . . . . . . . . . . . . . . . | | | 90 |
| 6. Glycoprotein biosynthesis . . . . . . . . . . . . . . . . . . . . . . . . . . | | | 92 |
| 6.1. The reaction and its partners . . . . . . . . . . . . . . . . . . . . . | | | 93 |
| | 6.1.1. | Activated forms of sugar donors. . . . . . . . . . . . . . . | 93 |
| | | 6.1.1.1. Glycosyl nucleotides . . . . . . . . . . . . . . . . | 93 |
| | | 6.1.1.2. Lipid intermediates . . . . . . . . . . . . . . . . | 94 |
| | 6.1.2. | Glycosyltransferase systems. . . . . . . . . . . . . . . . . | 97 |
| | | 6.1.2.1. The enzymes . . . . . . . . . . . . . . . . . . . . | 97 |
| | | 6.1.2.2. Cellular localization of glycosyltransferases . . | 99 |
| | 6.1.3. | Acceptors . . . . . . . . . . . . . . . . . . . . . . . . . . . . . | 100 |
| 6.2. Control of glycoprotein biosynthesis. . . . . . . . . . . . . . . . . . | | | 101 |
| | 6.2.1. | General mechanisms. . . . . . . . . . . . . . . . . . . . . . | 101 |

    6.2.1.1. Genetic control.................... 101
    6.2.1.2. Control of substrate availability.......... 101
    6.2.1.3. Involvement of glycosidases............. 102
    6.2.1.4. Control by the peptide-chain conformation .. 103
  6.2.2. Control of the biosynthesis of N-glycosylproteins .... 107
    6.2.2.1. Glycan primary structure as a guide for glycosylation ........................ 107
    6.2.2.2. Maturation of glycans................. 110
  6.2.3. Biosynthesis of O-glycosylproteins .............. 113
    6.2.3.1. Mucin and mucin-like glycoproteins ....... 113
    6.2.3.2. Proteoglycans...................... 115
    6.2.3.3. Collagen-type glycoproteins............. 118
    6.2.3.4. Extensin-type glycoproteins ............ 119
7. Glycoprotein catabolism and pathology.................... 119
 7.1. Enzymes involved in the catabolism of glycoproteins ....... 120
  7.1.1. Proteolytic enzymes......................... 120
  7.1.2. Specific enzymes involved in degradation of glycans. .. 120
 7.2. Catabolic pathway of glycoproteins .................. 123
  7.2.1. Uptake and catabolism of glycoproteins........... 125
  7.2.2. Traffic of lysosomal glycosidases ............... 126
 7.3. Genetic disorders of glycoprotein catabolism ............ 129
8. Molecular biology and role of glycoprotein glycans ........... 136
 8.1. Induction of protein conformation................... 142
 8.2. Protection of proteins against proteolytic attack .......... 145
 8.3. Control of membrane permeability................... 145
 8.4. Control of protein secretion from the cell .............. 146
 8.5. Glycans as recognition signals — Membrane lectins......... 147
 8.6. Concluding remarks............................. 151
9. Conclusions ..................................... 154
10. Acknowledgements ................................ 157
References........................................ 158

*Chapter 2. Immunoglobulins and Histocompatibility Antigens.*
*Their Structure, Function and Metabolism*
by DAVID I. STOTT and ALAN R. WILLIAMSON

Immunoglobulins and histocompatibility antigens ............... 189
1. Immunoglobulins ................................. 191
 1.1. Introduction................................. 191
 1.2. Structure of immunoglobulins ...................... 192
  1.2.1. Polypeptide chain structure................... 192
    1.2.1.1. IgG........................... 193
    1.2.1.2. IgA........................... 193
    1.2.1.3. IgM........................... 198
    1.2.1.4. IgD and IgE ..................... 199

1.2.2. Primary structure ........................... 199
    1.2.2.1. Variable and constant regions ........... 199
    1.2.2.2. Variable region families, groups and subgroups . 200
    1.2.2.3. Hypervariable regions ................. 201
    1.2.2.4. The constant region .................. 205
    1.2.2.5. Light-chain types .................... 205
    1.2.2.6. Heavy-chain classes and subclasses ........ 207
    1.2.2.7. Carbohydrate ....................... 208
    1.2.2.8. Allotypes .......................... 209
    1.2.2.9. Idiotypes .......................... 211
1.2.3. The antigen-binding site ...................... 214
    1.2.3.1. Valency ............................ 214
    1.2.3.2. Location of the antigen-binding site ....... 215
    1.2.3.3. Size and shape of the antigen-binding site. ... 215
    1.2.3.4. Affinity labelling .................... 217
1.2.4. Tertiary structure ........................... 221
    1.2.4.1. Electron microscopy .................. 221
    1.2.4.2. X-Ray crystallography ................ 228
    1.2.4.3. Model building and spectroscopic methods. ... 235
1.3. Function ......................................... 241
    1.3.1. Introduction ............................... 241
    1.3.2. Antigen binding ............................ 243
    1.3.3. Class of antibody and the immune response ........ 245
    1.3.4. Membrane transmission ...................... 245
    1.3.5. Binding to cell membranes .................... 247
        1.3.5.1. Lymphocytes ....................... 247
        1.3.5.2. Monocytes and macrophages ............ 247
        1.3.5.3. Neutrophils ........................ 248
        1.3.5.4. Basophils and mast cells .............. 248
    1.3.6. Complement fixation ........................ 248
1.4. Biosynthesis of immunoglobulins ....................... 251
    1.4.1. Introduction — Cells involved in antibody formation .. 251
    1.4.2. Polyribosomes synthesising immunoglobulin chains ... 252
    1.4.3. Cell-free synthesis .......................... 253
    1.4.4. Addition of carbohydrate ..................... 254
    1.4.5. Assembly of four chain structures ............... 255
        1.4.5.1. Non-covalent assembly ............... 255
        1.4.5.2. Covalent assembly ................... 256
    1.4.6. Assembly of polymeric immunoglobulin ........... 257
        1.4.6.1. IgM .............................. 257
        1.4.6.2. IgA .............................. 260
    1.4.7. Synthesis of immunoglobulin during the cell cycle ... 260
    1.4.8. Hybrid cells ............................... 261
    1.4.9. Variable and constant region genes .............. 262
    1.4.10. Cell membrane immunoglobulins ................ 267
        1.4.10.1. General properties .................. 267
        1.4.10.2. Structure ......................... 270
        1.4.10.3. Biosynthesis and turnover ............ 273

    1.5. Catabolism.................................... 274
        1.5.1. Definitions and methodology.................. 274
        1.5.2. Catabolic rates of immunoglobulin classes and subclasses 276
        1.5.3. Control of immunoglobulin catabolism ........... 278
2. Histocompatibility antigens............................ 279
    2.1. Introduction................................... 279
    2.2. Genetics and serology of the major histocompatibility complex . 281
        2.2.1. HLA and H-2............................. 281
        2.2.2. Polymorphism ............................ 284
        2.2.3. Linkage disequilibrium ..................... 287
    2.3. Structure of the products of the major histocompatibility locus . 288
        2.3.1. Major transplantation antigens................. 288
        2.3.2. Immune associated (Ia) antigens................ 292
        2.3.3. Amino acid sequences ...................... 296
    2.4. Functions of MHC products ...................... 301
        2.4.1. Functions as histocompatibility antigens .......... 301
        2.4.2. MHC restriction .......................... 302
        2.4.3. Recognition of self ........................ 304
        2.4.4. Prevalence of allo-reactive T cells .............. 305
        2.4.5. Immune response (Ir) genes................... 307
        2.4.6. Ia antigen functions....................... 310
3. Conclusions ........................................ 313
Note added in proof................................... 314
References........................................... 315


## Chapter 3. Regulation of the Messenger RNA for Hepatic Tryptophan 2,3-Dioxygenase
### by LOIS A. KILLEWICH and PHILIP FEIGELSON

Abstract .............................................. 331

References............................................ 344


## Chapter 4. Human Haemoglobin
### by H. LEHMANN and R. CASEY

Haemoglobin ......................................... 347
1. Distribution of haemoglobin .......................... 347
2. Haemoglobin structure and function .................... 349
    2.1. Introduction.................................... 349
    2.2. The structure of haemoglobin ...................... 351
        2.2.1. Haem structure........................... 351
        2.2.2. Protein structure.......................... 352

    2.2.2.1. Primary structure .................... 352
    2.2.2.2. Secondary structure................... 356
    2.2.2.3. Tertiary structure ................... 356
    2.2.2.4. Quaternary structure ................. 359
  2.3. Haemoglobin function in relation to structure............ 365
    2.3.1. $O_2$ affinity.............................. 365
    2.3.2. Co-operativity........................... 368
    2.3.3. The Bohr effect........................... 372
    2.3.4. Carbon dioxide binding .................. 374
    2.3.5. 2,3-Diphosphoglycerate binding............... 374
    2.3.6. Allostery ................................ 375
    2.3.7. Methaemoglobin.......................... 377
    2.3.8. Haemichrome............................ 377
    2.3.9. Superoxide dismutase...................... 378
3. Variation in haemoglobin structure....................... 379
  3.1. Genetic variability................................ 379
    3.1.1. Single point mutants ........................ 379
    3.1.2. Termination errors (elongated subunits)............ 381
    3.1.3. Frame shift variants......................... 381
    3.1.4. Deletions and insertions in phase ............... 383
    3.1.5. Nonsense mutants........................... 383
    3.1.6. Fusion variants............................ 384
  3.2. Non-genetic variation.............................. 386
  3.3. Distribution of different haemoglobins in man ........... 386
4. The effect of mutations on haemoglobin structure and function and the concept of "molecular pathology".................... 387
  4.1. Mutations which affect the stability of the tertiary or quaternary structure ....................................... 388
    4.1.1. Substitutions in the haem pocket ............... 388
    4.1.2. Replacement of an important non-polar residue by a polar residue............................... 388
    4.1.3. Replacements by proline..................... 389
    4.1.4. Replacements at tightly packed regions of the tetramer 389
  4.2. Mutations which affect the oxygen affinity ............. 390
    4.2.1. Haemoglobins M........................... 390
    4.2.2. Low-affinity variants ........................ 391
      4.2.2.1. Haemoglobins Kansas and Titusville ....... 391
      4.2.2.2. Haemoglobin Agenogi ................. 392
    4.2.3. High-affinity variants ....................... 393
      4.2.3.1. $\alpha_1\beta_2$ Interface mutations................ 393
      4.2.3.2. Haem pocket mutations ............... 393
      4.2.3.3. $\beta$-$\beta$ Interface mutations.................. 393
      4.2.3.4. 2,3-DPG-binding site mutations .......... 393
  4.3. Sickle-cell haemoglobin (HbS) ($\beta$6 Glu → Val) ........... 394
5. Haemoglobin biosynthesis................................ 396
  5.1. The genetics of normal human haemoglobin.............. 396
  5.2. Globin gene structure.............................. 397

5.3. Ontogeny of haemoglobin synthesis .................. 399
5.4. The control of globin synthesis ..................... 400
   5.4.1. Control at the gene and transcriptional level ........ 400
   5.4.2. Control at the translational level. ............... 401
   5.4.3. Control at the level of tetramer assembly .......... 402
5.5. Post-translational modification of haemoglobin .......... 403
6. The molecular basis of thalassaemia ..................... 403
   6.1. α-Thalassaemia ................................ 404
   6.2. β-Thalassaemia ................................ 405
      6.2.1. $β^+$-Thalassaemia ........................... 405
      6.2.2. $β^°$-Thalassaemia ........................... 406
   6.3. Haemoglobin Lepore ($α_2δβ_2$) ...................... 406
   6.4. Hereditary persistence of foetal haemoglobin (HPFH) ...... 406
   6.5. β-Thallassaemia due to structural variants ............. 407
7. The evolution of haemoglobin ......................... 407
Addendum (December 1980 and October 1981) ............... 409

References ............................................ 411

## Chapter 5. The Plasma Lipoproteins — their Formation and Metabolism
### by J. PAUL MILLER and ANTONIO M. GOTTO JR.

1. Introduction. ...................................... 419
2. Lipoprotein composition ............................. 420
   2.1. Chylomicrons ................................. 422
   2.2. Very low density lipoproteins ..................... 423
   2.3. Intermediate density lipoproteins .................. 423
   2.4. Low density lipoproteins ......................... 423
   2.5. High-density lipoproteins ........................ 424
   2.6. Lp(a) lipoproteins .............................. 424
   2.7. Lipoprotein-X. ................................ 426
3. The apolipoproteins. ................................ 428
   3.1. Apolipoprotein A-I ............................. 429
   3.2. Apolipoprotein A-II. ............................ 430
   3.3. Apolipoprotein B .............................. 430
   3.4. Apolipoprotein C .............................. 431
   3.5. Thin-line polypeptide ........................... 432
   3.6. Apolipoprotein E .............................. 432
   3.7. Proline-rich protein ............................ 433
4. Enzymes and exchange proteins ....................... 433
   4.1. Plasma triglyceride hydrolases .................... 433
      4.1.1. Lipoprotein lipase .......................... 433
      4.1.2. Hepatic triglyceride lipase ................... 436
   4.2. Lecithin: cholesterol acyltransferase ................ 437
   4.3. Cholesteryl ester exchange protein ................. 442

- 4.4. Triglyceride exchange protein ... 443
- 4.5. Phospholipid exchange proteins ... 443
- 5. Lipoprotein formation and secretion ... 444
  - 5.1. The intestine ... 444
    - 5.1.1. Chylomicrons and VLDL ... 445
    - 5.1.2. LDL and HDL ... 448
    - 5.1.3. Apolipoprotein secretion ... 448
      - 5.1.3.1. Apolipoprotein A-I ... 448
      - 5.1.3.2. Apolipoprotein A-II ... 449
      - 5.1.3.3. Apolipoprotein A-IV ... 449
      - 5.1.3.4. Apolipoprotein B ... 449
      - 5.1.3.5. C-apolipoproteins ... 450
      - 5.1.3.6. Apolipoprotein E ... 451
  - 5.2. The liver ... 451
    - 5.2.1. Very low density lipoproteins ... 451
    - 5.2.2. LDL and HDL ... 452
    - 5.2.3. Apolipoprotein secretion ... 453
      - 5.2.3.1. Apolipoprotein A-I ... 453
      - 5.2.3.2. Apolipoprotein A-IV ... 453
      - 5.2.3.3. Apolipoprotein B ... 453
      - 5.2.3.4. C-apolipoproteins ... 454
      - 5.2.3.5. Apolipoprotein E ... 454
    - 5.2.4. Molecular biology of apolipoprotein synthesis ... 454
- 6. Lipoprotein metabolism ... 455
  - 6.1. Triglyceride-rich lipoproteins ... 455
    - 6.1.1. Remnant formation ... 459
    - 6.1.2. Remnant removal by the liver ... 460
  - 6.2. Formation of low density lipoproteins ... 462
  - 6.3. The fate of plasma low density lipoproteins ... 463
  - 6.4. Kinetics of apolipoprotein B metabolism ... 466
  - 6.5. Metabolism of high density lipoproteins ... 468
  - 6.6. Kinetics of apolipoprotein A-I and A-II metabolism ... 471
  - 6.7. Metabolism of the C-apolipoproteins ... 473
  - 6.8. Metabolism of apolipoprotein E ... 474
  - 6.9. Metabolism of lipoproteins by cultured cells ... 476
    - 6.9.1. Low density lipoproteins ... 476
      - 6.9.1.1. LDL binding ... 477
      - 6.9.1.2. LDL uptake and degradation ... 479
      - 6.9.1.3. Regulation of cellular cholesterol metabolism ... 480
      - 6.9.1.4. The LDL pathway in vivo ... 482
      - 6.9.1.5. Non-specific uptake of LDL ... 483
    - 6.9.2. High density lipoproteins ... 484
    - 6.9.3. Interaction between high and low density lipoproteins ... 485
    - 6.9.4. Very low density lipoproteins ... 485
- 7. Conclusion ... 486

8. Acknowledgements ................................. 488
References ........................................ 489

## Chapter 6. The Chromosomal Proteins
by JOHN M. WALKER, GRAHAM H. GOODWIN, BRYAN J. SMITH and ERNEST W. JOHNS

1. Introduction. ................................. 507
   1.1. Historical. ............................... 507
   1.2. Definition ............................... 511
2. The histones. ................................ 512
   2.1. Occurrence and specificity ................ 512
   2.2. Preparation ............................. 514
   2.3. Amino acid composition. .................. 516
   2.4. Primary structures. ...................... 517
       2.4.1. Histone H4. ....................... 518
       2.4.2. Histone H3. ....................... 519
       2.4.3. Histone H2A ...................... 521
       2.4.4. Histone H2B. ..................... 523
       2.4.5. Histone H1. ...................... 523
       2.4.6. Histone H5. ...................... 526
       2.4.7. Histone variants ................. 527
   2.5. Post-synthetic modifications ............. 528
       2.5.1. Acetylation ..................... 528
       2.5.2. Methylation ..................... 529
       2.5.3. Phosphorylation ................. 529
       2.5.4. Poly ADP-ribosylation ........... 530
   2.6. Function ............................... 532
3. The HMG proteins. ........................... 535
   3.1. Introduction. .......................... 535
   3.2. Isolation .............................. 538
   3.3. Properties of the calf proteins ......... 539
       3.3.1. Amino acid analyses. ............ 539
       3.3.2. Microheterogeneity and modifications. 540
       3.3.3. Secondary and tertiary structures. 543
       3.3.4. Interactions with DNA. .......... 543
   3.4. Occurrence and specificity ............. 544
   3.5. Primary structure ...................... 547
       3.5.1. Calf thymus HMG14 ............... 547
       3.5.2. Calf thymus HMG17 ............... 549
       3.5.3. Trout testis protein H6. ........ 552
       3.5.4. Trout testis HMG-T .............. 552
       3.5.5. Calf thymus HMG1 and 2 .......... 554
       3.5.6. Sequence homologies witnin the HMG proteins. 554
       3.5.7. Other HMG proteins .............. 556
   3.6. Chromosomal and cellular localisation. .. 556

|         3.7. Functions ................................. 557
|      4. Other non-histone proteins ........................ 559
|         4.1. Protein A24 ................................ 559
|         4.2. Nucleolar protein C23 ....................... 560
|         4.3. Ubiquitin .................................. 561
|         4.4. Protein BA ................................. 563
|         4.5. Protein C14 ................................ 563
|         4.6. Nucleosome assembly factor ................... 563
|         4.7. Component 10 ............................... 564
|         4.8. Contractile proteins ......................... 564
|         4.9. DNA and RNA polymerases .................... 564
|              4.9.1. DNA polymerases ..................... 564
|              4.9.2. RNA polymerases ..................... 564
| References .......................................... 565
| *Subject Index* ........................................ 575

# COMPREHENSIVE BIOCHEMISTRY

*Section I — Physico-Chemical and Organic Aspects of Biochemistry*
Volume  1. Atomic and molecular structure
Volume  2. Organic and physical chemistry
Volume  3. Methods for the study of molecules
Volume  4. Separation methods

*Section II — Chemistry of Biological Compounds*
Volume  5. Carbohydrates
Volume  6. Lipids — Amino acids and related compounds
Volume  7. Proteins (Part 1)
Volume  8. Proteins (Part 2) and nucleic acids
Volume  9. Pyrrole pigments, isoprenoid compounds, phenolic plant constituents
Volume 10. Sterols, bile acids and steroids
Volume 11. Water-soluble vitamins, hormones, antibiotics

*Section III — Biochemical Reaction Mechanisms*
Volume 12. Enzymes — general considerations
Volume 13. (third edition). Enzyme nomenclature (1972)
Volume 14. Biological oxidations
Volume 15. Group-transfer reactions
Volume 16. Hydrolytic reactions; cobamide and boitin coenzymes

*Section IV — Metabolism*
Volume 17. Carbohydrate metabolism
Volume 18. Lipid metabolism
Volume 19. Metabolism of amino acids, proteins, purines, and pyrimidines
Volume 20. Metabolism of cyclic compounds
Volume 21. Metabolism of vitamins and trace elements

*Section V — Chemical Biology*
Volume 22. Bioenergetics
Volume 23. Cytochemistry
Volume 24. Biological information transfer
Volume 25. Regulatory functions — Mechanisms of hormone action
Volume 26. Part A. Extracellular and supporting structures
Volume 26. Part B. Extracellular and supporting structures (continued)
Volume 26. Part C. Extracellular and supporting structures (continued)
Volume 27. Photobiology, ionizing radiations
Volume 28. Morphogenesis, differentiation and development
Volume 29. Part A. Comparative biochemistry, molecular evolution
Volume 29. Part B. Comparative biochemistry, molecular evolution (continued)

*Section VI — A History of Biochemistry*
Volume 30. *Part I.* Proto-biochemistry
*Part II.* From proto-biochemistry to biochemistry
Volume 31. *Part III.* Identification of the sources of free energy
Volume 32. *Part IV.* Early studies on biosynthesis
Volume 33. *Part V.* The unravelling of biosynthetic pathways
Volume 34. *Part VI.* History of molecular interpretations of physiological and biological concepts, and of the origins of the conception of life as the expression of a molecular order

*Chapter 1*

# Glycoproteins

JEAN MONTREUIL

*Laboratoire de Chimie Biologique et Laboratoire Associé au C.N.R.S. No. 217, Université des Sciences et Techniques de Lille I, Villeneuve d'Ascq (France)*

In Memoriam Alfred Gottschalk

## 1. Introduction

Glycoproteins* which, together with glycolipids, constitute the class of glycoconjugates, were defined in 1908 by the Committee on Protein Nomenclature of the American Physiological Society and the American Society of Biochemists as *compounds of the protein molecule with a substance or substances containing a carbohydrate group, other than nucleic acid*. They result from the classical processes of translation of the genome message at the polysomal level, leading to the birth of a peptide chain to which carbohydrate moieties, called "glycans", are linked in a second step through covalent bonds.

Glycosylation of proteins represent one of the most important, if not the most important, of the post-translational events because of the universality of the phenomenon. In fact, most proteins are glycosylated, and the glycoproteins are widely distributed in animals, plants, microorganisms and viruses where they account for a major portion of the carbohydrates present in polymer form.

---

*For books and general reviews, see [1—79].

It is, therefore, not surprising that this class of components of living matter was discovered in the early age of Biochemistry, and that the curiosity of scientists was aroused by substances they systematically encountered in the most varied biological media. However, the advancement of our knowledge of glycoproteins was very slow. In 1925, that is to say about 60 years after the isolation and characterisation of the first glycoproteins, Levene [1] gave free course to his impatience in writing:

In no chapter of chemistry or even in the special branch of tissue components is there so much confusion as in that of glucoproteins and mucoproteins. No attempt has been made to co-ordinate the older information with the new, no effort has been made to apply the newer methods of physical chemistry to the study of the molecules as a whole. The subject of the chemical structure of these tissue elements seems to be a very neglected branch of science.

Even at the present time, one must confess that, in spite of the considerable advances realised in the last few years, the knowledge of glycoproteins is far behind that of nucleic acids and proteins.

The reasons for this slower development could be related, first, to the difficulties encountered in the isolation and study of glycoproteins, and second, to the fact that, for a long time, carbohydrates have been regarded as reserve substances, energy-storage compounds, and support structures. But, during the last decade, the chemistry and biochemistry of glycoproteins have acquired an importance as great as that of proteins and nucleic acids because of the tenacity and trust of a few scientists who were firmly convinced that the knowledge of the metabolism, of the molecular biology and pathology of glycoproteins is based on the knowledge of the primary structure of glycans. Thanks to their works, the complete primary structure of numerous glycans has been established by the improvement of chemical, physical and enzymic approaches they developed.

Moreover, a series of discoveries and observations pointed to the role of the glycans and we now know that (1) glycoconjugates are cell-surface antigens and their structure and function are modified in transformed cells and in cancer cells; (2) glycans play an important role in intracellular adhesion and recognition and also in cell-contact inhibition; (3) they are receptor sites for viruses, proteins and hormones; (4) they regulate the catabolism of circulating pro-

# HISTORICAL DEVELOPMENT

teins by different tissues and thus determine the lifetime of proteins and cells; (5) they protect the protein moiety against proteolytic attack; (6) they may influence the conformation of peptidic chains: (7) they probably control the permeability of cell membrane; (8) due to a lack of lysosomal enzymes, their catabolism is dramatically perturbed in the case of the so-called "glycoproteinoses" or "glycanoses".

Thus, in a few years the foundations of the molecular biology and pathology of the glycoconjugates in general, and of the glycoproteins, in particular, were established. But, long was the way which led from the first discoveries to the present concepts and I consider that the epic of glycoproteins is worthwhile to be retold, even briefly, since the sources of current knowledge are in the past. Moreover, in this way we can pay homage to the first "conquistadores" of the glycoprotein continent.

## 2. Historical development of the knowledge and concepts of glycoproteins

The history of glycoproteins (for reviews, see [1, 80, 81]) can be divided into three periods of very unequal length. During the first, which seems to begin, as far back as one can go, at the end of the eighteenth century, scientists discovered the glycoconjugates and tried to answer the question: what is a glycoprotein? This long period is characterized by a discontinuous evolution of the knowledge and by sporadic discoveries often separated by long years of silence during which research floundered, waiting for new and more efficient analytical methods. In my opinion, the pioneers' work terminated in 1925. It is perfectly summarized in the fascinating book *Hexosamines and Mucoproteins* edited at that time by P. A. Levene [1] and which is a wonderful museum piece as well as a treasure of information. During the second period, the physicochemical properties of pure glycoproteins were described and the first structures of glycans were established. At the same time research into the metabolism of these compounds began. The last period started around 1968 and corresponds to the birth of both molecular biology and molecular pathology of glycoproteins.

*References p. 158*

## 2.1. The early history

One can state that the early history of glycoproteins is closely linked to that of mucins, polysaccharides of connective tissue, serum glycoproteins and egg white glycoproteins. The reasons for such a strange choice could be easily explained. In fact, at that time, analytical methods related to carbohydrates required great quantities of substances. Before 1888, monosaccharides were characterized only by their reductive and fermentative behaviour and by the properties of their osazones. The coloured reactions of furfural derivatives with xylidine acetate according to Schiff (1880) and with $\alpha$-naphthol according to Molisch (1886) were used for the first time by Seegen [82] in 1886 and by von Udranszky [83] in 1888 for characterising the sugars in glycoproteins such as albumin, globulin, casein and fibrin. This explains why the first authors described only glucose as a constituent of glycoproteins, characterised by its osazone.

Of course, we now know that mannose and glucosamine also give glucosazone. But, to be fair, we have to remember that, before 1878, glucose, galactose, sorbose and fructose were the only known monosaccharides and that the discovery of new sugars occurred later. In fact, *glucuronic acid* was the first monosaccharide to be identified in a glycan, chondroitin sulfuric acid. This finding made in 1891 by Schmiedeberg [84] was facilitated by the fact that, at that time, the author himself [85] had already isolated and described the compound he obtained by acidic hydrolysis of the glucuronic acid-derivative of camphor which accumulated in the urine of dogs having ingested camphor. *"Chitosamine"*, discovered by Ledderhose [86] in 1878 as a component of chitin was characterized for the first time in glycoproteins only in 1898. This finding was a consequence of results obtained by several authors, leading to the conclusion that the reducing substance obtainable on acid hydrolysis of mucins and mucoids was not glucose but a nitrogenous compound. This view was confirmed by the isolation, in 1896—98, of crystalline "chitosamine" hydrochloride from ovalbumin by Seemann [87] and from mucin of the bronchial mucus by Müller [88]. Moreover, it is worthwhile mentioning that the "chitosamine" was identified as D-glucosamine only in 1939 by Haworth et al. [89] and Cox et al. [90]. *"Chondrosamine"* was discovered in chondroitin sulfate by Levene and La Forge [91]

# EARLY HISTORY

in 1914 and identified as D-galactosamine by James et al. [92], 31 years later. D-*Mannose*, isolated by Reiss in 1889 from mannans, was first identified in a glycoprotein, ovomucoid, only in 1927 by Fraenkel and Jellinek [93]. D-*Galactose*, discovered by Louis Pasteur, and L-*fucose*, obtained by Maquenne and Tollens from algae in 1909, were characterized for the first time in serum glycoproteins in 1929—30 by Bierry [94]; with regard to *sialic acids*, their discovery is more recent. In fact, the term sialic acid, to which are attached the names of Blix and Svennerholm, Gottschalk, Klenk and Faillard, Kuhn and Brossmer, Roseman, and Schauer, was first employed by Blix et al. [95] to describe a substance previously isolated by Blix [96] in 1936 as a crystalline degradation product of bovine submaxillary mucin. In 1941, Klenk [97] isolated neuraminic acid from gangliosides in the form of the methoxy derivative. Gottschalk [98], in 1955, and Klenk et al. [99], in 1956, proposed the first formulae for neuraminic acid (Fig. 1).

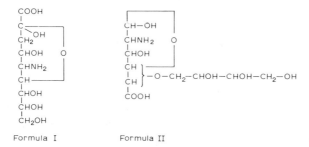

Fig. 1. First proposed structures for neuraminic acid. Formula I: Gottschalk [98]; Formula II: Klenk et al. [99].

Concerning these structures Klenk [100] wrote in 1958

The experimental finding obtained so far confirm the correctness of formula I proposed by Gottschalk, in which neuraminic acid is an aldol condensate of pyruvic acid and N-acetylhexosamine. Formula II which we suggested proved to be erroneous, as it cannot explain the appearance of N-acetylglucosamine among the degradation products of N-acetylneuraminic acid, as found by Kuhn and Brossmer, in 1956. The final decision for the one or the other of the two formulae will not be possible until further details of the chemical constitution of the substance are known.

*References p. 158*

The "final decision" came from the work of Comb and Roseman [101] in 1958. These authors demonstrated by using the aldolase from *Clostridium perfringens*, that N-acetylmannosamine was the amino sugar incorporated into N-acetylneuraminic acid. However, how true is the latin axiom "Nihil novum sub sole!" In 1871, Obolenski [102] obtained in Hoppe Seyler's laboratory, on heating a submaxillary mucin with very dilute sulfuric acid followed by neutralisation with barium hydroxide, a product which readily dissolved in water, displayed strong reducing power, and which had a nitrogen content of 4.42%. Indubitably, Obolenski had isolated a sialic acid.

Reading and analysis of the older literature led to the conclusion that the first works on glycoproteins were concerned with mucins which were, as a matter of fact, initially described under the name of mucus. Mucus were originally characterised as chemical entities by Bostock [103] in 1805 on the basis of their behaviour toward heat: mucus, in contrast to albumin, is not coagulated by heating. Rapidly, the criterion "behaviour toward heat", was completed by the criterion "behaviour toward acetic acid and alcohol" and this led to the definition of mucins and, later, of mucoids (Fig. 2). On the basis of these criteria, Berzelius [104], in 1828, extracted from gliadin, a protein originally prepared from wheat by Taddei in 1819. He described this material as "schleimige Materie" for which De Saussure proposed in 1835 the word "Muzin".

These observations on plant substances leading to a distinction between "proteins", non-precipitable by acetic acid, and "mucins", acid-precipitable, were paralleled by similar ones on animal fluids. Gmelin [105], in 1826, isolated an intestinal mucus by precipitation with acetic acid in the cold. In the same way, Scherer [106]

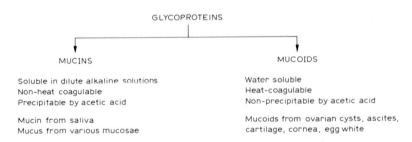

Fig. 2. Hammarsten's classification of glycoproteins (1891).

obtained in 1845 a mucin from human bronchial secretions and defined the *mucins* by their property to be acid-precipitable in the cold, in contrast to *mucinoids*.

1865 is an important date since it can be considered as the year of birth of glycoproteins. In fact, at that time, Eichwald [107], a young Russian Doctor of Medicine, working in Scherer's laboratory, provided the first evidence for the presence of a carbohydrate, assumed to be glucose, in various mucins. Accordingly he defined the latter substances as "conjugated single compounds consisting of a moiety with all properties of a genuine protein and a moiety released under certain conditions as a sugar". This opinion was definitively confirmed in 1888 by Hammarsten [108] who prepared pure mucin, applying a method which is still the basis of the actual most commonly used methods for purification of mucus glycoproteins, and who concluded that, in mucin, carbohydrate and protein were not a mixture but were attached together by a firm chemical linkage.

A further advance in the knowledge of the chemical structure of mucoproteins was made when Landwehr and Hammarsten provided the first proof that the sugar moiety of mucoproteins was not a single monosaccharide, as believed until this moment, but a polysaccharide. In fact, Landwehr [109] isolated in 1882 a polysaccharide from the alkali-treated mucin of *Helix pomatia* and, because of its resemblance to vegetable gums, he called this substance "animal gum". However, Landwehr failed to see any relationship between polysaccharide and mucoprotein and considered the latter merely as a mechanical mixture of animal gum and protein. Hammarsten [108, 110] confirmed Landwehr's finding and further demonstrated that polysaccharide and protein were joined together by a chemical bond. He proposed to call such compound "Glycoproteiden". So the term "glycoproteins" was born.

Concomitant with the gradual demonstration of sugars firmly linked to proteins in mucins, other types of glycoproteins were extracted from the three biological sources I mentioned above: egg white, blood serum and connective tissue. *Ovalbumin* was obtained from hen egg white in crystalline form by Hofmeister [111] in 1890 and *ovomucoid* was discovered by Neumeister [112] in 1890 and isolated by Mörner [113] in 1894. They were recognised as glycoproteins in 1898 by Eichholz [114] and by Hofmeister [115] who demonstrated the presence of a sugar iden-

*References p. 158*

tified as glucose. Thus the finding of Pavy [116—118] was confirmed who had shown the presence of a reducing sugar giving glucosazone after sulfuric acid hydrolysis of egg proteins he called "glycosides". In this connection, and in order to show how slowly knowledge progressed, it is interesting to note that glucosamine and mannose were characterised in ovalbumin in 1898 by Seeman [87] and in 1927 by Fraenkel and Jellinek [93]. Moreover, it was not until 1938 that Neuberger [119] definitely determined the sugar composition of this glycoprotein. At the same time the realisation that serum was rich in high-molecular-weight carbohydrate-containing substances came through the works of Zanetti who isolated from boiled horse serum a glycoprotein which, because of its similarity to ovomucoid, he called *seromucoid* [120] and in which he demonstrated the presence of glucosamine [121]. We now know that this compound contains as the major component the $\alpha_1$-acid glycoprotein isolated 50 years later in a pure form by Weimer et al. [122] and Schmid [123].

Almost simultaneously with these findings, another type of protein-linked carbohydrate was characterised in the ground substance of connective tissue. In fact, the presence of sugars in cartilage was demonstrated as far back as 1854 by Boedeker [124], but chondroitin sulfuric acid was prepared for the first time only in 1884 by Krukenberg [125] and obtained in a pure form by Mörner [126] in 1889. In 1891, Schmiedeberg [84] demonstrated that it was possible to remove sulfate by acid hydrolysis of this compound and to obtain a non-reducing substance he named "chondroitin", introducing at the same time the term "chondroitin sulfuric acid" to designate the native component. Moreover, Schmiedeberg showed that deacetylation of chondroitin led to a disaccharide he called "chondrosin" and he considered this sugar as a reducing disaccharide consisting of glucuronic acid and hexosamine, later identified and named chondrosamine by Levene and LaForge [91], in 1914 and then galactosamine by James et al. [92] in 1945. On the basis of his experiments, Schmiedeberg [127] proposed an audacious representation of chondroitin sulfuric acid given in Fig. 3. In this formula, the glucuronic acid residue was conjugated to the amino group of hexosamine, assumed to be chitosamine, and three acetyl groups were believed to be combined with carbon C-1 of the latter. It must be mentioned here that Schmiedeberg was the first to determine the complete composition of a carbo-

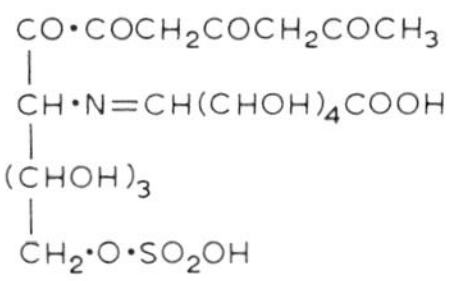

# EARLY HISTORY

$$\begin{array}{l}\text{CO·COCH}_2\text{COCH}_2\text{COCH}_3\\|\\\text{CH·N=CH(CHOH)}_4\text{COOH}\\|\\\text{(CHOH)}_3\\|\\\text{CH}_2\text{·O·SO}_2\text{OH}\end{array}$$

Fig. 3. Representation of chondroitin sulfuric acid proposed by Schmiedeberg in 1920.

hydrate group of a mucoprotein and to propose a tentative structure for this type of molecule.

However, there was a sudden decline of interest in glycoproteins since the difficulties encountered in the isolation and analysis of glycans were considerable. Only a few groups continued their exploration with obstinacy and the period between 1900 and 1925 is essentially marked by the works of Levene's and Schmiedeberg's groups. The situation in 1925 is clearly detailed in Levene's book *Hexosamines and Mucoproteins* [1] which marks the end of an era, that of the pioneers who failed to find the truth due to the lack of means, but who opened the way.

In 1925, the basic concepts on glycoproteins were dominated by Levene's concepts and dogmas which can be summarised as follows:

(1) The carbohydrate group of all mucoproteins is conjugated with sulfuric acid. The group is built up of four components in equimolar proportions. The components are sulfuric acid, acetic acid, hexosamine and glucuronic acid.

(2) In all mucoproteins analysed the carbohydrate component is either chondroitin or mucoitin sulfuric acid.

Levene divided all mucoproteins into two groups (Table I), namely, the *chondroproteins*, containing chondrosamine and constituted of the association of chondroitin sulfuric acid with proteins; the *mucins* and *mucoids* are constituted by the association of proteins with mucoitin sulfuric acid which Levene and Lopez Suarez [128] isolated in 1916 from pig gastric mucosa and from funis mucin*. Discrimination between mucoids (first sub-group)

---

*We now know that this compound does not exist and that it was probably sulfohyaluronic acid.

*References p. 158*

## TABLE I
### Classification of mucoproteins according to Levene [1]

| | | Mucoitin sulfuric acid | |
|---|---|---|---|
| | | Type A | Type B |
| Composition | Acetic acid<br>Sulfuric acid<br>Glucuronic acid<br>Chondrosamine | Acetic acid<br>Sulfuric acid<br>Glucuronic acid<br>Chitosamine | |
| Nature of the disaccharide unit | Chondrosine | Mucosin | |
| Nature of prosthetic group | Chondroitin sulfuric acid | Mucoitin sulfuric acid Type A | Mucoitin sulfuric acid Type B |
| Properties of the prosthetic groups | Intermediate between mucoitin sulfuric acids A and B | More soluble in water<br>Precipitated from aqueous solutions by an excess of acetic acid | Greater insolubility in water<br>Precipitated from aqueous solutions by dilute acetic acid |
| Type of mucoprotein | Chondroproteins | Mucoids | Mucins |
| Origin | Cartilage<br>Tendons<br>Aorta<br>Sclera | Mucin of gastric mucosa<br>Serum mucoid<br>Ovomucoid | Funis mucin<br>Vitreous humor<br>Cornea |

and mucins (second sub-group) was based, on the one hand, on the existence of two types of mucoitin sulfuric acids (i.e. types A in mucoids and B in mucins) presenting distinct differences in their physical properties and, on the other hand, on the solubility of the glycoproteins in acetic acid:

Both are very soluble in water and are precipitated from aqueous solution only by a large excess of glacial acetic acid, and when thus precipitated, substances of first sub-group come down in the form of very light floccules, in contradistinction to the substance of the second sub-group. These, under the same conditions, form a heavy gelatinous mass.

(3) Carbohydrate and protein are conjugated by ester linkages.

# EARLY HISTORY

In the light of the work on mucoitin and chondroitin sulfuric acid the formulation of the mode of union between the carbohydrate radicle and the protein is quite simple. From the experimental part ... it was seen that the union between the two parts is comparatively labile. Alkali of a dilution too low to bring about a disruption of the protein molecules, or of a disruption of the carbohydrate radicle, was sufficient to sever the union between protein and carbohydrate. Thus the simplest assumption is that the union between protein and carbohydrate is in the nature of an ester linkage.

(4) Chondroitin sulfuric acid is a non-reducing tetrasaccharide, built symmetrically of two acetylchondrosin sulfuric acid moieties, as expressed in Fig. 4.

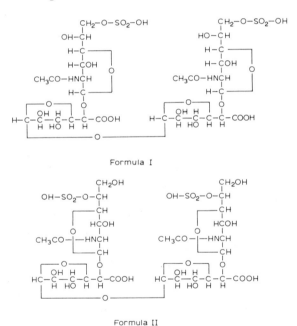

Fig. 4. Structure proposed by Levene [1] for chondroitin sulfuric acid (Formula I) and for mucoitin sulfuric acid (Formula II).

Thus, Levene restricted his choice of possible structures to only two or three. He was conscious of this restrictive aspect of his proposals and wrote:

*References p. 158*

Mucoproteins are complex proteins. The protein parts undoubtedly differ widely from tissue to tissue, from species to species. Thus, for mucoproteins, the same thing is true as for nucleoproteins, namely, the chief variable component is not the carbohydrate complex but the protein group.

Of course we are astonished by this concept, but we have to consider that at that time no great biological importance was attached to glycoproteins, but only a biochemical significance. In fact, from 1885 to 1900 they were regarded as the precursors of sugars and glycogen. This theory came from the demonstration in 1877 by the French physiologist Claude Bernard [129] that food proteins are converted into liver glycogen. Bernard's findings were confirmed by many physiologists and clinicians and led, independently of the work on mucoproteins performed by physicochemists, to the theory of the glucosidic nature of proteins which is one of the rare instances when physiological concepts were the basis of a new concept of chemical structures. To account for the conversion of proteins into sugars, it was logical to assume that most of the proteins contained preformed sugars and that they were of a "glycosidic nature". Strong support for this concept of "glucosidoproteins" was provided by the finding, in the meantime, that almost all the protein preparations contained sugars [82, 83]. That explains why so many people like Pavy, a British physician who worked with Claude Bernard around 1852, were still ardent defenders of the glycosidic structure of proteins at the end of the century. Having demonstrated the positive sugar reactions and the reducing power of the acid hydrolysates of material obtained from egg albumin, serum proteins, casein and fibrin by treatment with alkali, Pavy was convinced of "the glycosidic nature of proteins" and claimed that the formation of sugars from proteins was caused by the liberation of the carbohydrate moieties present in glucosidoproteins [116—118]. At a time when the structure of proteins was not known at all, it was natural to assume that the sugar molecules were preformed in proteins. However, due to the progress in the knowledge of protein structure on the one hand, and since Graham Lusk and coworkers demonstrated that sugars were in fact formed from amino acids, on the other hand, the physiological basis for the glucosidic structure of proteins and the theory of glucosidoproteins playing a metabolic role in being the precursors of sugars both vanished. After that time and during half a cen-

# EARLY HISTORY

tury glycoproteins will be considered as protective and supporting substances. In this connection, the following quotation of Levene, is particularly significant.

The occurrence of mucoproteins either in connective tissue or in the linings of organs such as mucous membranes indicate either an architectonic or perhaps a cementing function. On the other hand, as components of secretions they exercise a special protecting function against mechanical irritation by foreign bodies, and there are indications that in certain instances they serve as a protecting agent against bacterial infection. Also from the chemical structure of mucoproteins some deduction may be made regarding their biological significance. In the plants, cellulose is the principal architectonic and protecting tissue component. In the lower animals chitin plays a similar part. Chitin is a polysaccharide, consisting of chitosamine and acetic acid. Both these components also enter into the molecule of mucoproteins. The latter are in fact polysaccharides containing in their molecule the components of chitin to which are joined other carbohydrates and which finally combine with proteins. Thus genetically they are related to chitin. May they not then be related to chitin also functionally?".

This concept obviously does not require a wide variety of specific structures.

However, Levene was in a constant doctrinal conflict with several scientists, particularly Schmiedeberg, not only about the structure of chondroproteins, but chiefly about the prosthetic groups of other glycoproteins. In fact, Schmiedeberg isolated carbohydrate moieties from several glycoproteins and called them "hyaloidins" (Fig. 5), in the formulae of which appeared for the first time neutral sugars: hyaloidin of type I from ovalbumin, ovomucoid, mucoid of ovarian cyst and mucin of the gastric mucosa; hyaloidin of type II from fibrin and serum globulin, the only point of variation being that the latter contains a fructose residue in place of the acetyl group. The presence of the ketose is justified by a positive Seliwanoff reaction which was probably due to sialic acids.

Thus, the past could be summarised as follows. During this long period, the research workers investigated all aspects of glycoproteins: isolation, composition, structure, metabolism and function. Apart from the identification of monosaccharides, they were wrong. In fact, they were mostly dealing with impure preparations and they did not have the sophisticated methods we possess now. That is why they did not succeed in defining the true structure of gly-

*References p. 158*

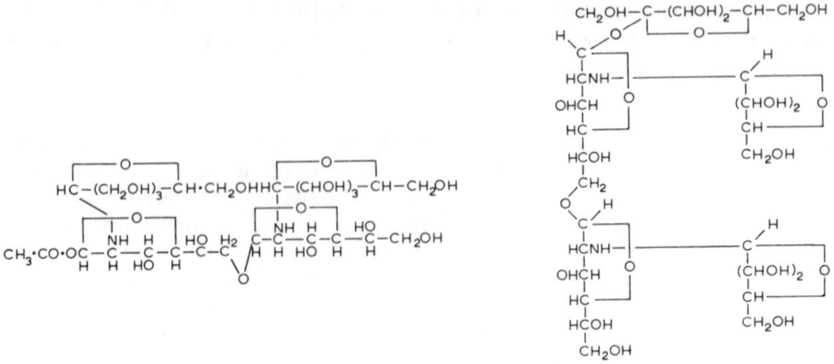

Fig. 5. Structure of hyaloidins of types I and II proposed by Schmiedeberg [127] in the 1920s.

coproteins. Lacking the structural basis, they were unable to ask the relevant question and to obtain the subsequent information on the metabolism and role of glycoproteins. Too many pieces of the puzzle were missing to have any idea of the complete picture, but as anyone knows, the most difficult pieces to fit into a puzzle are always the first ones. In addition, they did not have all the pieces.

## 2.2. Middle Ages and Renaissance of glycoproteins

The period 1920—1930 is poor in important discoveries in the field of glycoproteins. This with the exception, however, of the demonstration that concanavalin A reacted with glycogen [130], the isolation of hyaluronic acid [131], the characterisation of galactose, mannose and fucose in glycoproteins, and the discovery of the sialic acids. The latter did not create much of a stir at the time and the importance of the sialic acids in the molecular biology of glycoproteins was not recognised for another 20 years.

The reasons for this lack of interest in glycoproteins are multiple. First, in the 1920s scientists were still convinced that specific information was carried only by proteins, that carbohydrates did not play any important biological role and that glycoproteins were devoid of any interesting practical application. This is why the efforts of carbohydrate chemists turned to sugars presenting some industrial interest, such as starch, cellulose and gums, the structures of which were actively explored. Moreover, the erroneous and deeply restrictive dogmas of Levene we described before strongly influenced ideas about the nature of glycans and were probably not entirely irrelevant to the stagnation of research on glycoproteins. Second, due to the lack of proper methods for the isolation and structural analysis of glycans, verification of the smallest fact required efforts out of proportion. Moreover, the existence of a carbohydrate-protein linkage was not yet unambiguously demonstrated and glycoproteins were often contested as entities.

Fortunately, a series of events ended this long and sterile period which lasted from 1900 about 1940 and followed the flourishing period at the end of the last century. These following events are at the origin of the Renaissance period of glycoproteins.

(1) First of all, Neuberger [119] provided in 1938 an unequivocal demonstration of the presence of covalently bound carbohydrate in ovalbumin.

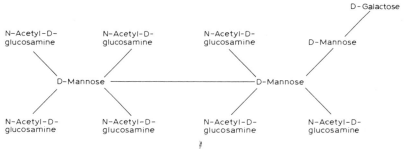

Fig. 6. Structure of ovomucoid glycan proposed by Stacey and Woolley [132].

(2) Stacey [132] introduced in 1940—1942 the methylation procedure to the study of glycoproteins and proposed for ovomucoid the structure depicted in Fig. 6. This structure comprises galactose and *N*-acetylglucosamine residues in external positions and a mannotriosidic core, in an internal position, as verified 30 years later. In this connection, it is astonishing to read in Stacey's paper the following sentence, which still remains essentially true:

*References p. 158*

From the identification of these products it was possible to give a means of portraying the type of structure (Fig. 6) present in the carbohydrate molecule. In this, by glycosidic attachment, seven N-acetylglucosamine units and one D-galactose unit radiate from *a central core of three* D-*mannose units* ... It is considered that (this compound) does not represent a repeating unit but rather that it depicts the whole molecule as being that of a hendecasaccharide having a type of structure which appears to be unique in carbohydrate chemistry ... In regard to the structure of ovomucoid itself, of which the carbohydrate forms 20%, it would appear that *the peptide constituents are mainly attached to the N-acetyl-*D*-glucosamine terminal residues* ...

(3) The end of 1930s and the beginning of 1940s saw the advent of the refined methods that the "glycoproteinists" needed to progress: analytical ultracentrifugation, chromatography and electrophoresis, radioactive labelling, ion-exchange resins and celluloses, immunoelectrophoresis, enzymic and chemical cleavage of glycosidic bonds.

Application of these sophisticated methods for isolation and study of glycoproteins, as well as a series of observations of biochemical, biological and pathological order incited more and more scientists to go into the field of glycoprotein research. The consequence was that this field has developed exponentially and that an enormous sum of results accumulated that it would be useless to even attempt to make a resumé. I will thus limit myself to citing the most outstanding findings or those which were at the origin of new research on glycoproteins. In the first place the development of fractionation procedures allowed the resolution of starting materials and the preparation of pure glycoproteins such as $\alpha_1$-acid glycoprotein, isolated by Weimer et al. [122] and crystallised by Schmid [123]. This also applies for most of the acid mucopolysaccharides: chondroitin sulfates A, B and C [133] and their protein complexes, heparan sulfate [134] and keratan sulfate [135]. In addition, the perfection of analytical procedures permitted the characterisation of glycoproteins in very diverse biological environments and the demonstration that they were widely distributed.

In the field of glycan primary structure, one can assume that the development of the methodology started with the work on milk oligosaccharides in Kuhn's [136] and Montreuil's [137] groups and on blood group substances in Morgan and Watkin's and in Kabat's laboratories (for early reviews see [138]). In fact, for the first time, well defined and pure heterosaccharides were encountered and new methods for structural analysis had to be devised. The development of the methods for partial hydrolysis and for methylation and of the use of glycosidases began at that time. The first determinations of complete glycan primary structures concerned those of ovine submaxillary mucin glycan, by Gottschalk [139] and those of acidic mucopolysaccharides, chiefly thanks to the work of groups led by Dorfman, Jeanloz, Jorpes, Masamune, Meyer, Ogston and Rodén. This can be explained by the relative simplicity of the structures of mucin glycans which are disaccharides, and of mucopolysaccharides which are linear chains of repeating disaccharide units. However, progress was much slower in the case of other glycoproteins, the $N$-glycosylproteins in particular. This sluggish progress was in part due to the structural complexity of branched glycans and especially to the existence of variants of the glycoproteins and to the microheterogeneity of the glycans.

For example, using starch-gel electrophoresis K. Schmid characterised 7 to 9 variants [140] of $\alpha_1$-acid glycoprotein and demonstrated that each one carried 5 structurally different glycan structures [141]. The results obtained by structural studies were thus statistical and consequently uninterpretable. Moreover, the structures proposed for the $\alpha_1$-acid glycoprotein glycans were numerous and rather far from reality, as can be seen when comparing the structures proposed by Wagh et al. [142] (Fig. 7) with those cited in Table XIII and Fig. 26. However, the further improvement of fractionation methods now permits the resolution of this thorny

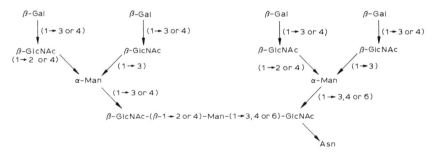

Fig. 7. Structure of glycopeptide V of $\alpha_1$-acid glycoprotein proposed by Wagh et al. [142].

*References p. 158*

problem and in this particular quoted case K. Schmid has isolated 22 glycopeptides [143] of $\alpha_1$-acid glycoprotein of which the complete primary structures were determined in 1978—1979 by combined NMR and methylation techniques [144, 145].

For the above reasons, at the beginning of the 1960s the progress in the structural analysis of the sugar moiety of glycoproteins remained very slow. That is why many qualified workers in glycan chemistry stopped their efforts and turned their attention to other aspects of glycoprotein research, principally to the biosynthesis and to the biology of the carbohydrate groups. Progress in this area has been spectacular due to the definition, in the meantime, of the "machinery" for biosynthesis of oligo- and polysaccharides thanks to the fundamental work of Leloir's school. Leloir and coworkers [146] discovered UDP-D-glucose in 1949—1950 and demonstrated that this compound was a precursor in the glucosylation processes and that it could be enzymatically converted to UDP-D-galactose [147]. The way was then open for the isolation of all the sugar nucleotides involved in the biosynthesis of glycans: UDP-D-GlcNAc by Cabib et al. [148], UDP-D-GlcUA by Dutton et al. [149] in 1953, GDP-D-Man by Cabib et al. [150] in 1954, UDP-D-GalNAc by Pontis [151] in 1955, CMP-NeuAc by Comb et al. [152] in 1958 and GDP-L-Fuc by Ginsburg [153] in 1960. On the basis of the methodology defined by Leloir's work, Dorfman [154] and Glaser and Brown [55] reported for the first time the biosynthesis of a glycan, that of hyaluronic acid. However, the first intensive work on the biosynthesis of glycans really started at the beginning of the 1960s with experiments carried out on the biosynthesis of blood plasma glycoproteins in the liver [156—159], of thyroglobulin in thyroid gland [160] and of blood-group substances in hog gastric mucosa [161]. The results obtained indicated that all monosaccharide residues were incorporated into glycans by membrane-linked enzymes, termed "glycosyltransferases". The precise localisation of the enzymes in the endoplasmic reticulum was rapidly achieved. In 1965—1967 Molnar et al. [158, 162] and Schachter et al. [163] demonstrated that mannose and N-acetylglucosamine residues were incorporated in the rough reticulum, whereas the incorporation of N-acetylglucosamine, galactose, fucose and sialic acid residues occurred in the smooth reticulum and Golgi apparatus. At the same time, investigation into the specificity of glycosyltransferases and on the mechanism of sugar chain elongation led Roseman [164, 165] to the definition of the multiglycosyltransferase (MGT) system which differs from chain elongation in proteins and nucleic acids in that the latter are formed by the polymerisation of different monomers by a single enzyme system. However, nothing is ever simple in the mechanisms of life. In fact, 15 years after the discovery of sugar nucleotides, a new class of activated carbohydrate derivatives was discovered in bacteria and identified as sugar-phosphate or sugar-diphosphate-polyprenols which act as intermediates in the synthesis of oligo- and polysaccharides (for review see [166, 167]). In 1963, Hemming's group [168] discovered this kind of "lipid intermediates" in mammalian tissues, identified as dolichols. The way was open for the demonstration, realised about 10 years later, that glycans did not result from the direct transfer of the first monosaccharide residue onto the protein, followed by a progressive sugar chain elongation resulting from the addition of monosaccharide residues one after the other. On the contrary, a part of the glycan is transferred "en bloc" onto the peptide chain from lipid intermediates.

Parallel to this work on the primary structure and metabolism of glycoproteins very active research developed in the domains of Pathology and Biology. This brought a new interest to glycoproteins and encouraged many research workers to venture into the field. The discovery of disorders of glycoprotein metabolism in various illnesses such as cancer, inflammatory syndromes, mucoviscidosis and lysosomal storage diseases stimulated numerous lines of research into the molecular pathology of glycoproteins (for reviews, see [35, 68, 72, 74, 75, 78, 169—175]). However, it was the observations made on membrane glycoproteins (for review see [176]) which is stimulated a revival of research on glycoproteins. The discovery of conjugated sugar in the plasma membrane should be considered as the principal motif of the attraction exerted on researchers by glycoproteins from 1960. It seems that the first chemical demonstration of the presence of sugars in the cell plasma membrane should be attributed to Cook et al. [177], who in 1960 characterised glycoproteins among the products of the action of proteases on cells. Direct evidence has involved the isolation of pure membranes in which the presence of carbohydrates was demonstrated by sugar analysis, for the first time in rat hepatocytes by Herzenberg and Herzenberg [178] in 1961.

*References p. 158*

The use of light and electron microscopy provided further evidence for the presence of carbohydrates in plasma membranes [179]. Parallel to this, important research developed on lectins and on their interactions with membrane sugars. Already in 1888–1889, Stillmark [180] had demonstrated that *Ricinus* and *Abrus* bean extracts strongly agglutinate red blood cells. This observation led to the isolation of concanavalin A by Sumner [181] in 1919 and to the demonstration in 1936 by Sumner and Howell [130] that this lectin reacted with glycogen. In the 1940s the studies of Renkonen [182] and of Boyd and Regnera [183] led to the discovery of lectins specific for some human blood group antigens. On the basis of these observations and with the establishment of the structure of oligosaccharide determinants of blood group activity the idea was born that haemagglutination could be the result of an interaction between lectins and glucidic structures at the red blood cell surface. The extraordinary development which followed in the research of new lectins, and their specificity towards glycan structures is well known (for reviews see [184–199]. These findings were at the beginning of the discovery that the carbohydrate structures were profoundly modified in transformed cells and cancer cells. The ensemble of these observations radically modified the concept of glycoproteins. They came at the moment when the procedures for exploring the primary structures of glycans, without having achieved perfection, allowed us to view with optimism the solution of the problem of the determination of glycan primary structure. After one century of gestation the birth of the Molecular Biology of Glycoconjugates thus occurred.

*2.3. The revolution of 1968 and the birth of the molecular biology of glycoconjugates*

If the years 1967–1969 did not cause important modifications in the social structure of man, they induced profound changes in our knowledge of the chemical structure of glycoproteins. This sudden change was due to a series of discoveries and observations which clearly revealed that complete knowledge of the primary structure and conformation of the glycans was necessary, first, to establish the relationships between structure and specific biological activities, and second, to understand the metabolism and to explain the pathology of the glycoproteins. For this reason, research on the structure of glycans of O-glycosyl and of N-glycosylproteins became intensive and the efforts displayed in numerous laboratories led in a few years to a great expansion of the knowledge of the structure of a considerable number of glycans.

If the attention of the scientific community was suddenly focused on the carbohydrate group of glycoproteins and if these latter received their "lettres de noblesse", this was due to the discovery that the glycans, which had been thought to have neither a role nor any biological significance are, on the contrary, extremely important. Burger and Goldberg, and Inbar and Sachs had just demonstrated that the glycoconjugate structure and the architecture of the cell membranes are profoundly altered in cancer cells, as revealed by the use of lectins [200]. The idea then arose that this molecular change could be related to the appearance of surface neoantigens and could be a factor in cancer induction and metastatic diffusion at the same time as it released an immune reaction against the neoplastic cells. This observation, added to the knowledge that the biological activity of many glycoproteins are diminished or abolished by artificial modification of the glycan moieties (see Section 8) led to the view that glycans are not metabolic accidents, born at random from the fantasy of glycosyltransferases, but that they could be recognition signals. Thus began the rush towards the glycoproteins.

It is, in this regard, surprising that this view came so late. More than 30 years ago, findings of Avery, together with those of Heidelberger and Goebel demonstrated that carbohydrate structures isolated from bacteria carried serological specificity as antigens or haptens. In 1940, Landsteiner and Harte [201] proved that the purified blood group A substance consisted of a polysaccharide to which was firmly attached a peptide chain. In 1952, Morgan and Watkins in London, provided the earliest indication, that ABH blood group specificity was carried by sugars (reviews on blood group substances in [65, 138, 202—208]: $N$-acetyl-$\alpha$-D-galactosamine is the immunodeterminant of blood group A, $\alpha$-D-galactose of blood group B and $\alpha$-L-fucose of blood group O. Enzymic removal by specific $\alpha$-D-glycosidases of $N$-acetylgalactosamine and of galactose from red blood cells of the types A and B, respectively, converted both to the type O. In the years 1948—1960, mainly

*References p. 158*

thanks to the efforts of Burnet, Gottschalk, Blix and Klenk, it was demonstrated very clearly that the influenza virus binds to the erythrocytes through sialic acid residues present on a cell surface receptor, so inducing haemagglutination. Removal of sialic acid from the red blood cell surface by neuraminidases prevents the fixation of influenza virus on them and inhibits the haemagglutination (reviews in [16, 209—211]). In 1963, Aub et al. [212] observed, as later did Burger and Goldberg [199] and Inbar and Sachs [200], that the agglutinability of cancer cells by some lectins is greatly modified if compared with normal cells. In 1964, Gesner and Ginsburg [213] provided an additional example of sugars as determinants of specificity on cell surfaces in showing that rat lymphocytes treated with a fucosidase migrate to the liver instead of the spleen, their normal destination. The fucose present at the lymphocyte surface thus constitutes a recognition signal of these cells by a receptor present on the spleen cell-membranes. In 1968, Ashwell and coworkers (for reviews see [214—216] completed the demonstration that the glycan moieties of glycoconjugates are endowed with a certain "biological intelligence" in establishing that carbohydrates regulate the clearance and thus determine the life-time of blood glycoproteins. In fact, the enzymic elimination of the terminal sialic acid residues from $N$-glycosylproteins (Section 4.3) induces the disappearance of these compounds in less than half an hour from the plasma of the animal into which they have been injected. The terminal galactosyl residues, unmasked by removal of sialic acid residues, are thus the recognition signals of the asialoglycoproteins for the hepatocytes.

Convinced of the fundamental importance of the glycoproteins, principally of the membrane glycoproteins and receptors, many scientists turned their attention to these exciting compounds so that research in this field exploded in a few years leading to a fabulous accumulation of findings which was unforeseeable ten years earlier. These findings will be developed in the next chapters. However, we can summarise the principal achievements as follows.

(1) Until around 1968, it was assumed that specific information was carried by nucleic acids and proteins. Since 1968, this concept has been enlarged to include carbohydrates. Indeed, as claimed by Nathan Sharon

We know now that the specificity of many natural polymers is written in terms of sugar residues, not of amino acids or nucleotides.

(2) Most of the primary structures of glycans from O-glycosylproteins and all from N-glycosylproteins have been determined in the past 10 years: that of an oligomannosidic type glycan from Taka amylase A [217] in 1971 and that of N-acetyllactosaminic type glycans in 1973—74, those from human serotransferrin [218—220] and of immunoglobulins [221].

(3) The metabolic pathways of glycoproteins have been almost completely elucidated.

(4) A glycoprotein pathology that is due to a lack of lysosomal glycosidases have been discovered, and the terms "glycoproteinosis" and "glycanosis" have been coined.

## 3. Definition and classification of glycans and glycoproteins

Fewer terms related to biochemical compounds have undergone so many changes in their meaning as those concerning the glycans and glycoproteins. Till recently the nomenclature and definition of the latter substances remained in a chaotic state. This is not surprising since the large number of proposed classifications and definitions reflects perfectly the progress made in our knowledge of glycoproteins over more than a century as well as the evolution of the concepts on the structure of these compounds. In fact, classifications were first based on the physical characteristics of glycoproteins (see Fig. 2, for example), then on the chemistry of the carbohydrate moiety and, finally on the structural details of the glycans.

The first attempts to classify the glycoproteins on the basis of the nature of the carbohydrate moiety was made in 1938, then in 1945 and again in 1953 by Meyer [2, 3, 222] who included only hexosamine-containing compounds. He proposed to term "mucopolysaccharides" the "high-molecular weight polysaccharides containing hexosamine" and to classify the hexosamine-containing compounds as described in Table II.

More simple was the classification, proposed by Stacey [4] in 1946, into *mucopolysaccharides* as "compounds having a low but significant protein content and giving reactions which are pre-

TABLE II

Classification of hexosamine-containing compounds according to Meyer [222]

| | | |
|---|---|---|
| (I) | Mucopolysaccharides: | |
| | (A) Neutral mucopolysaccharides. Example: Chitin | |
| | (B) Acid mucopolysaccharides | |
| | (1) Simple acid mucopolysaccharides. Example: Hyaluronic acid | |
| | (2) Complex acid mucopolysaccharides. Example: Chondroitin sulfates | |
| (II) | Mucoproteins: | Salts resulting from dissociable ionic binding to a protein of an acid mucopolysaccharide containing more than 4% hexosamine. Example: Chondroproteins |
| (III) | Mucoids: | Compounds resulting from covalent linkage between a protein and a mucopolysaccharide containing more than 4% hexosamine. Examples: Ovomucoid, serum mucoids, ovomucin, submaxillary mucins |
| (IV) | Glycoproteins: | Compounds resulting from covalent linkage between a protein and a mucopolysaccharide containing less than 4% hexosamine. Examples: Ovalbumin, serum globulins |

dominantly carbohydrate", *mucoproteins* as "compounds with relatively high protein or peptide content, the chemical reactions of which are predominantly protein", and *mucolipids*. This classification was similar to the one used presently.

Other classifications were proposed [5, 223, 224] but none were retained because the criteria upon which they were based were somewhat arbitrary. In fact, in the 1940s—1950s little was known about the composition and primary structure of glycans and about sugar-protein linkage. However, little by little these parameters became better defined and the nomenclatures and definitions that were further proposed by Jeanloz in 1960 and by Gottschalk in 1962 are still wholly or partly accepted and used.

Jeanloz [17] proposed to subdivide the substances containing sugars of low or high molecular weight into five groups: (1) Pure polysaccharides or *glycosaminoglycans;* (2) *polysaccharide-protein complexes* (chondroitin sulfate protein complexes, etc.) in which the "glycan" is bound to the protein through a weak linkage such as ionic or hydrogen bonding; (3) *glycoproteins, glycopolypeptides* and *glycopeptides* containing a carbohydrate group firmly conjugated with the peptide chain through a covalent linkage and comprising blood-group substances, mucins, glycoproteins of blood, etc.; (4) *glycolipids* and (5) *glycolipoproteins* resulting from the association of glycans with lipids or lipoproteins, respectively.

Gottschalk [225, 226] criticised all the classifications and definitions using the type of linkage between sugar and protein as a criterion because the number of examples for which this information was available at that time was small. Moreover, following Muir's [227] findings that serine could be engaged in a glycosidic bond with chondroitin 4-sulfate, it became less and less evident that the acid mucopolysaccharides were solely attached to proteins by ionic linkages. Therefore, Gottschalk proposed to classify the sugar-protein conjugates only on the basis of the structure of their carbohydrate moiety and to distinguish the two following families: (1) *Polysaccharide-protein complexes* defined as homo- and hetero-polysaccharides of high molecular weight and of linear structure composed of small repeating units, probably covalently bound to serine and/or threonine. Balasz [228] proposed in 1967 to call this kind of glycoconjugates *proteoglycans* and this term is now generally accepted. (2) *Glycoproteins* which comprise conjugated proteins containing as a prosthetic group one or more heterosaccharides, usually branched with a relatively low number of sugar residues, lacking a serially repeating unit and bound covalently to the polypeptide chain. Thus, in Gottschalk's classification, the term glycoprotein loses its general significance and this is not satisfactory from an etymological point of view. Moreover, the distinction between proteoglycans and glycoproteins and the differences in the structure and metabolism of both classes become more and more blurred as the knowledge of glycoconjugates advances. A demonstrative example is given by the keratosulfates (see Table XII) as compared to the glycans of rat sublingual glycoprotein (Fig. 21), of human lactotransferrin (Fig. 33), Chinese hamster ovary cell glycoprotein (Fig. 34) and poly(glycosyl) proteins of human erythrocyte membrane (Figs. 32 and 35) in which repeating sequence $\beta$-Gal-$(1 \rightarrow 4)$-$\beta$-GlcNAc-$(1 \rightarrow 3)$ of variable

*References p. 158*

length has been demonstrated. On the other hand, keratan sulfates I and II contain sialic acid and mannose residues as well as the above glycoproteins and are conjugated to proteins in the same manner: the first one, through the $N$-acetylglucosaminyl-asparagine linkage as found in $N$-glycosylproteins; the second one through an $N$-acetyl-$\alpha$-galactosaminyl-serine or threonine as in many $O$-glycosylproteins. Moreover, metabolic studies have shown that glycoproteins and proteoglycans have in common similar biosynthetic mechanisms to conjugate the carbohydrate moiety to the peptide chain.

For these reasons, it seems better to return to the definition of 1908 (see Introduction) and to consider as "glycoprotein" all compounds resulting of the covalent association of carbohydrate(s) with protein(s) through $O$- or $N$-glycosylic linkage(s), irrespective of the nature and structure of the sugar moiety. On the basis of this fundamental definition, I proposed [78] the tentative classification of Fig. 8 which is based both on the nature of glycosidic bonds between glycan and protein and on the structure of glycans. In a certain manner this classification resembles that of Marshall

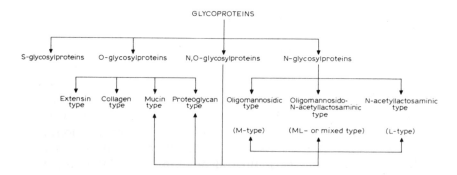

Fig. 8. Tentative classification of glycoproteins proposed by Montreuil [78]. In all $N$-glycosylproteins, the sugar-peptide linkage is of the $N$-acetylglucosaminyl-asparagine type. The sub-groups of $O$-glycosylproteins are differentiated by the nature of the monosaccharide-amino acid bond: $\beta$-xylosyl serine in "proteoglycan type", $N$-acetyl-$\alpha$-galactosaminyl serine or threonine in "mucin type", $\beta$-galactosyl hydroxylysine in "collagen type", $\beta$-arabinosyl hydroxyproline in "extensin type". The glycans of the $N$-acetyl-lactosaminic type (L-type), of the oligomannosidic type (M-type) and of the oligomannosido-$N$-acetyl-lactosaminic type (ML- or mixed type) are also designed as complex glycans, mannose-rich glycans and hybrid structures, respectively.

METHODS

and Neuberger [34] in which glycoproteins were divided into three groups: Group A: glycoproteins with N-glycosidic linkage involving the amide group of asparagine residues; Group B: glycoproteins with O-glycosidic bonds involving the hydroxyl group of serine and/or threonine residues; Group C: glycoproteins whose carbohydrate-peptide bond involves δ-hydroxylysine.

## 4. Structure and conformation of glycoprotein glycans

Thanks to the progress in chemical, physical and enzymatic methodologies in the 1960s—1970s, the determination of the primary structure of glycans has become easier and relatively rapid. As a result glycan structures are known now by the dozen. Thus, where several years were necessary for determining the first complete and definitive structures of glycans, such as those of serotransferrin and of immunoglobulins, a few weeks were sufficient to define structures of 18 glycopeptides [144] isolated by Karl Schmid from proteolytic hydrolysates of $\alpha_1$-acid glycoprotein.

### 4.1. Methods

The problem of the determination of glycan primary structure could be considered as virtually solved due to the improvements of an impressive series of sensitive and efficient methods for the isolation of glycoprotein variants and of each of their glycan residues, the splitting of glycosidic bonds either by chemical cleavage or by enzymic hydrolysis due to the preparation of many monospecific exo- and endo-glycosidases and the definition of their activity in relation to oligosaccharide structure, and the development of physical methods such as mass fragmentometry and high-resolution nuclear magnetic resonance.

Because these methods are reported in detail in numerous reviews or books [35, 46, 58, 61, 66, 69, 75, 76, 229—236] the present paragraph will be restricted to a rapid survey and to a short account of the most up-to-date techniques.

*4.1.1. Isolation of glycoproteins and glycans.* One of the most difficult problems to solve before applying procedures for the determination of the primary structure of the carbohydrate moieties of glycoproteins is the isolation in a pure state of each of the glycans present in the latter. Indeed, many glycoproteins which appear as homogeneous by physicochemical and immunological criteria are in fact complex mixtures of variants. The existence of these variants is essentially related to a glycan microheterogeneity which is discussed in Section 4.4.

*4.1.1.1. Isolation of glycoproteins.* Glycoproteins cannot be isolated by a uniform procedure. In fact, each of them is a particular entity due to the variability in amount and in properties of the carbohydrate moieties it carries. In general, the techniques applied are the classical methods for isolating the proteins [46, 237]. Moreover, an additional and efficient procedure has been developed in the last few years which is based on the use of lectins and which uses with advantage the presence of the carbohydrate part.

The term "lectins" proposed by Boyd and Sharpleigh [238] defines "sugar-binding proteins or glycoproteins of non-immune origin which agglutinate cells and/or precipitate glycoconjugates" [239]. They are found predominantly in plants, in particular in leguminosae, but also in animals. From a biological point of view, lectins exhibit the remarkable property of being able to agglutinate cells and, for certain of them, to be mitogenic in that they stimulate the transformation of resting lymphocytes into lymphoblasts. In this connection, the lectins represent a powerful tool for exploring the carbohydrate structures present in cell plasma membranes. For reviews, see [184—199].

The interaction of lectins with saccharides can be inhibited specifically by simple sugars and the use of the method of fixation-site saturation has led to the definition of the nature of the monosaccharide recognised by each lectin. In Table III are reported the sugar specificity of some of the most widely utilised lectins.

Originally, it was believed and stated that lectins recognised monosaccharides in non-reducing external position. However, findings of Kornfeld and Ferris, and Baenziger and Fiete [240] and of Debray et al. [241] in our laboratory led to the conclusion that in most cases complex oligosaccharides are several thousandfold more potent as "haptens" towards lectins than monosaccharides themselves. So, the notion of recognition of a monosaccharide residue in non-reducing external position by a given lectin, was

# METHODS

TABLE III

Sugar recognition by lectins [241]

| Lectin | Dominant inhibitory monosaccharides[a] | N-Acetyl-lactosaminic type structures the most inhibitory for haemagglutination |
|---|---|---|
| Concanavalin A | α-D-Man<br>α-D-Glc<br>α-D-GlcNAc | β-GlcNAc-(1 → 2)-α-Man $\begin{array}{l}(1\to 3)\\ \phantom{xx}\\(1\to 6)\end{array}$ β-Man-(1 → 4)-β-GlcNAc<br>β-GlcNAc-(1 → 4) ―――― 0-1<br>β-GlcNAc-(1 → 2)-α-Man |
| Lens escularis | α-D-Man<br>α-D-Glc<br>α-D-GlcNAc | α-NeuAc-(2 → 6)-β-Gal-(1 → 4)-β-GlcNAc-(1 → 2)-α-Man $\begin{array}{l}(1\to 3)\\ \phantom{xx}\\(1\to 6)\end{array}$ β-Man-(1 → 4)-β-GlcNAc-(1 → 4)-β-GlcNAc-(1 → )-Asn<br>α-NeuAc-(2 → 6)-β-Gal-(1 → 4)-β-GlcNAc-(1 → 2)-α-Man $\quad\quad\quad\quad\quad\quad\quad\quad\quad\quad\quad\quad\quad\quad\quad\quad\quad\quad$ $\mid$ (1 → 6)<br>$\quad\quad\quad\quad\quad\quad\quad\quad\quad\quad\quad\quad\quad\quad\quad\quad\quad\quad\quad\quad\quad\quad\quad\quad\quad\quad\quad\quad\quad$ α-Fuc |
| Vicia faba | α-D-Man<br>α-D-Glc<br>α-D-GlcNAc | α-NeuAc-(2 → 6)-β-Gal-(1 → 4)-β-GlcNAc-(1 → 2)-α-Man $\begin{array}{l}(1\to 3)\\ \phantom{xx}\\(1\to 6)\end{array}$ β-Man-(1 → 4)-β-GlcNAc-(1 → 4)-β-GlcNAc-(1 → )-Asn<br>β-Gal-(1 → 4)-β-GlcNAc-(1 → 2)-α-Man $\quad\quad\quad\quad\quad\quad\quad\quad\quad\quad\quad\quad\quad\quad\quad\quad\quad\quad\quad\quad\quad$ $\mid$ (1 → 6)<br>$\quad\quad\quad\quad\quad\quad\quad\quad\quad\quad\quad\quad\quad\quad\quad\quad\quad\quad\quad\quad\quad\quad\quad\quad\quad\quad$ α-Fuc |
| Wax bean | — | β-GlcNAc-(1 → 2) $\begin{array}{l}(1\to 3)\\ \phantom{xx}\\(1\to 6)\end{array}$ β-Man-(1 → 4)-GlcNAc<br>β-GlcNAc-(1 → 2) |
| Ricinus communis I | β-D-Gal | β-Gal-(1 → 4)-β-GlcNAc |
| PHA | — | β-Gal-(1 → 4)-β-GlcNAc-(1 → 2)-Man |
| Lotus | α-L-Fuc | α-Fuc-(1 → 6)-GlcNAc |
| Ulex I | α-L-Fuc | α-Fuc-(1 → 6)-GlcNAc-(1 →  )-Asn |

[a] Cited in descending order of affinity.

replaced by the concept of recognition of oligosaccharidic sequences including a dominant monosaccharide.

Lectins specifically precipitate oligosaccharides and glycoproteins and this property is widely used for the preparation and purification of glycoconjugates and for the study of their primary structure. It also led to the development of the fractionation and isolation of glycans, glycopeptides and glycoproteins by affinity chromatography on immobilised lectin columns. For reviews, see [188, 197, 198, 242—244].

*4.1.1.2. Isolation of glycans.* Fractionation and isolation of glycans are relatively easy, although these are often present as complex mixtures showing a marked microheterogeneity. The wide variety of procedures for the separation of oligosaccharides by the classical methods of chromatography and electrophoresis, to which one can add the use of immobilised lectins, allow practically all the problems posed by glycan isolation to be solved. These fractionation procedures are usually applied to two types of compounds: glycopeptides or glycans themselves.

Glycopeptides or glycoasparagines are obtained by the action of various proteases, the most efficient of them being pronase, an enzyme from *Streptomyces griseus*.

Glycans are removed either chemically or enzymatically. The chemical liberation of the carbohydrate moieties of glycoproteins or glycopeptides can be achieved in different ways. The action of alkali [245—248] in the presence of sodium or potassium borohydride liberates, by a $\beta$-elimination mechanism elucidated by Hartley and Jevons [249] and by Montreuil et al. [250] the glycans which are *O*-glycosidically linked to serine and threonine. *N*-Glycosically linked glycans can be removed by hydrazinolysis [251, 252], and *N*- and *O*-glycosically linked glycans by trifluoroacetolysis [253]. In both cases the native carbohydrate moieties can be re-*N*-acetylated before being further fractionated or investigated.

The enzymic removal of glycans can be performed by using endoglycosidases which are described in the next paragraph.

*4.1.2. Determination of the primary structure of glycans.* The complete determination of the primary structure of a glycan involves (1) the determination of its centesimal and molar composition with respect to simple sugars, (2) the sequential arrangement of monosaccharide units as well as the linkage between these units and the definition of the anomeric configuration of glycosidic bonds, (3) the definition of the nature of linkage between the glycan and the peptide chain.

*4.1.2.1. Sugar composition of glycans.* The centesimal composition of a glycoprotein or of a glycan in terms of neutral sugars, amino sugars, uronic acids and sialic acids, is usually determined by colorimetric procedures which have been reviewed in [229—231] and [254—256]. The identification and quantitative estimation of the principal sugars found in glycoproteins, D-mannose, D-galactose, D-glucose, D-xylose, L-arabinose, L-fucose, N-acetyl-D-glucosamine, N-acetyl-D-galactosamine and sialic acids can be performed by paper chromatography, ion-exchange chromatography and gas-liquid chromatography and does not present any particular problem (for reviews, see [46, 61, 254]). On the contrary, the quantitative liberation of monosaccharides, which must be realised before the chromatographic analysis, is still a problem of extreme complexity. In fact, all the glycosidic bonds must be split without any destruction of the sugars, and these bonds vary in stability depending on the nature of the monosaccharide and on the type of its glycosidic linkage. That is to say there is no single procedure for the quantitative liberation of monosaccharides and the principal techniques used are methanolysis and hydrolysis associated with gas-liquid chromatography. In the first case, methylglycosides are directly analysed. In the second case, the liberated monosaccharides are reduced to polyols and peracetylated prior to analysis on g.l.c. According to Marshall and Neuberger,

the effects of acid on glycoproteins still pose the most formidable problem encountered in the carbohydrate analysis of glycoproteins [255].

*4.1.2.2. Methods for the determination of the primary structure of glycans.* Numerous methods for establishing the primary structure of glycans which have become classical are reviewed or described in detail in [35, 46, 58, 61, 66, 69, 75, 76, 229—236].

*Chemical methods.* A series of chemical procedures has been described (i) for partial degradation of glycans with subsequent analysis of sugar fragments by chromatography and mass spectro-

*References p. 158*

metry: hydrolysis [257, 258], acetolysis [258—261], hydrazinolysis-nitrous acid deamination which specifically splits the N-acetylhexosamine bonds [251, 252, 262, 263] and periodic acid oxidation [264]; (ii) for complete methylation of carbohydrate, Hakomori's procedure [265] being the most efficient method, and for identification of methylated sugars by gas-liquid chromatography coupled to mass spectrometry and mass fragmentometry [266—270] being most satisfactory.

*Enzymic methods.* A second series of methodologies is based on the use of specific enzymes: exo- and endoglycosidases. Exoglycosidases hydrolyse the glycosidic bonds of monosaccharides in terminal non-reducing position and are specfic for $\alpha$- or $\beta$-glycosidic linkages. Thus, they yield information on the sequence of monosaccharides applying the method of sequential degradation, and in addition, they allow the nature of the anomeric linkage of each conjugated monosaccharide to be determined. At the present time, the use of exoglycosidases is developing very rapidly due, on the one hand, to improvements of the methods for the preparation of pure and "monospecific" enzymes by affinity chromatography and, on the other hand, to a better definition of their activity which depends on the origin of the glycosidases and on the nature and structure of the sugar to which the monosaccharide to be removed is attached (for general and technical reviews, see [233, 235, 236, 266, 271—275].

In addition, the use of endoglycosidases has been recently introduced and rapidly developed because of the very promising performance of this type of enzymes which release oligosaccharides from conjugated glycans. In fact, for a long time, the lack of endoglycosidases put the "glycoproteinists" in an inferior position as compared with the protein structuralists who possessed numerous endopeptidases. At the present time, several endoglycosidases have been described which are powerful tools in exploring the structure of glycans. They are listed in Table IV and reviewed in references 64, 235, 236 273—277. They could be divided in two classes. In Class I, the enzymes split the monosaccharide-amino acid or monosaccharide-peptide linkage, removing the complete glycan. In Class II, the enzymes cut sugar-sugar glycosidic bond, liberating part of the glycan. Endoglycosidases of Class I are the most interesting because they release the entire glycan available for further structural analysis. To this kind of enzyme belongs the

## METHODS

### TABLE IV
Sources and specificity of endoglycosidases

| Enzymes | Class of enzyme | Sources | Specificity[a] |
|---|---|---|---|
| 4'-L-Aspartylglycosylamine amidohydrolase | | Sheep epididymis [277] <br> Hen oviduct [278] <br> Hog kidney [279] | R-GlcNAc-(1 → ↓)-Asn |
| | | | Glycoasparagines of oligomannosidic and of N-acetyl-lactosaminic type |
| | | Almond emulsin [280] | Bromelain glycan (see Fig. 42) |
| Endo-N-acetyl-α-D-galactosaminidase | I | Clostridium perfringens [281] <br> Diplococcus pneumoniae [282, 283] | β-Gal-(1 → 3)-α-GalNAc-(1 → 3)↓Ser(Thr) |
| Endo-N-acetyl-β-D-glucosaminidase | II | | R-β-Man-(1 → 4)-β-GlcNAc-(1 → 4)-β-GlcNAc-(1 → ↓)-Asn |
| D-enzyme <br> H-enzyme | | Diplococcus pneumoniae [284] <br> Streptomyces griseus and plicatus [285, 286] | |
| C-I and C-II enzymes | | C. perfringens [287] <br> Mammalian kidney, liver and spleen [288] | Glycoasparagines, -peptides and/or -proteins of the mixed type and/or the oligomannosidic type |
| F-I and F-II enzymes | | Fig [289, 290] <br> Hen oviduct [291] | |
| L-enzyme | | Streptomyces griseus [285] | β-Man-(1 → 4)-β-GlcNAc-(1 → 4)↓β-GlcNAc-(1 → )-Asn |
| B-enzyme | | Basidiomyces [292] | Asialo- or monosialoglycoasparagines, -peptides and -proteins of the biantennary N-acetyl-lactosaminic type (see Table XIII) |
| Endo-β-D-galactosidase | II | Diplococcus pneumoniae [293] | α-GalNAc-(1 → 3)- or α-Gal-(1 → 3)-[α-Fuc-(1 → 2)]-β-Gal-(1 → 4)↓β-GlcNAc-(1 → )-R |
| | | Escherichia freundii [294] | (1 → 3)-[β-Gal-(1 → 4)-β-GlcNAc(SO₃ → )] |

[a] Arrows indicate the points of enzymic cleavage.

oldest one, the 4'-L-aspartylglycosylamine amido hydrolase (for review see [276]) which hydrolyses the N-acetylglucosaminyl-asparagine linkage of glycoasparagines only, whereas the amidase is inactive on glycopeptides [295]. On the other hand, the structure and size of carbohydrate moieties do not seem to affect the enzyme activity. In fact, it seems that all glycoasparagines of oligomannosidic and N-acetyllactosaminic type are split.

Another type of amidase has been prepared from almond emulsin by Takahashi [280]. This enzyme can cleave the N-acetylglucosaminyl-asparagine linkage in glycoasparagines and glycopeptides.

Also belonging to the Class I is an endo-N-acetyl-α-D-galactosaminidase from *Clostridium perfringens* [281] and *Diplococcus pneumoniae* [282, 283] which releases only β-Gal-(1 → 3)-GalNAc disaccharide α-(1 → 3)- linked to serine or threonine residues of glycoproteins.

The second class of endoglycosidases contains chiefly enzymes from various sources that act as endo-N-acetyl-β-D-glucosaminidases on the β-1,4-N-acetylglucosaminide bond of the NN'-diacetylchitobiose residue present in N-glycosylproteins. Five enzymes, endo-N-acetylglucosaminidases D, H, $C_I$, $C_{II}$ and L have been prepared at present from various sources. All act in different ways on glycan structures of the oligomannosidic type: D- and $C_I$-enzymes hydrolyse only oligosaccharides in which the α-Man-(1 → 3) moiety is not substituted (structures 8 to 10 in Fig. 23, for example) even if the N-acetylglucosamine residue of the glycosylamine linkage is fucosylated. H-enzyme is active on glycans of the oligomannosidic type and of the mixed type, like those of Fig. 23 and 36 and inactive on fucosylated glycoasparagines, while $C_{II}$-enzyme splits only the glycans of the oligomannosidic type, like those of structures 8—10 in Fig. 23. Thus, both enzymes can be used to discriminate between the oligomannosidic type glycans and the mixed type. L-Enzyme is active on the structures β-GlcNAc-(1 →   )-Asn and β-Man-(1 → 4)-β-GlcNAc-(1 → 4)-β-GlcNAc-(1 →   )-Asn only (reviews in [235, 236]).

Recently a novel endo-N-acetyl-β-D-glucosaminidase has been characterised by our group [292] in a culture filtrate of Basidiomycetes and called B-enzyme. This enzyme hydrolyses biantennary glycoasparagines and peptides or proteins of the N-acetyl-lactosaminic type (see Table XIII) provided that at least one β-galactose residue is in a non-reducing terminal position, even if the N-acetyl-

glucosamine residues of the glycosylamine linkage and of the
N-acetyl-lactosamine residues are fucosylated (see for example
structures 43 to 45 in Table XIII).
An interesting endoglycosidase has been found in the culture
fluid of D. pneumoniae which may bring forth new information
on the distribution and function of the ABO blood group determinants in human tissues [293] (see Table IV). Another endo-β-
D-galactosidase has been described in culture fluid of E. freundii
active on keratosulfate [294].

*Physical methods.* Carbohydrate moieties of glycopeptides and
of glycoproteins have been submitted to investigation by physical
methodologies at an early date, e.g. optical rotatory dispersion
and circular dichroism, mass spectrometry and X-ray diffraction
(reviews in [46]). Until recently, the application of these methods
has produced only poor results. But lately, a very efficient and
sensitive method for determining the complete primary structure
of glycans of the N-acetyl-lactosaminic type on the basis of
methylation and 360 MHz [144, 145, 296—306] or 500 MHz
[307] $^1$H-NMR spectroscopy only has been introduced. This procedure has been extended to the study of glycans of the oligomannosidic type [308] and of carbohydrate moieties of mucins
[309]. This very promising technique which requires only about
1 mg of total sugars is likely to be widely applied in the near
future to the study of the structure of cell membrane glycoproteins and of the oligosaccharides newly formed by the glycosyltransferases using glucidic acceptors. Therefore, I should like
to briefly summarise the principal results obtained by the application of this new procedure to glycans of the N-acetyl-lactosaminic
type to show its performance.

The use as reference of native and enzymically degraded serotransferrin glycoasparagines and oligosaccharides obtained by
chemical degradation of glycoprotein glycans or isolated from the
urine of patients with lysosomal storage diseases (see Figs. 61—68)
allowed us to translate the NMR message of human serotransferrin
glycan (Table V). The results obtained [296] entirely confirmed
the primary structure of this glycan (see structure No. 13 in Table
XIII) previously elucidated by Spik et al. [218—220] who had
applied chemical and enzymic procedures. In particular, integration
showed that 9 anomeric protons were present, in accordance with
the proposed number of monosaccharide units. Integration of peak

*References p. 158*

TABLE V

¹H-NMR data[a] for anomeric protons, mannose H-2 protons and N-acetyl methyl protons[b] of human serotransferrin asialoglycan (from Ref. 296).

| H-1 of residue | | | | | | | | H-2 of residue | | | NAc of residue | | | |
|---|---|---|---|---|---|---|---|---|---|---|---|---|---|---|
| 1 | 2 | 3 | 4 | 4' | 5 | 5' | 6 | 6' | 3 | 4 | 4' | 1 | 2 | 5 | 5' |
| 5.07 | 4.61 | 4.77 | 5.12 | 4.93 | 4.58 | 4.58 | 4.47 | 4.47 | 4.24 | 4.18 | 4.11 | 2.01 | 2.08 | 2.05 | 2.05 |
| (9.6) | (7.6) | (0.9)[c] | (1.3)[c] | (1.7)[c] | (8.0) | (8.0) | (8.0) | (8.0) | | | | | | | |

[a] Chemical shifts (δ) are given in ppm downfield from sodium 4,4-dimethyl-4-silapentane-1-sulphonate; the values in parentheses are coupling constants ($J_{1,2}$) in Hz.
[b] The numbering of the monosaccharide units in the reference compounds corresponds to that in the asialoglycan-Asn (see fundamental structure in Tabel XIII).
[c] Values from a convolution-difference spectrum.

areas is an accurate method for the determination of the molar ration of monosaccharides constituting a glycan.

On the basis of the serotransferrin asialoglycan data, the influence of substitution of this basic structure by mono- or oligosaccharides has been explored. The principal results obtained may be summarised as follows. (i) The di-, tri- and tetra-antennary structures may be distinguished [299, 304, 307] on the basis of the chemical shifts of H-1 and H-2 of the three mannose residues, namely, 3, 4 and 4' (see basic structure of Table XIII). An additional N-acetylglucosamine residue linked to the mannose-3 residue gives rise [297, 304] to changes in the chemical shifts of the three mannose residues (see Table VI). (ii) The occurrence of N-acetylneuraminic acid residues in $\alpha$-(2 → 3)- or $\alpha$-(2 → 6)-linkages, or both, to galactose residues induces chemical shifts [301, 304] of various anomeric and N-acetyl protons that depend on the nature of the sialyl linkage (see Table VII). Particularly interesting is the fact that $\alpha$-(2 → 6) sialyl linkages are able to induce shifts of the anomeric protons of mannose-4 and -4' residues, which are not affected at all by $\alpha$-(2 → 3) sialyl linkages. This action is probably related to the rigidity of $\alpha$-(2 → 3)-linkages in comparison with the free rotation of $\alpha$-(2 → 6)-linkages, as demonstrated by the use of molecular models. Thus, it has become possible to determine rapidly the relative positions of $\alpha$-(2 → 3) and $\alpha$-(2 → 6) sialyl residues on the different antennae in a glycan molecule. (iii) The presence of a fucose residue $\alpha$-(1 → 6)-linked to the N-acetylglycosamine-1 residue can be detected. In fact, the attachment of fucose to C-6 of the N-acetylglucosamine residue gives rise to changes in the chemical shifts for H-4, H-5 and H-6 of the aminosugar residue, as compared to the shifts of the $\beta$-GlcNAc-(1 → )-Asn residue (see Table VIII) [144, 298, 304]. In the same way, substitution by fucose on C-3 of N-acetylglucosamine of N-acetyl-lactosamine residues, as in $\alpha_1$-acid glycoprotein for example (see Fig. 26), can be demonstrated.

So, on the basis of well defined values, it is possible to use high-resolution $^1$H-NMR spectroscopy as a "finger-print method" enabling a rapid and sure determination of glycan structures. NMR spectroscopy allows the determination of the monosaccharide molar ratio [296, 304] and the complete primary structure of glycans [296, 304], the number of antennae [297, 299, 304, 307]

*References p. 158*

TABLE VI

Chemical shift data for mannose H-1 and H-2 protons in di-, tri- and tetraantennary structures [299, 304]

| Type of structure | Examples | H-1 of residue | | | H-2 of residue | | |
|---|---|---|---|---|---|---|---|
| | | 3 | 4 | 4' | 3 | 4 | 4' |
| Man⁴ — Man³ <br> Man₄' — | Human serum transferrin (see Structure 13 in Table XIII) | 4.77 | 5.12 | 4.93 | 4.24 | 4.18 | 4.11 |
| Man⁴ — Man³ <br> Man₄' — | α₁-acid glycoprotein (see Fig. 26) | 4.76 | 5.12 | 4.93 | 4.21 | 4.21 | 4.11 |
| Man⁴ — Man³ <br> Man₄' — | α₁-acid glycoprotein (see Fig. 26) | 4.77 | 5.12 | 4.86 | 4.22 | 4.22 | 4.09 |
| Man — Man <br> Man — | Human IgG (see Structure 48 in Table XIII) | 4.70 | 5.06 | 5 | 4.18 | 4.25 | 4.15 |

## TABLE VII

Influence of α-(2 → 3)-linked and α-(2 → 6)-linked terminal N-acetylneuraminic acid residues on the chemical shift of anomeric and N-acetyl protons of other monosaccharide residues present in a diantennary glycan[a] [301, 304]

| NeuAc linkage | Residues influenced | Chemical shift of | |
|---|---|---|---|
| | | Asialo chain (reference) | Sialo chain (observed) |
| α-(2 → 3) | Gal-6, -6' | 4.470 | 4.548 ± 0.003 |
| | NAc-5, -5' | 2.047 | 2.030 ± 0.001 |
| α-(2 → 6) | Gal-6, -6' | 4.470 | 4.445 ± 0.002 |
| | GlcNAc-5, -5' | 4.581 | 4.602 ± 0.005 |
| | Man-4 | 5.119 | 5.136 ± 0.003 |
| | Man-4' | 4.926 | 4.947 ± 0.005 |
| | NAc-5, -5' | 2.047 | 2.030 ± 0.001 |

[a] See footnotes [a] and [b] in Table V.

## TABLE VIII

Chemical shifts for β-GlcNAc-(1 → )-Asn and α-Fuc-(1 → 6)-β-GlcNAc-(1 → )-Asn [144, 298, 304]

| Proton of GlcNAc residue | Chemical shift of | |
|---|---|---|
| | 1[a] | 2[b] |
| 1 | 5.09 | 5.09 |
| 2 | 3.83 | 3.84 |
| 3 | 3.62 | 3.62 |
| 4 | 3.48 | 3.53 |
| 5 | 3.53 | 3.67 |
| 6a | 3.89 | 3.98 |
| 6b | 3.75 | 3.73 |

[a] β-GlcNAc-(1 → )-Asn; [b] α-Fuc-(1 → 6)-β-GlcNAc-(1 → )-Asn.

*References p. 158*

and the number and position of sialic acid [301, 304] and fucose [298, 304] residues.

This procedure has been successfully applied to the determination of the following structures: glycans of rabbit serotransferrin [300], horse pancreatic ribonuclease [302], $\alpha_1$-acid glycoprotein [144, 145], hog submaxillary glycoprotein [309], oligosaccharides from urine of sialidosis [301], mannosidosis [308], fucosidosis [298] patients and patients having Sandhoff's disease [297]. Moreover, the fact that the application of this method allowed Van den Eijnden et al. [310] to localise the position of sialic acid and fucose residues transferred by sialyl- and fucosyltransferase onto glycan acceptors. Similarly Debray et al. [305] succeeded in determining the complete primary structure of a biantennary glycoasparagine isolated from rat hepatocyte membrane with a 200 µg sample and this gives an idea of the precision and of the sensitivity of the NMR analysis of glycans.

### 4.2. Nature of linkages between glycans and protein

It is now firmly established that glycans are conjugated to peptide chains by two types of primary covalent linkages (reviews in [1, 13, 23, 25, 40, 43, 48, 58, 65, 70, 71, 77, 311—313]): *O*-glycosyl and *N*-glycosyl linkages, leading to the definition of two classes of glycoproteins: *O*-glycosylproteins and *N*-glycosylproteins (see Table IX). However, in the period between the demonstration given in 1885 by Hammarsten [110] that carbohydrate and protein were joined together by a firm chemical bond, on the one hand, and the first determination in 1961 of a monosaccharide-amino acid linkage achieved in Neuberger's laboratory [314], on the other hand, the concepts concerning the sugar-amino acid bonding have considerably evolved.

Initially, it was claimed by Levene, because of the great lability towards alkali of carbohydrate-protein linkages in numerous mucoproteins that the union was in the nature of an ester bond ([1], pp. 124—130). On the basis of the action of alkaline solutions and of lithium borohydride, Gottschalk [210] drew the same conclusion about the ovine submaxillary mucin and postulated a linkage between the disaccharide $\alpha$-NeuAc-(2 → 6)-GalNAc to the carboxyl groups of aspartic and glutamic acid residues. This pro-

# NATURE OF GLYCAN-PROTEIN LINKAGES

## TABLE IX
### Carbohydrate-peptide linkages in glycoproteins

| Amino acid and sugar residues generally engaged in the linkage | | Type of linkage | Nature of the linkage | Stability to alkali | Representative macromolecule | First ref. |
|---|---|---|---|---|---|---|
| Amino acid | Corresponding monosaccharide | | | | | |
| L-Asparagine | 2-acetamido-2-deoxy-D-glucopyranose | N-glycosidic | β-D-GlcNAcp-(1 → )-L-Asn | + | N-glycosylproteins | 314, 317—320 |
| L-Serine; L-threonine | 2-acetamido-2-deoxy-D-galactopyranose | O-glycosidic | α-D-GalNAcp-(1 → 3)-L-Ser or -L-Thr | — | Mucins | 321—325 |
| L-Serine; L-threonine | D-mannopyranose | O-glycosidic | ?-D-Manp-(1 → 3)-L-Ser or -L-Thr | — | Worm and plant glycoproteins | 336, 340 |
| L-Serine | D-xylopyranose | O-glycosidic | β-D-Xylp-(1 → 3)-L-Ser | — | Animal proteoglycans | 227, 326—330 |
| L-Serine | D-galactopyranose | O-glycosidic | α-D-Galp-(1 → 3)-L-Ser | — | Worm and plant glycoproteins | 334—339 |
| L-Threonine | D-xylopyranose | O-glycosidic | β-D-Xylp-(1 → 3)-L-Thr | — | Plant glycoproteins | 332, 333 |
| Hydroxy-L-lysine | D-galactopyranose | O-glycosidic | β-D-Gal-(1 → 5)-L-Hyl | + | Collagens | 345—348 |
| Hydroxy-L-proline | L-arabofuranose | O-glycosidic | β-L-Araf-(1 → 4)-L-Hyp | + | Plant cell wall glycoproteins | 355 |
| | D-galactopyranose | O-glycosidic | β-D-Galp-(1 → 4)-L-Hyp | + | Plant soluble glycoproteins | 359 |
| | | | | | | 364, 365 |

*References p. 158*

posal was withdrawn by Gottschalk himself when he found that dimethoxyborane, a decomposition product of lithium borohydride, can reduce free carboxyl groups [315].

Irvine and Hynd (cited in [1], p. 129) conceived the structure of mucoproteins in a different manner. In order to explain the peculiar resistance of the glycosides of chitosamine towards the action of mineral acids, the authors assumed a hydantoin-like structure of these glycosides and formulated the structure of mucoproteins in the tentative scheme of Fig. 9.

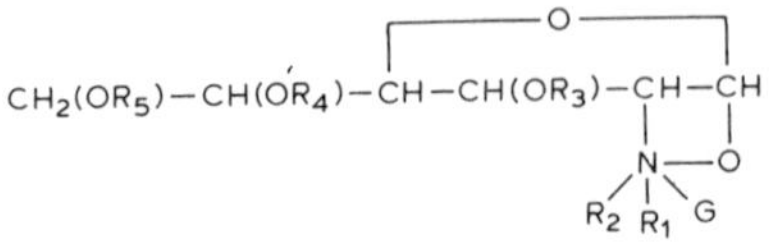

$$CH_2(OR_5)-CH(OR_4)-CH-CH(OR_3)-CH-CH$$
$$\underset{R_2\ R_1\ G}{\underset{|}{N}-O}$$
(with O bridging the two right-most CH groups)

Fig. 9. Hydantoin-like bond between glucosamine and amino acids in mucoproteins, proposed by Irvine and Hynd (quoted by Levene [1]). R stands for the radicles of peptides as well as of amino acids, and G stands for an amino acid or for a peptide.

Ether bonds were assumed by Masamune [316] to bind the peptide and carbohydrate in a blood group substance mucopolysaccharide from pig stomach mucus. The author drew this conclusion after having isolated from partial hydrolysates a crystallised acetylglucosamine-galactoside serine compound to which they ascribed the structure of Fig. 10.

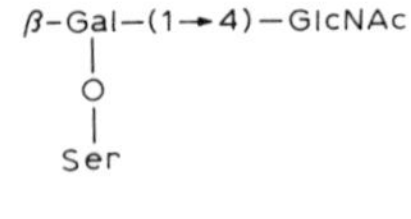

$$\beta\text{-Gal}-(1\rightarrow 4)-\text{GlcNAc}$$
$$|$$
$$O$$
$$|$$
$$\text{Ser}$$

Fig. 10. Structure of the $N$-acetylglucosamine galactoside serine ether proposed by Masamune [316].

Ionic bonds were also suggested for example between acidic mucopolysaccharides and proteins and some classifications were founded on the existence of this type of linkage, such as that of Karl Meyer (see above).

The first unambiguous demonstration of the existence of carbohydrate covalently bound to proteins was provided by Neuberger [119] in 1938. In fact, the author showed that the carbohydrate content of crystalline ovalbumin became constant on recrystallisation and that it is impossible to separate the carbohydrate from the protein by denaturation or ultrafiltration. Moreover, Neuberger isolated from a proteolytic digest of crystalline ovalbumin a pure glycopeptide in about 95% overall yield. The molecular weight estimation gave a value of about 1200 which allowed the conclusion that all the carbohydrate in ovalbumin, i.e. mannose and N-acetylglucosamine, was present as a single glycan group. This discovery led in 1961 to the first elucidation by Johansen et al. [314] of the structure of a monosaccharide-amino acid group, N-acetylglucosaminylasparagine. This was rapidly followed by the identification of the sugar-amino acid O-glycosyl linkages.

*4.2.1. N-Glycosidic linkages (glycosylamine linkages).* The first proposal of an N-glycosidic bond was made in 1955 by Masamune [316] who described an N-acetylglucosamine-aspartic acid linkage in a group substance mucopolysaccharide from pig stomach mucus. But, the attachment involved the α-amino group of aspartic acid, not the β-amino group of asparagine (Fig. 11) as was later established in a series of papers by Neuberger and coworkers [314, 317, 318], and almost simultaneously by others [319, 320]. In fact, these authors described the "binding unit" between the sugar and the peptidic chain of ovalbumin as a 2-acetamido-1-(L-β-aspartamido)-1,2-dideoxy-β-D-glucopyranose or 2-acetamido-1-N-(4'-L-aspartyl)-2-deoxy-β-D-glucopyranosylamine (Fig. 12) resulting from the binding of an N-acetylglucosamine residue by a β-linkage to the amido group of L-asparagine. This structure was confirmed by organic synthesis [245]. This structural entity is commonly termed N-acetylglucosaminyl-asparagine although it can be viewed as a condensation product of the glycosylamine of N-acetylglucosamine.

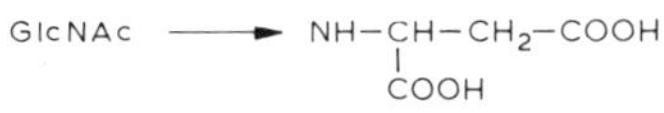

$$GlcNAc \longrightarrow NH-CH-CH_2-COOH$$
$$\phantom{GlcNAc \longrightarrow NH-}|$$
$$\phantom{GlcNAc \longrightarrow NH-}COOH$$

Fig. 11. N-glycosidic bond in group substance from pig stomach mucus according to Masamune [316].

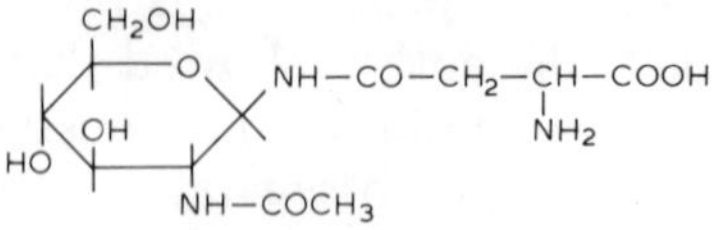

Fig. 12. 2-Acetamido-1-(L-β-aspartamido)-1,2-dideoxy-β-D-glucopyranose or 2-acetamido-1-N-(4′-L-aspartyl)-2-deoxy-β-D-glucopyranosylamine linkage, also called N-acetylglucosaminyl-asparagine.

Until now, the only N-glycosidic bond characterised in glycoproteins regardless whether it originates from animals, plants, microorganisms or viruses, is the N-acetylglycosaminyl-asparagine. In the case of mucopeptides, the linkage between the N-acetylmuramic acid and D-alanine residues is of the amide type.

*4.2.2. O-Glycosidic linkages.* In contrast to the N-glycosidic type, which is of a single type throughout the evolutionary scale, the O-glycosidic type offers a very wide variety of linkages which may be subdivided as follows.

*4.2.2.1. Linkages to L-serine and L-threonine.* Unambiguous proof for the existence in glycoproteins of glycosidic linkages of glycans to serine and threonine residues was provided almost simultaneously by several authors in 1963—1964. The evidence was based on the alkali-lability of the bond and on the fact that the combined losses of serine and threonine residues were almost equivalent to the amount of glycan residues which appeared in the medium. The first experiments concerned the salivary mucins but these were rapidly extended to many other glycoproteins. At the present time, the well defined linkages between L-serine, L-threonine and monosaccharides are the following:

*Linkage between* N*-acetyl-*D*-galactosamine and* L*-serine or* L*-threonine.* The N-acetyl-D-galactosaminyl-L-serine and L-threonine bond, in which the sugar is of the α-D-configuration (Fig. 13) was

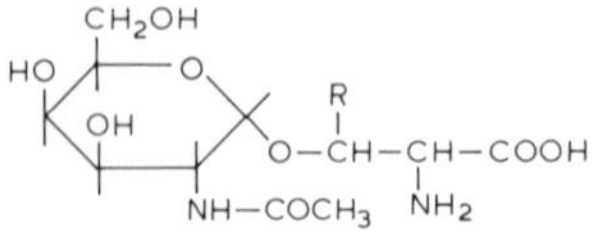

Fig. 13. 2-Acetamido-2-deoxy-O-α-D-galactopyranosyl-(1 → 3)-L-serine (R = H) or -L-threonine (R = CH$_3$) linkage.

first demonstrated in the years 1962—1963 in the mucins [321—325] and was then found in very numerous glycoproteins of which the structure of the glycosidic fraction is defined in Tables X, XI and XII. Evidently, this type of linkage is widely distributed in nature.

*Linkage between D-xylose and L-serine or L-threonine* (Fig. 14). The O-glycosidic carbohydrate-serine bond was first recognized in 1958 by Helen Muir [227] on the basis of amino acid analysis of the proteolytic digestion products of chondroitin-4-sulfate and of the alkali lability of the linkage. This proposal was more conclusively proven by the isolation of glycopeptides which contained only serine [326]. The involvement of xylose in the polysaccharide-protein linkage, was demonstrated in 1964 by Lindahl and Rodén who obtained xylosyl-serine from glycopeptides isolated from heparin [327, 328] and chondroitin 4-sulfate-protein complex [329]. Evidence that the glycosidic linkage has a β-configuration was obtained by Brendel et al. [330]. At present, the β-D-xylopyranosyl-(1 → 3)-L-serine linkage is considered as characteristically confined to animal proteoglycans (review in [213, 331]).

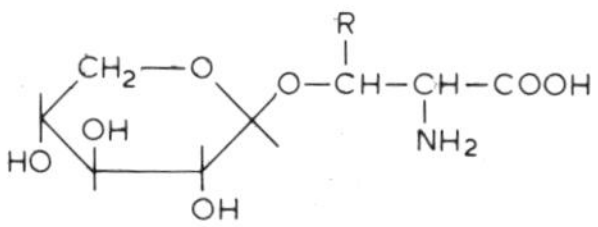

Fig. 14. O-β-D-Xylopyranosyl-(1 → 3)-L-serine (R = H) or -L-threonine (R = $CH_3$) linkage.

The β-xylosyl-L-threonine bond has been demonstrated in red algae [332] and in maize root slime cap [333].

*Linkage between D-galactose and L-serine.* A linkage of this type, in which the D-galactose is of the α-configuration [334], occurs within the collagen of *Lombricus* [335, 336] and *Nereis* [337] in tomato "extensin" [338] and in potato lectin [334, 339].

*Linkage between D-mannose and L-serine or L-threonine.* This kind of linkage is present in *Lombricus* and *Nereis* collagens [336], in the mannoproteins of *Saccharomyces cerevisiae* [340], *Cryptococcus laurentii* [341], *Penicillium melinii* [342] and *Torulopsis candida* [343], and in rat brain chondroitin sulfate proteoglycan [344].

*References p. 158*

*4.2.2.2. Linkages to 5-hydroxy-L-lysine.* Recognition of the involvement of galactose and hydroxy-L-lysine in sugar-protein linkage came through studies carried out in 1966 by Butler and Cunningham [345] on guinea pig skin tropocollagen and in 1967 by Spiro [346—348] on bovine renal glomerular basement membrane. The β-D-configuration of the galactose residue was demonstrated by Spiro [346–348] leading to the structure of Fig. 15. It seems that this linkage moiety is present in basement membrane of kidney glomeruli [348] and in collagen and collagen-like polymers (review in [349]) such as the tropocollagen from guinea pig skin [346] and collagens from the anterior lens capsule [350], from the cuttlefish [351] and from body-wall glycoproteins of the leech [352]. However, it has been reported to be also present in the C1q component of the complement system which contains amino acid sequences of collagen-like structure [353].

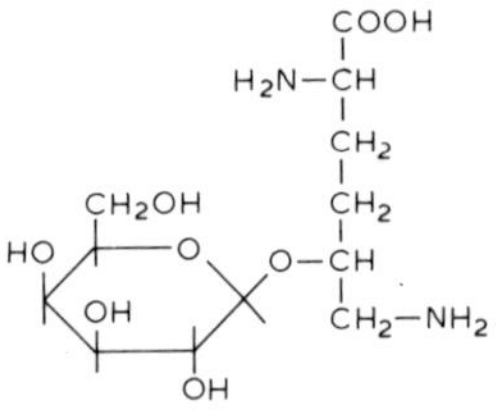

Fig. 15. *O*-β-D-Galactopyranosyl-(1 → 5)-hydroxyl-L-lysine linkage.

In all of these glycoproteins, the sugar moiety linked to the 5-hydroxy-L-lysine residues is present either as galactose only, or as the disaccharide α-D-Glc-(1 → 2)-β-D-Gal [354].

The galactosylhydroxylysine bond is very stable in strong alkali, much more so than peptide linkages, as would be anticipated from its *O*-glycosidic nature.

*4.2.2.3. Linkage to hydroxy-L-proline.* The *O*-L-arabofuranosyl-(1 → 4)-hydroxy-L-proline alkali-stable linkage (Fig. 16) was re-

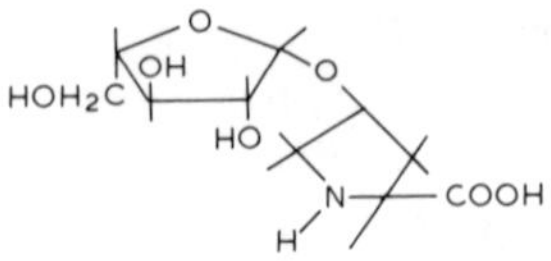

Fig. 16. *O*-β-L-Arabofuranosyl-(1 → 4)-hydroxy-L-proline linkage.

cognised in 1967 by Lamport [335] in sycamore "extensin". The β-configuration of the arabinosyl linkage was determined independently by Akiyama and Kato [356] in cell wall preparations and by Allen et al. [334, 357] in potato lectin. Since this discovery, it has been identified in a very wide variety of higher- and lower-plant cell tissues (reviews in [54, 70, 356]), such as tomato [358], tobacco [356] and green algal [359] cell walls, sandal leaves [360], arabinogalactan proteins from plant tissue culture medium [361], potato lectin [338, 339] and from seeds, leaves, stems and fruits of a variety of plants [362]. In apple fruit glycoproteins and in rice bran arabino-galactan protein, the anomeric configuration of L-arabinofuranose is α [363].

Another type of linkage involving hydroxy-L-proline, the O-β-D-galactopyranosyl-(1 → 4)-hydroxy-L-proline, has been found in green algal cell wall [359], and in arabinogalactan proteins from wheat endosperm [364, 365].

*4.2.2.4. Unusual linkages.* In addition to the linkages just described, the literature reports other, novel and unusual types of carbohydrate-peptide bonds: Linkage between L-fucose and L-threonine in a glycopeptide from human urine [366] and rat tissues [367] having the structure β-D-Glc-(1 → 3)-α-L-Fuc-(1 → 3)-L-Thr; linkage between L-fucose and L-serine α-Fuc-(1 → 3)-Ser, in rat tissues [367]; 1-thioglycosidic alkali-labile linkage between the SH group of cysteine and either a galactose disaccharide unit as in glycopeptide isolated from normal urine and originating from the kidney membrane [368] or a galactose trisaccharide unit as in glycopeptides from erythrocyte membranes [369]; linkage between the phenolic group of L-tyrosine and N-acetylneuraminic acid in chicken ovomucoid [370].

*4.2.3. Concluding remarks.* In conclusion, I should like to cite this sentence of N. Sharon [77]:

The recognition of the presence in glycoproteins of covalent bonds linking certain sugar residues to specific amino acids can be considered as the starting point of modern research in the field of glycoproteins.

It is true that the covalent linkage of saccharides to the peptide chain represents a central aspect of glycoprotein structure, the sugar-amino acid bond being generally used to classify the glycoproteins. Moreover, and above all, the perfect definition of the nature and of the properties of the carbohydrate-protein linkages, particularly of their stability towards chemical agents, led to the perfection of methods for the isolation of the glycan moieties making the study of their structure much easier. In addition, the definition of the nature of sugar-amino acid linkages laid the foundations for new, more rigorous classifications and so eliminated those based on the existence of ionic bonds between carbohydrates and proteins. Finally, it has led to the definition of rules concerning the nature of monosaccharides and amino acids which are partners in the glycan-protein linkage, in the same way as the knowledge of the primary structure of numerous glycans now allows us to define structural rules (see Section 4.3). The following facts are quite striking, from an evolutionary point of view.

(i) In almost all the glycoproteins of the animal kingdom the monosaccharide involved in the carbohydrate-protein linkage is an amino sugar, $N$-acetyl-$\beta$-D-glucosamine in the glycosylamine linkage, $N$-acetyl-$\alpha$-D-galactosamine in $O$-seryl and $O$-threonyl linkage.

(ii) The $N$-acetylglucosaminyl-asparagine is the only structural entity bonding the glycan and the peptide chain which has been found, up to now, in the $N$-glycosylproteins from animals, plants, microorganisms and viruses.

(iii) The xylosyl-serine linkage is specific for the connective tissue proteoglycans and has not yet been demonstrated in other classes of compounds, as are the galactosyl-hydroxylysine linkages, specific for basement membranes and collagens, and the arabinosyl-hydroxyproline for plant cell walls and glycoproteins. That is to say, on the basis of the sugar and amino acid composition, each of these compounds is highly distinctive, on the one hand, and of the relative labilities of linkages towards alkali, on the other hand, it is now possible to define as a first approximation the general class to which a glycoprotein belongs.

(iv) Although there is no apparent similarity between the amino acids involved in the sugar-protein linkage, certain homologies became evident when the codons for these amino acids are compared. In fact, each of the two codons for L-asparagine (AAU and AAC) can give rise by single nucleotide-substitution to the codons for L-serine (AGU and AGC), L-threonine (ACU and ACC) and L-lysine (AAA and AAG), this latter amino acid been known to be hydroxylated after incorporation into the peptidic chain. On the other hand, the codons for L-glutamine (CAA and GAG) would require two base substitutions of the codons of L-asparagine. From these data, Jett and Jamieson [371] inferred that the glycosylamine linkage between $N$-acetylglucosamine and L-asparagine is the most primitive and that the other amino acids, except 4-hydroxy-L-proline, involved in the carbohydrate-protein linkages, could be derived from L-asparagine by single-substitution of the original codon. The fact that the $N$-acetylglucosaminyl-asparagine linkage is the most widely distributed in living matter as well as the endo-$N$-acetyl-$\beta$-D-glucosaminidases (see Table IV) is in favour of this interesting view.

*4.3. Primary structure of glycoprotein glycans*

*4.3.1. Concepts and rules.* The knowledge we possess of the structures of numerous glycans raises a very exciting problem from a comparative biochemical point of view and entirely confirms the concepts developed in 1962—65 and 1974—75 [64]. In fact, glycan structures are not randomly constructed. On the contrary they are subject to laws which already appeared in the biosynthesis of the sugar-protein linkage and the bases of which are found in the specificity of glycosyltransferases and in a conservative evolution of these enzymes. In fact, the survey of the glycan structures described in Figs. 19 to 36 shows that they may be divided into families, within each of which, structures are very similar and present common oligosaccharide sequences, whether they originate from animals, microorganisms, plants or viruses. Consequently the following series of classes, concepts and rules now seems to be firmly established.

*4.3.1.1. n-Glycans and isoglycans.* Glycans are present either in $N$-glycosylproteins or in $O$-glycosylproteins in two distinct forms, for which the following terminology was proposed [64]: linear glycans or $n$-glycans, as typified by acid glycosaminoglycans (Fig. 22); branched glycans or isoglycans which are more complex and present branching points like the glycans of $N$-glycosylproteins.

*4.3.1.2. The concept of the common "inner-core" and of the*

*References p. 158*

inv *fraction of glycans*. The carbohydrate moieties of N- and O-glycosylproteins derive from the substitution of oligosaccharide structures common to numerous glycans and which are consequently non-specific. These structures are conjugated to the peptide chain and constitute the most "internal" part of glycans, namely, the "inner-core" (Fig. 17).

$\beta$-Gal-(1→3)-$\alpha$-GalNAc-(1→3)-Ser (Thr)    A

$\beta$-Gal-(1→3)-$\beta$-Gal-(1→4)-$\beta$-Xyl-(1→3)-Ser    B

$$\begin{array}{c} \phantom{xx}4 \\ \alpha\text{-Man} \\ \phantom{xx}\diagdown (1\to 3) \\ \phantom{xxxxx}\beta\text{-Man-}(1\to 4)\text{-}\beta\text{-GlcNAc-}(1\to 4)\text{-}\beta\text{-GlcNAc-}(1\to\phantom{x})\text{-Asn} \\ \phantom{xxxxxxxxxx}3 \phantom{xxxxxxxxxxxxxxxxx} 2 \phantom{xxxxxxxxxxxxx} 1 \\ \diagup (1\to 6) \\ \alpha\text{-Man} \\ \phantom{x}4' \end{array} \quad \text{C}$$

Fig. 17. Oligosaccharide inner cores of glycoproteins.

Core A exists in all the O-glycosylproteins in which the carbohydrate-protein linkage is of the N-acetyl-$\alpha$-D-galactosaminyl-(1 → 4)-L-serine of threonine type (see Fig. 13). However, a novel core structure has been recently reported by Hounsell et al. [372] which was found in one of the main oligosaccharide fractions obtained from Ii active gastric mucin of sheep. It consists of the tri-N-acetylhexosamine core $\beta$-GlcNAc-(1 → 3)-[$\beta$-GlcNAc-(1 → 6)]-N-acetylgalactosamine O-glycosidically linked to serine or threonine through the GalNAc residue.

Core B constitutes the terminal sequence of almost all the glycosaminoglycans of proteoglycans (Table IX and Fig. 22).

Core C is, with a few rare exceptions, common to all N-glycosylproteins. It results from the association, by a $\beta$-glycosidic linkage, of a mannotriose residue with a di-N-acetylchitobiose residue, itself linked to an asparagine residue through a glycosylamine linkage.

It has been proposed to term *inv*-fraction this common and non-specific "inner-core" which constitutes the invariant fraction of glycans [64]. The concept of the "core" may be extended to the carbohydrate part of glycosphingolipids in which, except in the glycans of gluco- and galactocerebrosides, the "inner-core" is composed of a lactose residue.

4.3.1.3. *The concept of the antenna and of the* var-*fraction of glycans*. The glycan structures derive from the substitution of the *inv* inner-cores by a very wide variety of glycosidic structures that confer specificity to the glycans and which thus constitute the variable fraction of *var*-fraction of the latter. On the basis both of the spatial conformation of the glycans (see Fig. 45), and of the hypothesis that the *var* glycosidic structures are the recognition signals carried by glycoproteins, it was proposed to use the term antennae for the outer branches substituting the inner-cores [64].

*4.3.1.4. Oligomannosidic and N-acetyl-lactosaminic type structures of glycans of the N-glycosylproteins.* The *N*-glycosylproteins may be divided into two families according to the nature of the carbohydrate that is linked to the pentasaccharidic inner-core C and that represents the specific structural characteristic of the glycans. In the first family (Fig. 18A) the glycans contain mannose and *N*-acetylglucosamine only and they result from the addition to the pentasaccharidic core uniquely of mannose residues. These glycans have been called glycans of the oligomannosidic type or M-type by Montreuil [64], or high-mannose type glycans by Kornfeld and Kornfeld [66]. In the second family (Fig. 18B), the sugar composition of glycans is more complex. In these glycans, galactose, fucose and sialic acids are found as constituents in addition to mannose and *N*-acetylglucosamine. However, the resulting structures (see Figs. 26—35) present evident analogies and derive fundamentally from the addition to the pentasaccharidic core of a

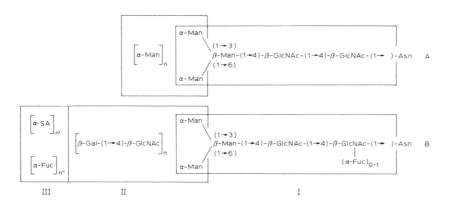

Fig. 18. General structural schemes of *N*-glycosidically linked glycans. (A) oligomannosidic type (M-type) or mannose-rich type; (B) *N*-acetyl-lactosaminic type (L-type) or complex type.

*References p. 158*

variable number of N-acetyl-lactosamine β-Gal-(1 → 4)-GlcNAc, sialic acid and/or fucose residues. These structures have been called glycans of the N-acetyl-lactosaminic type or L-type by Montreuil [64] and of the complex-type by Kornfeld and Kornfeld [66]. More recent work on the structure of glycopeptides from ovalbumin, human myeloma IgM and bovine rhodopsin (see Fig. 36) led to the definition of a third family of N-glycosylproteins in which the glycans simultaneously have structures of the oligomannosidic and of the N-acetyl-lactosaminic type. Thus they belong to the so-called oligomannosidic-N-acetyl-lactosaminic type glycans (ML-type; mixed type) [64] or hybrid type glycans [373].

*4.3.1.5. Substitution rules.* The survey of the structures described in Tables XI to XIII and in Figs. 19—36 shows that the substitution of monosaccharides conforms to a certain orthodoxy which is very restrictive [78, 304]. The possibilities of substitution of a given sugar are limited to 1, 2 or 3 well-defined monosaccharides, generally conjugated by a unique type of glycosidic linkage. The action of glycosyltransferases is thus guided by substitution rules which depend on the nature of the glycans.

For example, until now, the monosaccharide residues of the disaccharide basic structure β-Gal-(1 → 3)-GalNAc of numerous O-glycosylproteins are substituted as follows (see Table XI and Fig. 19): GalNAc residue by sialic acid or by N-acetylglucosamine in the C-6 position, rarely in the C-3 position; Gal residue by sialic acid in the C-3 position, by fucose in the C-2 position or by N-acetylglucosamine in the C-3 position.

In the case of the N-glycosically conjugated glycans, the substitution rules of the basic structure of Table XIII are as follows: GlcNAc-1 by α-fucose in the C-6 position (Table XIII and Figs. 29, 30 and 33); GlcNAc-5 and/or -5' by α-fucose in the C-3 positions, excluding the presence of sialic acid residues on Gal-6 and/or 6' (see structures 43 to 45 and 47 in Table XIII); Man-3 by N-acetyl-β-glucosamine in the C-4 position (see structures 46 to 48 in Table XIII and Fig. 31); Man-4 by α-mannose in the C-2 position (Fig. 23), by N-acetyl-β-glucosamine in the C-2 and/or in the C-4 position (Table XIII and Figs. 26, 27, 39, 40); Man-4' by α-mannose in the C-3 and/or in the C-6 position (Figs. 23 and 36), by N-acetyl-β-glucosamine in the C-2 and/or in the C-6 position (Table XIII and Figs. 26 and 29); Gal-6 and/or -6' by sialic acid in the C-3

PRIMARY STRUCTURE 53

or in the C-6 position (Table XIII and Figs. 27—30 and 33), by α-fucose in the C-2 or in the C-6 position (Fig. 63), by β-galactose in the C-3 position (see Structure 42 in Table XIII and structures 7 to 12 in Fig. 65, and [374—376], by N-acetyl-β-glucosamine in the C-3 position (Figs. 32—35).

These substitution rules raise very interesting questions of comparative biochemistry and of phylogeny and, in addition, allow the foundations to be laid for the genetics of glycosyltransferases. However, they must not be accepted as rigid and definitive rules as several "non-orthodox" structures have been described (see Figs. 37—43) on the basis of which other substitution rules could be established. However, these structures are unusual and even unique and do not detract from the generalisation made above.

*4.3.2. Description of glycan structures present in O-glycosylproteins.* The family of glycoproteins described in Table X is the only one to have glycans consisting of only one monosaccharide residue. However, this kind of monosaccharidic structure is so far restricted to a very small number of glycoproteins and, generally, the glycan structures of O-glycosylproteins are more complicated and are classified in structural groups depending on the nature of the carbohydrate-protein linkage.

TABLE X

O-Glycosylproteins with sugar moieties constituted of only one monosaccharide residue

| Origin | Structure | Ref. |
|---|---|---|
| Porcine submaxillary mucin | α-GalNAc-(1 → 3)-Ser(Thr) | 309, 377—379 |
| Human serum IgA$_1$ | α-GalNAc-(1 → 3)-Ser(Thr) | 380 |
| Tn-reactive erythrocytes | α-GalNAc-(1 → 3)-Ser(Thr) | 381 |
| Epiglycanin of TA3-Ha cells | α-GalNAc-(1 → 3)-Ser(Thr) | 382 |
| Rat brain chondroitin sulfate proteoglycan | Man-(1 → 3)-Ser(Thr) | 344 |
| Extensin | α-Gal-(1 → 3)-Ser | 338 |
| Potato lectin | α-Gal-(1 → 3)-Ser | 339 |
| Collagens | β-Gal-(1 → 5)-Hyl | 354 |

*References p. 158*

TABLE XI

Structure of oligosaccharidic glycans O-glycosylically conjugated to protein through the linkage α-GalNAc-(1 → 3)-Ser or Thr

| Designation of structures | Origin of glycans | Fundamental structure: β-Gal-(1 → 3)-α-GalNAc-(1 → 3)-Ser(Thr) | | Ref. |
|---|---|---|---|---|
| | | Substitution of Gal residue by | Substitution of GalNAc residue by | |
| 1 | Submaxillary mucins | no galactose residue | α-SA$^a$-(2 → 6) | 377, 378, 383–385 |
| 2 | Human erythrocytes | no galactose residue | α-NeuAc-(2 → 6) | 386 |
| 3–5 | Monkey cervical mucus | no galactose residue | α-NeuAc-(2 → 6) | 387 |
| | | | β-Gal-(1 → 3) | |
| | | | β-GlcNAc-(1 → 3) | |
| 6–7 | Human bronchial mucus | no galactose residue | β-GlcNAc-(1 → 3) | 388 |
| | | no galactose residue | βGal-(1 → 4)-β-GlcNAc-(1 → 3) | |
| 8 | Bovine cartilage keratan sulfate | | | 389 |
| 9 | Anti-freeze glycoprotein from antarctic fish | | | 390 |
| 10 | Porcine submaxillary mucin | | | 309, 378, 379, 391 |
| 11 | Human chorionic gonadotropin β-subunit | | | 392 |
| 12 | Human ovarian cyst fluid | | | 393 |
| 13 | Human serum IgA$_1$ | none | none | 380 |
| 14 | Epiglycanin of TA3-Ha cells | | | 382, 394 |
| 15 | T-reactive erythrocytes | | | 386, 395 |
| 16 | Rat brain glycoproteins | | | 396 |
| 17 | Human bronchial mucus | | | 388 |
| 18 | Herring egg | | | 397 |
| 19 | Rat small-intestinal mucin | | | 398 |
| 20 | Milk fat globules | | | 395, 399 |
| 21 | Human milk sIgA | | | 400, 401 |
| 22 | Bovine synovial fluid | | | 402 |
| 23 | Canine submaxillary mucin | none | α-NeuAc-(2 → 6) | 385 |
| 24 | Rat brain glycoproteins | | α-NeuAc-(2 → 6) | 396 |

# PRIMARY STRUCTURE

| # | Source | | | References |
|---|---|---|---|---|
| 25 | Herring egg | | α-NeuAc-(2 → 6) | 397 |
| 26 | Porcine submaxillary mucin | none | α-NeuGl-(2 → 6) | 378, 379 |
| 27 | Trout egg | | α-NeuGl-(2 → 8)-α-NeuGl-(2 → 6) | 397 |
| 28 | Human glycophorin | α-NeuAc-(2 → 3) | | 403 |
| 29 | Bovine kappa casein | α-NeuAc-(2 → 3) | | 404—406 |
| 30 | Fetuin | α-NeuAc-(2 → 3) | | 336, 407 |
| 31 | Epiglycanin of TA3-Ha cells | α-NeuAc-(2 → 3) | | 382, 394 |
| 32 | Rat brain glycoproteins | α-NeuAc-(2 → 3) | none | 396 |
| 33 | Bovine plasma kininogen | α-NeuAc-(2 → 3) | | 408 |
| 34 | Human milk sIgA | α-NeuAc-(2 → 3) | | 400, 401 |
| 35 | Bovine synovial fluid | α-NeuAc-(2 → 3) | | 402 |
| 36—37 | N-Blood group | α-NeuAc-(2 → 3) | none | 395, 403, 409 |
|  |  | α-NeuAc-(2 → ?) | β-Gal-(1 → ?) | 410, 411 |
| 38—39 | M-Blood group | α-NeuAc-(2 → 3) | α-NeuAc-(2 → 6) | 395, 403, 409 |
|  |  | α-NeuAc-(2 → ?) | α-NeuAc-(2 → ?)-β-Gal-(1 → ?) | 410, 411 |
| 40 | Human glycophorin | α-NeuAc-(2 → 3) | α-NeuAc-(2 → 6) | 403 |
| 41 | Milk fat globules | α-NeuAc-(2 → 3) | α-NeuAc-(2 → 6) | 395, 399 |
| 42 | Human chorionic gonadotropin β | α-NeuAc-(2 → 3) | α-NeuAc-(2 → 6) | 392, 412 |
| 43 | Fetuin | α-NeuAc-(2 → 3) | α-NeuAc-(2 → 6) | 336, 407 |
| 44 | Rat brain glycoproteins | α-NeuAc-(2 → 3) | α-NeuAc-(2 → 6) | 396, 413 |
| 45 | Bovine κ casein | α-NeuAc-(2 → 3) | α-NeuAc-(2 → 6) | 405 |
| 46 | Lymphocyte plasma membrane | α-NeuAc-(2 → 3) | α-NeuAc-(2 → 6) | 414 |
| 47 | Bovine plasma kininogen | α-NeuAc-(2 → 3) | α-NeuAc-(2 → 6) | 408 |
| 48 | Human plasminogens 1 and 2 | α-NeuAc-(2 → 3) | α-NeuAc-(2 → 6) | 415 |
| 49 | Epiglycanin of TA3-Ha cells | α-NeuAc-(2 → 3)-β-Gal-(1 → 4)-β-GlcNAc-(1 → 3) | α-NeuAc-(2 → 6) | 382 |
| 50 | Hen egg ovomucin | α-NeuAc-(2 → 3) | $SO_3^-$-(→ 6) | 416 |
| 51 | Porcine submaxillary mucin (H-active) | α-Fuc-(1 → 2) | none | 309, 377—379, 391 |
| 52 | Rat small intestinal mucosa | α-Fuc-(1 → 2) | none | 398 |
| 53 | Human bronchial mucus | α-Fuc-(1 → 2) | none | 388 |
| 54—55 | Human erythrocyte membrane type O | α-Fuc-(1 → 2) | none | 417 |
|  |  |  | α-NeuAc-(2 → 6) |  |
| 56 | Porcine and canine submaxillary mucin | α-Fuc-(1 → 2) | α-NeuGl-(2 → 6) | 309, 377—379, 385 |
| 57 | Porcine gastric mucin | α-Fuc-(1 → 2) | α-Fuc-(1 → 2)-β-Gal-(1 → 4)-β-GlcNAc-(1 → 6) | 418 |

*References p. 158*

TABLE XI (continued)

Fundamental structure: $\beta$-Gal-$(1 \rightarrow 3)$-$\alpha$-GalNAc-$(1 \rightarrow 3)$-Ser(Thr)

| Designation of structures | Origin of glycans | Substitution of Gal residue by | Substitution of GalNAc residue by | Ref. |
|---|---|---|---|---|
| 58–59 | Porcine submaxillary mucin (A active) | $\alpha$-Fuc-$(1 \rightarrow 2)$; $\alpha$-GalNAc-$(1 \rightarrow 3)$ | none | 309, 377–379 |
| 60 | Canine submaxillary mucin | $\alpha$-Fuc-$(1 \rightarrow 2)$; $\alpha$-GalNAc-$(1 \rightarrow 3)$ | $\alpha$-NeuGl-$(2 \rightarrow 6)$ | 378, 379 |
|  |  | $\alpha$-Fuc-$(1 \rightarrow 2)$; $\alpha$-Fuc-$(1 \rightarrow 2)$-$\beta$-Gal-$(1 \rightarrow ?)$-[$SO_4$]-$\beta$-GlcNAc-$(1 \rightarrow 6)$ |  |  |
| 61 | Human bronchial mucus | $\beta$-GlcNAc-$(1 \rightarrow 3)$ | none | 385 |
|  |  |  | none | 388 |
| 62 | Porcine gastric mucin | $\alpha$-GlcNAc-$(1 \rightarrow 4)$ | $\beta$-GlcNAc-$(1 \rightarrow 4)$-$\beta$-Gal-$(1 \rightarrow 4)$-$\beta$-GlcNAc-$(1 \rightarrow 6)$ | 419 |
| 63–64 | Human gastric mucin and core region of human and hog blood group substances | $\beta$-Gal-$(1 \rightarrow 3)$-$\beta$-GlcNAc-$(1 \rightarrow 3)$ | $\beta$-Gal-$(1 \rightarrow 4)$-$\beta$-GlcNAc-$(1 \rightarrow 6)$ | 420, 421 |
|  |  | $\beta$-Gal-$(1 \rightarrow 4)$-$\beta$-GlcNAc-$(1 \rightarrow 3$ and/or $6)$ | $\beta$-Gal-$(1 \rightarrow 4)$-$\beta$-GlcNAc-$(1 \rightarrow 6)$ | 422, 423 |
| 65 | Human milk sIgA | none | $\beta$-GlcNAc-$(1 \rightarrow 6)$ | 400, 401 |
|  |  | none | $\beta$-Gal-$(1 \rightarrow 4)$-$\beta$-GlcNAc-$(1 \rightarrow 6)$ |  |

[a] SA, sialic acid.

# PRIMARY STRUCTURE

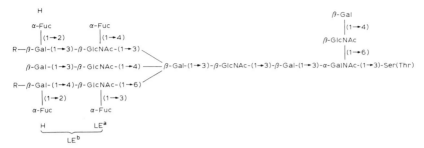

Fig. 19. Composite structure proposed for the megaloglycan from A, B, H, Le$^a$ and Le$^b$ substances by Lloyd and Kabat [420, 424]. R: α-GalNAc-(1 → 3), α-Gal-(1 → 3) or none in A, B and H substances, respectively.

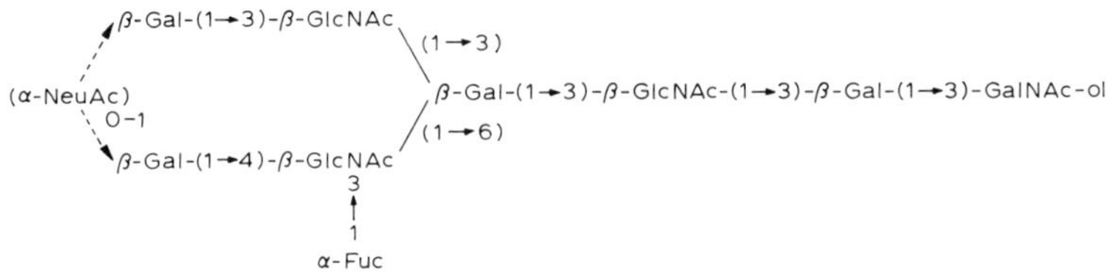

Fig. 20. Structure of O-glycosidically linked nona- and decasaccharides located in the hinge region of human milk IgA [400]. Structure of other glycans from the hinge region is given in Table XI (Structures 21, 34 and 65).

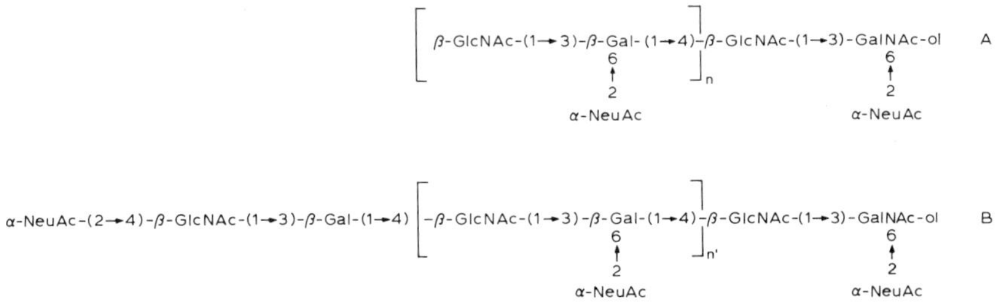

Fig. 21. Structure of the acidic oligosaccharides of rat sublingual glycoprotein [425]. Structure A: n = 4, 3 and 1 in oligosaccharides I, III and V, respectively. Structure B: n' = 2 and 1 in oligosaccharides II and IV, respectively. It is interesting to compare these structures with that of keratan sulfate II (Table XII).

*References p. 158*

*4.3.2.1. Glycans conjugated through an N-acetyl-β-D-galactosaminyl-L-serine or L-threonine linkage.* This group comprises principally structures consisting of mucins and these glycans are often designated as "mucin-like" structures, even if hey are present in plasma cell membranes and in glycoproteins from biological fluids. All the structures described until now have been listed in Table XI and described in Figs. 19—21. The main observation which could be made concerning these structures is the lack in structural specificity. This point will be discussed in Section 8.

*4.3.2.2. Glycans conjugated through a β-D-xylosyl-L-serine linkage.* This second group consists of the acid mucopolysaccharides or glycosaminoglycans. The terms "acid mucopolysaccharides" and "glycosaminoglycans" introduced by Karl Meyer [2] in 1938 and by Roger Jeanloz [17] in 1960, respectively, concern a homogeneous group of high-molecular-weight carbohydrates which constitute the major part of the ground substance of connective tissue, also called the intercellular matrix, in which the fibroblasts are embedded. All are linear polymers made up of disaccharide repeating units consisting of a repeating sequence of uronic acid, D-glucuronic or/and L-iduronic acid, or a galactose residue with a hexosamine, either glucosamine or galactosamine, generally *N*-acetylated. Moreover, sulfate groups are present as *O*-sulfate groups substituting on C-4 or C-6 of hexosamine and galactose residues of most of the mucopolysaccharides. Heparin makes an exception to this definition in two ways. Half of its L-iduronic acid residues are, in addition, sulfated on C-2 and none of its glucosamine residues is *N*-acetylated but these are *N*-sulfated.

For a long time it was believed that acid mucopolysaccharides occurred in nature as free polysaccharides or as protein salts, since the early methods of isolation performed in alkaline conditions induced the removal of carbohydrate moieties by a mechanism of β-elimination yielding compounds which generally contained no protein. But, it has been demonstrated (except for hyaluronic acid

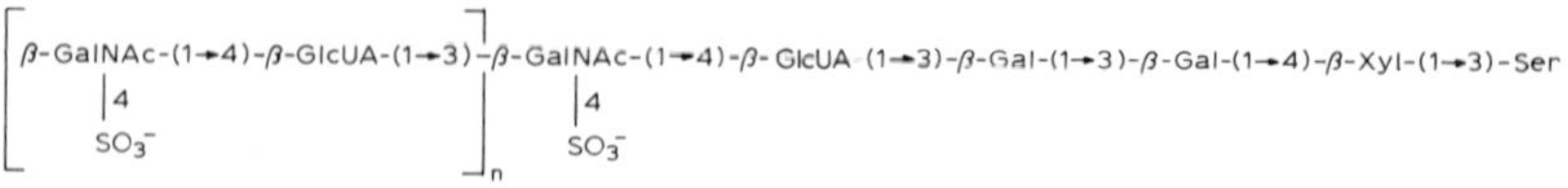

$$\left[ \beta\text{-GalNAc-}(1{\rightarrow}4)\text{-}\beta\text{-GlcUA-}(1{\rightarrow}3) \atop {\underset{SO_3^-}{|4}} \right]_n \beta\text{-GalNAc-}(1{\rightarrow}4)\text{-}\beta\text{-GlcUA }(1{\rightarrow}3)\text{-}\beta\text{-Gal-}(1{\rightarrow}3)\text{-}\beta\text{-Gal-}(1{\rightarrow}4)\text{-}\beta\text{-Xyl-}(1{\rightarrow}3)\text{-Ser} \atop {\underset{SO_3^-}{|4}}$$

Fig. 22. Primary structure of seryl-chondroitin 4-sulfate. n = 12 to 20.

## TABLE XII
### Properties, structure and occurrence of acid mucopolysaccharides (glycosaminoglycans)

| Mucopolysaccharide | Molecular weight | Monosaccharides in repeating unit | Sulfate groups | |
|---|---|---|---|---|
| | | | Per disaccharide unit | Position |
| Hyaluronic acid | $4-8 \cdot 10^6$ | D-GlcNAc<br>D-GlcUA | 0 | —<br>— |
| Chondroitin 4-sulfate<br>(Chondroitin sulfate A) | $5-50 \cdot 10^3$ | D-GalNAc<br>D-GlcUA | $0.1-1.3$ | C-4<br>— |
| Chondroitin 6-sulfate<br>(Chondroitin sulfate C) | $5-50 \cdot 10^3$ | D-GalNAc<br>D-GlcUA | $0.1-1.3$ | C-6<br>— |
| Dermatan sulfate<br>(Chondroitin sulfate B) | $15-40 \cdot 10^3$ | D-GalNAc<br>L-IdUA<br>D-GlcUA | $1-3$ | C-4<br>—<br>— |
| Heparin | $6-25 \cdot 10^3$ | D-GlcNH$_2$<br>L-IdUA<br>D-GlcUA | $1.6-3$ | N;C-3;C-6<br>C-2<br>— |
| Heparan sulfate<br>(Heparitin sulfate) | $2-10 \cdot 10^3$ | D-GlcNAc<br>D-GlcNH$_2$<br>L-IdUA<br>D-GlcUA | $0.4-2$ | C-3;C-6<br>N<br>C-2<br>— |
| Keratan sulfate I | $4-19 \cdot 10^3$ | D-GlcNAc<br>D-Gal[c] | $0.9-1.8$ | C-6<br>C-6 |
| Keratan sulfate II | $4-19 \cdot 10^3$ | D-GlcNAc<br>D-Gal[c] | $0.9-1.8$ | C-6<br>C-6 |

[a] The existence of hyaluronic acid in the tissues as a proteoglycan has not yet been demonstrated unequivocally.
[b] β-Gal-(1 → 3)-β-Gal-(1 → 4)-β-Xyl-(1 → 3)-Ser.
[c] Keratan sulfates, in addition, contain sialic acids, mannose and fucose.
[d] N-acetylglucosaminyl-asparagine linkage.
[e] α-GalNAc-(1 → 3)-Ser(Thr).

TABLE XII (continued)

| Sugar-sugar linkage | Sugar-protein linkage | Occurrence |
|---|---|---|
| β-(1 → 4)<br>β-(1 → 3) | a | Cartilage, cock's comb, skin, synovial fluid, umbilical cord, vitreous humor, mammalian cell surface, bacteria |
| β-(1 → 4)<br>β-(1 → 3) | Gal → Gal → Xyl → Ser[b] | Bones, cartilage, chondrosarcoma, cornea, granulation tissues, umbilical cord, urine |
| β-(1 → 4)<br>β-(1 → 3) | Gal → Gal → Xyl → Ser | Cartilage, heart valves, nucleus pulposus, saliva, skin, tendon |
| β-(1 → 4)<br>α-(1 → 3)<br>β-(1 → 3) | Gal → Gal → Xyl → Ser | Arterial wall, heart valves, skin, tendon |
| α-(1 → 4)<br>α-(1 → 4)<br>β-(1 → 4) | Gal → Gal → Xyl → Ser | Liver, lung mast cells, skin |
| α-(1 → 4)<br>α-(1 → 4)<br>α-(1 → 4)<br>β-(1 → 4) | Gal → Gal → Xyl → Ser | Arterial wall, cell surfaces, lung |
| β-(1 → 3)<br>β-(1 → 4) | GlcNAc → Asn[d] | Cornea |
| β-(1 → 3)<br>β-(1 → 4) | GalNAc-Ser(Thr)[e] | Cartilage |

| Designation of structure | Glycan structure | Occurrence |
|---|---|---|
| 1 | 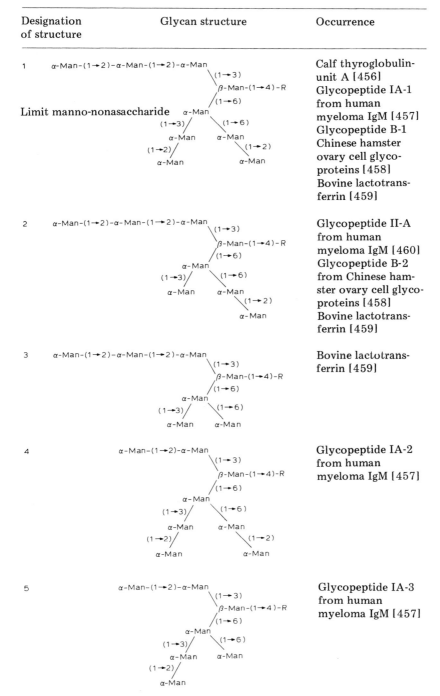 Limit manno-nonasaccharide | Calf thyroglobulin-unit A [456] Glycopeptide IA-1 from human myeloma IgM [457] Glycopeptide B-1 Chinese hamster ovary cell glycoproteins [458] Bovine lactotransferrin [459] |
| 2 | | Glycopeptide II-A from human myeloma IgM [460] Glycopeptide B-2 from Chinese hamster ovary cell glycoproteins [458] Bovine lactotransferrin [459] |
| 3 | | Bovine lactotransferrin [459] |
| 4 | | Glycopeptide IA-2 from human myeloma IgM [457] |
| 5 | | Glycopeptide IA-3 from human myeloma IgM [457] |

*References p. 158*

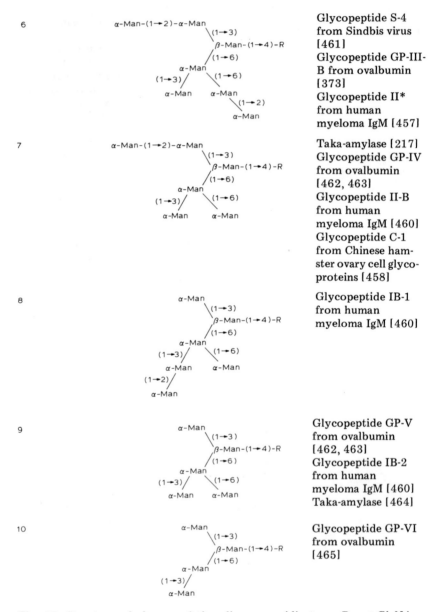

Fig. 23. Structure of glycans of the oligomannosidic type. R = β-GlcNAc-(1 → 4)-β-GlcNAc-(1 → )-Asn.

PRIMARY STRUCTURE 63

which has not yet been found conjugated to proteins) that the mucopolysaccharide moieties are covalently linked to the peptidic chains, the bond involving serine and xylose. However, in the case of keratan sulfates the linkage is somewhat different. In fact, in cartilage keratan sulfate II, N-acetylgalactosamine is attached to serine and threonine residues [389] and in corneal keratan sulfate I the linkage is of the N-acetylglucosaminyl-asparagine type [426, 427] (Table XII). As they differ significantly from other glycoproteins in their structure and physicochemical properties, the acid mucopolysaccharide-protein complexes are referred to as proteoglycans.

Since the structure of acid mucopolysaccharides has been previously reviewed in Comprehensive Biochemistry [428, 429] (see also [23, 39, 203, 312, 430—455]) we have limited the description to the principal characteristics of these compounds (Table XII). We have represented only one primary structure (Fig. 22), because almost all other glycosaminoglycans are built on the same model.

Recently, glycans conjugated through a D-mannosyl-serine or threonine linkage have been found in a chondroitin sulfate proteoglycan from rat brain [344]. The proposed structures for mannitol containing oligosaccharides obtained by mild alkaline-borohydride treatment are $\beta$-GlcNAc-(1 → 3)-Man-ol and $\beta$-Gal-(1 → 4)-[$\alpha$-Fuc-(1 → 3)]-$\beta$-GlcNAc-(1 → 3)-Man-ol.

*4.3.2.3. Glycans conjugated through a $\beta$-D-galactosyl-hydroxy-L-lysine linkage* (see Section 4.2.2.2).

*4.3.2.4. Glycans conjugated through a $\beta$-L-arabofurannosyl-hydroxy-L-proline linkage.* Until now this type of glycan has been found only in plants. The best known structures are those of oligosaccharides isolated from the cell-wall of tobacco cells and established as [$\beta$-L-Ara$f$-(1 → 3)]$_{0-1}$-$\beta$-L-Ara$f$-(1 → 2)-$\beta$-L-Ara$f$-(1 → 2)-$\beta$-L-Ara$f$-(1 → 5)-Hyp [357].

*4.3.3. Description of glycan structures present in N-glycosylproteins*

*4.3.3.1. Structure of glycans of the oligomannosidic type (or mannose-rich type).* The glycans of the oligomannosidic type constitute a very homogeneous group. In fact, almost all of them (Fig. 23) can be considered as being derived from a "limit mannononasaccharide structure" (Structure 1 in Fig. 23) and are common to numerous glycoproteins of different origins and roles.

*References p. 158*

Thus, they do not present any structural specificity and the interpretation of this observation in relation to their biological and biochemical significance will be discussed in Section 8. In addition to the structures in Fig. 23, glycans of the soybean lectin [466] and of mannoproteins from *S. cerivisiae* [467—469], respectively, are described in Figs. 24 and 25. The first is a glycan of plant origin and this could explain why its structure is different from those of Fig. 23. The second one is of particular interest because it shows that the pentasaccharidic core is present in a megaloglycan forming part of a micro-organism *N*-glycosylprotein.

### 4.3.3.2. Structure of glycans of the N-acetyl-lactosaminic type (or complex type).
Up to now, the glycan structures that have

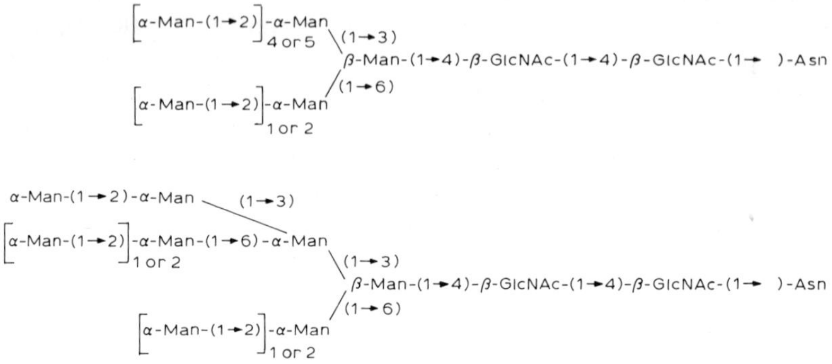

$$\begin{bmatrix}\alpha\text{-Man-}(1\to 2)\end{bmatrix}\text{-}\alpha\text{-Man}_{4\text{ or }5}\diagdown (1\to 3)$$
$$\beta\text{-Man-}(1\to 4)\text{-}\beta\text{-GlcNAc-}(1\to 4)\text{-}\beta\text{-GlcNAc-}(1\to\ )\text{-Asn}$$
$$\begin{bmatrix}\alpha\text{-Man-}(1\to 2)\end{bmatrix}\text{-}\alpha\text{-Man}_{1\text{ or }2}\diagup (1\to 6)$$

$$\alpha\text{-Man-}(1\to 2)\text{-}\alpha\text{-Man}\diagdown (1\to 3)$$
$$\begin{bmatrix}\alpha\text{-Man-}(1\to 2)\end{bmatrix}\text{-}\alpha\text{-Man-}(1\to 6)\text{-}\alpha\text{-Man}$$
$$\begin{matrix}\ _{1\text{ or }2}\end{matrix}\diagdown (1\to 3)$$
$$\beta\text{-Man-}(1\to 4)\text{-}\beta\text{-GlcNAc-}(1\to 4)\text{-}\beta\text{-GlcNAc-}(1\to\ )\text{-Asn}$$
$$\begin{bmatrix}\alpha\text{-Man-}(1\to 2)\end{bmatrix}\text{-}\alpha\text{-Man}_{1\text{ or }2}\diagup (1\to 6)$$

Fig. 24. Proposed structure of the glycan units of soybean lectin [466]*.

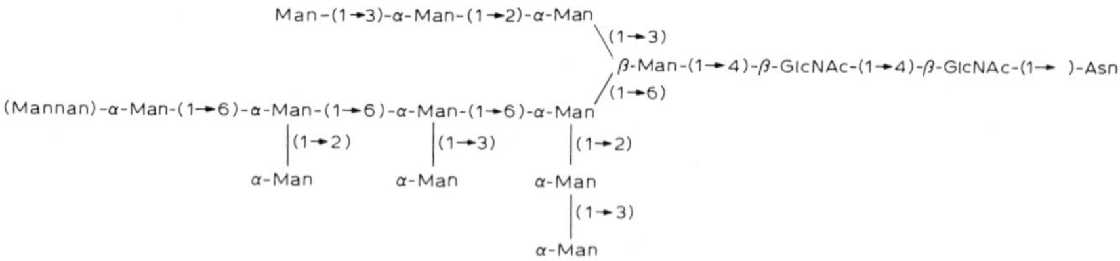

$$\text{Man-}(1\to 3)\text{-}\alpha\text{-Man-}(1\to 2)\text{-}\alpha\text{-Man}\diagdown (1\to 3)$$
$$\beta\text{-Man-}(1\to 4)\text{-}\beta\text{-GlcNAc-}(1\to 4)\text{-}\beta\text{-GlcNAc-}(1\to\ )\text{-Asn}$$
$$(\text{Mannan})\text{-}\alpha\text{-Man-}(1\to 6)\text{-}\alpha\text{-Man-}(1\to 6)\text{-}\alpha\text{-Man-}(1\to 6)\text{-}\alpha\text{-Man}\diagup (1\to 6)$$
$$|(1\to 2)\quad\quad |(1\to 3)\quad\quad |(1\to 2)$$
$$\alpha\text{-Man}\quad\quad\ \ \alpha\text{-Man}\quad\quad\ \ \alpha\text{-Man}$$
$$|(1\to 3)$$
$$\alpha\text{-Man}$$

Fig. 25. Inner-core glycoasparagine of *Saccharomyces cerevisiae* cell wall mannoproteins [467—469]. The mannoproteins themselves derive from the substitution of the core by a branched or by a linear mannan in X2180 wild-type and in X2180 *mnn2* mutant, respectively.

---

* See note (1) on p. 157.

# PRIMARY STRUCTURE

## TABLE XIII

Structures of biantennary glycans N-glycosydically linked to protein through the linkage $\beta\text{-GlcNAc-}(1 \rightarrow\ )\text{-Asn}$

**Fundamental structure**

$$\begin{array}{c}
\overset{6}{\beta\text{-Gal-}(1 \rightarrow 4)\text{-}\beta\text{-GlcNAc-}(1 \rightarrow 2)\text{-}\alpha\text{-Man}} \\
\overset{(1 \rightarrow 3)}{} \\
\beta\text{-Man-}(1 \rightarrow 4)\text{-}\beta\text{-GlcNAc-}(1 \rightarrow 4)\text{-}\beta\text{-GlcNAc-}(1 \rightarrow\ )\text{-Asn} \\
\overset{3}{} \\
\overset{(1 \rightarrow 6)}{} \\
\beta\text{-Gal-}(1 \rightarrow 4)\text{-}\beta\text{-GlcNAc-}(1 \rightarrow 2)\text{-}\alpha\text{-Man} \\
\overset{6'}{\ }\quad\overset{5'}{\ }\quad\overset{4'}{\ }
\end{array}$$

| Designation of structures | Origin of glycans | Substitution by additional monosaccharides of the following residues: | | | | | | Ref. |
|---|---|---|---|---|---|---|---|---|
| | | Gal-6 | Gal-6' | GlcNAc-5 | GlcNAc-5' | Man-3 | GlcNAc-1 | |
| 1 | Bovine colostrum IgG | — | — | none | | | | 470 |
| 2 | Waldenström IgM J-chain | — | — | | | | | 471 |
| 3 | Rabbit serotransferrin | α-NeuAc-(2 → 6) | — | — | — | — | — | 300 |
| 4 | Human complement subcomponent C₁q | α-NeuAc-(2 → 6) | — | — | — | — | — | 472 |
| 5 | Human plasminogen 1 | α-NeuAc-(2 → 6) | — | — | — | — | — | 415 |
| 6 | Bovine cold insoluble globulin | α-NeuAc-(2 → 6) | — | — | — | — | — | 473, 474 |
| 7 | Waldenström IgM J-chain | α-NeuAc-(2 → 6) | — | — | — | — | — | 471 |
| 8 | Human α₁-microglobulin | α-NeuAc-(2 → 6) | — | — | — | — | — | 475 |
| 9 | Bovine colostrum IgG | — | α-NeuAc-(2 → 6) | — | — | — | — | 470 |
| 10 | Rabbit serotransferrin | — | α-NeuAc-(2 → 6) | — | — | — | — | 300 |
| 11 | Human plasminogen 1 | — | α-NeuAc-(2 → 6) | — | — | — | — | 415 |
| 12 | Human α₁-microglobulin | — | α-NeuAc-(2 → 6) | — | — | — | — | 475 |
| 13 | Human serotransferrin | α-NeuAc-(2 → 6) | α-NeuAc-(2 → 6) | — | — | — | — | 218—220 |
| 14 | Rabbit serotransferrin | α-NeuAc-(2 → 6) | α-NeuAc-(2 → 6) | — | — | — | — | 300 |
| 15 | Human α₁-acid glycoprotein | α-NeuAc-(2 → 6) | α-NeuAc-(2 → 6) | — | — | — | — | 476 |
| 16 | Thyroxin-binding globulin | α-NeuAc-(2 → 6) | α-NeuAc-(2 → 6) | — | — | — | — | 477 |
| 17 | Rat liver plasma membrane | α-NeuAc-(2 → 6) | α-NeuAc-(2 → 6) | — | — | — | — | 305 |
| 18 | Waldenström IgM J-chain | α-NeuAc-(2 → 6) | α-NeuAc-(2 → 6) | — | — | — | — | 471 |
| 19 | Bovine prothrombin | α-NeuAc-(2 → 6) | α-NeuAc-(2 → 6) | — | — | — | — | 478 |
| 20 | Human plasminogen 1 | α-NeuAc-(2 → 6) | α-NeuAc-(2 → 6) | — | — | — | — | 415 |
| 21 | Rat α-lactalbumin | SA-(2 → 6) | SA-(2 → 6) | — | — | — | — | 479 |
| 22 | Human chorionic gonadotropin | α-NeuAc-(2 → 3) | — | — | — | — | — | 480 |

*References p. 158*

TABLE XIII (continued)

Fundamental structure

$$\begin{array}{c} 6 \\ \beta\text{-Gal-}(1 \to 4)\beta\text{-GlcNAc-}(1 \to 2)\alpha\text{-Man} \\ \phantom{xxxxxxxxxxxxxxxxxxxxxxxxxxxxxxxxxx} {}^{(1 \to 3)}\searrow \\ \phantom{xxxxxxxxxxxxxxxxxxxx} \beta\text{-Man-}(1 \to 4)\beta\text{-GlcNAc-}(1 \to 4)\beta\text{-GlcNAc-}(1 \to \phantom{x})\text{Asn} \\ \phantom{xxxxxxxxxxxxxxxxxxxxxxxxxxxxxxxxxx} {}^{(1 \to 6)}\nearrow \phantom{xxx}^{3}\phantom{xxxxxxxxxxxxxxxxxx}^{2}\phantom{xxxxxxxxxxxxx}^{1} \\ \beta\text{-Gal-}(1 \to 4)\beta\text{-GlcNAc-}(1 \to 2)\alpha\text{-Man} \\ {}^{6'}\phantom{xxxxxxxxx}{}^{5'}\phantom{xxxxxxxxxxxxx}{}^{4'} \end{array}$$

| Designation of structures | Origin of glycans | Substitution by additional monosaccharides of the following residues: | | | | | | Ref. |
|---|---|---|---|---|---|---|---|---|
| | | Gal-6 | Gal-6' | GlcNAc-5 | GlcNAc-5' | Man-3 | GlcNAc-1 | |
| 23 | Human chorionic gonadotropin | α-NeuAc-(2 → 3) | α-NeuAc-(2 → 3) | — | — | — | — | 480, 481 |
| 24 | Sindbis virus | α-NeuAc-(2 → 3) | α-NeuAc-(2 → 3) | — | — | — | — | 461 |
| 25 | Bovine serum IgG | — | — | — | — | — | α-Fuc-(1 → 6) | 482 |
| 26 | Bovine colostrum IgG | — | — | — | — | — | α-Fuc-(1 → 6) | 470 |
| 27 | Hamster embryo fibroblast galactoprotein a | — | — | — | — | — | α-Fuc-(1 → 6) | 483 |
| 28 | Human serum IgE | α-NeuAc-(2 → 6) | — | — | — | — | α-Fuc-(1 → 6) | 221, 380 |
| 29 | Human complement sub-component C₁q | α-NeuAc-(2 → 6) | — | — | — | — | α-Fuc-(1 → 6) | 472 |
| 30 | Human milk sIgA | α-NeuAc-(2 → 6) | — | — | — | — | α-Fuc-(1 → 6) | 484 |
| 31 | Human lactotransferrin | α-NeuAc-(2 → 6) | — | — | — | — | α-Fuc-(1 → 6) | 485 |
| 32 | Bovine colostrum IgG | — | α-NeuAc-(2 → 6) | — | — | — | α-Fuc-(1 → 6) | 470 |
| 33 | Human chorionic gonadotropin | α-NeuAc-(2 → 3) | — | — | — | — | α-Fuc-(1 → 6) | 480 |
| 34 | Hamster embryo fibroblast galactoprotein a | — | α-NeuAc-(2 → 3) | — | — | — | α-Fuc-(1 → 6) | 483 |
| 35 | Human lactotransferrin | α-NeuAc-(2 → 6) | α-NeuAc-(2 → 6) | — | — | — | α-Fuc-(1 → 6) | 218, 485 |
| 36 | Human serum IgA and IgE | α-NeuAc-(2 → 6) | α-NeuAc-(2 → 6) | — | — | — | α-Fuc-(1 → 6) | 221, 380 |
| 37 | Rat α-lactalbumin | SA-(2 → 6) | SA-(2 → 6) | — | — | — | α-Fuc-(1 → 6) | 479 |
| 38 | Horse pancreatic ribonuclease | α-NeuAc-(2 → 6) | α-NeuAc-(2 → 3) | — | — | — | α-Fuc-(1 → 6) | 302 |
| 39 | Sindbis virus | α-NeuAc-(2 → 3) | α-NeuAc-(2 → 3) | — | — | — | α-Fuc-(1 → 6) | 461 |
| 40 | Human chorionic gonadotropin | α-NeuAc-(2 → 3) | α-NeuAc-(2 → 3) | — | — | — | α-Fuc-(1 → 6) | 480, 481 |

# PRIMARY STRUCTURE

| | | | | | | |
|---|---|---|---|---|---|---|
| 41–42 | Human milk sIgA | α-NeuAc-(2 → 6) | α-Fuc-(1 → 6) | — | — | α-Fuc-(1 → 6) | 484 |
| | | α-NeuAc-(2 → 6) | β-Gal-(1 → 3) | — | — | α-Fuc-(1 → 6) | 484 |
| 43–44 | Human milk sIgA | — | — | α-Fuc-(1 → 3) | — | α-Fuc-(1 → 6) | 484 |
| | | α-NeuAc-(2 → 6) | — | α-Fuc-(1 → 3) | — | α-Fuc-(1 → 6) | 484 |
| 45 | Human lactotransferrin | α-NeuAc-(2 → 6) | — | α-Fuc-(1 → 3) | — | α-Fuc-(1 → 6) | 485, 486 |
| 46 | Human serum IgA₁ | α-NeuAc-(2 → 6) | — | — | β-GlcNAc-(1 → 4) | — | 380 |
| 47 | Human milk sIgA | — | — | α-Fuc-(1 → 3) | β-GlcNAc-(1 → 4) | α-Fuc-(1 → 6) | 484 |
| 48 | Human serum IgG | α-NeuAc-(2 → 6) | α-NeuAc-(2 → 6) | — | β-GlcNAc-(1 → 4) | α-Fuc-(1 → 6) | 221, 380 |

*References p. 158*

been found in this group, associated with the pentasaccharidic core C, are the following: (i) substitution on C-2 of the mannose-4 and -4' residues of the "basic structure" of Table XIII by 2 residues of N-acetyl-lactosamine, leading to the biantennary glycans described in Table XIII and in Fig. 26A; (ii) substitution by 3 residues of N-acetyl-lactosamine, giving triantennary glycans, either on C-2 and C-4 of mannose-4 and on C-2 of mannose-4' (Figs. 26B and BF, 27 and 28), or on C-2 of mannose-4 and on C-2 and C-6 of mannose-4' (Figs. 29 and 30); (iii) substitution by four residues of N-acetyl-lactosamine on C-2 and C-4 of the mannose-4 and on C-2 and C-6 of the mannose-4' which makes up the tetra-antennary glycans (Figs. 26C and CF); (iv) substitution by 5 residues of N-acetyl-lactosamine on C-2 and C-4 of the mannose-4 and on C-2, C-4, and C-6 of the mannose-4' leading to *penta-antennary glycans* (Fig. 28); (v) substitution on C-4 of the mannose-3 by an N-acetyl-glucosamine residue or "intersecting N-acetylglucosamine" (Structures 46 to 48 in Table XIII; Fig. 31); (vi) substitution on C-6 of the N-acetylglucosamine-1 by a fucose residue (Structures 25 to 45, 47 and 48 in Table XIII; Figs. 29, 30 and 33); (vii) substitution by fucose residue(s) on C-3 of the N-acetylglucosamine-5 and -5', or both (Structures 43 to 45 and 47 in Table XIII and Fig. 33) or of the N-acetylglucosamine-7 (Figs. 26 BF and CF).

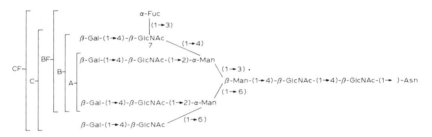

Fig. 26. Primary structures of classes A, B, BF, C and CF asialo-glycans of human plasma $\alpha_1$-acid glycoprotein [144, 145]. The 23 glycopeptides isolated from pronase hydrolysate of desialylated $\alpha_1$-acid glycoprotein are distributed as follows: Class A: GP-I-8, GP-II-6; Class B: GP-I-3, GP-I-6, GP-I-7, GP-II-5, GP-III-3, GP-III-7, GP-IV-7; Class C: GP-II-3, GP-III-2, GP-III-6, GP-IV-3, GP-IV-6, GP-V-4; Class BF: GP-I-2, GP-I-5, GP-II-4; Class CF: GP-III-5, GP-IV-4, GP-IV-5, GP-V-2, GP-V-3. For nomenclature of GP-glycopeptides from $\alpha_1$-acid glycoprotein, see [142] and Table XIV. In the native glycoprotein galactose residues are substituted in C-3 and C-6 position by N-acetylneuraminic acid residues.

## PRIMARY STRUCTURE

```
α-NeuAc-(2→3)-β-Gal-(1→4)-β-GlcNAc
                                    \(1—4)
α-NeuAc-(2→6)-β-Gal-(1→4)-β-GlcNAc-(1→2)-α-Man
                                                \(1→3)
                                                 β-Man-(1→4)-β-GlcNAc-(1→4)-β-GlcNAc-(1→  )-Asn
                                                /(1→6)
α-NeuAc-(2→3)-β-Gal-(1→4)-β-GlcNAc-(1→2)-α-Man
```

Fig. 27. Structure of calf fetuin glycan [407].

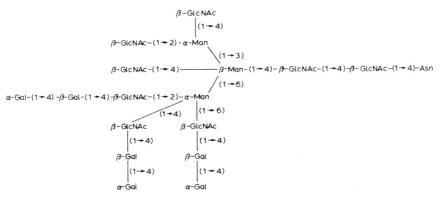

Fig. 28. Structure of glycopeptide PR-5 from turtle-dove ovomucoid possessing a P-serologic activity [487].

```
β-Gal-(1→4)-β-GlcNAc-(1→2)-α-Man
                                 \(1→3)
                                  β-Man-(1→4)-β-GlcNAc-(1→4)-β-GlcNAc-(1→  )-Asn
                                 /(1→6)                              |(1→6)
α-NeuAc-(2→6)-β-Gal-(1→4)-β-GlcNAc-(1→2)-α-Man                        α-Fuc
                                       /(1→6)
α-NeuAc-(2→6)-β-Gal-(1→4)-β-GlcNAc
```

Fig. 29. Structure of Unit-B type of glycopeptide GP-3 of porcine thyroglobulin [488, 489].

```
α-NeuAc-(2→3)-β-Gal-(1→4)-β-GlcNAc-(1→2)-α-Man
                                              \(1→3)
                                               β-Man-(1→4)-β-GlcNAc-(1→  )-Asn
                                              /(1→6)                |(1→6)
α-NeuAc-(2→3)-β-Gal-(1→4)-β-GlcNAc-(1→2)-α-Man                       α-Fuc
                                         /(1—6)
α-NeuAc-(2→3)-β-Gal-(1→4)-β-GlcNAc
```

Fig. 30. Structure of vesicular stomatitis virus membrane glycoprotein glycan [490].

*References p. 158*

```
β-GlcNAc
         \(1→4)
β-GlcNAc-(1→2)-α-Man
                    \(1→3)
β-GlcNAc-(1→4) ———— β-Man-(1→4)-β-GlcNAc-(1→4)-β-GlcNAc-(1→  )-Asn
                    /(1→6)
β-GlcNAc-(1→2)-α-Man
```

Fig. 31. Structure of hen ovotransferrin glycan [306, 491]. Compare with structure 6 of Fig. 36.

Occasionally, the branchings are incomplete and the formation of the N-acetyl-lactosamine residues is only started in outline, as in the glycans of hen ovotransferrin (Fig. 31) and of ovalbumin (Fig. 36). In other instances, glycan structures are enriched with supplementary monosaccharide residues: for example, the occurrence of disialyl groups α-NeuAc-(2 → 8)-α-NeuAc in different tissues and cell membranes [374, 375, 492, 493] and of β-Gal-(1 → 3) residues linked to terminal galactose residues in calf-thymocyte membranes [376] and in human milk sIgA [484] (see Structure 42 in Table XIII). Recently, structures containing linear oligo- or poly-N-acetyl-lactosaminyl sequences have been found and named poly(glycosyl)-peptides by Krusius et al. [494—496].

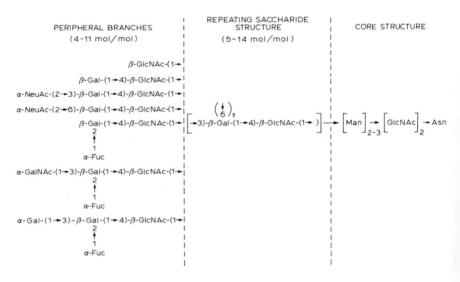

Fig. 32. Proposed general structure of the poly(glycosyl) peptides of human erythrocyte membrane [496].

Fig. 33. Structure of the minor glycopeptide from human lactotransferrin. Structure of other glycans of lactotransferrin are given in Table XIII (Structures 31, 35 and 45) [485].

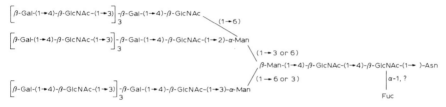

Fig. 34. Proposed structure of the major glycopeptide isolated from pronase digests of Chinese hamster ovary cell glycoproteins [497]. Tetraantennary glycans of the same type have also been characterized in the pronase hydrolysate, as well as glycans of the oligomannosidic type (see Structures 1, 2 and 7 of Fig. 23).

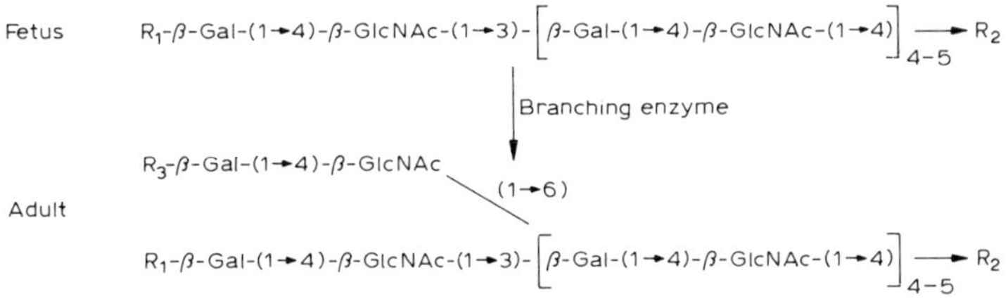

Fetus      $R_1$-β-Gal-(1→4)-β-GlcNAc-(1→3)-$\left[\text{β-Gal-(1→4)-β-GlcNAc-(1→4)}\right]_{4-5}$ ⟶ $R_2$

|Branching enzyme

           $R_3$-β-Gal-(1→4)-β-GlcNAc
                                      \(1→6)
Adult
           $R_1$-β-Gal-(1→4)-β-GlcNAc-(1→3)-$\left[\text{β-Gal-(1→4)-β-GlcNAc-(1→4)}\right]_{4-5}$ ⟶ $R_2$

Fig. 35. Proposed structures of Band 3 glycan from fetus and adult human erythrocyte membrane [498]. $R_1$: H; α-Fuc-(1 → 2); α-NeuAc-(2 → 3 or 6). $R_2$: oligosaccharide $(Man)_3(GlcNAc)_{4-5}$, the exact structure of which is unknown; $R_3$: the same as $R_1$ or $R_1 \to [(\beta\text{-Gal-}(1 \to 4)\text{-}\beta\text{-GlcNAc-}(1 \to 3)]_n$.

*References p. 158*

| Designation of structure | Glycan structure | Occurrence |
|---|---|---|
| 1 | β-GlcNAc-(1→2)-α-Man<br>　　　　　　　　　＼(1→3)<br>　　　　　　　　　　β-Man-(1→4)-R<br>　　　　　　　　　／(1→6)<br>　　　　　　　α-Man | Glycopeptide A from bovine rhodopsin [500] |
| 2 | β-GlcNAc-(1→2)-α-Man<br>　　　　　　　　　＼(1→3)<br>　　　　　　　　　　β-Man-(1→4)-R<br>　　　　　　　　　／(1→6)<br>　　　　　　　α-Man<br>　　　(1→3)／<br>　　α-Man | Glycopeptide B from bovine rhodopsin [500] |
| 3 | β-GlcNAc-(1→2)-α-Man<br>　　　　　　　　　＼(1→3)<br>　　　　　　　　　　β-Man-(1→4)-R<br>　　　　　　　　　／(1→6)<br>　　　　　　　α-Man<br>　　(1→3)／　＼(1→6)<br>　α-Man　　α-Man | Glycopeptide C from bovine rhodopsin [500]<br>Glycopeptide IA-4 from human myeloma [457] IgM |
| 4 | β-GlcNAc-(1→2)-α-Man<br>　　　　　　　　　＼(1→3)<br>β-GlcNAc-(1→4)————β-Man-(1→4)-R<br>　　　　　　　　　／(1→6)<br>　　　　　　　α-Man<br>　　(1→3)／　＼(1→6)<br>　α-Man　　α-Man | Glycopeptide GP-III-A from ovalbumin [373, 463] |
| 5 | β-GlcNAc<br>　　　　＼(1→4)<br>β-GlcNAc-(1→2)-α-Man<br>　　　　　　　　　＼(1→3)<br>β-GlcNAc-(1→4)————β-Man-(1→4)-R<br>　　　　　　　　　／(1→6)<br>　　　　　　　α-Man<br>　　(1→3)／　＼(1→6)<br>　α-Man　　α-Man | Glycopeptide GP-II-B from ovalbumin [501] |
| 6 | β-GlcNAc<br>　　　　＼(1→4)<br>β-GlcNAc-(1→2)-α-Man<br>　　　　　　　　　＼(1→3)<br>β-GlcNAc-(1→4)————β-Man-(1→4)-R<br>　　　　　　　　　／(1→6)<br>　　　　　　　α-Man<br>　　(1→3)／<br>　α-Man | Glycopeptide GP-III-C from ovalbumin [373] |

# PRIMARY STRUCTURE

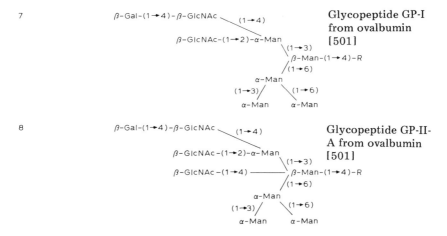

Fig. 36. Structure of glycans of the oligomannosido-N-acetyl-lactosaminic type (ML-type, mixed type or hybrid type). R = β-GlcNAc-(1 → 4)-β-GlcNac-(1→)-Asn.

These authors have isolated from human erythrocyte membranes, glycopeptides containing a repeating β-Gal-(1 → 4)-GlcNAc-(1 → 3) structure with branch points at the C-3 and C-6 positions of the galactose-6 and 6' residues of the "basic structure" of Table XIII (Fig. 32). The saccharide chains are terminated with N-acetylglucosaminyl, galactosyl, α-2,3 and/or α-2,6-N-acetylneuraminyl and α-1,2-fucosyl residues. They also contain blood group A and B determinants. These results suggest that protein-bound carbohydrates occur in a novel type of "megaloglycans" containing 20—70 monosaccharide residues and this raises the possibility that membrane blood group antigens could be carried by glycoproteins and not only by glycolipids. Similar poly-N-acetyl-lactosaminyl residues have been observed in human lactotransferrin [485] (Fig. 33), in Chinese hamster ovary cell glycoprotein [497] (Fig. 34) and in Band 3 glycoprotein of human erythrocyte membrane [498, 499] (Fig. 35).

In contrast to the glycans just described, glycans of the N-acetyl-lactosaminic type present a wide variety of structures. However, several biantennary structures are common to different glycoproteins. This latter point will be developed in Section 8.

*4.3.3.3. Structure of glycans of the mixed (or hybrid) type.* Find-

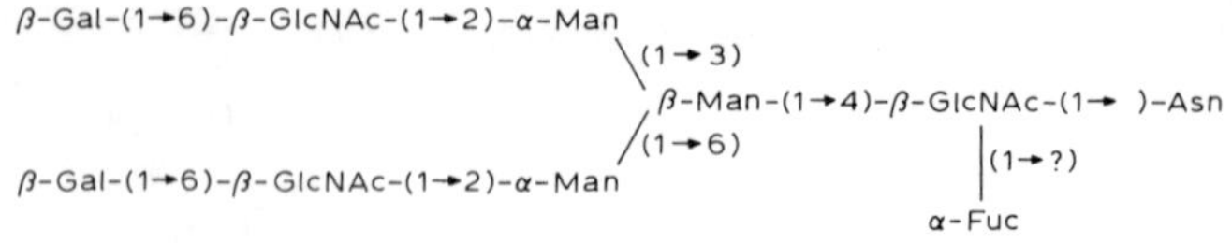

```
β-Gal-(1→6)-β-GlcNAc-(1→2)-α-Man
                                  \(1→3)
                                   β-Man-(1→4)-β-GlcNAc-(1→  )-Asn
                                  /(1→6)            |(1→?)
β-Gal-(1→6)-β-GlcNAc-(1→2)-α-Man                    α-Fuc
```

Fig. 37. Structure of human myeloma IgM glycan [502]. Unusual structures: only one GlcNAc residue in the inner-core; iso-$N$-acetyl-lactosaminic structure.

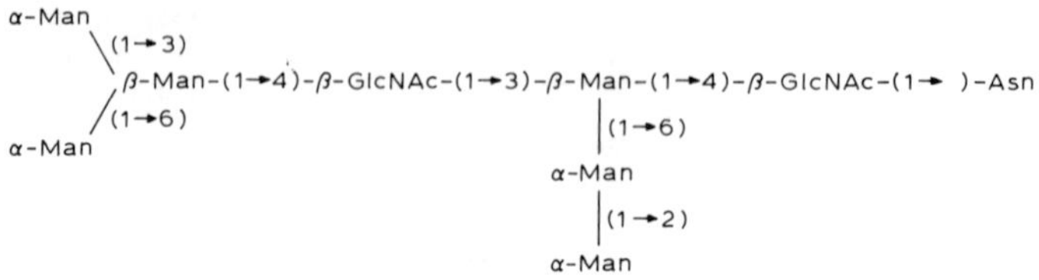

```
α-Man
     \(1→3)
      β-Man-(1→4)-β-GlcNAc-(1→3)-β-Man-(1→4)-β-GlcNAc-(1→  )-Asn
     /(1→6)                              |(1→6)
α-Man                                    α-Man
                                         |(1→2)
                                         α-Man
```

Fig. 38. Structure of C-1 glycopeptide of human myeloma IgM [221]. Unusual structure: inserted mannotriose structure into the pentasaccharidic core.

ings of Tai et al. [373, 462] and of Conchie et al. [463] on ovalbumin glycopeptides (Fig. 36) led to the definition of a third group of $N$-glycosylproteins in which the glycans simultaneously present structure of the oligomannosidic and of the $N$-acetyl-lactosaminic type, and thus belong to the "mixed" or "hybrid" type. Similar results have been obtained with human myeloma IgM [457] and bovine rhodopsin [500]. The biological significance and the origin of this kind of structure will be discussed in Section 8.

4.3.3.4. *Unusual structures.* In Figs. 37—43 are described glycan structures which do not follow the substitution rules mentioned above. These structures differ from the "orthodox" structures in the occurrence of a single $N$-acetylglucosamine residue, instead of two, at the end conjugated to asparagine (Figs. 37 and 38), by the replacement of $N$-acetyl-lactosamine residues by structures of the "iso-$N$-acetyl-lactosamine type" (Figs. 37, 39 and 40) and by the presence of unusual monosaccharide or oligosaccharide sequences and linkages which are discussed in the legends to Figs. 37—43. Could one consider these structures as "non-orthodox"? The answer is negative and it is preferable and prudent to use the term "unusual". In fact the rules previously mentioned cannot be accepted as dogma in the present state of the knowledge of glycoprotein structure. On the other hand, it is interesting to note that some of these structures belong to glycoproteins of pathological

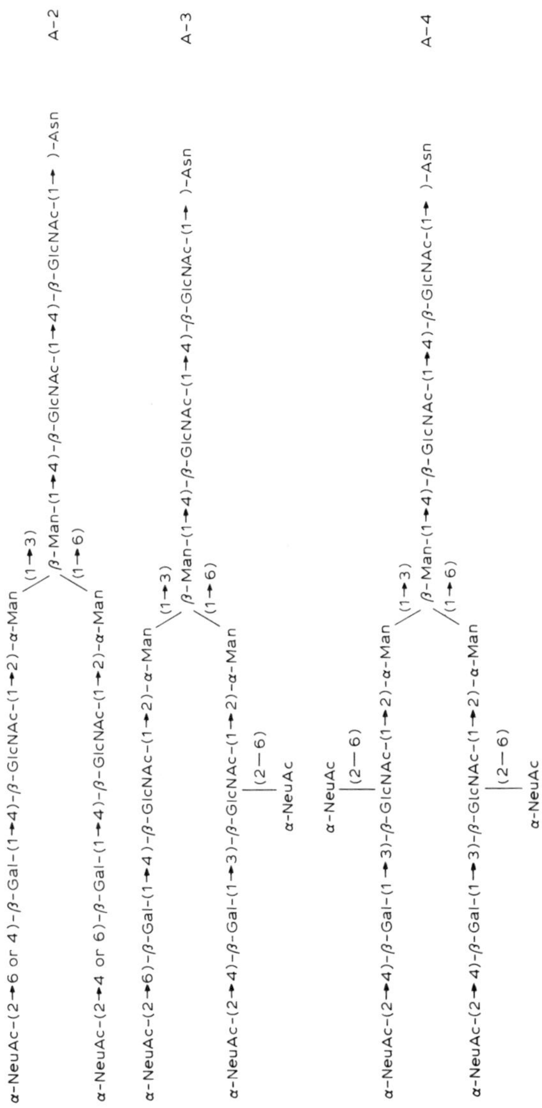

# PRIMARY STRUCTURE

α-NeuAc-(2→6 or 4)-β-Gal-(1→4)-β-GlcNAc-(1→2)-α-Man  
                                                                    (1→3)  
                                                                     β-Man-(1→4)-β-GlcNAc-(1→4)-β-GlcNAc-(1→)-Asn    A-2  
α-NeuAc-(2→4 or 6)-β-Gal-(1→4)-β-GlcNAc-(1→2)-α-Man  
                                                                    (1→6)

α-NeuAc-(2→6)-β-Gal-(1→4)-β-GlcNAc-(1→2)-α-Man  
                                            (1→3)  
                                                β-Man-(1→4)-β-GlcNAc-(1→4)-β-GlcNAc-(1→)-Asn    A-3  
α-NeuAc-(2→4)-β-Gal-(1→3)-β-GlcNAc-(1→2)-α-Man  
             |(2—6)                                           (1→6)  
            α-NeuAc

α-NeuAc  
     |(2—6)  
α-NeuAc-(2→4)-β-Gal-(1→3)-β-GlcNAc-(1→2)-α-Man  
                                            (1→3)  
                                                β-Man-(1→4)-β-GlcNAc-(1→4)-β-GlcNAc-(1→)-Asn    A-4  
α-NeuAc-(2→4)-β-Gal-(1→3)-β-GlcNAc-(1→2)-α-Man  
             |(2—6)                                           (1→6)  
            α-NeuAc

Fig. 39. Proposed structures of the glycans A-2, A-3 and A-4 of bovine cold insoluble globulin [473, 474]. Unusual structures: α-NeuAc-(2 → 4)-Gal and α-NeuAc-(2 → 6)-GlcNAc linkages.

*References p. 158*

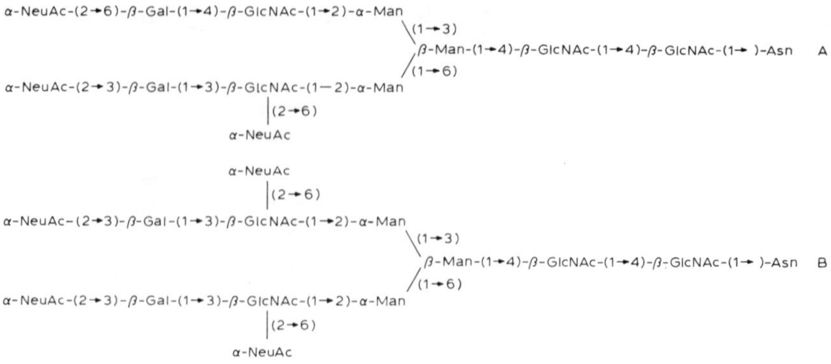

```
α-NeuAc-(2→6)-β-Gal-(1→4)-β-GlcNAc-(1→2)-α-Man
                                              \(1→3)
                                              /β-Man-(1→4)-β-GlcNAc-(1→4)-β-GlcNAc-(1→ )-Asn   A
α-NeuAc-(2→3)-β-Gal-(1→3)-β-GlcNAc-(1—2)-α-Man/(1→6)
                                              |(2→6)
                                           α-NeuAc

                                           α-NeuAc
                                              |(2→6)
α-NeuAc-(2→3)-β-Gal-(1→3)-β-GlcNAc-(1→2)-α-Man
                                              \(1→3)
                                              /β-Man-(1→4)-β-GlcNAc-(1→4)-β-GlcNAc-(1→ )-Asn   B
α-NeuAc-(2→3)-β-Gal-(1→3)-β-GlcNAc-(1→2)-α-Man/(1→6)
                                              |(2→6)
                                           α-NeuAc
```

Fig. 40. Structure of glycans A and B of bovine prothrombin [478]. Unusual structures: β-Gal-(1 → 3)-GlcNAc and α-Neu-(2 → 6)-GlcNAc linkages.

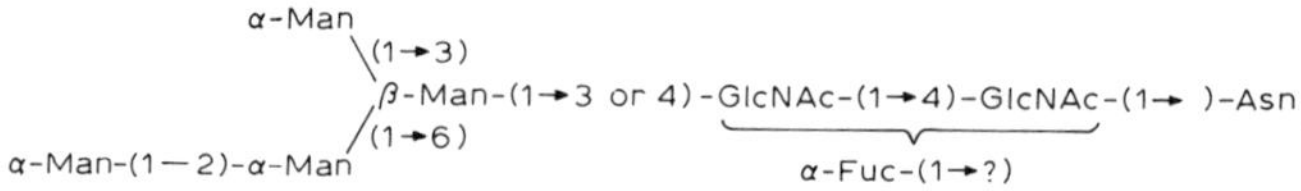

```
                    α-Man
                       \(1→3)
                       /β-Man-(1→3 or 4)-GlcNAc-(1→4)-GlcNAc-(1→ )-Asn
                       /(1→6)                ⎴⎴⎴⎴⎴⎴⎴⎴⎴⎴⎴⎴⎴⎴⎴⎴
α-Man-(1—2)-α-Man                              α-Fuc-(1→?)
```

Fig. 41. Tentative structure of the glycan of the lima bean lectin [503]. Unusual structures: α-Man-(1 → 2)-α-Man-(1 → 6) sequence and presence of fucose.

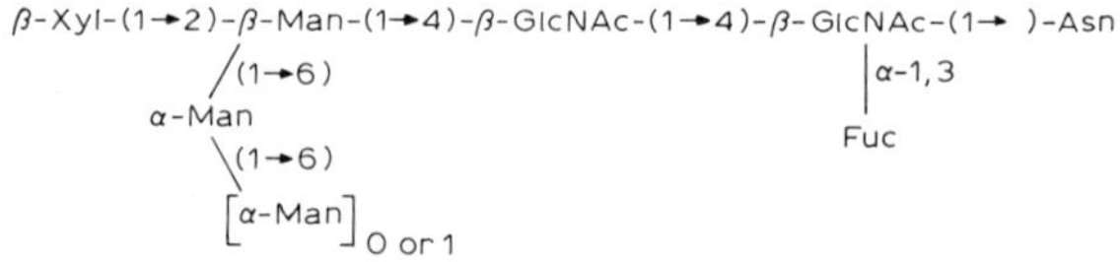

```
β-Xyl-(1→2)-β-Man-(1→4)-β-GlcNAc-(1→4)-β-GlcNAc-(1→ )-Asn
           /(1→6)                              |α-1,3
           α-Man                               Fuc
           \(1→6)
           [α-Man]_0 or 1
```

Fig. 42. Proposed structures of glycans 1 and 2 from bromelain [504]. Glycans 1 and 2 contain 3 and 2 mannose residues respectively. Unusual structures: presence of α-1,3-linked fucose.

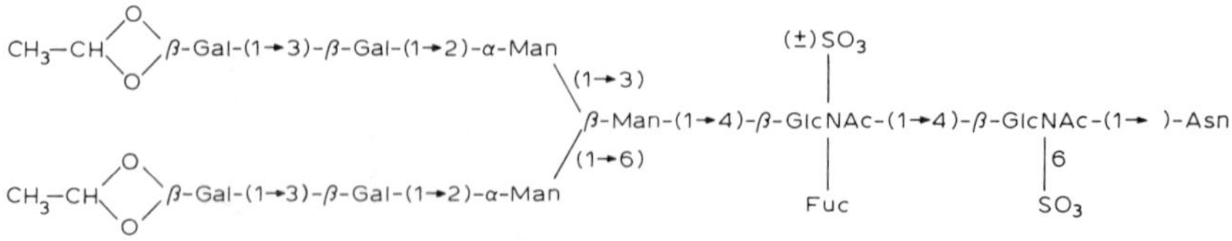

```
         O
        / \
CH₃—CH    /β-Gal-(1→3)-β-Gal-(1→2)-α-Man              (±)SO₃
        \O/                          \(1→3)              |
                                     /β-Man-(1→4)-β-GlcNAc-(1→4)-β-GlcNAc-(1→ )-Asn
         O                           /(1→6)              |              |6
        / \                                              Fuc            SO₃
CH₃—CH    /β-Gal-(1→3)-β-Gal-(1→2)-α-Man
        \O/
```

Fig. 43. Structure of the glycan of Paramyxovirus SV-5 grown in bovine kidney cell [505]. Unusual structures: presence of acetal groups; lack of N-acetyllactosamine residue; linkage of the fucose residue on N-acetylglucosamine-2.

origin (Figs. 37 and 38) and they are, perhaps, evidence of the molecular lesion being part of the nature of the disease; others are from plants (Figs. 41 and 42) or of virus (Fig. 43) origin. In addition, the isolation, from the urine of an Angus calf having mannosidosis, of a pentasaccharide having a tri-$N$-acetylchitobiose residue in the terminal reducing position (Fig. 44) suggests the existence of glycan cores containing more than two residues of $N$-acetylglucosamine [506].

$$\begin{matrix} & \beta\text{-Man-}(1\to 4)\text{-}\beta\text{-GlcNAc-}(1\to 4)\text{-}\beta\text{-GlcNAc-}(1\to 4)\text{-GlcNAc} \\ & \diagup(1\to 6) \\ \alpha\text{-Man} & \end{matrix}$$

Fig. 44. Structure of the oligosaccharide I isolated from urine of bovine mannosidosis [506].

## 4.4. Microheterogeneity of glycans

In addition to genetically determined variants expressed as variations in their polypeptide chains, almost all glycoproteins reveal another form of polymorphism associated with their carbohydrate residues. In fact, a given glycan located at a given amino acid in a glycoprotein, often presents a structural heterogeneity which is produced by partial substitution of sugar residues on a basically similar core structure. This type of diversity is termed "microheterogeneity" or "peripheral heterogeneity" because it involves the number and positions of the most external monosaccharides in the glycan residues. It must be distinguished from the presence along the same peptide chain of glycans belonging to different types of structure and/or linkage (see Fig. 49). This other type of diversity is often termed "central heterogeneity", because it is located in the core of the carbohydrate groups.

In some cases, the polymorphism of glycoproteins due to the carbohydrate can be recognized by electrophoresis or by anion-exchange chromatography of the intact molecules due to differences in charge, if the variability is related to the number and positions of sialic acid residues. In most cases, the diversity becomes apparent only from a study of glycopeptides or oligosaccharides obtained by enzymatic or chemical methods.

Since the microheterogeneity has been extensively reviewed [35, 37, 40, 50, 58, 65, 66, 507, 508] and because it is found in almost

*References p. 158*

all glycoproteins, I have limited this review to the study of only two glycoproteins taken as the most representative and illustrating both the nature and the significance of the microheterogeneity carried by their carbohydrate groups: human $\alpha_1$-acid glycoprotein and hen egg ovalbumin.

Native and pure $\alpha_1$-acid glycoprotein from human serum is homogeneous as shown by chemical analysis of the composition in amino acids and sugars, in $N$- and C-terminal amino acids, ultracentrifugation, immuno-electrophoresis and starch gel electrophoresis under conventional conditions (pH 8.6). However, it displays polymorphism near pH 2.9: 7 bands are observed with the pooled glycoprotein [140] and 5, 6, 7 or 8 bands with the glycoprotein isolated from individuals [509]. After removal of sialic acid by neuraminidase, the preparations show either a fast or a slow moving band, or both which correspond to variants of the protein part which are genetically transmitted [509]. On the basis of this observation, Schmid et al. [140] and then Yamauchi and Yamashina [510] succeeded in separating some of the polymorphic forms in a pure state by chromatography on DEAE-cellulose and demonstrated that they were identical in amino acid composition but different in their monosaccharide molar ratio and, more particularly, in their content of fucose and sialic acid. Finally, the heterogeneity of the carbohydrate moiety of $\alpha_1$-acid glycoprotein was firmly established by Schmid et al. [141]. These investigators elucidated the complete amino acid sequence of the protein part and thereby demonstrated that $\alpha_1$-acid glycoprotein possessed five glycosylation sites (see Fig. 46). Secondly, they prepared by proteolytic digestion of asialo $\alpha_1$-acid glycoprotein, 22 pure glycopeptides each derived from one of the five glycosylation sites [142]. The results described in Table XIV show that (i) the glycopeptides are different in their sugar composition; (ii) each glycosylation site of $\alpha_1$-acid glycoprotein possesses carbohydrate units with different structures. Recently, the final solution came from the complete elucidation of the primary structure of the glycan moieties of all the 22 asialo glycopeptides (see Fig. 26) [144, 145]. The glycosylation of the protein thus leads to a great variety of structures of $\alpha_1$-acid glycoprotein.

The microheterogeneity may occur in glycoproteins containing a single carbohydrate unit as well as in those containing multiple units. In this connection, hen egg ovalbumin is a remarkable exam-

## TABLE XIV

The carbohydrate composition of glycopeptides derived from asialo-$\alpha_1$-acid glycoprotein [141, 143—145]

| Site of glycosylation | Glycopeptide | Monosaccharide residues (expressed in mol/mol of glycopeptide)[a] | | | | Class of glycopeptide[b] |
|---|---|---|---|---|---|---|
| | | Fuc | Gal | Man | GlcNAc | |
| I | GP-I-2 | 0.60 (1) | 2.60 (3) | 3.00 | 5.10 (5) | BF |
| | 3 | 0 | 3.10 (3) | 3.00 | 4.80 (5) | B |
| | 5 | 0.70 (1) | 2.90 (3) | 3.00 | 4.50 (5) | BF |
| | 6 | 0 | 3.10 (3) | 3.00 | 4.80 (5) | B |
| | 7 | 0 | 2.80 (3) | 3.00 | 4.90 (5) | B |
| | 8 | 0 | 2.10 (2) | 3.00 | 3.70 (4) | A |
| II | GP-II-3 | 0 | 3.64 (4) | 3.00 | 5.94 (6) | C |
| | 4 | 0.71 (1) | 2.82 (3) | 3.00 | 5.10 (5) | BF |
| | 5 | 0 | 2.71 (3) | 3.00 | 5.17 (5) | B |
| | 6 | 0 | 1.92 (2) | 3.00 | 3.97 (4) | A |
| III | GP-III-2 | 0 | 3.72 (4) | 3.00 | 5.94 (6) | C |
| | 3 | 0 | 2.64 (3) | 3.00 | 4.93 (5) | B |
| | 5 | 0.85 (1) | 4.09 (4) | 3.00 | 5.64 (6) | CF |
| | 6 | 0 | 3.82 (4) | 3.00 | 5.99 (6) | C |
| | 7 | 0 | 2.77 (3) | 3.00 | 4.94 (5) | B |
| IV | GP-IV-5 | 0.94 (1) | 4.09 (4) | 3.00 | 5.64 (6) | CF |
| | 6 | 0 | 3.75 (4) | 3.00 | 5.96 (6) | C |
| | 7 | 0 | 2.60 (3) | 3.00 | 4.73 (5) | B |
| V | GP-V-2 | 0.91 (1) | 3.52 (4) | 3.00 | 6.21 (6) | CF |
| | 3 | 1.12 (1) | 4.06 (4) | 3.00 | 5.82 (6) | CF |
| | 4 | 0 | 3.54 (4) | 3.00 | 5.73 (6) | C |
| | 5 | 0 | 2.99 (3) | 3.00 | 4.99 (5) | B |

[a] The number of mannose residues per carbohydrate chain was assumed to be 3.00.
[b] See Fig. 26.

ple. This glycoprotein which can be easily obtained very pure and crystalline [119] consists of a single polypeptide chain to which is conjugated in a unique location [511] a single glycan group composed of mannose and N-acetylglucosamine in the molar ratio 5 : 2. Such preparations which appear homogeneous by physicochemical and immunological criteria, are, however, heterogeneous. In fact, Cunningham et al. [511] and Huang et al. [512] isolated from protease digests by ion-exchange chromatography a series of five glycopeptides which differed in their sugar composition. These authors proposed a general structure of a "whole" glycan from which the five glycopeptides could all be derived. In reality, this structure was wrong and only the improvements in the glycopeptide fractionation procedures and in the structural methodologies later developed have allowed several authors [373, 462—465, 501] to give another and more correct view of the microheterogeneity of ovalbumin (see Figs. 23 and 36).

The knowledge we now possess, on the one hand, of the structure of glycanic variants of polydisperse glycoproteins and, on the other hand, of the metabolism of glycans allows us to propose two mechanisms explaining the microheterogeneity of glycans. First, one can imagine that in the endoplasmic reticulum and Golgi apparatus where the proteins are glycosylated, all the glycosyltransferases have not the opportunity to act on a growing carbohydrate unit, depending on the rapidity of the passage of the nascent glycoprotein through the membrane network and of its excretion out of the cell. If completion of the carbohydrate units does not take place before the glycoprotein leaves the cell, molecules at varying stages of synthesis will be exported. This might be the case, for example, for glycans from bovine colostrum IgG (see structures 1, 9, 26 and 32 in Table XIII) and Waldenström IgM J-chain (see structures 2, 7 and 18 in Table XIII). However, one can not exclude the possibility that these apparently "unfinished" structures are really complete, and correspond to a finality and to recognition signals decided by the cell.

On the other hand, one could assume that glycosidases are responsible for the glycan microheterogeneity. These enzymes are widely distributed in all cells and not only in the lysosomes, so that they could in practice degrade the carbohydrate moieties of glycoproteins during their passage in or out of the cell. This hypothesis of the trimming of glycans by glycosidases has been recently supported by the work on the maturation of glycans of the $N$-acetyl-lactosaminic type. It has been demonstrated (see Section 8) that the glycans of the oligomannosidic and of the mixed type are the maturation forms of the former. Thus, the structures described in Figs. 23 and 36, including those of ovalbumin glycan variants, are metabolic intermediates in the biosynthesis of more complex glycans of the $N$-acetyl-lactosaminic type.

### 4.5. Spatial conformation of glycans

Until now the conformation of glycans remains unknown and one

# SPATIAL CONFORMATION OF GLYCANS

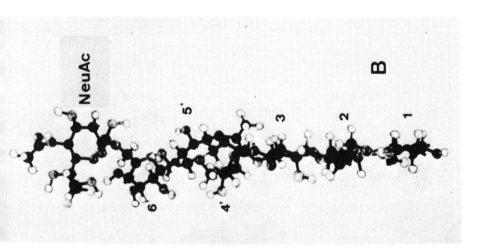

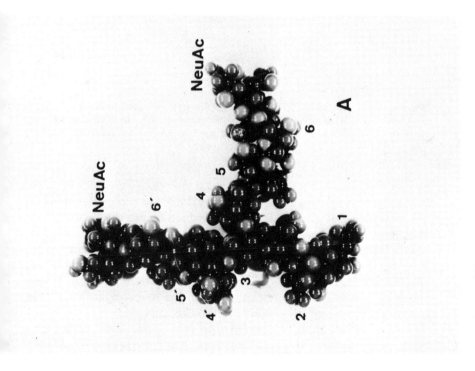

(legend on p. 82)

*References p. 158*

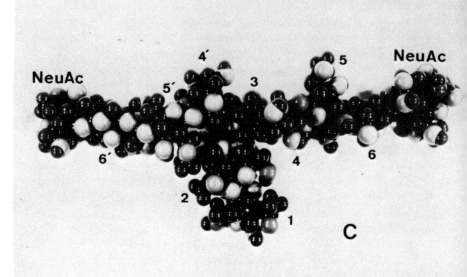

Fig. 45. Molecular model (A) and lateral view (B) of the biantennary glycan of human serotransferrin in the T-conformation, (C) "bird-conformation". Numbers correspond to the numbering used in Table XIII (see fundamental structure).

can only anticipate future results by building molecular models. Thus an image of glycans is obtained which allows the definition of the question to be posed to experimenters. The construction of the biantennary glycan molecule of human serotransferrin (structure 13 in Table XIII), creating thermodynamically possible hydrogen-bonds, led to the conformation inllustrated in Fig. 45. The essential observations that can be provided by examination of the molecular model are the following [64, 78, 304].

The structure may be divided into two parts. The first is compact: it is constituted by the *inv* part of the inner-core of mannotriosido-di-$N$-acetylchitobiose. The second is looser: it is made up of the trisaccharidic antennae $\alpha$-NeuAc-(2 → 6)-$\beta$-Gal-(1 → 4)-$\beta$-

GlcNAc-(1 → 2), the *var* structure attached to the core.

The terminal trisaccharide β-Man-(1 → 4)-β-GlcNAc-(1 → 4)-β-GlcNAc-(1 → 4) is flat (Fig. 45B). It is also rigid, due to hydrogen bonds joining the GlcNAc-1 and 2-residues, and the GlcNAc-2 and Man-3 residues.

The antennae have a helix shape (Fig. 45B).

The position of Man-4 residue is such that the antenna which it supports is disposed almost perpendicularly in relation to the plane of the trisaccharide β-Man-(1 → 4)-β-GlcNAc-(1 → 4)-β-GlcNAc-(1 → 4) leading to the T-conformation depicted in Fig. 45A. The X-ray diffraction pattern of crystals of the trisaccharide α-Man-(1 → 3)-β-Man-(1 → 4)-GlcNAc isolated from mannosidosis urine (see Fig. 61) is in good accordance with the proposed conformation. In fact the experimental results obtained [513] confirmed the proposition that the disaccharide β-Man-(1 → 4)-GlcNAc is planar, that a hydrogen bond exists between the oxygen atom of the oxygen bridge of the Man-3 residue and the OH group on C-6 of the GlcNAc-2 residue, and that the position of Man-4 is perpendicular to the plane of the disaccharide β-Man-(1 → 4)-GlcNAc.

More recently, the conformation illustrated in Fig. 45C has been adopted, that I proposed to call "bird-conformation" [514]. This new type of conformation was accepted on the basis of a series of concordant results obtained by different ways: (a) crystallographic structural studies of Fc fragment from human IgG based on a Fourier map at 2.9 and 3.5 Å resolution (Huber [515]; Deisenhofer, personal communication); (b) analysis by freeze-fracture electron microscopy and low-angle X-ray scattering of synthetic amphipathic block polymers obtained by coupling glycopeptides of ovomucoid with hydrophobic peptide block: the γ-benzyl-L-glutamate block [516]; (c) by questioning computers that are ideal for calculating which structures are sterically feasible and which conformation is energetically the most favourable (Perez, Warin and Montreuil, unpublished results).

At first sight, the bird-conformation seems to be far from the concept of recognition signal. A few years ago, one believed that lectins, including membrane lectins, were able to recognise and bind only monosaccharides in terminal non-reducing position. We now know that lectins, and also antibodies, could recognise and bind not only "terminally" but also "laterally" mono- and oligosaccharides located in an internal position in the glycan moieties.

*References p. 158*

On the other hand, the bird-conformation is more favourable in creating interactions between glycans and proteins in glycoproteins. In addition, regarding the biosynthesis of glycans, the bird-conformation is very satisfactory since all the glycosylable hydroxyl groups of monosaccharides are more accessible by glycosyltransferases. Moreover, this new concept allows us to raise an interesting point: that of the relative dimensions of glycans towards the protein moiety [514]. For instance, the dimensions of human serotransferrin biantennary glycan are: $55 \times 24 \times 5$ Å, while those of the protein moiety are $95 \times 60 \times 50$ Å. In the case of $\alpha_1$-acid glycoprotein, because of the presence of 5 glycan moieties which are essentially of the tetra-antennary type, the protein is enveloped by the carbohydrate moieties which are in an "umbrella-conformation", each covering an area of 3000 Å$^2$ [514]. In this way the resistance towards proteases of numerous glycoproteins as well as their weak antigenicity, glycans acting as shields, could be explained.

Glycan structure should not be regarded as fixed in the conformation shown in Fig. 45. Of course, certain sequences are "rigid" and solidly maintained by hydrogen bonds, but they also have points of flexibility leading to a relative mobility of antennae. One of these is located at the $\alpha$-$(1 \rightarrow 6)$-bond linking the Man-4' residue to the Man-3 residue, conferring a certain freedom of rotation in the space of the antenna carried by this linkage. The existence of such an internal degree of freedom in the glycan structure has been recently confirmed by an ESR study carried out by Davoust et al. [517] on human serotransferrin glycoasparagine. These investigators, using these compounds covalently spin-labelled on their sialic acid residues, demonstrated spin-spin interactions originating from a collision effect between probes which are decreased by concanavalin A. Moreover, Gallot et al. [516] demonstrated that the bird- and T-conformations are interconvertible, high concentration of glycopeptides favouring the bird-conformation.

Other points of flexibility involve the two $N$-acetylneuraminic acid residues which are able, due to their $\alpha$-$(2 \rightarrow 6)$-linkage, to take up most varied positions in space, even to fall back on the GlcNAc-5 and 5' residues. But they adopt a more fixed position when conjugated by an $\alpha$-$(2 \rightarrow 3)$ bond. This property may explain the shifts of $^1$H of the Man-4 and 4' residues, observed by NMR spectroscopy, only when the galactose-6 and 6' residues are substituted in C-6 position by $N$-acetylneuraminic acid (see Table VII).

The double character of rigidity and flexibility that the glycans possess is compatible with the role of recognition signals which is ascribed to them. They may be imagined as solidly planted on the proteins by a rigid arm constituted of the terminal core trisaccharide which is perhaps able, on account of its planar conformation, to penetrate into the protein. In this connection, it is of interest to note that the terminal trisaccharide of the inner-core is relatively hydrophobic because of the presence of the two or sometimes three-fold substituted Man-3 residue and of the two $N$-acetylglucosamine residues. The hydrophobicity of this structure is still more enhanced when the GlcNAc-1 is fucosylated. It is therefore not unreasonable to assume that this hydrophobic part of glycans could interact with hydrophobic peptide inner-sequences to better anchor the flexible antennae. The latter, because of their unimpeded mobility, could rapidly adapt themselves to the receptor sites and interact with them.

## 5. Some selected glycoproteins

Glycoproteins are widely distributed in nature and we know now that most proteins are, in fact, glycoproteins. They differ (1) in the amount of sugars which varies from 0.5% in collagen to 85% in blood group substances, (2) in the number (1 in ovalbumin, 800 in ovine submaxillary mucin), the size (1 monosaccharide in glycoprotein reported in Table X, 250 in acidic mucopolysaccharides), and the structure (see Fig. 8) of oligosaccharide chains, (3) in the linkage of glycan moieties to peptides (see Fig. 8) producing $O$- and $N$-glycosylproteins. Thus, the hundreds of glycoproteins we now know present a considerable diversity. Moreover, recent literature abounds with reports on the occurrence and the physico-chemical properties of glycoproteins (reviews in [12, 15, 21, 26, 34, 35, 39, 40, 46, 47, 52, 56, 58, 65, 68, 69, 72—74, 77, 176, 518, 519]).

For these reasons, I chose to limit this Section to the description of the main characteristics of the five following compounds: human serotransferrin, as an $N$-glycosylprotein in which the two glycans of the $N$-acetyl-lactosaminic type are identical and devoid of any microheterogeneity; $\alpha_1$-acid glycoprotein, as an $N$-glycoprotein with 5 glycosylation sites occupied by glycans of the $N$-acetyl-lactosaminic type presenting a wide microheterogeneity; thyroglobulin as an example of glycoprotein carrying together glycans of the oligomannosidic and of the $N$-acetyl-lactosaminic type; proteoglycans as an example of $O$-glycosylproteins; glycophorin, as a glycoprotein in which $O$- and $N$-glycosidically conjugated glycans coexist, and as a model of a membrane glycoprotein.

## 5.1. Human serotransferrin

"Transferrins", are glycoproteins which bind reversibly two atoms of iron ($Fe^{3+}$) and which affect the transport and distribution of the metal to and from the body organs (reviews in [46 and 69]). Human transferrin or siderophilin is the iron-binding protein of serum (about 3 g/l). It has an $M_r$ = 78 000 ± 2000, a single peptide chain the primary sequence of which is known [520] and contains 5.9% carbohydrates distributed over two identical biantennary glycans of the $N$-acetyl-lactosaminic type (see structure 13 in Table XIII) [218–220, 296]. A fascinating aspect of the role of serotransferrin is the transfer of the metal into the reticulocyte. Thus, it has been demonstrated that saturated serotransferrin is recognized and fixed by a specific receptor present at the surface of reticulocytes and that, after the iron has been transferred into the cell, the apotransferrin separates from the membrane [521]. It has not yet been shown that both glycans are recognition signals for the reticulocyte membrane receptor.

The family of transferrins represents a very interesting problem of the comparative biochemistry of glycans. Thus, the carbohydrate moieties of transferrins of different species are entirely different. Rabbit serotransferrin glycan (Structure 3 in Table XIII) is identical with those found in human serotransferrin, but there is only one per molecule. Like human serotransferrin, human milk lactotransferrin possesses two glycans of the $N$-acetyl-lactosamine type but they are different from the former because they are fucosylated and more complex (see Structures 31, 35 and 45 in Table XIII, and Fig. 33). Bovine milk lactotransferrin in contrast to human lactotransferrin, contains one glycan of the $N$-acetyl-lactosamine type, the structure of which has not yet been determined, and another one of the oligomannosidic type (see Structure 1 in Fig. 23). The structure of the only hen ovotransferrin glycan differs from the preceding glycans by the fact that it bears four incomplete and immature branches devoid of galactose residues (see Fig. 31).

### 5.2. $\alpha_1$-Acid glycoprotein

$\alpha_1$-Acid glycoprotein or "orosomucoid" (reviews in [522, 523]) is a globulin from human plasma (0.5—0.7 g/l) having an $M_r$ 40 000 ± 2000, characterised by the following properties: (1) it has a single peptide chain, the primary structure [141] of which is given in Fig. 46, (2) it has a very high carbohydrate (45%) and sialic acid (11—12%) content giving the glycoprotein a very acidic isoelectric point, (3) it shows an important microheterogeneity of the glycan moieties which are distributed at five glycosylation sites (Sites I to V) located at the $N$-terminus of the peptide chain. The microheterogeneity of $\alpha_1$-acid glycoprotein, which is the most important known until now, has been discussed in Section 4.4 and has been explained owing to the complete determination of the structure of all glycans substituting the five glycosylation sites (see Table XIV and Fig. 26). The role of glycans and the significance of the high microheterogeneity they present as well as the polymorphism of $\alpha_1$-acid glycoprotein are not yet interpretable. The question will remain obscure until the role of the latter is known.

### 5.3. Thyroglobulin

Thyroglobulin (reviews in [58, 524, 525]) the major constituent of the thyroid gland, is a high $M_r$ (670 000) iodinated glycoprotein (10% total sugars) with approx. 300 monosaccharide residues distributed among two types of carbohydrate units [526]: unit A of the oligomannosidic type (see Structure 1 in Fig. 23) and unit B of the tri-antennary $N$-acetyl-lactosaminic type (see Fig. 29). The repartition of units A and B varies from one thyroglobulin to another: 5 units A and 14 units B, 7 units A and 22 units B in calf and in human proteins, respectively. Moreover, human thyroglobulin, in contrast to the other thyroglobulins, contains $N$-acetylgalactosamine linked to serine and threonine and forming

*References p. 158*

```
  1
<Glu - Ile - Pro - Leu - Cys - Ala - Asn - Leu - Val - Pro - Val - Pro - Ile - Thr - Asn*- Ala - Thr - Leu - Asp - Arg - Ile - Thr - Gly - Lys - Trp - Phe - Tyr - Ile - Ala - Ser -
                                                 10                          15                       Gln
                          38                                         50                                                    54
Ala - Phe - Arg - Asn - Glu - Glu - Tyr - Asn*- Lys - Ser - Val - Glu - Glu - Ile - Gln - Ala - Thr - Phe - Phe - Tyr - Phe - Thr - Pro - Asn*- Lys - Thr - Glu - Asp - Thr - Ile -
      Ala                         40                                                                                                                                         60
                                                       Ile        75                 Thr                                                                 85
Phe - Leu - Arg - Glu - Tyr - Gln - Thr - Arg - Gln - Asp - Gln - Cys - Tyr - Asn*- Ser - Thr - Tyr - Leu - Asn - Val - Gln - Arg - Glu - Asn*- Gly - Thr - Val - Ser - Arg -
                                         70                Phe       Ser                                                                      Ile         90
                                                100                                                              Phe - Gly - Ser - Tyr - Leu - Asp                  Cys
Tyr - Val - Gly - Gly - Gln - Glu - His - Val - Ala - His - Leu - Leu - Ile - Leu - Arg - Asp - Thr - Lys - Thr - Leu - Met - Phe - Gly - Ser - Tyr - Leu - Asp - Asp - Glu - Lys -
      Glu             Arg            Phe                                                                    Tyr - Ala - Phe - Asp                  Val                 150
                          130                                                   140                                                                             Cys
Asn - Trp - Gly - Leu - Ser - Phe - Tyr - Ala - Asp - Lys - Pro - Glu - Thr - Thr - Lys - Glu - Gln - Leu - Gly - Glu - Phe - Tyr - Glu - Ala - Leu - Asp - Cys - Leu - Arg -
                              Val                                                                                                                               170
Pro - Arg - Ser - Asp - Val - Met - Tyr - Thr - Asp - Trp - Lys - Lys - Asp - Cys - Glu - Pro - Leu - Glu - Lys - Gln - His - Glu - Lys - Glu - Arg - Lys - Gln - Glu - Glu - Gly -
      Lys                     Val                      160                                                                                                               180
Ser - COOH
```

Fig. 46. Complete amino acid sequence of $\alpha_1$-acid glycoprotein [141]. The five glycosylation sites I to V are in position 15, 38, 54, 75 and 85, respectively.

part of the glycans of units C [527]. These findings show how complex the architecture can be of a glycoprotein in which coexist different types of linkages and glycan structures. As with the $\alpha_1$-acid glycoprotein, the biological and biochemical significance of these structures is an enigma.

## 5.4. Proteoglycans

There is now general agreement that the acidic mucopolysaccharides (see Section 4) are covalently linked to protein, except hyaluronic acid, the proteoglycan nature of which remains uncertain. The various proteoglycans differ in protein content, molecular size, and number of polysaccharide chains per molecule. The current model for the structure of proteoglycan monomer molecules isolated from hyaline cartilages is that of a central "protein core" [528], with an average $M_r$ = 200 000, to which approx. 100 chondroitin sulfate and 50 keratan sulfate side chains are covalently bound. Recently Heinegård and Axelsson [529] have proposed a model (Fig. 47) for the proteoglycan monomer in which they

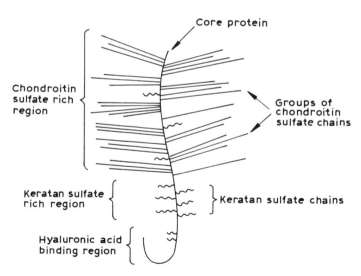

Fig. 47. Tentative model for the proteoglycan monomer structure [529]. Length of "core protein": 300—400 nm, of chondroitin sulfate chains: approx. 40 nm [530].

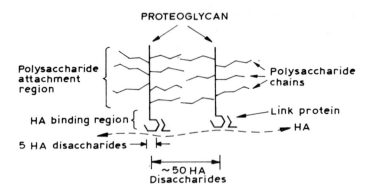

Fig. 48. Postulated model for the structure of proteoglycan aggregates. HA: hyaluronic acid. Length of the hyaluronic acid molecule: 1200 nm. Distance between two proteoglycan chains: 20—50 nm [530].

distinguished three domains; (1) a keratan-sulfate-enriched region containing about 60% of the keratan sulfate in the monomer, but only 10% of the chondroitin sulfate; (2) a chondroitin sulfate-enriched region located at the other end of the monomer containing 90% of the chondroitin sulfate chains, but only about 20% of the total number of keratan sulfate chains; (3) a "hyaluronic acid binding region" which contains 20% of the keratan sulfate chains and which carries the binding site of proteoglycan monomer core protein for hyaluronic acid.

In the intercellular matrix of cartilage, most of the proteoglycan exists in the form of aggregates [531], the molecular architecture of which is shown in Fig. 48. These aggregates are formed by the association of many proteoglycan monomers with hyaluronic acid. A low $M_r$ (= 45 000) glycoprotein, called "link protein" is responsible for promoting proteoglycan-hyaluronic acid aggregation and for stabilising the complex. Such aggregates appear to be the predominant way in which the proteoglycans are organised in cartilage extracellular matrices (reviews in [75] pp. 329—412, and [532]).

## 5.5. Glycophorin A

Glycophorin A [533, 534] is the major sialoglycoprotein of human erythrocyte membrane (review in [535]). This glycoprotein has an

## GLYCOPHORIN A

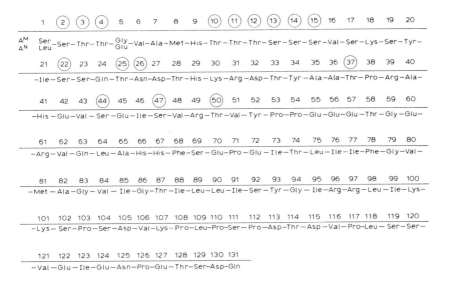

Fig. 49. Amino acid sequence of human erythrocyte glycophorin A [536]. Circled numbers indicate the position of glycans. Amino acid-26 is $N$-glycosylated. All the other conjugated amino acids are $O$-glycosylated. $A^M$, $A^N$: glycophorin A isolated from red blood cells of individuals which are homozygous for the blood group antigens of MN system.

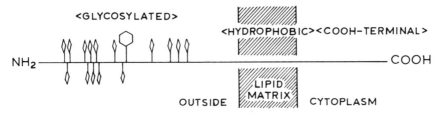

Fig. 50. Schematic representation of the order of peptides in human erythrocyte glycophorin A and of its integration in the membrane [537].

$M_r$ = 31 000 and contains about 60% carbohydrate and is particularly rich in $N$-acetylneuraminic acid (25%). This protein is the major carrier of cell surface carbohydrate in the erythrocyte and represents a remarkable model of integration of glycoproteins into cellular membranes. Its complete amino acid sequence and the sites of glycosylation have been determined [536] and are

*References p. 158*

shown in Fig. 49. This transmembrane glycoprotein can be divided into three domains [537] (Fig. 50): The hydrophilic N-terminal sequence of 70 amino acids is oriented toward the exterior of the plasma membrane. This fragment contains 15 O-glycosidically linked glycans and only one N-glycosidically bound carbohydrate moiety. The structure of this latter is not yet well defined and the only information we now possess favours a sialylated and fucosylated triantennary glycan of the N-acetyl-lactosaminic type. By contrast, a few structures of O-glycosidically linked glycans are known (see Structures 28 and 40 in Table XI). Moreover, some authors assume that glycophorin carries the human erythrocyte blood group-M and -N antigens (see Structures 36 to 39 in Table XI); the second domain is highly hydrophobic and comprises the amino acids residues 71 to 92. It is associated with the hydrophobic membrane lipid bilayer; finally, the C-terminal of the protein is exposed to the cytoplasm and can interact with cytoplasmic elements.

This model of a glycoprotein integrated in a cell membrane could be generalised to other membrane glycoproteins. It conforms to the present concept of the role played by membranar glycoconjugates: reception and transmission of environmental stimuli, cell-cell contact, and membrane transport.

From a structural point of view, glycophorin is an interesting glycoprotein. It is an O,N-glycosylprotein presenting an asymmetry in the distribution of oligosaccharides which are located at the N-terminus of the protein chain, nine of them densely substituting tri- and hexapeptidic sequence of β-hydroxy-amino acids.

## 6. Glycoprotein biosynthesis
(with the collaboration of André Verbert)

The glycosyltransferase general reactions may schematically be represented as follows:

$$\text{Sugar-donor + Acceptor} \xrightarrow{\text{Glycosyltransferase}} \text{Glycosylated acceptor + De-glycosylated donor.}$$

In the last decade a tremendous amount of information has been obtained about these reactions and very much has been written about the metabolic pathway in which these compounds participate (reviews in [38, 40, 43, 44, 47, 49, 52, 61, 65, 75—79, 176, 538 – 560]).

Rather than being a detailed review, the following pages, apart from giving general information, illustrate the fact that the better knowledge of structure has made possible the recent investigations on the metabolism and this will allow further development in this field. The precise knowledge of structure as a basis for a better understanding of function is obvious for nucleic acid or protein synthesis for which a template is required, but it is still more relevant to glycoprotein synthesis where the primary structure of the glycan itself will take part in the control of the formation of the carbohydrate chain.

## 6.1. The reaction and its partners

Glycosyltransferases are enzymes which transfer monosaccharides or oligosaccharides from activated sugars to various acceptors. Before looking at the glycosyltransferases themselves, attention must first be given to the two substrates: the activated sugar donors, and the acceptors.

*6.1.1. Activated forms of sugar donors.* The activated sugar donors are always phosphorylated derivatives in which the carbohydrate moiety is connected through the anomeric carbon atom to either a phosphate or pyrophosphate. The latter units are linked either to a nucleoside moiety to form a nucleotide sugar, also called glycosyl nucleotide, or to a polyprenol lipid in the case of "lipid intermediates".

*6.1.1.1. Glycosyl nucleotides.* All sugars present in the carbohydrate moieties of glycoproteins have nucleotide sugars as precursors either directly or *via* the lipid intermediates, and it is important to emphasize the work of Leloir and collaborators, who discovered the nucleotide sugars [146]. Schemes outlining the interconversion pathways from the D-glucose to the formation of all the nucleotide sugar donors are reviewed in [542, 561]. The site of synthesis of sugar precursors is of importance as the availability of the substrate is one of the first points of control in the synthesis of glycoprotein. Nucleotide sugars appear to be found in the cytosol, as the pyrophosphorylases which catalyse the final activation step and conjugation of sugar to the nucleotide are mostly cytoplasmic enzymes. It is worthwhile mentioning an exception concerning the formation of CMP-NeuAc, which occurs in the nuclei and for which no real explanation has been produced.

During the transfer reaction, breakage of the high-energy phosphate or phosphodiester bonds will supply the energy required for the formation of the glycosidic bond. This is not the only advantage of such and activated form. The transfer reaction appears to proceed through a nucleophilic substitution of the second order ($SN_2$) which leads to inversion of configuration. As all sugars are bound through an α-linkage to a phosphodiester bond, with the exception of CMP-NeuAc and GDP-Fuc which are β-linked, the anomeric configuration of the conjugated sugar may be predicted. We shall see later how lipid intermediates by doubling the transfer reaction, will allow two successive inversions, i.e. to restore the initial anomeric linkage. It would be hazardous to state that this rule is always respected, but it fits perfectly well with the anomeric linkages formed via lipid-linked intermediates encountered in the N-glycosylproteins so far. However, it cannot be argued that, because the anomeric configuration in the product has the same configuration as the nucleotide donor, the glycosylation must involve a lipid-linked intermediate. Indeed examples will be seen in Section 6.1.1.2 of α-mannosidic linkages being formed from GDP-α-mannose without any involvement of lipid intermediates.

A special comment has to be made about sialic acids which are N-acetyl and N-glycolyl derivatives of neuraminic acid often with additional O-acetyl groups (review in [562]). It has been shown that these further modifications of N-acetyl neuraminic acid may occur after its incorporation into glycoprotein. However, as synthetases are capable of reacting CTP with N-acetyl, N-glycolyl and other N,O-acetyl neuraminic acids to make the corresponding CMP-sialic acids, it may be assumed that these donors could be used to transfer any of the series of neuraminic acid derivatives to glycoproteins. In the first case, the different ratios of individual sialic acids found in mucus glycoprotein would mainly depend on the activity of hydroxylating and acetylating enzymes, in the second case it would rather depend on the activity of the various CMP-sialic acid synthetases or sialyltransferases.

*6.1.1.2. Lipid intermediates.* The second type of activated sugar donors are lipid intermediates which intervene mainly in the N-

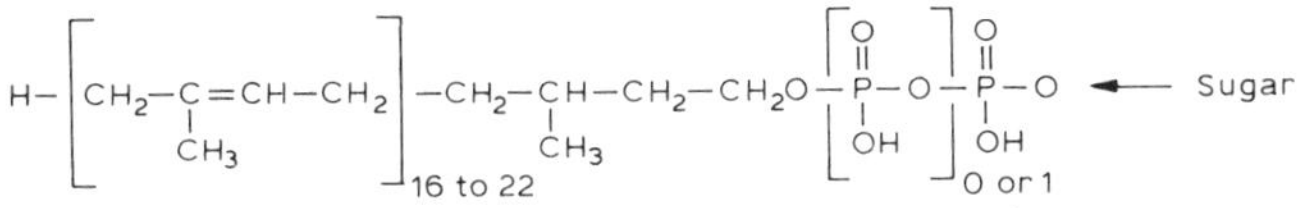

Fig. 51. General structure of dolichol mono-, and diphosphate sugars.

glycosylprotein biosynthesis. This has been one of the most important discoveries in the field of oligosaccharide synthesis and here again, most of the early investigations are due to Leloir and his group [563].

A family of long chain isoprenyl alcohols, the dolichols, containing between 16 to 22 isoprene units with the terminal unit saturated (Fig. 51) has been found in animal tissues as carriers of sugars activated via a phosphate or phosphodiester bond. These dolichol derivatives are mainly observed for glucose, mannose and N-acetylglucosamine. They have also been found to be linked to xylose, galactose [564] and the disaccharide β-glucuronyl-(1 →4)-N-acetylglucosamine (general reviews in [75, 565—574]). They are formed from their respective sugar nucleotides transferring their sugar moiety to dolichol-phosphate. The dolichol-phosphate sugars observed are dolichol pyrophosphate-α-GlcNAc, dolichol monophosphate-β-Man and dolichol monophosphate-β-Glc. The production of these compounds occurs in a large variety of animal tissues from which they can be extracted by a chloroform—methanol (2 : 1 by vol.) mixture.

Another major contribution from Leloir's group was the discovery of dolichol-pyrophosphate oligosaccharides. These products were shown to be insoluble in water, trichloroacetic acid and in chloroform—methanol (2 : 1 by vol.) but were soluble in chloroform—methanol—water (10 : 10 : 3 by vol.) as shown first by Behrens et al. [575]. The dolichol pathway which leads to the dolichol-pyrophosphate oligosaccharides has been very widely investigated [566, 568, 571, 576—581] and is illustrated in Fig. 52.

It has to be noted that all reactions are catalysed by specific glycosyltransferases. However, the very first one is not a transfer of the GlcNAc-1-P moiety of UDP-GlcNAc. So, it resembles more the reverse action of a phosphodiesterase. Another remark has to be made about the stereochemical consideration resulting from the choice of the donor. This is particularly well illustrated with the mannose. The glycosidic linkage formed by transfer of mannose to

*References p. 158*

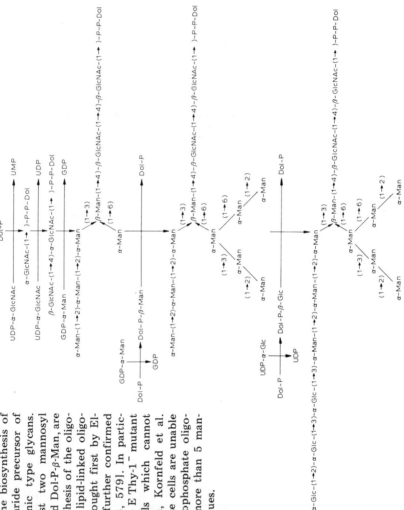

Fig. 52. Pathway of the biosynthesis of lipid-linked oligosaccharide precursor of the N-acetyl-lactosaminic type glycans. Evidence that at least two mannosyl donors: GDP-α-Man and Dol-P-β-Man, are involved in the biosynthesis of the oligomannosidic moiety of lipid-linked oligosaccharide has been brought first by Elbein et al. [577] and further confirmed by other authors [578, 579]. In particular, by using the Class E Thy-1⁻ mutant mouse lymphoma cells which cannot synthesize Dol-P-β-Man, Kornfeld et al. demonstrated that these cells are unable to make dolichol pyrophosphate oligosaccharide containing more than 5 mannose residues.

the lipid-oligosaccharide intermediate will be of the α-configuration if the donor is the dolichol monophosphate sugar. However, when the donor is GDP-mannose the configuration can be either α- or β-. Presumably the α-form results from mannosylation of some unknown intermediate, possibly part of the enzyme molecules. The anomeric configuration of the glucosyl residues in the glucosylated dolichol diphosphate oligosaccharide is not yet certain, but it is probably α.

Besides dolichol, the derivates of another polyprenol, retinol or vitamin A, has been investigated by the group of De Luca [582]. It may be involved in mammalian protein glycosylation. Retinol is a cyclised form of tetrahydrotetraprenol with all-trans double bonds.

At the moment, no precise role for the polyprenyl moiety of lipid intermediates has been given and the mechanism of its action is not yet known.

### 6.1.2. Glycosyltransferase systems

*6.1.2.1. The enzymes.* The glycosyltransferases are widely distributed in nature and have been described in bacteria, plants and higher organisms. Few of them have been thoroughly purified mainly because most of them are membrane bound, extraction of which requires detergent treatment which often alters enzyme structure and conformation. However, the glycosyltransferases of biological fluids which are in a soluble form can be more readily purified. Use of affinity chromatography has been useful to obtain pure enzymes as for example with the galactosyltransferase from bovine milk which was purified by immobilised α-lactalbumin [583] and with sialyltransferase from bovine colostrum for which a CMP-agarose column [584] was used.

The glycosyltransferases show a generally broad spectrum of activity toward temperature and pH, generally around physiological values. They have cation requirements for which the manganese cation appears the most common ion and is sometimes essential. The glycosyltransferases show a very strict dependence for the nucleotide sugars and this specificity extends to both the nucleotide and the monosaccharide moieties: any substitution leads to a drop of the transfer activity to less than a few percent of what is usually obtained with the correct precursor. Until recent years, as far as the transfer activity was only based on the amount of sugar transferred it was believed that acceptor specificity was very loose and many investigators used various acceptors to assay their enzyme activity. However, with the increasing development of methodologies allowing precise determination of structure, it becomes possible to characterise the nature of the glycosidic bond formed and to determine the structure of the different products. The results obtained so far lead to the obvious concept of "one monosaccharide-one glycosyltransferase" but complemented by the "one linkage-one glycosyltransferase". For example, Roseman's group [585—588] has characterised a number of sialyltransferase activities that differ from one another in their substrate specificities or in the nature of the glycosidic linkages formed between sialic acid and the carbohydrate acceptor, suggesting that different transferases are responsible for forming the different sialyl linkages. This point was confirmed by Paulson and coworkers [589, 590] who isolated a CMP-$N$-acetylneuraminate: $\beta$-D-galactoside $\alpha$-(2 → 6)-sialyltransferase from bovine colostrum responsible for incorporating $N$-acetylneuraminic acid onto the sequence $\beta$-Gal-(1 → 4)-GlcNAc. It has a strict substrate specificity, and forms only the $\alpha$-(2 → 6)-sialyl linkage. $N$-Acetyl-lactosamine and asialoglycoproteins of the $N$-acetyl-lactosaminic type are the best acceptor substrates. Isomers of $N$-acetyllactosamine having $\beta$-(1 → 3)- or $\beta$-(1 → 6)-glycosidic linkages, as well as lactose, are poor acceptors. Similar results have been observed more recently by Hill's group [591] who succeeded in the isolation from porcine submaxillary glands of two highly specific sialyltransferases: $\beta$-D-galactoside $\alpha$-2,3- and $N$-acetyl-$\alpha$-galactosaminide $\alpha$-2,6-sialyltransferases. Both use the basic disaccharide (see Table XI) $\beta$-Gal-(1 → 3)-$\alpha$-GalNAc-(1 → 3)-Ser(Thr) as acceptor, but the former specifically transfers $N$-acetylneuraminic acid to the C-3 position of the galactose residue, while the latter transfers to the C-6 position of the $N$-acetylgalactosamine residue.

Importance has to be focused on the fact that specificity is just based on the various rates of transfer towards the acceptors whilst no attempt is made to discuss the $V_m$ or $K_m$ of the reaction, as the required conditions are rarely reached. Nevertheless, the acceptor specificity appears to be very much more restricted and controlled than was believed earlier (see Section 6.2).

*6.1.2.2. Cellular localization of glycosyltransferases.* It is now evident that most of glycosyltransferase machinery is located where the glycosylation process usually takes place during and after the synthesis of the protein moiety. That is why rough and smooth endoplasmic reticulum and the Golgi apparatus contain most of the total cell activities.

Attachment of sugar is considered to occur sequentially. For example, in the case of the $N$-glycosylproteins it has been well demonstrated that the biosynthesis of the glycoproteins starts in the rough endoplasmic reticulum, where all of the mannose and some of the $N$-acetylglucosamine residues are conjugated, and that it continues and is terminated in the smooth endoplasmic reticulum and the Golgi apparatus, where the remaining $N$-acetylglucosamine residues and all of the galactose, fucose, and sialic acid residues are transferred. It may thus be presumed that, for the $N$-acetyl-lactosaminic type of glycans, the pentasaccharide core common to all of these glycans (fragment I in Fig. 18) is synthesised, and conjugated to protein in the rough endoplasmic reticulum, and that the antennae (fragments II and III in Fig. 18) are formed in the smooth-membrane fraction of the cell. Such glycan structures as we know at present conform with the concepts we have of their biosynthesis, as the sequence of the monosaccharide residues in the glycan structures corresponds well with the chronological order of their conjugation.

This sequence of events implies that the specific glycosyltransferases are integrated in the membranes in a given order, as are the enzymes of the respiratory chains in mitochondrial *cristae*. This was first suggested by Roseman [165] and the "multiglycosyltransferase" (MGT)-system he proposed is still generally accepted*.

Besides these locations for which agreement is very wide (for recent review see [553]), it has to be noted that specific mannosyl and $N$-acetylglucosaminyltransferases have been described in the outer membrane of mitochondria (general reviews in [592—596]). The same group of authors has described the presence of five glycosyltransferases in rat liver nuclei [594, 597].

Finally, following the initial hypothesis of Roseman [165] numerous investigators have found glycosyltransferases at the cell surface (reviews in [598—600]). These findings have been very

---

* See note (2) on p. 157

*References p. 158*

controversial because of the difficulties of assigning ectoglycosyltransferase activities to intact cell activities [601]. In the last five years, attention has been focused on the causes of error and methodologies have been developed — some of them have come from our laboratory [600, 602—604] — to ascertain the presence of ectoglycosyltransferases. Due to their unique property of having their active site accessible from the outside of the cell, they have been assumed to be involved in a variety of phenomena of the "social life" of the cell, such as cell-cell recognition, cell adhesion, cell differentiation (recent review in [599]) on the basis of Roseman's hypothesis [165].

*6.1.3. Acceptors.* One has to keep in mind that most of the glycosyltransferase activity has been detected and studied with exogenous acceptors. They are native "incomplete" glycoproteins such as ovomucoid which, with its five terminal $N$-acetylglucosaminyl residues, is an excellent acceptor for the galactosyltransferase [605]. These incomplete glycoproteins may also be obtained by sequential "strip-tease" of glycoproteins as has been achieved with human $\alpha_1$-acid glycoprotein using sequentially neuraminidase, $\beta$-galactosidase and $N$-acetyl-$\beta$-glucosaminidase which expose in turn galactose, $N$-acetylglucosamine and mannose terminal residues as acceptors for the sialyl, galactosyl and $N$-acetylglucosaminyl transferases, respectively.

Glycopeptides and oligosaccharides are also used as acceptors. They are interesting tools in determining substrate or transfer specificity as their structures are well defined and may be reexamined after the glycosylation process.

We should also remember that all the previously described acceptors are "in vitro" acceptors. A more physiological approach is the characterisation of endogenous acceptors, but this is not often done. This approach has, however, been used in a simple system where a major glycoprotein is synthesized. This is the case for viral glycoprotein [606], where cellular protein synthesis is blocked, or for tissue where a major product is synthesized as, for example, the thyroglobulin for the thyroid [607] and ovalbumin in hen oviduct [608].

## 6.2. Control of glycoprotein biosynthesis

### 6.2.1. General mechanisms

*6.2.1.1. Genetic control.* The genetic control of glycan structure is exerted through the specificity of the glycosyltransferases involved in the biosynthesis. This hypothesis has been described by Roseman [165] as "one gene-one glycosidic linkage". The simplest interpretation of this genetic control is summarised as follows: each gene product is a specific glycosyltransferase that catalyses the formation of a glycosidic linkage by transfer from a particular donor to a preferred acceptor molecule. A new acceptor is produced that becomes the preferred substrate for another glycosyltransferase synthesized under the control of another gene. The synthesis may be growing until the newly formed carbohydrate chain is no more an acceptor for any glycosyl transferase present within the cell.

This concept has been applied very successfully to the interpretation of the inheritance of oligosaccharide structures carrying the human ABH, Le$^a$ and Le$^b$ serological determinants (complete reviews in [138 and 548]).

*6.2.1.2. Control of substrate availability.* One of the most direct controls of the glycosyltransferase reaction is the availability of the substrate. The control by the acceptor capacity of the nascent oligosaccharide chain will be discussed later in the paragraph concerning the role of the glycan moiety. The following section will briefly focus on the levels of regulation brought by control of the concentration of the various donors.

At the first level, that of nucleotide sugars, a control is exerted by the regulation of their own synthesis on one hand, but also by the availability near the glycosyltransferase active site. This regulation by compartmentalisation of the pool is obvious for the glycosyltransferases whose active sites are thought to be on the luminal face of the membranes, although the nucleotide sugars are mostly synthesised in the cytoplasm. The mode of passage through the cytoplasmic membrane is still very unclear. It has been suggested this role could be filled by the lipid carrier but this possibility seems to have been abandoned recently since there is no evidence for movement of dolichol phosphate sugar across membranes. Moreover, the flip-flop mechanism from one lipid leaflet of a membrane to the other is likely to be very slow. In the case of the

*References p. 158*

synthesis of Dol-P-P-di-*N*-acetyl chitobiose, it has been shown that the enzymatic system was at the cytoplasmic face of membranes, although the product itself was found on the luminal face. Hanover and Lennarz [609] have recently suggested that the transmembrane movement of the GlcNAc-1-P and GlcNAc units could be an integral part of the enzymatic reaction leading to their attachment to lipid. In addition, enzymes which degrade nucleotide sugars:

$$\text{XDP-Sugar} \longrightarrow \text{XMP} + \text{Sugar-1-P} \longrightarrow \text{Sugar} + P_i$$

have been detected in various biological systems [610—612]. They may play a role in controlling the synthesis of glycoproteins by adjusting the level of precursors. In this connection, Sela et al. [610] have shown that this type of enzyme was missing in cell lines transformed by SV40 and Rous sarcoma virus.

At the second level, that of control of the dolichol monophosphate sugars pool, the concentration of Dol-P, which is the common substrate for the formation of dolichol derivatives, appears to be very low in cells and is generally rate-limiting. Thus, it is plausible that changes in the membrane content of Dol-P could affect the rate of synthesis of glycoproteins by modulating the rate of assembly of the common oligosaccharidic core. Such a demonstration has been recently made by several groups [613—615] who showed that the glycosylation was directly related to the rate of synthesis of dolichol.

The third level is that of Dol-P-P-oligosaccharides which are the final donors before maturation is complete. It has been recently reported [616] that, in rat spleen lymphocytes, Dol-P-P-oligosaccharides could be degraded to phospho-oligosaccharides and/or oligosaccharides by specific enzymes (Fig. 58). Thus, the availability of Dol-P-P-oligosaccharides could be controlled through the action of a phosphodiesterase. In this connection, Wedgwood and Strominger [617] have reported, in the lymphocyte system, too, the existence of a phosphodiesterase acting on dolichol pyrophosphate.

*6.2.1.3. Involvement of glycosidases.* As the glycosyltransferase reaction does not appear to be reversible, a control must be possible when a transferred sugar has to be removed to give back the acceptor. This role could be assumed for specific exoglycosidases for which activities included in the MGT system have still to be proved. As reported in Section 4.4, the involvement of exoglycosidases has been proposed to explain the microheterogeneity of glycans.

Moreover, endoglycosidases could play a role in the control of glycoprotein synthesis by removing "en bloc" the glycan moieties. This will be discussed in Section 8.

*6.2.1.4. Control by the peptide-chain conformation.* Many investigations carried out in the last 10 years have clearly and unambiguously demonstrated that the primary structure and conformation of a protein plays an important role in its glycosylation. The existence of specific peptidic sequences around the glycan receptor amino acid, "sequons" [37], on the one hand, and the integration of these "sequons" in β-turns or loops of the peptide chain, on the other hand, are, at least, two prerequisites of glycosylation.

*"Sequon" and glycosylation* (reviews in [37, 40, 47, 58, 245, 618—620]). On the basis of an examination of the amino acid sequences of numerous N-glycosylproteins, many authors [37, 621] have demonstrated that the tripeptide sequence Asn-X-Ser(Thr), where X can be almost any amino acid, except proline [622, 623], is a requisite feature for N-glycosylation of proteins. In this regard, the most convincing example is given by the location of the single glycan in the bovine ribonuclease molecule. This enzyme contains 10 asparagine residues, and only 1, residue-34, is linked to a carbohydrate group in ribonuclease B because this asparagine residue is the only one which belongs to a "sequon", in this particular case the tripeptide Asn-Leu-Thr.

The concept of "sequon" has been extended to some O-glycosylproteins. For instance, the sequence Gly-X-Hyl-Gly-Y-Arg, in which X and Y are quite diverse amino acids, is the minimal "sequon" structure required for glycosylation of the hydroxylysine residues in vertebrate and invertebrate collagens and related proteins [624, 625]. Another example is given by the glycosylation of the threonine residue in position 98 in the basic protein of myelin. Young et al. [626] have tested the ability of nine synthetic peptides containing different sequences in the region of threonine-98, to be glycosylated by the UDP-GalNAc: mucin polypeptide N-acetylgalactosaminyl transferase from porcine submaxillary glands. They have observed that the glycosylation was greater for peptides containing the sequence Thr-Pro-Pro-Pro suggesting that the three proline residues C-terminal to the threonine residue provide a unique and suitable shape for the glycosylation, and that the minimum requirement for O-glycosylation is contained in the tetrapeptide

*References p. 158*

Thr-Pro-Pro-Pro. In the case of mucins or of "mucin-like" glycoprotein, it has been observed that specific peptide sequences are not required for the glycosylation which seems to depend only on the conformation of the peptidic chains. The same lack of specificity appears to occur in the sequence of amino acids around the carbohydrate-amino acid linkage in glycophorin (Fig. 49) and in the anti-freeze glycoprotein of Antartic fish the peptide chain of which contains the following repeating unit: Ala-Thr-Ala.

However, segments of the chain carrying the sugar moiety in some O-glycosylproteins, such as glycoproteins from mucus secretions, human chorionic gonadotropin β subunit, fetuin, $A_1$ protein from bovine myelin and rabbit IgG heavy chain, are rich in proline residues on the N-terminal side of each serine or threonine residue involved in the glycopeptide bond. So, these observations emphasise the possible role of proline in helping to determine the conjugation of N-acetylgalactosamine with serine or threonine residues in shaping the receptor region for glycosylation.

The use of synthetic peptides for studying the N-glycosylation [622, 623, 627] entirely confirmed the concept of the "sequon". Moreover, further information was obtained concerning the environment of the "key peptide sequence" which could be summarised as follows: (i) the tripeptide Asn-X-Ser(Thr) is the minimal peptide active as an acceptor of oligosaccharide, only if both its amino and carboxyl termini are blocked; (ii) the replacement of asparagine or serine (threonine) causes a complete loss of acceptor activity; (iii) the peptides with an Asn-Pro-Thr/Ser sequence cannot to be glycosylated; (iv) the affinity of the "oligosaccharide transferase" (Fig. 55) increases with the length of the acceptor peptide.

However, as "sequons" of certain peptide chains are sometimes not glycosylated, their presence is not a sufficient condition for glycosylation. Hunt and Dayhoff [621] performed a statistical analysis of the occurrence of the two tripeptides in proteins and found that the frequency of glycosylation was only about 65% of that expected. So, other factors such as the rate of folding of the polypeptide chain and its final conformation play a decisive role in glycosylation, and the accessibility of the "sequons" is the predominant structural feature of the protein that determines whether or not it is glycosylated.

*β-Turns and glycosylation.* Recently a second requirement for carbohydrate attachment to proteins has been defined: that glyco-

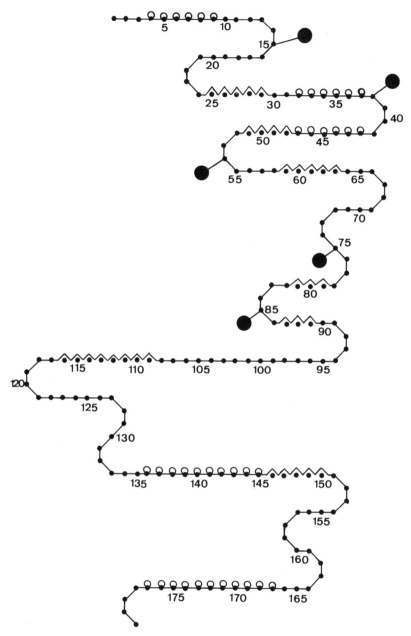

Fig. 53. Suggested secondary structure of $\alpha_1$-acid glycoprotein [633] according to Chou and Fasman [630]. Residues are represented in their respective conformational state: helical (○), β-sheet (△), coil (—), β-turn tetrapeptides ( ); Asn 15, 38, 54, 75 and 85 carry a carbohydrate moiety.

References p. 158

sylated regions of peptide chains could be associated with a specific, secondary structure. In fact, by applying predictive methods [628—630] to amino acid sequences adjacent to the glycosylated sites of numerous glycoproteins (9 O-glycosidically and 29 N-glycosidically linked glycans in [631—633] and 31 N-glycosylically linked glycans in [634], it has been demonstrated that glycans are located in amino acid sequences favouring turn or loop structures, as defined by Kuntz [628]; most are in tetrapeptide β-turns, and the rest are in other types of loops or turns of the peptide chain, such as "hairpins" and corners (Fig. 53 and [635]). Thus, the glycosylation may be favoured, as turn and loop conformations are generally located at the surface of globular proteins, making the asparagine residues present in these particular structures readily accessible to glycosyltransferases. This has been clearly demonstrated by Prasad et al. [636] who have sequenced the tryptic glycopeptide from rat α-lactalbumin which contains 13.4% carbohydrate, and demonstrated that Asn-45 linked to the glycan moiety takes part in an extended β turn, whereas Asn-45 in bovine α-lactalbumin, a poorly glycosylated protein is in a non-extended β-turn. In addition, proof has been obtained that the carbohydrate moieties are positioned on the outside of the glycoprotein molecules, and this result is in good agreement with the role of recognition signal played by glycans. Moreover, it is possible that protection against proteolytic attack may be due to the masking of turns and loops by the conjugated carbohydrate [634].

In relation to the conformation of protein, the question is whether the glycosylation step may occur "co-translationally, while the nascent polypeptide chain is still growing on the polysomal complex, or post-translationally, when its synthesis and folding have already been completed. The findings of investigators are generally consistent with the view that the initial glycosylation event occurs on the nascent polypeptide chain (e.g. [637—646]). Recently, Lennarz et al. [646] succeeded in defining precisely the minimum number of amino acid residues of nascent ovalbumin that must be translated after the addition of $Asn_{293}$ before glycosylation of this residue can occur. The results they obtained indicate that at least a 30 amino acid segment must be added before the acceptor site is glycosylated. This number of residues, independent of the length of polypeptide contained within the cleft, would be sufficient to allow the acceptor site to be accessible for glycosylation at the luminal face of the rough endoplasmic reticulum. In addition, the same authors have shown that ovalbumin contains a second potential glycosylation site at $Asn_{312}$ which is not glycosylated in vivo, although the two oligosaccharide acceptor sequences in ovalbumin are quite likely to form a $\beta$-turn. So, it is clear that the precise conformational features recognized by the oligosaccharide transferase leading to the structure B of Fig. 55 remain to be established. In this connection, the hypothesis of Weitzman et al. [645] according to which

the final structure of the processed oligosaccharide moiety probably results from interactions between the cellular glycosylating enzymes and aspects of the tertiary and quaternary structure of the finished protein

remains to be verified.

### 6.2.2. Control of the biosynthesis of N-glycosylproteins

*6.2.2.1. Glycan primary structure as a guide for glycosylation.* Examination of glycan structures shows that a series of "substitution laws" exist which have been defined in Section 4. It thus became desirable to understand, first, the basis of this determinism in the formation of glycosidic linkages, and second, the order in which the residues of monosaccharides are conjugated. Some results have now demonstrated that this double problem is linked simultaneously with the specificity of glycosyltransferases and the presence of well defined oligosaccharide structures in the glycan molecule. In this regard, some findings by Schachter and coworkers [560, 647—650] on the conjugation of monosaccharides at the branching points are of interest (Fig. 54).

In a first series of experiments, these authors demonstrated that the attachment of the fucose residue on the $N$-acetylglucosamine linked to asparagine, and the conjugation of $N$-acetylglucosamine residues on $\alpha$-mannose 4 and 4' (see the fundamental structure in Table XIII), occur in a well defined sequence. In fact, the conjugation, by the $N$-acetylglucosaminyltransferase I, of an $N$-acetylglucosamine residue to the mannose-4 is an essential prerequisite both for attachment by a fucosyltransferase of a fucose residue to the $N$-acetylglucosamine 1 and for the attachment by the $N$-acetylglucosaminyltransferase II of a second $N$-acetylglucosamine residue to the mannose-4' of the core.

More recently, Schachter et al. [560, 650] have demonstrated the important regulatory role in the maturation of glycans played by the $N$-acetylglucosamine residue $\beta$-(1 → 4)-linked by the $N$-

# GLYCOPROTEIN BIOSYNTHESIS

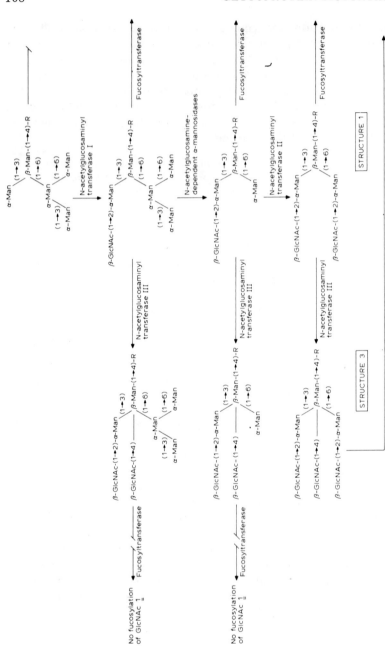

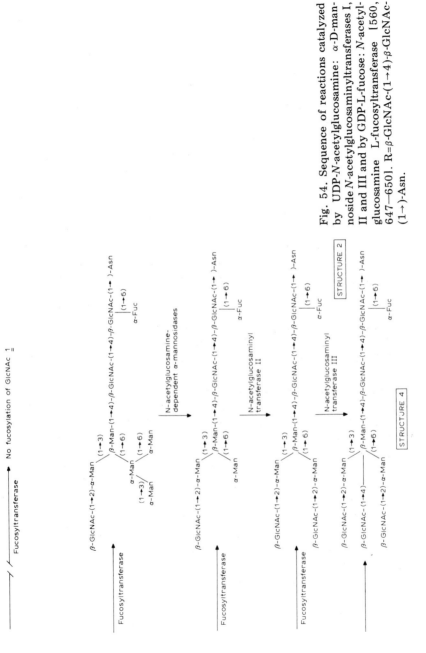

Fig. 54. Sequence of reactions catalyzed by UDP-*N*-acetylglucosamine: α-D-mannoside *N*-acetylglucosaminyltransferases I, II and III and by GDP-L-fucose: *N*-acetylglucosamine L-fucosyltransferase [560, 647–650]. R=β-GlcNAc-(1→4)-β-GlcNAc-(1→)-Asn.

acetylglucosaminyltransferase III, to the $\beta$-mannose 3 and called "intersecting $N$-acetylglucosamine". In fact, as illustrated by Fig. 54, the "intersecting $N$-acetylglucosamine residue" has three main effects: it turns off (i) the removal of any $\alpha$-mannose residue by the $N$-acetylglucosamine-dependent $\alpha$-mannosidase of Golgi membranes, (ii) the further action of $N$-acetylglucosaminyltransferase II, and (iii) the fucosylation of $N$-acetylglucosamine-1 residue linked to asparagine.

In the same way, galactose and sialic acid residues are also added in a defined order. Using asialo-galactoglycopeptide obtained by enzymic hydrolysis of a glycopeptide from IgG heavy chain, Rao and Mendicino [651] have given evidence for a mechanism which allows for the transfer of only a single galactosyl residue to one of the terminal $N$-acetylglucosamine units and they have shown that the rate of transfer of galactose to the second terminal $N$-acetylglucosamine residue decreases when galactose is already present on the other chain. The same kind of mechanism has been demonstrated by Van den Eijnden et al. [652] concerning the sialic acid conjugation which occurs preferentially on the galactose-6 residue (see fundamental structure of Table XIII). This could explain why so many monosialylated glycans are sialylated in the C-6 position of galactose (see structures in Table XIII).

Moreover, mechanisms of mutually exclusive glycosylation have been described. They show that the sialylation and fucosylation processes could control the degree of completion of oligosaccharide chains during synthesis. Indeed, as demonstrated by Hill's group [653] asialoglycan of human serotransferrin (structure 13 in Table XIII) can be either sialylated by a $\beta$-galactoside: $\alpha$-2,6 sialyltransferase leading to $\alpha$-NeuAc-(2 $\rightarrow$ 6)-$\beta$-Gal-(1 $\rightarrow$ 4)-GlcNAc structures, or fucosylated by an $N$-acetylglucosaminide: $\alpha$-1,3 fucosyltransferase to form $\beta$-Gal-(1 $\rightarrow$ 4)-[$\alpha$-Fuc-(1 $\rightarrow$ 3)]-$\beta$-GlcNAc structures. Sialylation blocks subsequent addition of fucose, and fucosylation prevents transfer of sialic acid. These results favour the concept that glycosylation by sialyl and fucosyl transferases is mutually exclusive and explain the occurrence of the structures 43 to 45 and 47 described in Table XIII.

*6.2.2.2. Maturation of glycans.* Nothing is ever simple even for $N$-glycosylation of proteins. In fact, it has been recently shown by Hunt's and Kornfeld's groups that glycoproteins of the $N$-acetyllactosaminic type are not biosynthesised directly in their definitive form, but that they pass through intermediate stages consisting of

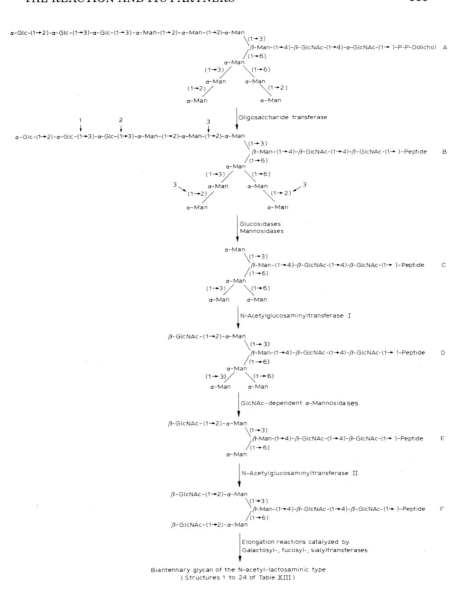

Fig. 55. Proposed sequence for the biosynthesis of biantennary glycans of the N-acetyl-lactosaminic type [657]. Elongation reactions could also occur with the structures 2, 3 and 4 of Fig. 54.

References p. 158

structures of the oligomannosidic type. The demonstration was brought about by associating the findings of Hunt et al. [654] who studied the formation of the envelope glycoprotein of vesicular stomatitis virus (VSV) and of Kornfeld's group [655—660] on the glycan biosynthesis of the VSV glycoprotein, of the IgG heavy-chain of the Chinese hamster ovary cells and of the Class E Thy-1 negative mouse lymphoma cells. The findings of these investigators are summarised in Fig. 55.

The first step in the biosynthesis of these $N$-glycosylproteins is the transfer "en bloc" of the glucosylated glycan conjugated to the dolichol pyrophosphate (Structure A) to the protein acceptor [577, 578, 580]. The reaction occurs within the luminal face of the rough endoplasmic reticulum [580] and is catalysed by an "oligosaccharide transferase" and leads to the formation of glycoprotein B. It is generally assumed that the glycosylation of the oligomannosidic part of the dolichol pyrophosphate oligosaccharide is the signal for this transfer [581].

In the following steps the glycan moiety of the newly formed glycoprotein is further "trimmed" to reach the $N$-acetyl-lactosaminic structure through the oligomannosidic and the mixed structures. First, the glucose residues are removed by specific microsomal $\alpha$-glycosidases [661] leading to an oligomannosidic type $N$-glycosylprotein containing 9 mannose residues. The further maturation steps of glycan are the following: after removal of the four $\alpha$-1,2-linked mannose residues by specific Golgi membrane mannosidases [662] the heptasaccharidic glycan (structure C of Fig. 55) is the acceptor [663] for the transfer of the $N$-acetylglucosaminyl residue on the $\alpha$-1,3-linked mannose (structure D) by the $N$-acetyl glucosaminyl transferase I [649] which has been called the "key enzyme", because no processing of the heptasaccharide may happen until this first transfer has occurred, as seen in the preceding paragraph. Once this is done, the two $\alpha$-1,3 and $\alpha$-1,6 "extra" mannose residues are cleaved off by Golgian-specific mannosidases [664] called $N$-acetylglucosamine-dependent $\alpha$-mannosidases [650], leading to structure E. Thus a further elongation step may take place by adding the second $N$-acetylglucosaminyl residue on the $\alpha$-1,6 mannose residue by the action of the $N$-acetylglucosaminyltransferase II [649] (structure F.). The completion of the chain is then achieved by adding galactosyl, fucosyl and sialyl residues, according to the substitution rules described above, leading to the biantennary structures 1 to 24 of

Table XIII. If the elongation and termination processes complete the structure 2, 3 and 4 of Fig. 54, glycans 25 to 45, glycan 46 and glycans 47 and 48 of Table XIII are synthesized, respectively. This complete processing is reviewed in detail by Schachter [549, 560] and by Gibson et al. [559]. As for the tri- and tetraantennary structures, neither the steps nor the mechanism of their formation are yet known. Anyhow, as pointed out by Hunt and coworkers, the finding of the maturation process of $N$-glycosylproteins

demonstrates an unprecedented mechanism that explains how the large oligomannosidyl structures observed for dolichol-linked oligosaccharides can be used to generate the smaller oligomannosyl structures found on glycoproteins. In addition, it suggests that the two major classes of oligosaccharide structures linked to L-asparagine in glycoproteins may share an intermediate.

Moreover, these recent advances solve the enigma of the complete identity of many oligomannosidic type glycans of very different origin. Thus the glycans of Taka-amylase A, GP-IV glycopeptides of ovalbumin, II-B glycopeptide of human IgM and glycopeptide C-1 from Chinese hamster ovary cell glycoprotein (Structure 7 of Fig. 23) are identical, as are those of GP-III-B of ovalbumin, Sindbis virus S-4 and glycopeptide II from myeloma IgM (Structure 6 of Fig. 23). The same applies to GP-V of ovalbumin and IB-2 of human IgM (Structure 9 of Fig. 23) and those of calf thyroglobulin Unit, bovine lactotransferrin, glycopeptide B-1 of CHO cells and glycopeptide IA-1 (Structure 1 of Fig. 23). In fact, one can assume that the structures of the oligomannosidic type are metabolic intermediates common to all cells and to all glycans of the $N$-acetyl-lactosaminic type. In this regard, the complete identity of Structure 1 of Fig. 23 with the deglucosylated oligosaccharidic moiety of Structure A of Fig. 55 is particularly convincing, as well as the identity of Structure 9 in Fig. 23 with the intermediary compound C in Fig. 55. Moreover, the origin of mixed structures of the oligomannosido-$N$-acetyl-lactosaminic type (compare structures D and E of Fig. 55 with Structures 3 and 1, respectively, of Fig. 36) is thus elucidated.

### 6.2.3. Biosynthesis of O-glycosylproteins

*6.2.3.1. Mucin and mucin-like glycoproteins.* Most of the work dealing with the synthesis of Ser(Thr)-GalNAc-linkage type oligosaccharides (reviews in [60, 72, 74, 556]) concerns the secretory glycoproteins called "mucins". As opposed to what has been de-

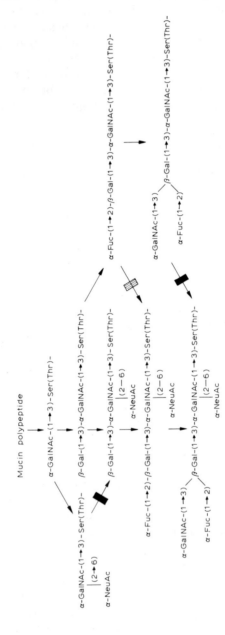

Fig. 56. Proposed pathway for the biosynthesis of the oligosaccharide chains of A⁺ porcine submaxillary mucin. The solid bars indicate that the reaction cannot proceed, and the hatched bar indicates that the reaction proceeds very slowly [665].

scribed about N-glycosylproteins, the incorporation of the oligosaccharide moiety into mucin occurs in the Golgi apparatus by the sequential addition of one sugar at a time, with no pre-assembly processing. No lipid intermediates seem to be involved in this type of biosynthesis, but as none of the specific glycosyltransferases have been purified to homogeneity, the possibility of a pathway which includes a lipid carrier may not be completely ruled out.

Fig. 56 summarises the biosynthetic path for the major oligosaccharide groups of porcine submaxillary mucin proposed by Beyer et al. [665]. The first step in the assembly process is the attachment of N-acetylgalactosamine residue to an hydroxy amino acid of the peptidic chain. It must be noted that the N-acetylgalactosaminyl transferase has a very high specificity for the protein chain of mucin. On the one hand large numbers of other hydroxy amino acid-containing proteins are completely ineffective as acceptors, on the other hand the high-molecular-weight structure seems to be required as pronase digestion of the carbohydrate-free ovine submaxillary mucin destroys its acceptor capacity (review in [550]).

The second step is a branching reaction as the galactosaminyl residue may be either sialylated or galactosylated. As in the case of the biosynthesis of N-glycosylprotein glycans, mechanisms of mutually exclusive glycosylation have been observed for O-glycosylprotein glycans. Thus, if sialylation occurs in C-6 position of the N-acetylgalactosamine residue linked to serine or threonine, any further carbohydrate addition is prevented (review in [558, 560]) and the OSM type glycan will be formed with the α-NeuAc-(2 → 6)-GalNAc linkage. But if galactose is transferred before the sialic acid to form the "basic disaccharide" β-Gal-(1→3)-α-GalNAc of Table XI, further growth of the oligosaccharide can occur by adding sialyl, fucosyl and N-acetylgalactosaminyl residues to form the PSM-type glycan. Thus, the key enzyme in the control of the synthesis of the mucin type glycan appears to be the UDP-Gal: GalNAc-mucin β-1,3 transferase.

In another type of mucin, the glycosylation in C-6 position of the N-acetylgalactosamine residue of the "basic disaccharide" β-Gal-(1 → 3)-α-GalNAc, by the sequence β-Gal-(1 → 3 or 4)-β-GlcNAc inhibits further growth of the chain.

The tetrasaccharide α-NeuAc-(2 → 3)-β-Gal-(1 → 3)-[α-NeuAc-(2 → 6)]-α-GalNAc-(1 → 3)-Ser(Thr) is assembled in a definite order: the galactose residue is incorporated first, then the α-(2 → 3)

*References p. 158*

and finally the α-(2 → 6)-linked sialic acid residues [666].

*6.2.3.2. Proteoglycans.* From the biosynthesis point of view (reviews in [39, 43, 65, 331, 429, 447, 448, 450, 542, 667—673]) proteoglycans may be divided into two main groups depending on the nature of the inner core (see Table XII). The first group includes hyaluronic acid, chondroitin sulfates, dermatan sulfate, heparin and heparan sulfate where the repeating saccharidic units are linked to a common core: Gal → Gal → Xyl → Ser. The second group consists of keratan sulfates where the repeating units are linked either to GlcNAc → Asn or to GalNAc → Ser(Thr). It must be immediately pointed out that both groups may be present in the same proteoglycan and the intriguing fact is that, in this latter situation, certain serine residues are substituted by chondroitin sulfate chains while others are linked to keratan sulfate chains and the rest remain unglycosylated. It can be assumed that, here again, the primary structure of the protein will play an important role in the control of the glycosylation. Nevertheless, the biosynthesis of proteoglycans of the first group obeys the general scheme illustrated in Fig. 57 [450]. The formation of the trisaccharide unit Gal → Gal → Xyl on the seryl residue is achieved stepwise. First, xylosylation occurs from UDP-xylose [674]. As for mucin-type glycoprotein, the initial glycosylation has a strict specificity for the high-molecular-weight substrate which gives support to the idea that the primary structure of the protein is of importance for xylosyl acceptor capacity. Then, the chain-initiating xylosyl transfer is followed by sequential addition of two galactosyl units. It has been demonstrated that two different galactosyltransferases are involved in the process and that the second galactosyltransferase has a very high degree of specificity toward the disaccharidic unit β-Gal-(1 → 4)-Xyl. Therefore, this strict specificity prevents any further addition of galactose on the trisaccharidic unit linked to serine and β-Gal-(1 → 3)-β-Gal-(1 → 4)-β-Xyl-(1 → 3)-Ser is formed. Next a glucuronic acid residue is added by the glucuronyltransferase I for which the terminal galactose residue is the acceptor. Further addition of the carbohydrate involved in the repeating units is achieved by alternating the action of the required glycosyltransferases whose specificity seems to be simply restricted to the correct terminal monosaccharide acceptor, whatever is the rest of the glycan moiety: *N*-acetylgalactosaminyl transferase is specific for terminal glucuronyl residue and glucuronyltransferase II is specific for terminal *N*-acetylgalactosamine residue. As a very

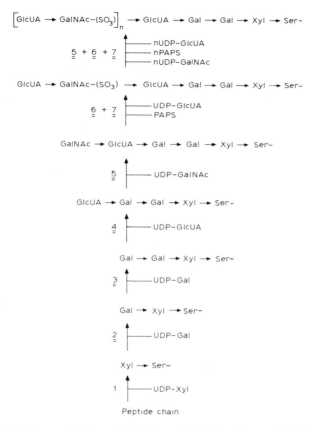

Fig. 57. Pathway of biosynthesis of a sulfate chondroprotein [450]. 1: xylosyltransferase, 2: galactosyltransferase I, 3: galactosyltransferase II, 4: glucuronyltransferase I, 5: N-acetylgalactosaminyltransferase, 6: sulfotransferase, 7: glucuronyltransferase II.

large number of such repeated sequences of glycosylation has to take place for the biosynthesis of a complete proteoglycan molecule, it has been proposed that highly organised enzymatic complexes should exist where the product of one enzymic reaction becomes the substrate for the next one and so on. Although such enzyme-enzyme complexes have been demonstrated in vitro [675], the existence in vivo of such a multienzymatic glycosyltransferase complex is still to be established. For completion, the chondroitin residues have then to be sulfated. Although UDP-N-acetylgalactosamine-4-sulfate has been described, it does not appear that

*References p. 158*

the sulfatation is obtained through this donor. In fact, Silbert et al. [676—678] have shown that sulfatation occurs on the formed polysaccharide during or immediately after polymerisation. The sulfate donor is 3'-phosphoadenyl-5'-phosphosulfate (PAPS) which is itself synthesized from ATP and sulfate is a two-step reaction. A similar mechanism operates for the insertion of L-iduronic acid into dermatan sulfate, heparin and heparin sulfate, for which the incorporation into the glycan is not caused by the transfer from UDP-L-iduronic acid (although this compound exists), but is caused by the epimerisation, at the polymer level, of the glucuronic acid by a glucuronosyl C-5 epimerase. The delineation of the above described pathway of proteoglycan biosynthesis has been chiefly obtained by studying in vitro the individual reactions. It has to be kept in mind that the situation in vivo may be somewhat different, namely there is a possibility that unknown intermediates participate in these reactions. Recently a xylose-containing lipid from hen oviduct has been characterised [679] and a glucuronyl-*N*-acetylglucosaminyl pyrophosphodolichol [680] from human lung fibroblasts for which involvement in the proteoglycan pathway is possible, although not definitively established.

The biosynthesis of the second group of proteoglycans, that of keratan sulfates, also occurs stepwise. In the case of keratan sulfate II (see Table XII) the formation of the GalNAc-Ser(Thr) is presumably formed by direct transfer from UDP-GalNAc in an analogous manner to that which has been described for the other *O*-glycosylated proteoglycans. However, with keratan sulfate I, it appears that the initiation of the polysaccharide biosynthesis requiring the binding of *N*-acetylglucosamine to asparagine is achieved through the lipid intermediate pathway. In fact, tunicamycin (see Section 8), a specific inhibitor of *N*-acetylglucosaminyl pyrophosphate dolichol utilisation, inhibits keratan sulfate biosynthesis. For both keratan sulfates, further glycosylation of the "initiating carbohydrate" occurs step by step by repeated action of galactosyl and *N*-acetylglucosaminyl-transferases.

*6.2.3.3. Collagen type glycoproteins.* It is obvious that before glycosylation of collagen (reviews in [39, 43, 65, 348, 349, 681]) can occur, the sites of attachment of the galactosyl residue must be formed by a post-translation hydroxylation of the lysine residues of the peptidic chain to give the hydroxylysine residues. This reaction is catalysed by the lysyl hydroxylase and is performed in the endoplasmic reticulum, while the peptidic chain is still growing (review in [682]) and prior to the formation of the triple-helical configuration of the collagen chain. After hydroxylation, a few hydroxylysine residues are galactosylated and some of these moieties are further glycosylated. All the glycosylation processes take place on the nascent polypeptide, after the modification of lysine residues, but before the folding of the peptidic chains into the triple helix.

It is worth mentioning that the amino acid sequence in the vicinity of the hydroxylysine residue plays a role in the control of the glycosylation as has been discussed in Section 6.2. However, it appears that all the potential sites are not glycosylated and that all the galactosyl residues are not further glycosylated. Thus, other specific environmental conditions of the neighbouring amino acid sequences are required as has been reported for the glycosylation of the Hyl-87 of four chains of various interstitial collagens which contain the similar sequence Gly-Leu(or Phe)-Hyp-Gly-X-Hyl-Gly-His-Arg. Since glycosylation ceases when the collagen is folded into the triple structure, it is possible that the nascent polypeptide possesses a special secondary structure which is lost with the rigidity obtained from the formation of the tertiary and quaternary structure. This may correctly expose some of the hydroxylysyl residues to glycosylation, as has been observed for the $N$-glycosylation.

*6.2.3.4. Extensin-type glycoproteins.* Extensins resemble collagens in their tertiary structure and in the mechanism of the post-translational modification of proline (reviews in [54, 70, 77, 355]). It is worth mentioning that both are structural glycoproteins and it is possible that the two proteins could share a common evolutionary origin [70]. Very little is known about the mechanism of linking arabinosyl residues to hydroxyproline. More than one enzyme has to be involved [357] as four different glycosidic linkages are present in the glycopeptide sequence: $\alpha$-L-Ara-(1 → 3)-$\beta$-L-Ara-(1 → 2)-$\beta$-L-Ara-(1 → 2)-$\beta$-L-Ara-(1 → 4)-Hyp and it is believed that the sequence required for glycosylation consists of four adjacent hydroxyprolines [338].

## 7. Glycoprotein catabolism and pathology
(With the collaboration of Geneviève Spik)

Experimental evidence accumulated until now indicates that lysosomes play an essential role in the degradation of glycoproteins of both intracellular and extracellular origin (reviews on glycoprotein catabolism in [43, 74, 450, 683—685]). Proof supporting this view has been obtained from biochemical findings and from pathological observations. Lysosomes contain the complete collection of glycosidases and proteases capable of extensive degradation of the carbohydrate moiety and protein core of all the glycoproteins. Thus, complete hydrolysis of these substances results in the formation of monosaccharides and amino acids that escape into the cytoplasm, by passing apparently freely through the lysosome membrane, where they are catabolised or recycled. By contrast, in storage diseases due to genetic defects in lysosomal glycosidases, the catabolism of glycoproteins is profoundly disturbed and their degradation results in oligosaccharide or in glycoasparagine fragments which may be retained within lysosomes, but which are small enough to be released from cells and excreted in urine. Study of the structure of these fragments has provided the basis for increasing our understanding of the way in which glycoproteins are catabolised and of the role of lysosomes in the degradation of these compounds.

### 7.1. Enzymes involved in the catabolism of glycoproteins

*7.1.1. Proteolytic enzymes.* Lysosomes are capable of completely digesting the peptide chain through action of the various proteases and peptidases they contain such as cathepsins (A and other catheptic carboxypeptidases, B, C, D and E), neutral proteases and a series of endo- and exopeptidases. The question is whether the complete degradation of the protein moiety up to glyco-amino acids is a prerequisite for the action of glycosidases. This problem has not yet been solved. However, taking into account the fact that activity of glycosidases towards glycopeptides decreases with the length of the peptide chain, one can assume that the action of glycosidases may start only after an extensive degradation of the protein moiety has been achieved.

*7.1.2. Specific enzymes involved in degradation of glycans.* Space limitation precludes very detailed discussion of this topic, abundantly reviewed in [26, 60, 274, 276, 683—687], in which the physicochemical properties and the enzymatic activity parameters of both exo- and endoglycosidases from microorganisms, plants, and vertebrate and invertebrate animals, are extensively described.

It is generally assumed that, at least in animal cells, the catabolism of glycoprotein glycans occurs essentially or entirely in lysosomes, and that it is performed by exoglycosidases. The organelles possess the glycosidase equipment indispensable for cleavage of all kinds of glycosidic linkages which are present in glycan structures: until now, the following lysosomal exoglycosidases, part of which are associated with the lysosome membrane, have been isolated or characterized: $\alpha$-2,3 and $\alpha$-2,6 neuraminidases, $\alpha$-D- and $\beta$-D-galactosidases, $N$-acetyl-$\beta$-D-hexosaminidases, -glucosaminidases and -galactosaminidases, $N$-acetyl-$\alpha$-D-galactosaminidases, $\alpha$-L-fucosidases, $\alpha$-D- and $\beta$-D-mannosidases, $\alpha$-D- and $\beta$-D-glucosidases, $\beta$-D-xylosidases, $\beta$-D-glucuronidases, $\beta$-L-iduronidase. In addition, $O$- and $N$-sulfatases with a narrow or wide specificity ensure the removal of sulfate residues from sulfated glycans (Fig. 59).

However, one cannot assert that glycoprotein catabolism is entirely performed by the lysosome and by lysosomal exoglycosidases. Exoglycosidases are not exclusively lysosomal in nature and non-lysosomal enzymes could play an important role in the degradation of glycoprotein glycans. For instance, neuraminidases have been found, apart from lysosomes, in plasma membranes, Golgi apparatus and cytosol [685, 688]. Another example is the glucosidase-mannosidase system located in the Golgi apparatus which is responsible for the trimming and maturation of $N$-glycosically linked glycans (Section 6). Moreover, one cannot exclude the possibility that endoglycosidases intervene in glycoprotein catabolism at an early stage. Carbohydrate chains could be initially degraded to oligosaccharides and subsequently cleaved by specific exoglycosidases. For instance, it is possible that the lysosomal hyaluronidase, an endoglycosidase, degrades hyaluronic acid and chondroitin sulfate isomers into oligosaccharides having glucuronic acid at the non-reducing end, that are further hydrolysed by exoglycosidases. Moreover, we described in Section 4.1 other endoglycosidases specific for $N$-glycosically linked glycans (Table IV) e.g. 4'-L-aspartylglycosylamine amidohydrolase or "glycoaspartamidase", a widely distributed lysosomal enzyme (review in [276]) that liberates complete glycans from glycoasparagines; endo-$N$-acetyl-$\beta$-D-glucosaminidases that split the di-$N$-acetylchitobiose residue of glycans of the $N$-acetyl-lactosamine type or of the oligomannosidic type. In this connection we must mention an interesting observation which has been recently reported by Pierce et al. [689, 690]. These authors have demonstrated that two types of endo-N-acetyl-β-D-glucosaminidases are located in the cytosol of rat liver and kidney, not in the lysosomes: (i) "manno-endo-N-acetyl-β-D-glucosaminidases" active on glycans of the oligomannosidic type; (ii) "galacto-endo-N-acetyl-β-D-glucosaminidases" active on glycans of the N-acetyl-lactosaminic type, on provided that at least a galactose residue is in a terminal non-reducing position. This result could be explained in two ways: (1) there are effectively no endo-N-acetylglucosaminidases in lysosomes and the N-glycosidically linked glycans are thus catabolised by the sequential action of exoglycosidases only; (2) the cytosolic endo-N-acetyl-β-glucosaminidases are not enzymes of the catabolism of cytoplasmic glycoproteins but belong to a control system of the biosynthesis of N-glycosylproteins which is proposed, as an hypothesis [691], in Fig. 58. When the glycan moiety of a glycoprotein is still immature (Section 7) it is possible to imagine that this glycoprotein is susceptible toward "manno" and "galacto" β-endo-N-acetyl-

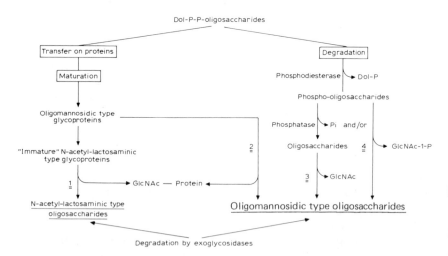

Fig. 58. Scheme proposed [691] for a catabolic pathway of dolichyl pyrophosphate-linked oligosaccharides and of "immature" glycoproteins (see Fig. 52). 1: endo-N-acetyl-β-D-glucosaminidase acting on glycans of the N-acetyl-lactosaminic type; 2, 3 and 4: endo-N-acetyl-β-D-glucosaminidase acting on glycans of the oligomannosidic type.

glucosaminidases. Thus, at any moment, the cell has the possibility to destroy the newly synthesised glycoproteins, if something triggers this choice. This concept has recently been extended in our laboratory [616] to the degradation of Dol-P-P-oligosaccharides (Fig. 58). It has been demonstrated that Dol-P-P-oligosaccharides may be degraded by a phosphodiesterase into dolichol-phosphate which is immediately available for recycling into the dolichol cycle, and into phospho-oligosaccharides. These substances are susceptible to alkaline phosphatase giving neutral-oligosaccharides and they may be cleaved by "manno"-endo-$N$-acetyl-$\beta$-D-glucosaminidase leaving GlcNAc-1-P and neutral-oligosaccharides. These data suggest that splitting of the phosphodiester bond of Dol-P-P-oligosaccharides, dephosphorylation and/or endo $N$-acetyl-$\beta$-D-glucosaminidase hydrolysis of the phosphorylated oligosaccharides could represent the beginning of the catabolic pathway of Dol-P-P-oligosaccharides. If this pathway really occurs in vivo, urinary oligosaccharides of the oligomannosidic type which are secreted by patients with a lysosomal enzyme deficiency such as mannosidosis (see Section 7.3) could originate not only from glycoprotein catabolism but also from the Dol-P-P-oligosaccharide catabolism.

## 7.2. Catabolic pathway of glycoproteins

In Figs. 59 and 69 examples are given of the schematic pathways of catabolism of heparan sulfate, a proteoglycan, by exoglycosidases and of an $N$-glycosidically linked glycan of the $N$-acetyl-lactosaminic type by the combined and successive actions, first of an $N$-acetyl-$\beta$-D-endoglycosidase and, then, of exoglycosidases. In both cases, one has to be aware of the complexity of a mechanism that involves an impressive number of enzymes: seven in the case of heparan sulfate to which have to be added three more enzymes, those hydrolysing the trisaccharidic core linking the polysaccharide chain to the protein (see Table XII): two $\beta$-D-galactosidases and one $\beta$-D-xylosidase; eight in the case of the sialofucoglycan of Fig. 69, plus the glycoaspartamidase (amido hydrolase).

However, the initiation of the catabolism of cellular glycoprotein as well as its control and regulation remain unclear. In particular, if the intracellular glycoproteins are really catabolised in lysosomes, we must know how they enter the organelle: are they in-

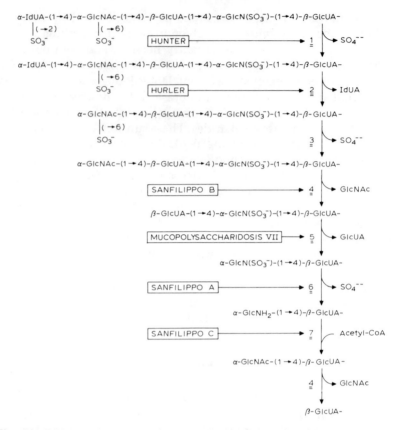

Fig. 59. Pathway for heparan sulfate catabolism by exoglycosidases [450, 692]. 1: L-iduronidase sulfatase, 2: α-L-iduronidase, 3: N-acetyl-D-glucosamine-6-sulfate sulfatase, 4: N-acetyl-α-D-glucosaminidase, 5: β-D-glucuronidase, 6: sulfamidase, 7: acetylCoA: α-glucosaminide N-acetyltransferase.

cluded in autophagosomes or do they penetrate into the lysosome after fixation on the cytoplasmic face of the lysosomal membrane? However, this last hypothesis can be accepted only if the existence of receptors located at the surface of lysosomes and recognising the monosaccharide end of glycoproteins is demonstrated. Findings of Ashwell's group led to the elucidation of the mechanism of the catabolic pathway of exocellular glycoproteins, particularly, that of serum glycoproteins. Thanks to a series of fascinating experiments, Ashwell and coworkers demonstrated how circulating glycoproteins could be specifically recognised by cell membranes, fixed by them and then endocytosed.

### 7.2.1. Uptake and catabolism of glycoproteins.

Ashwell et al. (for original publications and reviews: [214—216, 693—695]) demonstrated that the elimination of the terminal sialic acid residues from numerous circulating glycoproteins is equivalent to the death sentence for these compounds. In monitoring the evolution of radioactivity from such glycoproteins as $\alpha_1$-acid glycoprotein (see the glycan structure in Fig. 26) labelled with iodine-125, Ashwell and coworkers [696] demonstrated that asialo-glycoproteins, which thus possessed galactosyl groups in terminal position, disappeared in less than half an hour from the plasma of an animal into which they have been injected and that the injected radioactivity appeared in the hepatocytes. The elimination of the terminal galactosyl groups with $\beta$-D-galactosidase, which exposes $N$-acetylglucosamine groups in the terminal positions, slowed the clearance of asialo-agalacto-glycoproteins of the plasma. The terminal galactosyl groups are thus the recognition signals for uptake of these asialoglycoproteins for the hepatocytes. Ashwell's group proved that they are recognized at the hepatocyte membrane, by showing that the radioactivity of [$^{125}$I]asialo-glycoproteins are fixed very rapidly onto some hepatocyte-membrane proteins. Ashwell and coworkers [697] succeeded in isolating hepatocyte-membrane receptor for galactoglycoproteins by affinity chromatography. It consists of a sialylglycoprotein that loses its property of recognition of galactose if it is desialylated by neuraminidase. The proof was thus furnished that a "galactoglycan" carries the recognition signal for a glycoprotein receptor, itself a sialoglycoprotein, which is integrated in a plasma membrane.

These observations were extended to many other glycoproteins including ceruloplasmin, haptoglobin, thyroglobulin, fetuin, and prothrombin for which the circulatory life is dependent on sialic acid residues. So, the mechanism of extracellular glycoprotein degradation clearly occurs in steps: (1) binding of glycoprotein to cell plasma membrane receptors, (2) uptake by pinocytosis leading to the formation of pinocytic vesicles, (3) fusion of these latter with lysosomes in which the glycoproteins are completely degraded.

Further studies initiated by the findings of Ashwell's group have

*References p. 158*

TABLE XV

Carbohydrate recognition signals for adsorptive pinocytosis of glycoproteins (quoted in [695])

| Signal on ligand | Cell expressing the receptor |
| --- | --- |
| Gal, GalNAc, Glc | Mammalian hepatocyte |
| GlcNAc | Avian hepatocyte |
| $\alpha$-L-Fuc-(1 → 3)-GlcNAc | Mammalian hepatocyte |
| Man, GlcNAc, Glc | Reticuloendothelial cells |
| Mannose-6-phosphate | Fibroblasts |

shown that the hepatocyte was not the only cell possessing glycoprotein receptors and that galactose was not the only terminal monosaccharide to be recognised. By studying the effects of sugars as competitive inhibitors of the binding and of the uptake or clearance of glycoproteins or by using synthetic ligands called "neoglycoproteins" (review in [698]) in which sugars are attached to proteins, the carbohydrates listed in Table XV have been identified as recognition signals in adsorptive pinocytosis of glycoproteins.

*7.2.2. Traffic of lysosomal glycosidases.* One of the most interesting concepts to evolve in the story of receptor-mediated uptake of glycoproteins concerns the uptake and sequestration into fibroblast lysosomes of extracellular acid hydrolases, which are also glycoproteins.

Until recently there was fairly general agreement that lysosomal enzymes enter the vacuolar system of cells by primary lysosomes that originate in the Golgi apparatus and smooth endoplasmic reticulum. This view was called in question by observations made on I-cell disease or mucolipidosis type II (see below). This is an inherited disease [699] in which many lysosomal enzymes, including neuraminidases [700, 701] are absent or deficient from connective tissue-cells. Hickman and Neufeld [702] showed that cultured I-cell fibroblasts would pinocytose and retain exogenous lysosomal enzymes secreted by normal fibroblasts. However, the lysosomal enzymes secreted by I-cells were not taken up. They therefore proposed that the normal route by which lysosomal enzymes make their way from their site of synthesis to the vacuolar system is one that included exocytosis to the extracellular environment and pinocytic recapture by neighbouring cells (reviews in [216, 692—695 and 703]).

This concept received support from the fact that enzyme-deficient cells could be corrected by the addition to the culture medium of medium preincubated with normal cells, or of cells from any genotype other than that of the defective cell. The so-called "corrective factors" have been identified as the missing lysosomal glycosidases in the respective disorders. The recognition of lysosomal enzymes at the cell surface has been subjected to studies focused either on the nature and characteristics of the receptor at the cell surface [704] or on the structure of recognised moieties of the enzyme which are carbohydrates [705]. In 1977, the presence of phosphate groups in the recognised moiety was demonstrated [706]. Since then phosphate groups have been located on position C-6 of mannose residues of several lysosomal glycoenzymes [707]. The structure [708] and biosynthesis [709] of phosphooligomannosidic glycans of lysosomal glycoenzymes have recently been elucidated (Fig. 60).

However, this secretion-reuptake model has been contradicted by several experimental data [710] and it has been pointed out that this model can be modified to involve transfer of newly synthesised enzymes to lysosomes via the cell surface in a tightly bound form, by von Figura and Weber [710] or via intracellular receptor mediated recognition intracellularly by Sly and coworkers [694, 706]. The latter authors proposed that most of the lysosomal glycosidases reach lysosomes by intracellular route, i.e. without being first secreted. However, enzyme delivery in this model is still receptor-mediated.

According to these investigators and others [711], mannose-6-phosphate residues carried by glycans of the oligomannosidic type are the *intracellular* traffic signals which allow the segregation of the acid hydrolases from the secretory glycoproteins destined for export, by binding to receptors present on membranes of the endoplasmic reticulum. Enzyme lacking the recognition marker would fail to be segregated and would proceed through the Golgi where the glycans of the oligomannosidic type would be processed to the $N$-acetyl-lactosaminic type. Thus the acid hydrolases, which escape

*References p. 158*

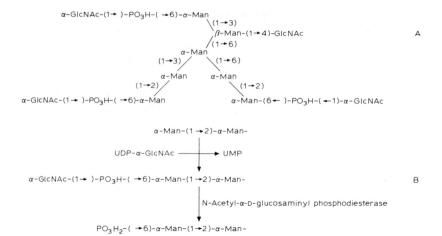

Fig. 60. (A) proposed structure for oligosaccharide II-2 obtained by the action of endo-N-acetyl-β-D-glucosaminidase H on N-acetyl-β-D-hexosaminidase and cathepsin D secreted by human skin fibroblasts [708]. Oligosaccharide I-2 differs from oligosaccharide II-A by the substitution with only 1 to 2 phosphate groups blocked by N-acetylglucosamine. (B) Proposed scheme for the phosphorylation of mannose residues of lysosomal enzyme glycans [709]. In this biosynthetic pathway, it is proposed that N-acetyl-α-glucosamine 1-phosphate residues are transferred from UDP-N-acetyl-α-glucosamine to mannose residues of the lysosomal enzyme. The N-acetylglucosamine residues are then removed by N-acetyl-α-glucosamine phosphodiesterase which has been recently characterized in smooth membranes of rat liver by Varki and Kornfeld [708]. This enzyme, by removing the N-acetylglucosamine residues, unmasks the targeting function of the phosphomannosyl residues of the glycans of newly synthesized lysosomal hydrolases.

the normal segregation process because they lack the mannose-6-phosphate marker, would be secreted with other soluble secretory glycoproteins which they resemble in their carbohydrate structure. Recently, Sly and coworkers [712] verified their concept of the existence of phosphomannosyl receptors on intracellular membranes in demonstrating that at least 80% of the total specific binding activity of β-glucuronosidase, a lysosomal "phosphomannosyl enzyme", are present on intracellular membranes from disrupted human fibroblasts and they concluded that

the total number of receptors (intracellular and cell surface) is insufficient to explain previously observed rates of enzyme pinocytosis without invoking receptor reutilization

and

the finding that the majority of receptors were on intracellular membranes is consistent with a role for the receptor in regulating intracellular traffic of newly synthesized acid hydrolases and directing them to lysosomes.

Other carbohydrate recognition systems mediate the clearance of acid hydrolases that enter the plasma. Lysosomal glycosidases, such as human and rat $\beta$-glucuronidase, rat $\beta$-galactosidase, $\alpha$-fucosidase, $\alpha$-mannosidase and $N$-acetylglucosaminidase [713] are promptly cleared from plasma by a specific recognition system largely associated with liver. After uptake, the above glycoenzymes are rapidly transported to non-parenchymal liver cell lysosomes [714]. The efficient capture of exocytosed enzymes was postulated to depend on interaction between a chemical moiety on the enzyme and a binding site on the plasma membrane. It has been shown that lysosomal clearance by liver non-parenchymal cells is mediated by a receptor that binds either $N$-acetylglucosamine or mannose residues associated with the enzymes [714].

## 7.3. Genetic disorders of glycoprotein catabolism

A number of various pathological conditions have been described in the last 15 years in which the degradation of compounds containing carbohydrates, including glycoproteins and glycolipids, appears to be disturbed. These diseases, generally fatal, are all characterised by the same symptoms: a red spot at the back of the eye, facial dysmorphia, and mental retardation, in particular. They result from deficiencies in lysosomal enzymes leading to the accumulation of oligosaccharides or/and of glycoasparagines in the tissues and urine of the patients (reviews in [60, 65, 74, 78, 79, 450, 683, 685, 692, 695, 715—718]).

The first description of this kind of disease affecting the glycoproteins, we propose to call "glycoproteinoses", concerned the disorder of acidic mucopolysaccharide degradation. The concept of the mechanism originated in the electron microscopic studies of Van Hoof and Hers who observed large vacuoles containing granular material in the hepatocytes of a Hurler patient (Fig. 59). They interpreted these vacuoles as lysosomes distended with undigested accumulated mucopolysaccharides. Later the diagnosis of this

*References p. 158*

disorder was established by the finding of elevated excretion in urine (100 mg/l) of mucopolysaccharide fragments of low molecular weight derived from dermatan sulfate and heparan sulfate.

At present, 25 glycosidase deficiencies have been related to genetic diseases of glycoconjugates. They are listen in Table XVI. 15 of them concern glycoprotein catabolism.

The description of the so-called "mucopolysaccharidoses" will be limited to Fig. 59 in which the normal and pathological catabolic pathways are described. But the disorders of the $N$-glycoproteins will be more extensively discussed because the investigation of oligosaccharides and glycoasparagines isolated from tissues and from urine of patients affected by diverse "$N$-glycoproteinoses" has furnished important information on the catabolism of $N$-glycosylproteins, and has allowed the reconstruction of the different steps of the latter process (reviews in [78, 304, 716, 718]).

Figs. 61 to 68 represent the structures of oligosaccharides and glycoasparagines extracted from the urine or tissues of patients affected by mannosidosis, sialidosis, fucosidosis, $GM_1$-gangliosidosis, Sandhoff's disease, and asparaginylglucosaminuria, diseases caused by deficiencies in, respectively, $\alpha$-D-mannosidase, $\alpha$-neuraminidase, $\alpha$-L-fucosidase, $\beta$-D-galactosidase, $N$-acetyl-$\beta$-D-hexosaminidase, and 4'-L-aspartylglycosylamine amidohydrolase.

It may be observed, first, that all of the oligosaccharides possess an $N$-acetylglucosamine residue in the terminal reducing position, to the C-4 of which is linked a $\beta$-mannose residue. In the second place they are fragments of glycans as they exist in numerous glycoproteins. For example, the tenth oligosaccharide depicted in Fig. 62 is found in numerous diantennary glycans of human origin (see Table XIII) and the oligosaccharides following in the same figure may be considered to be fragments of the tri- and tetraantennary glycans shown in Figs. 26 and 27. In the same way, the fifth glycan in Fig. 66 is part of the glycan of IgG and IgA shown in Table XIII (Structures 46—48). Particularly interesting is the glycoasparagine described in Fig. 68 because it is identical with the "limit nona-oligomannoside" of Structure 1 in Fig. 23 and of the deglycosylated Structure B in Fig. 55. In this connection, oligosaccharide 7 in Fig. 61 could be derived from the latter by the action of an endo-$N$-acetyl-$\beta$-D-glucosaminidase.

On the basis of these observations, the hypothesis has been put

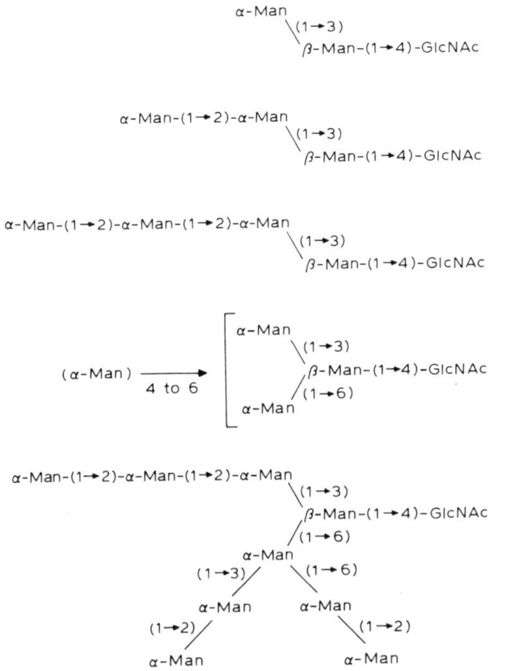

```
α-Man
    \(1→3)
     \β-Man-(1→4)-GlcNAc    1

α-Man-(1→2)-α-Man
              \(1→3)
               \β-Man-(1→4)-GlcNAc    2

α-Man-(1→2)-α-Man-(1→2)-α-Man
                         \(1→3)
                          \β-Man-(1→4)-GlcNAc    3

        ⎡ α-Man
        ⎢     \(1→3)
(α-Man) ──→  ⎢      β-Man-(1→4)-GlcNAc    4 to 6
  4 to 6 ⎢    /(1→6)
        ⎣ α-Man

α-Man-(1→2)-α-Man-(1→2)-α-Man
                         \(1→3)
                          \β-Man-(1→4)-GlcNAc    7
                          /(1→6)
                     α-Man
              (1→3)/      \(1→6)
                 α-Man    α-Man
           (1→2)/              \(1→2)
             α-Man              α-Man
```

Fig. 61. Structure of oligosaccharides isolated from urine of mannosidosis. Oligosaccharides 1 to 3 [719—721]; oligosaccharides 4 to 6 [721]; oligosaccharide 7 [308].

forward that the catabolism of glycans linked N-glycosylically to the protein starts by the action of endo-N-acetyl-β-D-glucosaminidases which split the residue of di-N-acetylchitobiose [64, 78, 304, 740, 741]. In splitting the residue of di-N-acetylchitobiose, this enzyme liberates the glycan minus a residue of N-acetylglucosamine. The oligosaccharide thus formed would then be degraded stepwise by lysosomal exoglycosidases. Fig. 68 illustrates the hypothetical mechanism of the normal and pathological catabolism of a glycan of N-glycosylproteins.

Thus, knowledge of the structure of the oligosaccharides that accumulate in the urine and tissues of patients suffering from glycoproteinoses has allowed the definition of the nature of the enzymic deficiencies that are a consequence of grave genetic dis-

*References p. 158*

## TABLE XVI
### Genetic diseases related to lysosomal glycosidase deficiencies

| Enzymatic deficit | Accumulated compounds | Nature of the disease |
|---|---|---|
| α-L-Iduronidase | Heparan sulfate and dermatan sulfate | Mucopolysaccharidosis $I_H$ (Hurler) Mucopolysaccharidosis $I_S$ (Scheie) |
| L-Iduronidase sulfatase | Heparan sulfate and dermatan sulfate | Mucopolysaccharidosis II (Hunter) |
| Sulfamidase | Heparan sulfate | Mucopolysaccharidosis $III_A$ (Sanfilippo A) |
| N-Acetyl-α-D-glucosaminidase | Heparan sulfate | Mucopolysaccharidosis $III_B$ (Sanfilippo B) |
| Acetyl CoA: α-glucosaminide transferase | Heparan sulfate | Mucopolysaccharidosis $III_C$ (Sanfilippo C) |
| N-Acetyl-D-galactosamine-6-$SO_4$ and galactose-6-$SO_4$ sulfatase | Keratan sulfate and chondroitin sulfate | Mucopolysaccharidosis IV (Morquio) |
| N-Acetyl-D-galactosamine-4-$SO_4$ sulfatase | Dermatan sulfate | Mucopolysaccharidosis VI (Maroteaux-Lamy) |
| β-D-Glucuronidase | Dermatan sulfate and heparan sulfate | Mucopolysaccharidosis VII |
| N-Acetyl-D-glucosamine-6-$SO_4$ sulfatase | Keratan sulfate and heparan sulfate | New type of mucopolysaccharidosis |
| Sphingomyelinase | Sphingomyelin | Nieman-Pick's disease |
| Ceramidase | Ceramide | Farber's disease |
| Cerebroside sulfatase | Cerebroside sulfate | Metachromatic leukodystrophy |
| Galactocerebrosidase | Galactosyl-sphingosine | Krabbe's disease |
| Glucocerebrosidase | Glucocerebroside | Gaucher's disease |
| α-D-Galactosidase | Ceramide trihexoside | Fabry's disease |
| β-D-Hexosaminidase A | Ganglioside GM 2 | Tay-Sachs disease |
| Ganglioside α-neuraminidase | Ganglioside GM 3 and GD 3 | Mucolipidosis IV |

| | | |
|---|---|---|
| UDP-GalNAc: GM 3 N-acetyl-galactosaminyltransferase (?) | Ganglioside GM 3 | GM 3 Gangliosidosis |
| β-D-Galactosidase | Ganglioside GM 1 and oligosaccharides with terminal galactose | GM 1 Gangliosidosis |
| β-D-Hexosaminidases A and B | Ganglioside GM 2, globoside and oligosaccharides with terminal N-acetylglycosamine | Sandhoff's disease |
| α-L-Fucosidase | Fucosyl-glycolipids, oligosaccharides with terminal fucose, and fucosyl-glycoasparagines | Fucosidosis |
| α-D-Mannosidase | Oligosaccharides with terminal mannose | α-Mannosidosis |
| 4′-L-Aspartylglycosylamine amido hydrolase | Glycoasparagines | Asparaginyl-N-acetylglucosaminuria |
| α-Neuraminidase | Sialyl-oligosaccharides | Sialidosis |

*References p. 158*

α-NeuAc-(2→3)-β-Gal-(1→4)-β-GlcNAc-(1→2)-α-Man
    \(1→3)
     β-Man-(1→4)-GlcNAc    1

α-NeuAc-(2→6)-β-Gal-(1→4)-β-GlcNAc-(1→2)-α-Man
    \(1→3)
     β-Man-(1→4)-GlcNAc    2

α-NeuAc-(2→6)-β-Gal-(1→4)-β-GlcNAc-(1→2)-α-Man
    \(1→3)
     β-Man-(1→4)-GlcNAc    3
    /(1→6)
α-Man

α-NeuAc-(2→3)-β-Gal-(1→4)-β-GlcNAc-(1→2)-α-Man
    \(1→3)
     β-Man-(1→4)-GlcNAc    4
    /(1→6)
β-Gal-(1→4)-β-GlcNAc-(1→2)-α-Man

α-NeuAc-(2→6)-β-Gal-(1→4)-β-GlcNAc-(1→2)-α-Man
    \(1→3)
     β-Man-(1→4)-GlcNAc    5
    /(1→6)
β-Gal-(1→4)-β-GlcNAc-(1→2)-α-Man

α-NeuAc-(2→3)-β-Gal-(1→4)-β-GlcNAc   (1→4)
α-NeuAc-(2→6)-β-Gal-(1→4)-β-GlcNAc-(1→2)-α-Man
    \(1→3)
     β-Man-(1→4)-GlcNAc    6

α-NeuAc-(2→3)-β-Gal-(1→4)-β-GlcNAc-(1→2)-α-Man
    \(1→3)
     β-Man-(1→4)-GlcNAc    7
    /(1→6)
α-NeuAc-(2→3)-β-Gal-(1→4)-β-GlcNAc-(1→2)-α-Man

α-NeuAc-(2→6)-β-Gal-(1→4)-β-GlcNAc-(1→2)-α-Man
    \(1→3)
     β-Man-(1→4)-GlcNAc    8
    /(1→6)
α-NeuAc-(2→3)-β-Gal-(1→4)-β-GlcNAc-(1→2)-α-Man

α-NeuAc-(2→3)-β-Gal-(1→4)-β-GlcNAc-(1→2)-α-Man
    \(1→3)
     β-Man-(1→4)-GlcNAc    9
    /(1→6)
α-NeuAc-(2→6)-β-Gal-(1→4)-β-GlcNAc-(1→2)-α-Man

α-NeuAc-(2→6)-β-Gal-(1→4)-β-GlcNAc-(1→2)-α-Man
    \(1→3)
     β-Man-(1→4)-GlcNAc    10
    /(1→6)
α-NeuAc-(2→6)-β-Gal-(1→4)-β-GlcNAc-(1→2)-α-Man

(Fig. 62; legend on p 135)

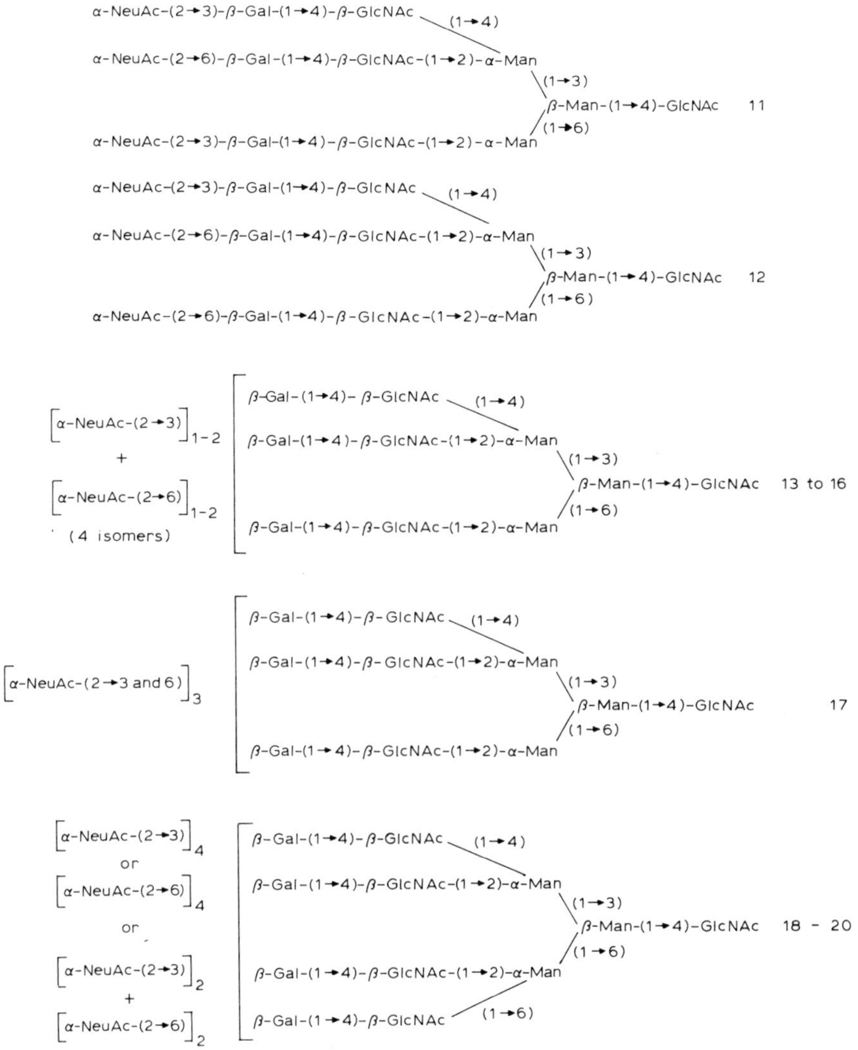

# GENETIC DISORDERS OF CATABOLISM

```
α-NeuAc-(2→3)-β-Gal-(1→4)-β-GlcNAc
                                    \(1→4)
α-NeuAc-(2→6)-β-Gal-(1→4)-β-GlcNAc-(1→2)-α-Man
                                               \(1→3)
                                               β-Man-(1→4)-GlcNAc    11
                                               /(1→6)
α-NeuAc-(2→3)-β-Gal-(1→4)-β-GlcNAc-(1→2)-α-Man


α-NeuAc-(2→3)-β-Gal-(1→4)-β-GlcNAc
                                    \(1→4)
α-NeuAc-(2→6)-β-Gal-(1→4)-β-GlcNAc-(1→2)-α-Man
                                               \(1→3)
                                               β-Man-(1→4)-GlcNAc    12
                                               /(1→6)
α-NeuAc-(2→6)-β-Gal-(1→4)-β-GlcNAc-(1→2)-α-Man
```

$\begin{bmatrix}\alpha\text{-NeuAc-}(2\to3)\end{bmatrix}_{1-2}$
$+$
$\begin{bmatrix}\alpha\text{-NeuAc-}(2\to6)\end{bmatrix}_{1-2}$
(4 isomers)

```
   β-Gal-(1→4)-β-GlcNAc
                        \(1→4)
   β-Gal-(1→4)-β-GlcNAc-(1→2)-α-Man
                                    \(1→3)
                                    β-Man-(1→4)-GlcNAc    13 to 16
                                    /(1→6)
   β-Gal-(1→4)-β-GlcNAc-(1→2)-α-Man
```

$\begin{bmatrix}\alpha\text{-NeuAc-}(2\to3\text{ and }6)\end{bmatrix}_3$

```
   β-Gal-(1→4)-β-GlcNAc
                        \(1→4)
   β-Gal-(1→4)-β-GlcNAc-(1→2)-α-Man
                                    \(1→3)
                                    β-Man-(1→4)-GlcNAc    17
                                    /(1→6)
   β-Gal-(1→4)-β-GlcNAc-(1→2)-α-Man
```

$\begin{bmatrix}\alpha\text{-NeuAc-}(2\to3)\end{bmatrix}_4$
or
$\begin{bmatrix}\alpha\text{-NeuAc-}(2\to6)\end{bmatrix}_4$
or
$\begin{bmatrix}\alpha\text{-NeuAc-}(2\to3)\end{bmatrix}_2$
$+$
$\begin{bmatrix}\alpha\text{-NeuAc-}(2\to6)\end{bmatrix}_2$

```
   β-Gal-(1→4)-β-GlcNAc
                        \(1→4)
   β-Gal-(1→4)-β-GlcNAc-(1→2)-α-Man
                                    \(1→3)
                                    β-Man-(1→4)-GlcNAc    18 - 20
                                    /(1→6)
   β-Gal-(1→4)-β-GlcNAc-(1→2)-α-Man
                                    /(1→6)
   β-Gal-(1→4)-β-GlcNAc
```

Fig. 62. Structure of oligosaccharides isolated from urine [301, 722—725] and from liver and brain [727] of sialidosis.

*References p. 158*

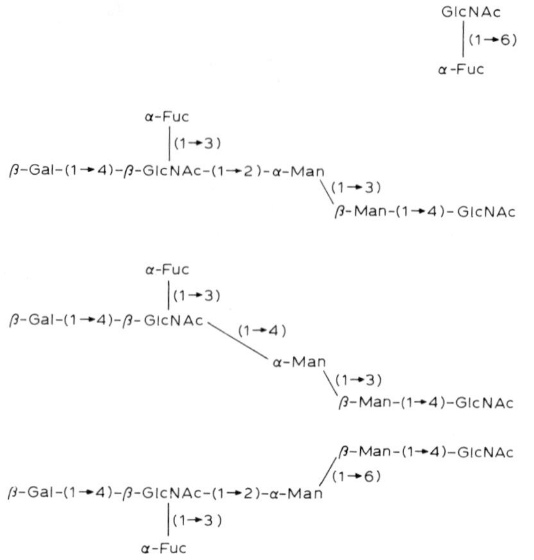

```
                                    GlcNAc
                                     │(1→6)    1
                                    α-Fuc

                α-Fuc
                 │(1→3)
  β-Gal-(1→4)-β-GlcNAc-(1→2)-α-Man
                                  \(1→3)
                                   β-Man-(1→4)-GlcNAc    2

                α-Fuc
                 │(1→3)
  β-Gal-(1→4)-β-GlcNAc
                      \(1→4)
                       α-Man
                            \(1→3)
                             β-Man-(1→4)-GlcNAc    3

                                      /β-Man-(1→4)-GlcNAc    4
                                     /(1→6)
  β-Gal-(1→4)-β-GlcNAc-(1→2)-α-Man
                                │(1→3)
                                α-Fuc
```

Fig. 63. Structure of oligosaccharides isolated from urine of fucosidosis [298, 726].

eases, and has led to perfection of methods for their antenatal diagnosis.

## 8. Molecular biology and role of glycoprotein glycans

The definition, in the last 5 years, of the primary structure of numerous glycoprotein glycans several of which are membrane constituents, has given rise to the hope that the mystery of the biological significance of the glycans would be rapidly resolved. However, despite all the intense work done in many laboratories the biological role of glycans remains unclear and the message they carry has not yet been decoded. In this connection, it suffices to consider the large number of hypotheses which have been put forward about the molecular biology of the glycans in the last 15 years to realise the extent of our ignorance. In fact, the greater its depth, the more the investigators' imagination, always unbounded, can run free. Furthermore, it has to be admitted that, to add to the confusion, many results are contradictory. I am thinking in

# GENETIC DISORDERS OF CATABOLISM

```
β-GlcNAc-(1→ )-Asn      GP-1                β-Gal-(1→4)-β-GlcNAc-(1→ )-Asn    GP-2
 |(1→6)                                      |(1→6)
 α-Fuc                                       α-Fuc
                           α-Fuc
                            |(1→3)
        β-Gal-(1→4)-β-GlcNAc-(1→2)-α-Man
                                \(1→3)
                                 β-Man-(1→4)-β-GlcNAc-(1→4)-β-GlcNAc-(1→ )-Asn    GP-3
                                                                    |(1→6)
                                                                    α-Fuc

α-Fuc-(1→2)-β-Gal-(1→4)-β-GlcNAc-(1→2)-α-Man
                                \(1→3)
                                 β-Man-(1→4)-β-GlcNAc-(1→4)-β-GlcNAc-(1→ )-Asn    GP-4
                                                                    |(1→6)
                                                                    α-Fuc

                                 /β-Man-(1→4)-β-GlcNAc-(1→4)-β-GlcNAc-(1→ )-Asn    GP-5
                                /(1→6)
                       α-Man
                                                                    |(1→6)
                                                                    α-Fuc

                                 /β-Man-(1→4)-β-GlcNAc-(1→4)-β-GlcNAc-(1→ )-Asn    GP-6
                                /(1→6)
        β-Gal-(1→4)-β-GlcNAc-(1→2)-α-Man
                                                                    |(1→6)
                                                                    α-Fuc

                                 /β-Man-(1→4)-β-GlcNAc-(1→4)-β-GlcNAc-(1→ )-Asn    GP-7
                                /(1→6)
        β-Gal-(1→4)-β-GlcNAc-(1→2)-α-Man
                                                                    |(1→6)
          |(1→3)                                                    α-Fuc
          α-Fuc

                                 /β-Man-(1→4)-β-GlcNAc-(1→4)-β-GlcNAc-(1→ )-Asn    GP-8
                                /(1→6)
α-Fuc-(1→6)-β-Gal-(1→4)-β-GlcNAc-(1→2)-α-Man
                                                                    |(1→6)
                                                                    α-Fuc

    ⎡      α-Fuc              ⎤
    ⎢       |(1→3)            ⎥     α-Man
    ⎢β-Gal-(1→4)-β-GlcNAc-(1→ )⎥         \(1→3)
    ⎣                         ⎦ 3-4      β-Man-(1→4)-β-GlcNAc-(1→4)-β-GlcNAc-(1→ )-Asn   GP-9 and 10
                                     /(1→6)
                                α-Man                               |(1→6)
                                                                    α-Fuc
```

Fig. 64. Structure of glycoasparagines isolated from urine of fucosidosis. GP-1 [298, 726, 728, 729]; GP-3 [730]; GP-2 and GP-4 to GP-10 [78, 298, 726, 729]. Some of these structures have been confirmed by Kobata et al. [731].

particular of those obtained by using tunicamycin, sometimes demonstrating the importance, sometimes the unimportance of glycans. Finally, the error is perhaps too often committed of wanting to attribute one specific role to glycan function. It is very possible that all the hypotheses will be confirmed, but that each explanation applies to a particular group of glycoproteins. It would be a

*References p. 158*

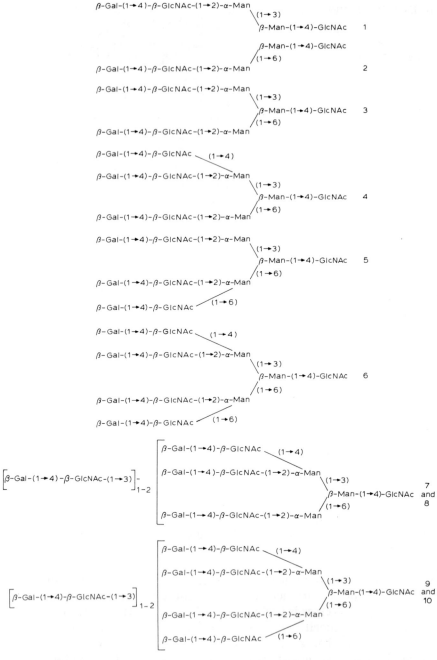

Fig. 65. Structure of oligosaccharides isolated from urine and liver of GM 1-gangliosidosis type I. Oligosaccharide 1 [732]; oligosaccharide 3 and 4: [733]; oligosaccharides 2 and 5 to 10 [78, 734, 741].

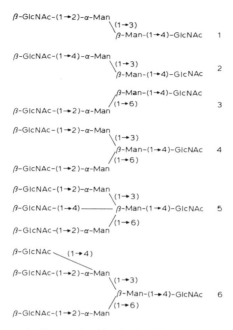

Fig. 66. Structure of oligosaccharides isolated from urine of Sandhoff's disease [297, 735, 736].

"mission impossible" to report or even to summarise the considerable body of results accumulated in the last 10 years, on the role of glycoproteins in the fields of molecular biology and pathology (reviews in [1, 29, 39, 43, 45—47, 50, 57, 64, 65, 76, 78, 165, 214—216, 235, 618, 741—747]). I will thus give one or two precise and demonstrative experimental examples for each hypothesis in the hope that the hundreds of other investigators will forgive me for not mentioning their work.

At the beginning of investigations into the biological role of glycoproteins numerous authors considered that glycans associated with proteins were metabolic accidents and that they played no important biological role. This "gadget hypothesis" was, in particular, upheld by Gottschalk [748] for whom the biosynthesis of the glycans took place when the following conditions were fulfilled: (1) the presence in the peptide chains of a "code sequence" of amino acids; (2) the presence in the cells of specific glycosyltransferases and of glycosylnucleotide precursors. Under these

*References p. 158*

```
β-GlcNAc-(1→ )-Asn                                              1

β-Gal-(1→4)-β-GlcNAc-(1→ )-Asn                                  2

β-Gal-(1→4)-β-GlcNAc-(1→ )-Asn                                  3
        |(1→6)
        α-Fuc

α-NeuAc-(2→3)-β-Gal-(1→4)-β-GlcNAc-(1→ )-Asn                    4

α-NeuAc-(2→6)-β-Gal-(1→4)-β-GlcNAc-(1→ )-Asn                    5

   β-Man-(1→4)-β-GlcNAc-(1→4)-β-GlcNAc-(1→ )-Asn                6
  /(1→6)
α-Man
```

Fig. 67. Structure of glycoasparagines isolated from urine of asparaginylglucosaminuria. Glycoasparagines 1 and 2 [737]; glycoasparagines 3 to 5 [78]; glycoasparagine 6 [738].

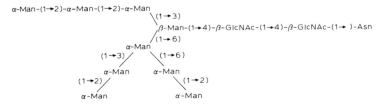

Fig. 68. Structure of a glycoasparagine isolated from urine of Gaucher disease [739].

conditions, the composition and structure of the glycans would, according to Gottschalk, depend on the relative concentrations of the sugar nucleotides. If such hypothesis were correct, the structure of the glycans would depend on chance, and would never be definite.

Then, little by little, on the basis of results I described in the Historical Section, and by the use of tunicamycin in exploring the effects of the non-glycosylation of proteins (recent reviews in [559, 749]), some progress was made. It has recently become feasible to study the role of glycan by employing tunicamycin, an antibiotic originally described by Tamura et al. [750] which blocks the first step in dolichol glycosylation involved in the assembly of the core N-glycosidically linked to protein [751]. Based on the

# GENETIC DISORDERS OF CATABOLISM 141

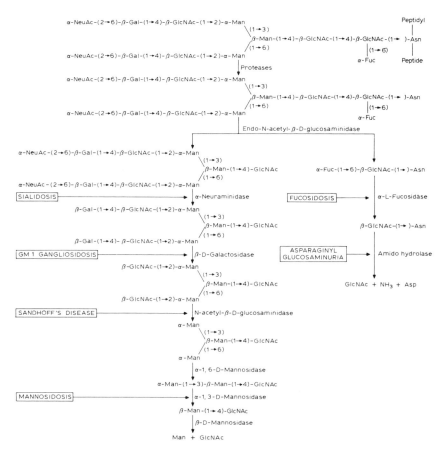

Fig. 69. Proposed pathway for catabolism of glycans of the N-acetyl-lactosaminic type [78, 304, 741].

known structure of tunicamycin, it has been proposed that the antibiotic acts either as an irreversible inhibitor of GlcNAc-1-phosphate transferase [752] hindering the formation of dolichol-diphosphate N-acetylglucosamine (Fig. 52) or as a multisubstrate analogue in which the branched hydrocarbon side chain mimics dolichol phosphate and the uracil-carbohydrate-N-acetylglucosamine backbone mimics UDP-N-acetylglucosamine [753] (Fig. 70). Hence, proteins which are normally N-glycosylated are synthesised devoid of glycan in the presence of tunicamycin. The

*References p. 158*

latter compound may therefore be used as a probe into the role of glycans. We may now summarise the hypotheses, which are based on experimental results that have been obtained on this subject up to the present time as follows.

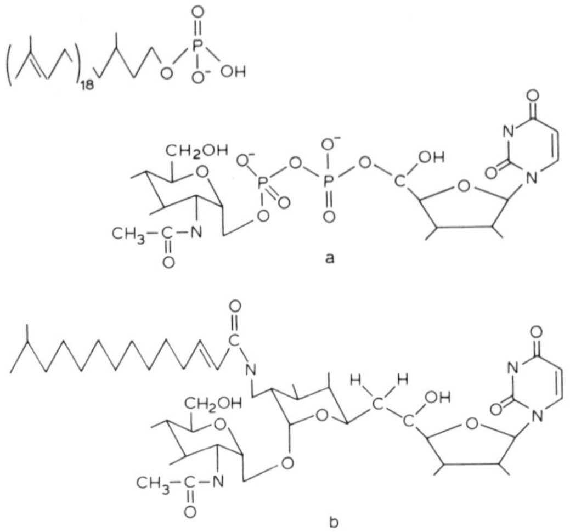

Fig. 70. Comparison of structures of (a) UDP-GlcNAc and dolichyl phosphate (upper left) with (b) tunicamycin according to Keller et al. [753].

## 8.1. Induction of protein conformation

This hypothesis, according to which the glycan moieties intervene in the folding and in the maintenance of protein conformation, is based on the glycan-glycan and glycan-protein interactions which bring into play ionic forces (either repulsive or attractive), hydrogen-bonds hydrophobic interactions, or both. A series of examples may demonstrate this view. The first one concerns one of the most important physico-chemical properties of mucins which is to give solution of high viscosity such as the secretions of mucous cells. This property is essentially linked to a rod-like conformation of structures having an elevated axial ratio. In the case of mucins, this conformation is maintained as a result of repulsive electronegative charges carried by the sialic acid residues present in the

800 disaccharide moieties which are almost regularly distributed along the long peptide chain [210]. Removal of sialic acid with a neuraminidase makes the glycoprotein globular, the solutions of which become weakly viscous.

The second interesting example is that of the participation of the two carbohydrate units of human $IgG_1$ immunoglobulins (Structure 48 in Table XIII) in the three-dimensional structure of the intact protein as determined by X-ray diffraction [754]. Glycans play a central role as the principal contact between the $C_H2$ domains and form, in addition, a large part of the interface between the Fc and Fab regions. By their action, the $C_H1$ domain is prevented from making close contacts with the $C_H2$ domain. The closest contact distances are, therefore, not between the $C_H1$ and the $C_H2$ domains directly, but through the carbohydrate moiety. The $C_L$ domain of the Fab fragment does not interact at all with the $C_H2$ domain or with the glycan. In addition, Huber and coworkers [755] have shown by X-ray diffraction analysis of crystallised Fc fragment that each carbohydrate moiety shields distinct, apolar regions of the polypeptide sheet and is also in intimate contact with the amino acids in the region functioning as the hinge in the antibody. According to these authors, the carbohydrate chains play a decisive role in the conversion of the antibody from the flexible Y to the rigid T shape induced by antigen binding. The function of the glycan chains seems to be "in screening apolar protein segments and keying in the final association between ordered protein domains". This view is supported by the experiments of Korde et al. [756] who, by employing glycosidases, confirmed the requirement of the glycans in the recognition events mediated by the Fc region without significantly affecting the binding ability of the antibody.

Moreover, the fact that certain glycoenzymes of the $N$-glycosylprotein type such as invertase, acid and alkaline phosphatases and carboxypeptidase Y from yeast are not produced in their active forms by tunicamycin-treated cells [757, 758] may signify that, as unglycosylated proteins, they cannot be correctly folded or processed. Additional support for this view comes from the observation on pepsinogen [759], a glycoprotein, the glycan moieties of which are released upon conversion of the molecule into active pepsin. As the active enzyme is denatured more easily than the zymogen, it can be concluded that the carbohydrate units stabilise the pepsinogen in a conformation in which the active site is protected. However, in the case of many glycoenzymes, it has been shown that the carbohydrate moiety is not involved in the enzyme activity and the reason for its presence is a real enigma. In fact, bovine pancreatic ribonuclease A, which is not glycosylated, has the same activity as ribonuclease B which contains a carbohydrate unit. The same observation has been made with the non-glycosylated and glycosylated forms of many other glycoenzymes: porcine ribonuclease, bovine β-glucuronidase, human parotid α-amylase and *A. niger* glucoamylase [56, 58].

In the same manner, the mechanism of milk casein clotting by rennin brings additional evidence on the role that certain glycans play in determining the final conformation and the physicochemical properties of glycoproteins [760]. Thus, native casein micelles might be constituted of insoluble α, β and γ caseins associated with the asialoglycoprotein, κ casein, the structures of some glycans of which are given in Table XI (Structures 29 and 45). The κ casein sialoglycans prevent the self-aggregation process and contribute to the maintenance of the micelles stable suspension in an aqueous medium. The rennin specifically splits one peptide linkage Phe-Met of the κ casein leading to the liberation of a soluble κ caseinoglycopeptide and of an insoluble κ paracasein. Devoid of its highly hydrophilic sialoglycans, the complex structure of the α, β and γ caseins, κ paracasein becomes insoluble and collapses.

Experiments carried out by Kornfeld and coworkers [761], using tunicamycin, on the effects of the non-glycosylation of G protein of the vesicular stomatitis virus envelope indicate that the carbohydrate moiety stabilises the polypeptide conformation. Thus, the non-glycosylated G protein is temperature sensitive and undergoes intracellular aggregation at elevated temperature, blocking the virus assembly. This proves that glycosylation influences the folding and ultimate conformation of the glycoprotein.

A last example is given by the carbohydrate units of proteoglycans which could play a role in the arrangement and stabilisation of collagen, which is due to the interaction between cationic groups of collagen fibers with anionic groups of the acid mucopolysaccharide moieties [455, 762]. Similarly the hydroxylysine-linked sugar moieties of collagens and basement membranes could act by regulating the packing arrangement of the peptide chains and, consequently affect the permeability of renal glomerular basement membrane [58].

## 8.2. Protection of proteins against proteolytic attack

The hypothesis of the protective effect of glycans against proteolytic attack is based on the observation that numerous glycoproteins lose their resistance to proteases on treatment with neuraminidases, this applies to mucins [763], enzymes [56] or intrinsic factor, a sialoglycoprotein secreted by the mucous membrane of the stomach. After removal of sialic acid, this retains its ability to bind vitamin $B_{12}$ but the absorption of the vitamin by the enterocyte is abolished. This loss of activity is due to destruction of the desialylated intrinsic factor by digestive proteolytic enzymes [764].

Using tunicamycin, Loh and Gainer [765] have confirmed this view. The experiments of these investigators are concerned with the common precursor of adrenocorticotropin, β-lipotropin, α-MSH and β-endorphin which is a glycoprotein synthesized by the neurointermediate lobe of the frog *Xenopus laevis*. These experiments support Beeley's hypothesis [634] according to which the addition of sugars at the β-turns of proteins would result in the masking of the turn conformation, and hence, protects the molecule from proteolysis. The authors have shown that the lack of glycosylation of the precursor by tunicamycin results in its degradation, and formation of atypical and inactive peptides.

## 8.3. Control of membrane permeability

The idea according to which the glycan moiety of glycoconjugates controls the permeability of cell membrane arose from the observation [766] that chemical or enzymic modification of the glycan β-Gal-(1 → 3)-α-GalNAc-(1 → 3)-Thr of the "antifreeze glycoprotein" of an antarctic fish abolishes the function of this protein, which is to lower the freezing point of the blood of this fish by 1.85°C. The glycan moiety most probably acts on the orientation of water molecules. On this basis, the hypothesis was put forward that the orientation and concentration of water molecules at the level of cell membranes and, in consequence, the movement of

*References p. 158*

mineral ions and organic substances of low molecular weight, are related to the glycans of membrane glycoconjugates. More particularly, they are linked to the relative number of the highly hydrophilic, sialic acid residues and of the relatively hydrophobic, fucose residues. Experimental evidence has recently been furnished by Olden et al. [767] to show that inhibition of protein glycosylation by tunicamycin results in defective membrane transport and glucose metabolism in chick embryo fibroblasts. These authors demonstrated that the carbohydrate moieties of glycoproteins are required for the normal transport of metabolites across the plasma membrane and their subsequent metabolism. This effect might occur directly on carrier function or indirectly by inhibiting insertion of transport molecules into the membrane or by increasing their degradation rate.

Thus, in controlling cell-membrane permeability, the glycans of surface glycoconjugates intervene powerfully in the regulation of the metabolism of the cell and of cell division. Any modification of their composition, of their structure and of their distribution at the cell surface might lead to metabolic disturbances, such as are observed in tranformed cells and cancer cells.

## 8.4. Control of protein secretion from the cell

Peterson and Leblond [768], in 1964, and Eylar [618] in 1965 proposed the "exit passport" hypothesis to explain why most extracellular proteins are glycosylated, whereas intracellular proteins are rarely so substituted. According to Eylar,

the lack of a specific function of the carbohydrate unit in biologically active glycoproteins suggests a more general role, and it is proposed that the carbohydrate acts as a chemical label which, upon interaction with a membrane receptor or carrier, promotes the transport of the newly synthesised glycoprotein into the extracellular environment.

He concluded

that the carbohydrate unit need to play no functional role in biological active proteins, that the carbohydrate unit in glycoproteins from the same cell would be identical or very similar, and that glycoprotein biosynthesis would involve similar primary or tertiary structures at those regions in the protein molecule where the carbohydrate unit is attached.

GLYCANS AS RECOGNITION SIGNALS    147

This hypothesis was much criticised and the recent finding obtained by using tunicamycin have not clarified the situation at all. In fact, Struck et al. [796] have demonstrated that, despite the inhibitory effect of the drug on glycosylation, secretion of transferrin, serum albumin and apoproteins A and F of VLDL by primary cultures of rat and chick hepatocytes was virtually unimpaired. Keller and Sivank [770] have made the same observation showing that ovalbumin is secreted by the oviduct even when glycosylation is blocked by tunicamycin. In the same way, Mizrahi et al. [771] studying the effects of the antibiotic on several properties of human immune interferon, induced in whole leukocyte cultures by a mitogen, phytohaemagglutinin-P, have observed that glycosylation was not necessary for interferon to be secreted by the cell or to express its antiviral function. Thus, it seems that, contrary to Eylar's hypothesis, glycosylation is not mandatory for the secretion of glycoproteins, even if recent results obtained by Tanzer et al. [772] lead to the conclusion that tunicamycin impairs glycosylation of procollagen, and that the decreased content of carbohydrate occurs concomitantly with a marked retardation of its secretion from human fibroblasts.

*8.5. Glycans as recognition signals — Membrane lectins*

In the 1970s, following the remarkable work of Ashwell's group the concept of "the glycans as recognition signals" developed considerably. These authors demonstrated that proteins could be recognised by cell membranes and become associated with these membranes owing to carbohydrate groups that the proteins carry and which play the role of "antennae" towards membrane receptors. So, the glycans fix the destiny of extracellular proteins and, in consequence, according to Winterburn and Phelps [742], the glycoproteins are synthesized *by* cells *for* cells. Such is the "target-cell hypothesis" in favour of which a series of experimental facts have been cited in previous sections (Sections 1.3 and 7.2).

Following the way that Ashwell and coworkers started with their work on mammalian liver, numerous investigators have systematically explored a wide variety of membranes and succeeded in characterising such membrane receptors in cells from bacteria, fishes, birds, mammals and plants (recent reviews [773]). These

*References p. 158*

receptors which are generally glycoproteins and which act as lectins are called "membrane lectins" (reviews in [216, 773, 774]). To this concept of glycans being involved in molecule—membrane interaction was rapidly added the concept of glycans involved in membrane—membrane interaction. Both contributed to a better approach to the function of membrane glycoconjugates.

It is now well established that the bulk of the conjugated carbohydrate of cell plasma membranes is extracellular in position and orientation and faces the external environment. But the role or roles that the glycoprotein (and glycolipid) glycans play is not yet well defined and is still a matter of speculation (reviews in [52, 176, 519, 775—786]). Perhaps the carbohydrate moiety is the essential determinant in the orientation and conformation of proteins within the membrane. But, if this were the only function of membrane glycans, the following question immediately arises: why does the cell build so many different complex structures for such a simple function? As a matter of fact, it is now well established that glycans of membrane glycoconjugates play a more specific biological role in modulating activity and function of cell membrane and in intervening in the social life of the cell.

In this connection, it is worth mentioning that the integrity of membrane proteins and lipids appears to be crucial for cell survival while, on the contrary, cells can survive profound structural change in the carbohydrates bound to their periphery. For instance, when a cell becomes malignant, marked and extensive changes occur in the membrane glycans and after transformation many glycoproteins, and also glycolipids, contain altered glycan units (reviews in [74, 775, 781, 785, 787, 788]). Despite these modifications, the cancer cell is alive and remains capable of dividing. However, the mechanisms for interacting with the environment and which depend on oligosaccharides located at the cell surface are altered e.g. lack of recognition, of binding and of response to soluble effectors and to membranes of other cells; modification of the reactivity towards lectins. These transformations result in continuous cell division, decrease of intercellular adhesiveness leading to metastatic diffusion, alteration of transfer mechanisms, loss of contact inhibition and of specialised functions, and alteration of immunogenicity. In the same way, the surface carbohydrate alterations of mutant mammalian cells resistant to plant lectins does not endanger the existence of this kind of transformed cells review in [789]). More-

GLYCANS AS RECOGNITION SIGNALS 149

over, cells in culture do not seem to suffer from treatment with glycosidases [235] and even if the rates of transport of various compounds are altered, the consequences are not lethal to the cells. Still more, in many cases, neuraminidase and protease treatment of resting cells stimulates mitosis (review in [790]). Another example is given by the AD6 cell, a mutant from Balb/c 3T3 cells which lacks the acetylation enzyme of glucosamine-6-phosphate and results in modifications of the biosynthesis of glycoproteins [791]. These changes in glycan structures are not lethal, but lead to alterations in adhesiveness to substratum, increase in cell surface microvilli, cell rounding, increase in agglutinability by concanavalin A and loss of directional mobility.

Thus, it appears that some of these phenomena are related to one of the roles that the glycans of membrane glycoproteins could play: that of cell receptor sites complementary to recognition sites carried by other environmental molecules in solution or by other cells. Related to this latter interaction, a number of observations have implicated surface membranes and, more specifically, the cell surface glycoconjugates in cell-cell recognition and adhesion phenomena (reviews in [76, 165, 176, 556, 773, 775, 777, 778, 781, 792—799]). It has been already noted (Section 1.3) that the fate of lymphocytes on incubation is greatly affected by removal of L-fucose from their surface. In the same manner, the survival of erythrocytes depends on the integrity of their cell surface sialyl residues [800, 801]. The desialylation of these cells leads to their rapid disappearance from the blood stream and to their destruction in the liver and the spleen which suggests they have to be first recognised by these specific tissues. This behaviour of desialylated erythrocytes may be similar to that of naturally aged erythrocytes which contain less membrane sialic acid than young erythrocytes. It has also been demonstrated that interaction between erythrocytes and enveloped virus, especially Myxovirus, is mediated by the presence of a viral lectin specific for the sialyl residues of glycophorin [802, 803].

These few examples are concerned with recognition between two different cell types or entities. It is rather an extension of Ashwell's concept in which the interaction between the receptor of one cell and the ligand bound to the other leads to phagocytosis. About the problem of cell-cell recognition and further adhesion to form tissue, there is still no description of the complete mole-

*References p. 158*

cular topography of any adhesion site and of the molecular mechanism, despite the number of investigations undertaken in many systems during the past decade. The methods used for the study of cell-cell adhesion are based on the assumption that complementary molecules cross-link cells and are usually directed towards identifying molecules that promote cell aggregation. The receptors of glycoconjugates could be glycosyltransferases, as suggested by Roseman [165] or lectin-like proteins.

In Roseman's concept (reviews in [792—795, 798]), it is suggested that the complementary molecules are cell-surface glycosyltransferases (ectoglycosyltransferases) (Section 6) and glycans. In the proposed mechanism an ectoglycosyltransferase — e.g. a sialyltransferase — on the surface of one cell may be able to bind to the appropriate carbohydrate acceptor on an opposing cell, a glycan possessing galactose residues in terminal non-reducing position, in the particular case of sialyltransferase.

The direct role of ectoglycosyltransferases in cell-cell adhesion has not yet been proved. But involvement of cell surface lectins is established, or at least strongly presumed in some cases, such as the interaction between slime mould cells, sponge cells and embryonic cells. Such a mechanism could also be involved in pollen-stigma and ovule—spermatozoid interactions, and in metastasis of cancer cells. It should be noted that in membrane lectins—ligand interaction, two cases may occur: either the complementary ligand is membrane bound and there is a direct association of cell to cell as it is for the slime mold aggregation in which "discoidin", a membrane lectin of protein nature of one cell binds to D-galactose and related sugars of cell surface glycoconjugates of the other cell (review in [804]); or a soluble extracellular bivalent ligand cross-links two cells as it would be for the species-specific multivalent glycoproteins that promote sponge-cell aggregation. In addition to these membrane-bound lectins, surface lectins — close to but not necessarily tightly apposed to the plasma membrane — may be responsible for making a bridge between the glycans of one cell to the glycans of another cell. For example, a retina specific aggregation promoting factor has been isolated as a pure protein from cultured chick neural retina cells [805]. Recently, a class of adhesive, high-molecular-weight glycoproteins, present on the surface of normal cultured fibroblast has been characterised. They bear various names: LETS (large external transformation-sensitive) protein, galactoprotein a, CSP (cell-surface protein) and fibronectin. They appear to be closely related or identical with the plasma cold insoluble globulin (reviews in [805–814]). These proteins help to maintain normal cell morphology, cell surface architecture, and cell interactions in cultured cells. Fibronectins may have an important role in cell-cell adhesion, and cell-substratum adhesiveness. Absent in transformed cells, fibronectins restore the normal phenotype when added to these transformed cells, i.e. they restore alignment, fibroblastic morphology and cell-surface morphology. The underglycosylated fibronectin produced by tunicamycin-treated cells [815] or by cells deficient in certain glycosyltransferases [816] remains active in mediating cell adhesion, making it unlikely that the glycan moiety of fibronectin (Table XIII, Structures 27 and 34) plays a role in the interactions with the cell surface. On the other hand, tunicamycin-treated cells (BHK fibroblasts) exhibit reduced amounts of surface-associated fibronectin and adhere relatively poorly to plastic or collagen surfaces [817]. This finding demonstrates that $N$-glycosically linked glycans of BHK cell surface glycoproteins play a role in interaction of cells with fibronectin.

In conclusion, it can be stated that the molecular mechanism of cellular associations processes, i.e. cell-substrate and cell-cell adhesion, is still not known and appears increasingly complex, due to the characterisation of more and more associating factors, such as proteoglycans, for example. However, direct or indirect evidence strongly suggests that carbohydrate-containing macromolecules on the surface of cells play a role in cell association, but the nature of adhesive bonds, as well as the conformation of the molecules implicated in adhesion and the structure of the association sites, remain undefined.

## 8.6. Concluding remarks

In conclusion, it appears more and more obvious that glycoproteins play two important roles. The first is of a physicochemical order. It concerns (1) the conformation of the peptide chain of glycoproteins and its protection against proteolytic attack, and (2) the orientation and concentration of water molecules, and also the movement of mineral ions and organic compounds of low molecular weight at the level of cell-surface membranes. This role could be essentially attributed to the O-glycosylproteins. In fact, the structures of the glycans of this type of glycoproteins do not present a wide diversity and the same structures can be found in glycoproteins of very different origins and roles (Table XI). Moreover, this kind of glycan is often located in "strategic" domains of peptidic chains, for example, in the hinge region of immunoglobulins (Fig. 71).

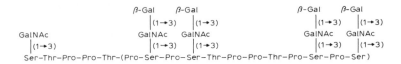

Fig. 71. Hinge region of immunoglobulin IgA$_1$ [380].

The second role is of a biological order. It is essentially based on the concept of a recognition signal carried by the glycans and in consequence, on their specific structure. This role could be essentially ascribed to the glycan structures of the $N$-acetyl-lactosaminic type because of the wide diversity they present and which will certainly be accentuated when the primary structures of cell membrane glycoproteins have yielded their secrets. We may perhaps discover that glycoconjugates were essentially invented to be membrane constituents.

As regards the glycans of the oligomannosidic and mixed types, on the basis (1) of our knowledge of their structures (Figs. 23 and 36) which reveal no specificity at all and (2) on the findings concerning the biosynthesis of the glycans of the $N$-acetyl-lactosaminic type (Section 6 and Fig. 55), they could be considered as only metabolic intermediates in the maturation of these proteins, playing secondary and accessory biological roles. However, this hypothesis must not be accepted as a dogma at all, as shown by recent findings of Baynes and coworkers [818] on the clearance of antibody-antigen complexes from the circulation and the role played by the oligomannosidic type glycans of IgM in the uptake of IgM-bovine serum albumin complexes. These authors demonstrated that the antigen induces conformational changes of the IgM molecules resulting in exposure of the glycans of the oligomannosidic type sterically inaccessible in the free immunoglobulin, which signal the clearance of the soluble immune complexes from the circulation via the mannose recognition system in hepatic non-parenchymal cells and other organs of the reticuloendothelial system. The clearance is inhibited by pre- or co-injection of mannans and ovalbumin, but not by asialofetuin. Moreover, IgM binds with concanavalin A only when complexed with antigen. Digestion of bovine serum albumin-IgM complexes with α-mannosidases abolishes both binding by concanavalin A and rapid clearance from rat circulation*.

However, whatever discoveries are made in the next few years on the role of glycans, there is one certainty: that of the enormous importance of two monosaccharides whose systematically external position designates them as recognition signals, i.e. sialic acids (reviews in [57, 779]) and fucose (review in [820]). They have often been evoked all through this review in numerous physicochemical, biological or pathological actions, generally in demonstrating that their elimination abolishes the activity of the glycoproteins carrying them. Nonetheless, even compared to fucose and all other monosaccharides, sialic acid possesses the most unusual and changeable personality: a "chameleon-sugar". In contrast to the other monosaccharides which have in practice a fixed and unitary structure (except for some modifications by sulfatation) sialic acid can present itself under diverse "disguises": N-acetyl or N-glycolylneuraminic acids; mono, di, tri or tetra-O-acetylated in position C-4, C-7, C-8 and C-9; O-glycolylated in position C-4; methylated in position C-8; lactylated in position C-9. This protean characteristic is in complete accord with the quality of recognition-sugar attributed to sialic acid.

---

*In the same way, the authors [819] have presented evidence that bovine serum albumin induces conformational changes in anti-bovine serum albumin IgG resulting in both recognition of galactose residues on IgG and rapid clearance of bovine serum albumin-IgG complexes from the circulation by Ashwell's receptor in hepatic parenchymal cells specific for galactose terminal glycoproteins.

*References p. 158*

## 9. Conclusions

The development of chemical, physical, and enzymic methods for investigating the primary structure of the carbohydrate fractions of glycoconjugates has, during the past 10 years, allowed the determination of the structures of numerous glycans. In this regard, the combination of the techniques of permethylation and 360- and 500-MHz, proton-NMR spectroscopy has afforded real progress in the rapid determination of the primary structure of oligosaccharide chains.

Our knowledge of the constitution of the latter is now sufficient to allow us to propose some "laws" relating to their structures, and to raise interesting questions of comparative biochemistry and phylogeny. In fact, there exist certain general rules regarding the structures of these glycans. All are derived by the substitution of common core oligosaccharides by variable oligosaccharide structures called "antennae" which are the basis of their biological specificity. This is due to the following variables: (a) the number and length of antennae substituting the core. In the case of glycans of the $N$-acetyl-lactosaminic type, for example, structures possessing 2,3 and 4 antennae have been found as well as repeating units of $N$-acetyl-lactosamine (Fig. 32); (b) the nature, number, distribution, and types of glycosidic linkage of monosaccharides in external position: sialic acid residues $\alpha$-(2 → 3) and/or $\alpha$-(2 → 6)-linked, fucose residues $\alpha$-(1 → 3) and/or $\alpha$-(1 → 6)-linked, or galactose residues $\beta$-(1 → 3)-linked; (c) the number of glycan residues per molecule of protein, and their structural type. The same peptidic chain may be substituted (a) by glycans of a single type: for example by the $N$-acetyl-lactosaminic or the oligomannosidic type, (b) by glycans of different type: for example by both the previous types, such as in thyroglobulin (Section 5.3), (c) by $O$- or $N$-glycosidically linked glycans, or by both as in glycophorin (Section 5.5). A large variety of specific carbohydrate loci therefore exist which amply covers the widely differing requirements of biological specificities.

The substitution rules that have been established on the basis of a certain "orthodoxy" of glycan primary structure facilitate the determination of this latter. In this connection, it even becomes possible to predict the glycan structure of human $N$-glycosylproteins which have not yet been discovered! As has been seen in

# CONCLUSIONS

Section 8, the knowledge of the structure of oligosaccharides from urine of glycoproteinoses allowed us to present the hypothesis according to which the catabolism of N-glycosylproteins would commence by the action of an endo-N-acetyl-β-glucosaminidase. This can split the residue of di-N-acetylchitobiose of the pentasaccharide core, and liberate oligosaccharides possessing an N-acetylglucosamine residue in the terminal, reducing position. We can therefore postulate [78, 304, 741] that the structures of the oligosaccharides of mannosidosis (Fig. 61), sialidosis (Fig. 62), and fucosidosis (Fig. 63) as well as those of certain oligosaccharides of $GM_1$-gangliosidosis (Fig. 65) pre-exist in the glycan structures of human N-glycosylproteins, even though they have not yet been found. These structures may be reconstituted by adding the sequence β-(1 → 4)-GlcNAc-β-(1 → 4)-Asn, α-fucosylated on C-6 of the 2-acetamido-2-deoxyglucose residue (Fig. 72) or not, to the

Fig. 72. General scheme of foreseeable glycan structures of human N-glycosylproteins [78]. See structure of oligosaccharides in Figs. 61 to 65.

reducing end of all of the oligosaccharides accumulated as a result of deficiencies in mannosidases, neuraminidases, fucosidases and galactosidases (enzymes that liberate the monosaccharides from terminal positions in native glycan structures). Certain of these reconstituted structures correspond to glycans already known. The others have not yet been discovered, probably because they exist in low concentrations in the human body. This is why we have hypothesised that these new structures are those of endocellular or membranous glycoproteins, or both. In glycoproteinoses, the oligosaccharides detached by the endo-N-acetyl-β-glucosaminidases are protected from any further action of exoglycosidases. Consequently, they accumulate, first in the tissues, and then in the urine, where they can be readily characterized. The glycoasparagines isolated from fucosidosis urines (Fig. 64), in asparaginyl-glycosaminuria (Fig. 67) and in Gaucher's disease (Fig. 68) reinforce the hypothesis of a glycoproteinic origin of the saccharidic structures accumulated in the urine, because, in the three pathological cases, the glycan moieties stay intact, with their residue of di-$N$-acetylchitobiose still attached to asparagine. It thus seems feasible to predict the existence in man of several dozens of glycan structures in glycoproteins that remain to be discovered, probably in cell membranes.

On the other hand, it is evident that the knowledge of the primary structure of $N$-glycosidically linked glycans has been decisive in the elucidation of the biosynthesis of $N$-glycosylproteins.

However, besides these fascinating discoveries in the field of glycan structure, the problem of the true biological role of glycoproteins is not yet solved. The reason is that, even now, we know too few glycan structures and the structures we know are for the most part derived from too limited a range of biological sources. The structures we know are essentially of glycoproteins from connective tissue, blood plasma, milk and egg white, which are "excreted" and not "cellular" compounds. One can believe that the best is yet to come and that the solution to the problem of the biological role of glycoproteins will be solved by studying membranes in which the most complicated and specific structures will be discovered. Glycosyltransferases have perhaps been invented by cells essentially to synthesise their plasma membrane glycoconjugates. However, we cannot exclude the hypothesis that the carbohydrate moiety of glycoproteins does not have any specialized role and that its function could be to favour a particular conformation of the protein and in this way inducing its particular activity.

On the whole, the role of glycans in glycoproteins is still obscure and still the subject of speculation. As pointed out by Neuberger [62] in 1974, it is still true that

we are now faced with the major problem of the biological function of the glycoproteins: in other words, we are asking the question how the activity of a protein, whatever it may be, is modified by the presence of sugar residues.

Nevertheless, the actual knowledge we possess of glycan primary structure allows us, on solid physicochemical and molecular grounds, to establish the study of the normal and pathological metabolism of glycans, to lay the foundations of the genetics and molecular biology of glycoproteins and to define better questions we have to ask of nature.

## 10. Acknowledgements

We would like to express our sincere gratitude to Prof. Albert Neuberger, Geneviève Spik, Brian M. Carden, Franck Hemming, William Sly and André Verbert for their critical comments of the manuscript. We wish to thank Miss Brigitte Mahieu and Mrs. Jocelyne Celen for excellent technical assistance in the elaboration of this manuscript.

## Notes added in proof

(1) Recently, Lis and Sharon (J. Biol. Chem., 256 (1981) 7708—7711) have demonstrated that the primary structure of the soybean lectin was identical to structure No. 1 of Fig. 23.

(2) The term "multienzymatic" is not exactly appropriate. In fact, it must not be taken in a strict sense defined as an assembly of enzymes in a particle which can be isolated as a whole, but must be considered as a topological arrangement of enzymes acting one after the other.

## REFERENCES

1 P. A. Levene (Ed.), Hexosamines and Mucoproteins, Longmans, Green, New York, 1925.
2 K. Meyer, Cold Spring Harbor Symp. Quant. Biol., 6 (1938) 91.
3 K. Meyer, Adv. Prot. Chem., 2 (1945) 249—275.
4 M. Stacey, Adv. Carbohyd. Chem., 2 (1946) 161—201.
5 G. Blix, in B. Flaschenträger and E. Lehnartz (Eds.), Physiologische Chemie, Vol. 1, Springer, Berlin, 1951, p. 751.
6 W. Pigman and R. M. Goepp (Eds)., Chemistry of the Carbohydrates, Academic Press, New York, 1952.
7 W. H. Cole (Ed.), Some Conjugated Proteins, Symposium, Rutgers, New Brunswick, 1953.
8 A. G. Ogston and J. E. Stanier, Disc. Faraday Soc., 13 (1953) 275.
9 R. L. Whistler and C. L. Smart (Eds), Polysaccharide Chemistry, Academic Press, New York, 1953.
10 P. W. Kent and M. W. Whitehouse (Eds.), Biochemistry of the Aminosugars, Butterworth, London, 1955.
11 G. F. Springer, Klin. Wochenschr., 33 (1955) 347.
12 J. Montreuil, Bull. Soc. Chim. Biol., 39, Suppl. III (1957) 3—92.
13 F. R. Bettelheim-Jevons, Adv. Prot. Chem., 13 (1958) 35—105.
14 G. E. W. Wolstenholme and M. O'Connor (Eds.), Ciba Found. Symp. Chemistry and Biology of Mucopolysaccharides, Churchill, London, 1958.
15 J. Montreuil, Les glycoprotéides, in M. Javillier, M. Polonovski, M. Florkin, P. Boulanger, M. Lemoigne, J. Roche and R. Wurmser (Eds.), Traité de Biochimie Générale, Vol. 1, Part 2, Masson, Paris, 1959, pp. 935—1002.
16 A. Gottschalk (Ed.), The Chemistry and Biology of Sialic Acids and Related Substances, Cambridge University Press, Cambridge, 1960.
17 R. W. Jeanloz, Arthr. Rheum., 3 (1960) 233.
18 V. Ginsburg and E. F. Neufeld, Annu. Rev. Biochem., 38 (1961) 371.
19 M. Stacey and S. A. Barker (Eds.), Carbohydrates of Living Tissues, Van Nostrand, London, 1962.
20 A. Delaunay, L. Robert and J. Polonovski, in P. Boulanger, M-F. Jayle and J. Roche (Eds.), Exp. Ann. Bioch. Méd., Vol. 24, Masson, Paris, 1963.
21 A. Gottschalk, in M. Florkin and E. H. Stotz (Eds.), Comprehensive Biochemistry, Vol. 8, Part 2, Elsevier, Amsterdam, 1963, pp. 17—37.
22 R. G. Spiro, New Engl. J. Med., 269 (1963) 566—573; 616—621.
23 P. T. Grant and J. L. Simkin, Annu. Reports Chem. Soc. (London), 61 (1964) 491—506.
24 R. W. Jeanloz, Medicine, 43 (1964) 363—369.
25 F. R. Jevons, in H. W. Schulz and A. F. Anglemier (Eds.), Symposium on Foods: Proteins and their Reactions, Avi Publishing, Westport, 1964, pp. 153—165.

# REFERENCES

26  E. A. Balasz and R. W. Jeanloz (Eds.), The Amino Sugars, Vol. 2A, Distribution and Biological Role, Academic Press, New York, 1965.
27  R. W. Jeanloz and E. A. Balasz (Eds.), The Amino Sugars, Vol. 1B, Glycosaminoglycans, Glycoproteins and Glycosaminolipids, Academic Press, New York, 1965.
28  R. J. Winzler, Clin. Chem., 11 (1965) 339—347.
29  E. A. Balasz and R. W. Jeanloz (Eds.), The Amino Sugars, Vol. 2B, Metabolism and Interactions, Academic Press, New York, 1966.
30  A. Gottschalk (Ed.), Glycoproteins. Their Composition, Structure and Function, Elsevier, Amsterdam, 1966.
31  A. Gottschalk and E. R. B. Graham, in H. Neurath (Ed.), The Proteins, Vol. 4, Academic Press, New York, 1966, pp. 95—151.
32  N. Sharon, Annu. Rev. Biochem., 35 (1966) 485—520.
33  P. W. Kent, Essays Biochem., 3 (1967) 105—151.
34  R. D. Marshall and A. Neuberger, in F. Dickens, P. J. Randle and W. J. Wheelan (Eds), Carbohydrate Metabolism and Its Disorders. Academic Press, New York, 1968, pp. 213—258.
35  E. Rossi and E. Stoll (Eds.), Biochemistry of Glycoproteins and Related Substances, Part II, Karger, Basel, 1968.
36  R. W. Jeanloz (Ed.), The Amino Sugars, Vol. 1A, Chemistry of Amino Sugars, Academic Press, New York, 1969.
37  A. Neuberger and R. D. Marshall, in H. W. Schultz, R. F. Cain and R. W. Wrolstad (Eds.), Carbohydrates and their Role, Avi Publishing, Westport, 1969, pp. 115—132.
38  R. G. Spiro, New Engl. J. Med., 281 (1969) 991; 1043.
39  E. A. Balasz (Ed.), Chemistry and Molecular Biology of the Intracellular Matrix, Academic Press, New York, 1970.
40  R. D. Marshall and A. Neuberger, Adv. Carbohydr. Chem., 25 (1970) 407—478.
41  W. Pigman and D. Horton (Eds.), The Carbohydrates. Chemistry, Biochemistry, Vol. 2B, Academic Press, New York, 1970.
42  R. Montgomery, in [41], pp. 628—709.
43  R. G. Spiro, Annu. Rev. Biochem., 39 (1970) 599—638.
44  E. C. Heath, Annu. Rev. Biochem., 40 (1971) 29—56.
45  G. A. Jamieson and T. J. Greenwalt (Eds.), Glycoproteins of Blood Cells and Plasma, Lippincott, Philadelphia, 1971.
46  A. Gottschalk (Ed.), Glycoproteins, Their Composition, Structure and Function, Elsevier, Amsterdam, 1972.
47  R. D. Marshall, Annu. Rev. Biochem., 41 (1972) 673—702.
48  M. Monsigny, D. Delay and F. Delmotte, Symbioses, 4 (1972) 39—72.
49  R. Piras and H. G. Pontis, Biochemistry of the Glycosidic Linkage, PAABS Symposium, Vol. 2, Academic Press, New York, 1972.
50  K. Schmid, Chimia, 26 (1972) 405—414.
51  V. Zambotti, G. Tettamanti and M. Arrigoni (Eds.), Glycolipids, Glycoproteins and Mucopolysaccharides of the Nervous System, Plenum, New York, 1972.
52  R. C. Hughes, Prog. Biophys. Mol. Biol., 26 (1973) 189—268.
53  J. F. Kennedy, Chem. Soc. Rev., 2 (1973) 355—395.

# REFERENCES

54 D. T. A. Lamport, in F. Loewus (Ed.), Biogenesis of Plant Cell Wall Polysaccharides, Academic Press, New York, 1973, pp. 149—164.
55 P. Louisot and J. Polonovski, in P. Boulanger, M-F. Jayle and J. Roche (Eds.), Exp. Ann. Biochim. Méd., Vol. 32, Masson, Paris, 1973.
56 J. H. Pazur and N. N. Aronson, Adv. Carbohyd. Chem., 27 (1973) 301—341.
57 R. Schauer, Angew. Chem. Int. Ed., 12 (1973) 127—138.
58 R. G. Spiro, Adv. Prot. Chem., 27 (1973) 349—467.
59 R. J. Winzler, in C. H. Li (Ed.), Hormonal Proteins and Peptides, Academic Press, New York, 1973, pp. 1—16.
60 R. Fricke and F. Hartmann (Eds.), Connective Tissues, Springer, Berlin, 1974.
61 J. Montreuil (Ed.), Méthodologie concernant la structure et le métabolisme des glycoconjugués, Actes du Colloque International du C.N.R.S. No. 221, Villeneuve d'Ascq, Juin 1973, C.N.R.S., Paris, 1974.
62 A. Neuberger, Biochem. Soc. Symp., 40 (1974) 1.
63 N. Sharon, in J. B. Pridham (Ed.), Plant Carbohydrate Biochemistry, Academic Press, New York, 1974, pp. 235—252.
64 J. Montreuil, G. Spik and A. Chosson, C.R. Acad. Sci. Paris, 255 (1962) 3493—3494; J. Montreuil, A. Adam-Chosson and G. Spik, Bull. Soc. Chim. Biol., 47 (1965) 1867—1880; J. Montreuil, Proc. VIIth Int. Symp. Carbohydr. Chem., Bratislava, August 1974, Pure and Appl. Chem., 42 (1974) 431—477.
65 N. Sharon (Ed.), Complex Carbohydrates, Their Chemistry, Biosynthesis and Functions, Addison-Wesley, Reading, 1975.
66 R. Kornfeld and S. Kornfeld, Annu. Rev. Biochem., 45 (1976) 217—237.
67 R. G. Brown and W. C. Kimmins, Int. Rev. Biochem., 13 (1977) 183—209.
68 G. G. Forstner (Ed.), Mucus Secretions and Cystic Fibrosis, Modern Problems in Paediatrics, Vol. 19, Karger, Basel, 1977.
69 M. Horowitz and W. Pigman (Eds.), The Glycoconjugates, Vol. 1, Academic Press, New York, 1977.
70 D. T. A. Lamport, Recent Adv. Phytochem., 11 (1977) 79—115; Biochem. Plants, 3 (1980) 501—541.
71 A. Neuberger, in F. Franks (Ed.), Characterisation of Protein Conformation and Function, Symposium Press, London, 1977, pp. 144—157.
72 J. R. Clamp (Ed.), Mucus, Vol. 34, Medical Department, British Council, London, 1978.
73 M. Horowitz and W. Pigman (Eds.), The Glycoconjugates, Vol. 2, Academic Press, New York, 1978.
74 E. F. Walborg (Ed.), Glycoproteins and Glycolipids in Disease Processes, ACS Symposium No. 80, Am. Chem. Soc. Publ., Washington, 1978.
75 J. D. Gregory and R. W. Jeanloz (Eds.), Glycoconjugate Research, Proc. IVth Int. Symp. Glycoconjugates, Woods Hole, Sept. 1977, Academic Press, New York, 1979.
76 R. Schauer, P. Boer, E. Buddecke, M. F. Kramer, J. F. G. Vliegenthart and H. Wiegandt (Eds.), Glycoconjugates, Proc. Vth Int. Symp. Glyco-

conjugates, Kiel, Sept. 1979, Thieme, Stuttgart, 1979.
77  N. Sharon and H. Lis, Biochem. Soc. Trans., 7 (1979) 783—799; in H. Neurath and R. L. Hill (Eds.), The Proteins, Vol. 5, Academic Press, New York, in press.
78  G. Strecker and J. Montreuil, Biochimie, 61 (1979) 1199—1246; J. Montreuil, Adv. Carbohyd. Chem. Biochem., 37 (1980) 157—223.
79  W. J. Lennartz (Ed.), The Biochemistry of Glycoproteins and Proteoglycans, Plenum, New York, 1980.
80  A. Neuberger, in [45], pp. 1—15.
81  A. Gottschalk, in [46], pp. 1—23.
82  J. Seegen, Zbl. Med. Wiss., 24 (1886) No. 44, 45.
83  L. v. Udránszky, Z. Physiol. Chem., 12 (1888) 377.
84  O. Schmiedeberg, Arch. Exp. Pathol. Pharmakol., 28 (1891) 355.
85  O. Schmiedeberg and H. H. Meyer, Z. Physiol. Chem., 3 (1878) 437.
86  G. Ledderhose, Z. Physiol. Chem., 2 (1878) 213.
87  J. Seemann, Inaugural Lecture Marburg, 1898; cf. J. Chem. Soc. (Abstracts), 76 (1899) 465.
88  F. Müller, Sitzungsber. Ges. Beförd Ges. Naturwiss., Marburg, (1896), No. 6, p. 53; (1898), No. 6, p. 117; Z. Biol., 42 (1901) 468.
89  W. N. Haworth, W. H. G. Lake and S. Peat, J. Chem. Soc. (1939) 271.
90  E. G. Cox and G. A. Jeffery, Nature, 143 (1939) 984.
91  P. A. Levene and F.B. LaForge, J. Biol. Chem., 18 (1914) 123.
92  S. P. James, F. Smith, M. Stacey and L. F. Wiggins, Nature, 156 (1945) 308.
93  S. Fraenkel and C. Jellinek, Biochem. Z., 185 (1927) 392.
94  H. Bierry, C. R. Soc. Biol., 101 (1929) 524; Compt. Rend., 191 (1930) 1381.
95  G. Blix, L. Svennerholm and I. Werner, Acta Chem. Scand., 6 (1952) 358.
96  G. Blix, Z. Physiol. Chem., 240 (1936) 43.
97  E. Klenk, Z. Physiol. Chem., 268 (1941) 50.
98  A. Gottschalk, Nature, 176 (1955) 881.
99  E. Klenk, H. Faillard, F. Weygand and H. H. Schöne, Z. Physiol. Chem., 304 (1956) 35.
100  E. Klenk, in [14], pp. 296—305.
101  D. G. Comb and S. Roseman, J. Am. Chem. Soc., 80 (1958) 497.
102  S. Obolenski, Arch. Ges. Physiol., 4 (1871) 336.
103  J. Bostock, J. Natl. Phil. Chem. Arts, 2nd Ser., 11 (1805) 244.
104  J. Berzelius, Jahresber. Fortschr. Phys. Wiss., 7 (1828) 231.
105  L. Gmelin, in F. Tiedemann and L. Gmelin (Eds.), Die Verdauung nach Versuchen, Vol. 1, Heidelberg and Leipzig, 1826.
106  J. J. Scherer, Ann. Chem. Pharmakol., 57 (1846) 196; Chemische und Mikroskopische Untersuchungen, Winter, Heidelberg, 1843, p. 93 ff.
107  E. Eichwald, Ann. Chem. Pharmakol., 134 (1865) 177.
108  O. Hammarsten, Z. Physiol. Chem., 12 (1888) 163—195.
109  H. A. Landwehr, Z. Physiol. Chem., 6 (1882) 74.
110  O. Hammarsten, Arch. Ges. Physiol., 36 (1885) 373.
111  F. Hofmeister, Z. Physiol. Chem., 14 (1890) 165.

112  R. Neumeister, Z. Biol., 27 (1890) 369.
113  C. T. Mörner, Z. Physiol. Chem., 18 (1894) 525.
114  A. Eichholz, J. Physiol., 23 (1898) 163.
115  F. Hofmeister, Z. Physiol. Chem., 24 (1898) 159.
116  F. W. Pavy, Proc. Roy. Soc. (Biol.), 54 (1893) 53.
117  F. W. Pavy (Ed.), Physiology of the Carbohydrate, Churchill, London, 1894.
118  F. W. Pavy (Ed.), Physiology of the Carbohydrate. An Epicriticism, Churchill, London, 1895.
119  A. Neuberger, Biochem. J., 32 (1938) 1435—1451.
120  C. U. Zanetti, Ann. Chim. Farm., 12 (1897) 1.
121  C. U. Zanetti, Gazz. Chim. Ital., 33 (1903) 160.
122  H. E. Weimer, J. W. Mehl and R. J. Winzler, J. Biol. Chem., 185 (1950) 561—568.
123  K. Schmid, J. Am. Chem. Soc., 75 (1953) 60—68.
124  C. Boedeker, cited in G. Fischer and C. Boedeker, Ann. Chem. Pharm., 117 (1854) 111.
125  C. F. W. Krukenberg, Z. Biol., 20 (1884) 307.
126  C. T. Mörner, Skand. Arch. Physiol., 1 (1889) 210.
127  O. Schmiedeberg, Arch. Exp. Pathol. Pharmakol., 87 (1920) 1; 31; 44; 47.
128  P. A. Levene and J. López-Suárez, J. Biol. Chem., 25 (1916) 511; 26 (1916) 373.
129  C. Bernard, Leçons sur le Diabète et la Glycogenèse Animale, Baillière, Paris, 1877.
130  J. B. Sumner and S. F. Howell, J. Bacteriology, 32 (1936) 227.
131  K. Meyer and J. W. Palmer, J. Biol. Chem., 107 (1934) 629.
132  M. Stacey and J. M. Woolley, J. Chem. Soc., (1940) 184; (1942) 550.
133  K. Meyer, E. A. Davidson, A. Linker and P. Hoffman, Biochim. Biophys. Acta, 21 (1956) 506.
134  J. E. Jorpes and S. Gardell, J. Biol. Chem., 176 (1948) 267.
135  K. Meyer, A. Linker, E. A. Davidson and B. Weismann, J. Biol. Chem., 205 (1953) 611; K. Meyer, P. Hoffman and A. Linker, Science, 128 (1958) 896.
136  R. Kuhn, Angew. Chem., 64 (1952) 493.
137  J. Montreuil, Bull. Soc. Chim. Biol., 42 (1960) 1399—1440; Ann. Nutrit. Aliment., 25 (1971) 1—37.
138  W. M. Watkins, in [46], pp. 830—891.
139  E. R. B. Graham and A. Gottschalk, Biochim. Biophys. Acta, 38 (1960) 513.
140  K. Schmid, J. P. Binette, S. Kamiyama, V. Pfister and S. Takahashi, Biochemistry, 1 (1962) 959—966.
141  K. Schmid, H. Kaufmann, S. Isemura, F. Bauer, J. Emura, T. Motoyama, M. Ishiguro and S. Nanno, Biochemistry, 12 (1973) 2711—2724.
142  P. V. Wagh, I. Bornstein and R. Winzler, J. Biol. Chem., 244 (1969) 658—665.
143  K. Schmid, R. B. Nimber, A. Kimura, H. Yamaguchi and J. P. Binette, Biochim. Biophys. Acta, 492 (1977) 291—302.

144 B. Fournet, G. Strecker and J. Montreuil; L. Dorland, J. Haverkamp and J. F. G. Vliegenthart; K. Schmid and J. P. Binette, Biochemistry, 17 (1978) 5206—5214; B. Fournet, G. Strecker, G. Spik and J. Montreuil; K. Schmid and J. P. Binette; L. Dorland, J. Haverkamp, B. L. Schut and J. F. G. Vliegenthart, in [75], pp. 149—156.
145 K. Schmid, J. P. Binette, L. Dorland, J. F. G. Vliegenthart, B. Fournet and J. Montreuil, Biochim. Biophys. Acta, 581 (1979) 356—359.
146 R. Caputto, L. F. Leloir, C. E. Cardini and A. C. Paladini, J. Biol. Chem., 179 (1949) 497—498; 184 (1950) 333—350.
147 L. F. Leloir, Arch. Biochem. Biophys., 33 (1951) 186.
148 E. Cabib, L. F. Leloir and C. E. Cardini, J. Biol. Chem., 203 (1953) 1055.
149 G. J. Dutton and I. D. E. Storey, Biochem. J., 53 (1953) XXXVII; 57 (1954) 275.
150 E. Cabib and L. F. Leloir, J. Biol. Chem., 216 (1955) 195.
151 H. G. Pontis, J. Biol. Chem., 216 (1955) 195.
152 D. G. Comb, F. Shimizu and S. Roseman, J. Am. Chem. Soc., 80 (1958) 653; 81 (1959) 5513.
153 V. Ginsburg, J. Biol. Chem., 235 (1960) 2196.
154 A. Dorfman, Pharmacol. Rev., 7 (1955) 1.
155 L. Glaser and D. H. Brown, Proc. Natl. Acad. Sci. USA, 41 (1955) 253.
156 E. J. Sarcione, Arch. Biochem. Biophys., 100 (1963) 516—521; J. Biol. Chem., 239 (1964) 1686.
157 E. J. Sarcione, M. Bohne and M. Leahy, Biochemistry, 3 (1964) 373.
158 J. Molnar, G. B. Robinson and R. J. Winzler, J. Biol. Chem., 240 (1965) 1438; 1882.
159 L. Helgeland, Biochim. Biophys. Acta, 101 (1965) 106.
160 R. G. Spiro and M. J. Spiro, J. Biol. Chem., 241 (1966) 1271.
161 S. Kornfeld, R. Kornfeld and V. Ginsburg, Arch. Biochem. Biophys., 110 (1965) 1.
162 J. Molnar and D. Sy, Biochemistry, 6 (1967) 1941—1947.
163 G. R. Lawford and H. Schachter, J. Biol. Chem., 241 (1966) 5308.
164 S. Roseman, in [35], pp. 244—269.
165 S. Roseman, Chem. Phys. Lipids, 5 (1970) 270—297.
166 L. F. Leloir, Science, 172 (1971) 1299—1303.
167 A. J. Parodi and L. F. Leloir, Biochim. Biophys. Acta, 559 (1979) 1—37.
168 J. Burgos, F. W. Hemming, J. F. Purnock and R. A. Morton, Biochem. J., 88 (1963) 470.
169 E. M. Greenspan, Adv. Int. Med., 7 (1955) 101—123.
170 M. F. Jayle and G. Boussier, Exp. Ann. Biochim. Méd., Vol. 17, Masson, Paris, 1955, 157—194.
171 J. Sonnet (Ed.), Les Glycoprotéines Sériques à l'Etat Normal et Pathologique, Arscia, Bruxelles, 1956.
172 G. Biserte, Bull. Soc. Chim. Biol., 39 (1957) 557—576.
174 H. Busch (Ed.), Biochemistry of the Cancer Cell, Academic Press, 1962, pp. 279—291; Z. Stary, Clin. Chim., 3 (1957) 557—576.
175 R. J. Winzler, in [45], pp. 204—218.

# REFERENCES

176  R. C. Hughes (Ed.), Membrane Glycoproteins, Butterworth, London, 1976.
177  G. M. W. Cook, D. H. Heard and G. V. F. Seaman, Nature, 188 (1960) 1011.
178  L. A. Herzenberg and L. A. Herzenberg, Proc. Natl. Acad. Sci. USA, 47 (1961) 762.
179  G. Gasic and T. Gasic, Proc. Natl. Acad. Sci. USA, 48 (1962) 1172; Proc. Soc. Exptl. Biol. Med., 114 (1963) 660.
180  H. Stillmark, Inaug. Diss. Univ. Dorpat, Germany, 1888; in Arbeiten des Pharmokologischen Institutes zu Dorpat, Vol. 3, Enke, Stuttgart, pp. 59—151.
181  J. B. Sumner, J. Biol. Chem., 37 (1919) 137.
182  K. O. Renkonen, Ann. Med. Exp. Biol. Fenn., 26 (1948) 66; 28 (1950) 45; 38 (1960) 26.
183  W. C. Boyd and R. M. Regnera, J. Immunol., 62 (1949) 333.
184  N. Sharon and H. Lis, Science, 177 (1972) 949—959.
185  H. Lis and N. Sharon, Annu. Rev. Biochem., 43 (1973) 541—574.
186  G. L. Nicolson, Int. Rev. Cytol., 39 (1974) 89—190.
187  A. M. C. Rapin and M. Burger, Adv. Cancer Res., 20 (1974) 1.
188  N. Sharon and H. Lis, Methods Memb. Biol., 3 (1975) 147—200.
189  I. E. Liener, Annu. Rev. Plant Physiol., 27 (1976) 291—319.
190  G. L. Nicolson, Biochim. Biophys. Acta, 457 (1976) 57.
191  H. Lis and N. Sharon, Lectins: their Chemistry and Application to Immunology, in M. Sela (Ed.), The Antigens, Vol. 4, Academic Press, New York, 1977, pp. 429—530.
192  H. Bittiger and H. P. Schnebli (Eds.), Concanavalin A as a Tool, Wiley, London, 1976.
193  N. Sharon, Sci. Am., 236 (1977) 108—119.
194  I. J. Goldstein and C. E. Hayes, Adv. Carbohydr. Chem. Biochem., 35 (1978) 127—340.
195  S. Kornfeld and R. Kornfeld, in [73], pp. 437—449; in [79], pp. 1—34.
196  K. D. Noonan, in C. Nicolau (Ed.), Virus-transformed Cell Membranes, Academic Press, New York, 1978, pp. 281—371.
197  I. J. Goldstein (Ed.), Carbohydrate—Protein Interaction, ACS Symposium Series No. 88, Am. Chem. Soc. Publ., Washington, 1979.
198  R. Lotan and G. L. Nicolson, Biochim. Biophys. Acta, 559 (1979) 329—376.
199  H. Lis and N. Sharon, Lectins in Higher Plants, in P. K. Stumpf and E. E. Cohn (Eds.), The Biochemistry of Plants: A Comprehensive Treatise, Vol. VI, Academic Press, New York, 1981, in press.
200  M. M. Burger and A. R. Goldberg, Proc. Natl. Acad. Sci. USA, 57 (1967) 359—366; M. M. Burger, Proc. Natl. Acad. Sci. USA, 62 (1969) 994—1001; M. Inbar and L. Sachs, Nature, 223 (1969) 710—712; Proc. Natl. Acad. USA, 63 (1969) 1418—1425.
201  K. Landsteiner and R. A. Harte, J. Exp. Med., 71 (1940) 551.
202  E. A. Kabat (Ed.), Blood Group Substances, Academic Press, New York, 1956.

# REFERENCES

203 J. S. Brimacombe and J. M. Webber (Eds.), Mucopolysaccharides, Elsevier, Amsterdam, 1964.
204 W. T. J. Morgan, in [14], pp. 200—215.
205 W. T. J. Morgan, in [35], pp. 170—184.
206 D. Aminoff (Ed.), Blood and Tissue Antigens, Academic Press, New York, 1970.
207 V. Ginsburg, Adv. Enzymol., 36 (1972) 131—149.
208 M. L. Horowitz, in [73], pp. 387—436.
209 A. Gottschalk, Biochim. Biophys. Acta, 23 (1957) 645; Physiol. Rev., 37 (1957) 66; [14], pp. 287—295.
210 A. Gottschalk, Nature, 186 (1960) 949—951.
211 A. Gottschalk, G. Belyavin and F. Biddle, in [46], pp. 1082—1096.
212 J. C. Aub, C. Tieslau and A. Lankester, Proc. Natl. Acad. Sci. USA, 50 (1963) 613.
213 B. M. Gesner and V. Ginsburg, Proc. Natl. Acad. Sci. USA, 52 (1964) 750—755.
214 G. Ashwell and A. G. Morell, in G. A. Jamieson and T. J. Greenwalt (Eds.), Glycoproteins of Blood Cells and Plasma, Lippincott, Philadelphia, 1971, pp. 173—189.
215 G. Ashwell and A. G. Morell, Adv. Enzymol., 41 (1974) 99—128.
216 G. Ashwell and A. G. Morell, Trends Biochem. Sci., 2 (1977) 76—78.
217 H. Yamaguchi, T. Ikenaka and Y. Matsushima, J. Biochem., 70 (1971) 587—594.
218 G. Spik, R. Vandersyppe, B. Fournet, B. Bayard, P. Charet, S. Bouquelet, G. Strecker and J. Montreuil, in [61], pp. 483—500.
219 G. Spik, B. Fournet, B. Bayard, R. Vandersyppe, G. Strecker, S. Bouquelet, P. Charet and J. Montreuil, Arch. Int. Physiol. Biochim., 82 (1974) 791.
220 G. Spik, B. Bayard, B. Fournet, G. Strecker, S. Bouquelet and J. Montreuil, FEBS-Lett., 50 (1975) 269—299.
221 J. Baenziger, S. Kornfeld and S. Kochwa, J. Biol. Chem., 249 (1974) 1889—1896; 1897—1903.
222 K. Meyer, in [7], p. 64.
223 H. Masamune, J. Japan. Biochem. Soc., 18 (1944) 247.
224 R. J. Winzler, in [14], pp. 245—263; in F. Putnam (Ed.), The Plasma Proteins, Vol. 1, Academic Press, New York, 1960, pp. 309—348.
225 A. Gottschalk, Perspect. Biol. Med., 5 (1962) 327.
226 A. Gottschalk, W. H. Murphy and E. R. B. Graham, Nature, 194 (1962) 1051.
227 H. Muir, Biochem. J., 69 (1958) 195.
228 E. A. Balasz, in [39], p. XXX.
229 D. Aminoff, W. W. Binkley, R. Schaffer and R. W. Mowry, in [41], pp. 740—807.
230 J. Montreuil, in P. Boulanger and J. Polonovski (Eds.), Problèmes Actuels de Biochimie Générale, Masson, Paris, 1972, pp. 175—269.
231 R. L. Whistler and J. N. BeMiller (Eds.), Methods in Carbohydrate Chemistry, Vol. 7, Academic Press, New York, 1976.
232 V. Ginsburg (Ed.), Methods in Enzymology, Vol. 28, Academic Press,

New York, 1972.
233 V. Ginsburg (Ed.), Methods in Enzymology, Vol. 50, Academic Press, New York, 1978.
234 E. Neufeld and V. Ginsburg, Methods in Enzymology, Vol. 8, Academic Press, New York, 1966.
235 H. M. Flowers and N. Sharon, Adv. Enzymol., 48 (1979) 29—95.
236 A. Kobata, Anal. Biochem., 100 (1979) 1—14.
237 M. I. Horowitz, in [69], pp. 15—34.
238 W. C. Boyd and E. Sharpleigh, Science, 119 (1954) 419.
239 I. J. Goldstein, R. C. Hughes, M. Monsigny, T. Osawa and N. Sharon, Nature, 285 (1980) 66.
240 R. Kornfeld and C. Ferris, J. Biol. Chem., 250 (1975) 2614—2619; J. V. Baenziger and D. Fiete, J. Biol. Chem., 254 (1979) 2400—2407.
241 H. Debray, D. Decout, G. Strecker, J. Montreuil and M. Monsigny, Proc. IXth Int. Symp. Carbohydr. Chem., London, 1978, pp. 385—386; Protides Biol. Fluids, 27 (1979) 451—454; H. Debray, D. Decout, G. Strecker, G. Spik and J. Montreuil, Eur. J. Biochem., 117 (1981) 41—55.
242 T. Krusius, J. Finne and H. Rauvala, FEBS Lett., 71 (1976) 117—120; T. Krusius and J. Finne, Eur. J. Biochem., 78 (1977) 369—379.
243 V. Ginsburg, Methods Enzymol., 28, Part B (1972); 50, Part C (1978).
244 S. Narasimhan, J. R. Wilson, E. Martin and H. Schachter, Can. J. Biochem., 57 (1979) 83—96.
245 A. Neuberger, A. Gottschalk, R. D. Marshall and R. G. Spiro, in [46], pp. 450—490.
246 W. Pigman, J. Moschera, F. Downs and T. Wakabayashi, in [61], pp. 231—247.
247 D. M. Carlson, in [61], pp. 249—262.
248 W. Pigman and F. Downs, in [69], pp. 80—82.
249 F. K. Hartley and F. R. Jevons, Biochem. J., 84 (1962) 134—139.
250 J. Montreuil, M. Monsigny and M-T. Buchet, C.R. Acad. Sci., Ser. D., 264 (1967) 2068—2071; in [61], pp. 245—247.
251 B. Bayard and J. Montreuil, in [61], pp. 209—218.
252 M. Isemura and K. Schmid, Biochem. J., 124 (1971) 591—604.
253 L. E. Franzén and S. Svensson, in [76], pp. 8—9; B. Nilsson and S. Svensson, Carbohydr. Res., 72 (1979) 183—190.
254 J. Montreuil and G. Spik (Ed.), Méthodes Colorimétriques de Dosage des Glucides totaux. Lab. Chim. Biol. Fac. Sci. Lille, 1963; Procédés de Dosages Chromatographiques et Electrophorétiques des Monosaccharides Constituant les Glycoprotéines, Lab. Chim. Biol. Fac. Sci. Lille, 1968.
255 R. D. Marshall and A. Neuberger, in [46], pp. 224—299.
256 J. R. Clamp, T. Bhatti and R. E. Chambers, in [46], pp. 300—321.
257 J. Montreuil, A. Adam-Chosson and G. Spik, Bull. Soc. Chim. Biol., 47 (1964) 1867—1880; A. Adam-Chosson and J. Montreuil, Bull. Soc. Chim. Biol., 47 (1965) 1881—1900.
258 B. Lindberg, J. Lönngren and S. Svensson, Adv. Carbohydr. Chem. Biochem., 31 (1975) 185—240.

## REFERENCES

259 B. Bayard and J. Montreuil, Carbohydr. Res., 24 (1972) 427—443.
260 B. Bayard, B. Fournet, S. Bouquelet, G. Strecker, G. Spik and J. Montreuil, Carbohydr. Res., 24 (1972) 445—456.
261 B. Fournet, J-M. Dhalluin. G. Strecker and J. Montreuil, Anal. Biochem., 108 (1980) 35—56.
262 B. Bayard and B. Fournet, Carbohydr. Res., 46 (1976) 75—86.
263 G. Strecker, A. Pierce-Crétel, B. Fournet, G. Spik and J. Montreuil, Analyt. Biochem., 111 (1981) 17—26.
264 J. A. Rothfus and E. L. Smith, J. Biol. Chem., 238 (1963) 1402.
265 S. I. Hakomori, J. Biochem., 55 (1964) 205—208.
266 R. D. Marshall and A. Neuberger, in [46], pp. 322—380.
267 B. Fournet, Y. Leroy and J. Montreuil, in [61], pp. 111—130.
268 B. Fournet and J. Montreuil, J. Chromatog., 75 (1973) 29—37; 92 (1974) 184—190.
269 B. Fournet, J-M. Dhalluin, Y. Leroy and J. Montreuil, J. Chromatog., 153 (1978) 91—99.
270 B. Lindberg and J. Lönngren, in [233], pp. 3—33.
271 A. Gottschalk and R. Drzeniek, in [46], pp. 381—398.
272 Y-T. Li and S-C. Li, in [231], pp. 221—225.
273 Y-T. Li and S-C Li, in [69], pp. 51—67.
274 F. Maley, A. L. Tarentino and R. B. Trimble, in [74], pp. 86—103.
275 Y-T. Li, in [75], pp. 3—15.
276 I. Yamashina, in [46], pp. 1187—1210.
277 E. H. Eylar and M. Murakami, in [234], pp. 597—600.
278 A. L. Tarentino, in [232], pp. 782—786.
279 M. Kohno and I. Yamashina, Biochim. Biophys. Acta, 257 (1972) 600; in [232], pp. 786—792.
280 N. Takahashi, Biochem. Biophys. Res. Commun., 76 (1977) 1194—1201; N. Takahashi and H. Nishibe, J. Biochem. (Tokyo), 84 (1978) 1467—1473.
281 C. C. Huang and D. Aminoff, J. Biol. Chem., 247 (1972) 6737.
282 Y. Endo and A. Kobata, J. Biochem. (Tokyo), 80 (1976) 1—8.
283 J. Umemoto, V. P. Bhavanandan and E. A. Davidson, J. Biol. Chem., 252 (1977) 8609—8614.
284 T. Muramatsu, J. Biol. Chem., 246 (1971) 5535—5537; N. Koide and T. Muramatsu, J. Biol. Chem., 249 (1974) 4897—4904.
285 A. L. Tarentino and F. Maley, J. Biol. Chem., 249 (1974) 811—817; A. L. Tarentino, T. H. Plummer Jr. and F. Maley, J. Biol. Chem., 248 (1973) 5547; 249 (1974) 818—824.
286 M. Arakawa and T. Muramatsu, J. Biochem. (Tokyo), 76 (1974) 307—317.
287 S. Ito, T. Muramatsu and A. Kobata, Arch. Biochem. Biophys., 171 (1975) 78—86.
288 M. Nishigaki, T. Muramatsu and A. Kobata, Biochem. Biophys. Res. Commun., 59 (1974) 638—645.
289 A. M. Ogata, T. Muramatsu and A. Kobata, J. Biochem. (Tokyo), 82 (1977) 611—614.
290 S. F. Chien, R. Weinburg, S. C. Li and Y. T. Li, Biochem. Biophys. Res.

Commun., 76 (1977) 317—323.
291  A. L. Tarentino and F. Maley, J. Biol. Chem., 251 (1976) 6537—6543; in [233], p. 580.
292  S. Bouquelet, G. Strecker, J. Montreuil and G. Spik, in [76], pp. 374—375; Biochimie, 62 (1980) 43—49.
293  S. Takasaki and A. Kobata, J. Biol. Chem., 251 (1976) 3603—3609.
294  M. N. Fukuda and G. Matsumura, J. Biol. Chem., 251 (1976) 6218.
295  M. Makino, T. Kojïma and I. Yamashina, Biochem. Biophys. Res. Commun., 24 (1966) 961.
296  L. Dorland, J. Haverkamp, B. L. Schut, J. F. G. Vliegenthart, G. Spik, G. Strecker, B. Fournet and J. Montreuil, FEBS Lett., 77 (1977) 15—20.
297  G. Strecker, M-C. Herlant-Peers, B. Fournet, J. Montreuil, L. Dorland, J. Haverkamp, J. F. G. Vliegenthart and J-P. Farriaux, Eur. J. Biochem., 81 (1977) 165—171.
298  L. Dorland, B. L. Schut, J. F. G. Vliegenthart, G. Strecker, B. Fournet, G. Spik and J. Montreuil, Eur. J. Biochem., 73 (1977) 93—97; G. Strecker, B. Fournet, J. Montreuil, L. Dorland, J. Haverkamp, J. F. G. Vliegenthart and D. Dubesset, Biochimie, 60 (1978) 725—734.
299  L. Dorland, J. Haverkamp, J. F. G. Vliegenthart, B. Fournet, G. Strecker, G. Spik and J. Montreuil, FEBS Lett., 89 (1978) 149—152.
300  D. Léger, V. Tordera and G. Spik; L. Dorland, J. Haverkamp and J. F. G. Vliegenthart, FEBS Lett., 93 (1978) 255—260.
301  L. Dorland, J. Haverkamp, J. F. G. Vliegenthart; G. Strecker, J. C. Michalski, B. Fournet, G. Spik and J. Montreuil, Eur. J. Biochem., 87 (1978) 323—329.
302  B. L. Schut, L. Dorland, J. Haverkamp, J. F. G. Vliegenthart and B. Fournet, Biochim. Biophys. Res. Commun., 82 (1978) 1223—1228.
303  L. Dorland, Structure Determination of the Complex Carbohydrate Chains of Glycoproteins by 360 MHz $^1$H-NMR Spectroscopy, Thesis, Utrecht, 1979.
304  J. Montreuil and J. F. G. Vliegenthart, in [75], pp. 35—78.
305  H. Debray, J. Montreuil, L. Dorland and J. F. G. Vliegenthart, in [76], p. 86; H. van Halbeek, L. Dorland, J. F. G. Vliegenthart, G. Spik, A. Chéron and J. Montreuil, Biochim. Biophys. Acta, 675 (1981) 293—296.
306  L. Dorland, J. Haverkamp, J. F. G. Vliegenthart, G. Spik, B. Fournet and J. Montreuil, Eur. J. Biochem., 100 (1979) 569—574.
307  H. van Halbeek, L. Dorland, J. F. G. Vliegenthart, K. Schmid, J. Montreuil, B. Fournet and W. E. Hull, FEBS Lett., 114 (1980) 11—16; H. van Halbeek, L. Dorland, G. A. Veldink, J. F. G. Vliegenthart, J. C. Michalski, J. Montreuil, G. Strecker and W. E. Hull, FEBS Lett., 121 (1980) 65—70; H. van Halbeek, L. Dorland, J. F. G. Vliegenthart, J. Montreuil, B. Fournet and K. Schmid, J. Biol. Chem., 256 (1981) 5580—5590.
308  H. van Halbeek, L. Dorland, G. A. Veldink, J. F. G. Vliegenthart, G. Strecker, J. C. Michalski, J. Montreuil and W. E. Hull, FEBS Lett., 121 (1980) 71—77; H. Debray, B. Fournet, J. Montreuil, L. Dorland and J. F. G. Vliegenthart, Eur. J. Biochem, 115 (1981) 559—563.

309 L. Dorland, H. van Halbeek, J. Haverkamp, G. A. Veldink, J. F. G. Vliegenthart, B. Fournet, J. Montreuil and D. Aminoff, in [76], pp. 29—30.
310 D. H. von den Eijnden, D. H. Joziasse, L. Dorland, H. van Halbeek, J. F. G. Vliegenthart and K. Schmid, Biochem. Biophys. Res. Commun., 92 (1980) 839—845.
311 R. S. Blacklow, A. P. Fletcher, R. D. Marshall and A. Neuberger, in [35], pp. 79—93.
312 B. Lindahl and L. Rodén, in [46], pp. 491—517.
313 A. B. Zinn, J. J. Plantner and Don M. Carlson, in [69], pp. 69—85.
314 P. G. Johansen, R. D. Marshall and A. Neuberger, Biochem. J., 78 (1961) 518—527.
315 A. Gottschalk and W. König, Biochim. Biophys. Acta, 158 (1968) 358.
316 H. Masamune, in C. Liébecq, Proc. 3th Int. Congress Biochem., Brussels 1955, Academic Press, New York, pp. 72—77.
317 A. P. Fletcher, G. S. Marks, R. D. Marshall and A. Neuberger, Biochem. J., 87 (1963) 265—273.
318 G. S. Marks, R. D. Marshall and A. Neuberger, Biochem. J., 87 (1963) 274—281.
319 R. H. Nuenke and L. W. Cunningham, J. Biol. Chem., 236 (1961) 2451—2460.
320 I. Yamashina and M. Makino, J. Biochem. (Tokyo), 51 (1962) 359—364.
321 Y. Hashimoto and W. Pigman, Ann. N.Y. Acad. Sci., 93 (1962) 541.
322 K. Tanaka, M. Bertolini and W. Pigman, Biochem. Biophys. Res. Commun., 16 (1964) 404.
323 B. Anderson, H. Seno, P. Sampson, J. G. Riley, P. Hoffman and K. Meyer, J. Biol. Chem., 239 (1964) PC 2716.
324 S. Harbon, F. Herman, B. Rossignol, P. Jollès and H. Clauser, Biochem. Biophys. Res. Commun., 17 (1964) 57.
325 V. P. Bhavanandan, E. Buddecke, R. Carubelli and A. Gottschalk, Biochem. Biophys. Res. Commun., 16 (1964) 333.
326 J. D. Gregory, T. C. Laurent and L. Rodén, J. Biol. Chem., 239 (1964) 3312—3320.
327 U. Lindahl and L. Rodén, Biochem. Biophys. Res. Commun., 17 (1964) 254—259.
328 U. Lindahl and L. Rodén, J. Biol. Chem., 240 (1965) 2821—2826.
329 L. Rodén and U. Lindahl, Fed. Proc., 24 (1965) 606.
330 K. Brendel and E. A. Davidson, Carbohydr. Res., 2 (1966) 42.
331 L. Rodén, in [35], pp. 185—202; in [79], pp. 267—371.
332 J. Heaney-Kieras, L. Rodén and D. J. Chapman, Biochem. J., 165 (1977) 1—9.
333 J. R. Green and D. H. Northcote, Biochem. J., 170 (1978) 599—608.
334 A. K. Allen, N. N. Desai, A. Neuberger and J. M. Creeth, Biochem. J., 171 (1978) 665—674.
335 L. Muir et Y. C. Lee, J. Biol. Chem., 244 (1969) 2343—2349.
336 R. G. Spiro and V. D. Bhoyroo, J. Biol. Chem., 249 (1974) 5704—5717;

255 (1980) 5347—5354.
337 R. G. Spiro and V. D. Bhoyroo, Fed. Proc., 30 (1971) 1223.
338 D. T. A. Lamport, L. Katona and S. Roerig, Biochem. J., 133 (1973) 125—131.
339 R. H. A. Muray and D. H. Northcote, Phytochemistry, 17 (1978) 623—629.
340 T. Nakajima and C. E. Ballou, J. Biol. Chem., 249 (1974) 7679-7684.
341 M. K. Raizada, J. S. Schutzbach and H. Ankel, J. Biol. Chem., 250 (1975) 3310—3315.
342 A. L. Rosenthal and J. H. Nordin, J. Biol. Chem., 250 (1975) 5295—5303.
343 J-B. Leleu, B. Fournet, J-P. Morilhat, R. Bonaly et J. Montreuil, Biochimie, 59 (1977) 687—692.
344 J. Finne, T. Krusius, R. K. Margolis and R. U. Margolis, J. Biol. Chem., 254 (1979) 10295—10300.
345 W. T. Butler et L. W. Cunningham, J. Biol. Chem., 240 (1965) PC 3449; 241 (1966) 3882—3888.
346 R. G. Spiro, J. Biol. Chem., 242 (1967) 1923—1932.
347 R. G. Spiro, J. Biol. Chem., 242 (1967) 4813-4823.
348 R. G. Spiro, in [39], pp. 195—215; in [46], pp. 964—999.
349 W. T. Butler, in [74], pp. 213—226.
350 R. G. Spiro and S. Fukushi, J. Biol. Chem., 244 (1969) 2049—2058.
351 M. Isemura, T. Ikenaka and Y. Matsushima, J. Biochem. (Tokyo), 74 (1973) 11—21.
352 T. Biswas and A. K. Mukherjee, Carbohydr. Res., 63 (1978) 173—181.
353 M. A. Calcott and H. J. Müller-Eberhard, J. Biol. Chem., 11 (1972) 3443—3450.
354 L. W. Cunningham and J. D. Ford, J. Biol. Chem., 243 (1968) 2390.
355 D. T. A. Lamport, Nature, 216 (1967) 1322—1324.
356 Y. Akiyama and K. Kato, Agric. Biol. Chem., 41 (1977) 79—81.
357 D. Ashford and A. Neuberger, Trends Biochem. Sci., 5 (1980) 245—248; D. Ashford, N. Desai, A. K. Allen, A. Neuberger, M. A. O'Neill and R. R. Selvendran, Biochem. J., 201 (1982) 1, 199—208.
358 D. T. A. Lamport, Biochemistry, 8 (1969) 1155—1163.
359 D. H. Miller, D. T. A. Lamport and M. Miller, Science, 176 (1972) 918—920; D. H. Miller, E. S. Mellman, D. T. A. Lamport and M. Miller, J. Cell Biol., 63 (1974) 420—429.
360 U. V. Mani and A. N. Radnakrishnan, Biochem. J., 141 (1974) 147—153.
361 D. G. Pope, Plant Physiol., 59 (1977) 894—900.
362 D. T. A. Lamport and D. H. Miller, Plant Physiol., 48 (1971) 454—456.
363 M. Knee, Phytochemistry, 14 (1975) 2181—2188; T. Yamagishi, K. Matsuda and T. Watanabe, Carbohyd. Res., 50 (1976) 63—74.
364 M. K. McNamara and B. A. Stone, Proc. IXth Int. Symp. Carbohydr. Chem., London, 1978, pp. 43—44.
365 G. B. Fincher, W. H. Sawyer and B. A. Stone, Biochem. J., 139 (1974) 535—545.
366 P. Hallgren, A. Lundblad et S. Svensson, J. Biol. Chem., 250 (1975)

5312—5314.
367 G. Larriba, M. Klinger, S. Sramek and S. Steiner, Biochem. Biophys. Res. Commun., 77 (1977) 79—85.
368 C. J. Lote and J. B. Weiss, Biochem. J., 123 (1971) 25; FEBS Lett., 16 (1971) 81—85.
369 J. B. Weiss, C. J. Lote and H. Bobinski, Nature, 234 (1971) 25—26.
370 M. A. Krysteva, I. N. Mancheva and I. D. Dobrev, Eur. J. Biochem., 40 (1973) 155—161.
371 M. Jett and G. A. Jamieson, Carbohydr. Res., 18 (1971) 466—468.
372 E. F. Hounsell, M. Fukuda, M. E. Powell, T. Feizi and S. I. Hakomori, Biochem. Biophys. Res. Commun., 92 (1980) 1143—1150.
373 T. Tai, K. Yamashita, S. Ito and A. Kobata, J. Biol. Chem., 252 (1977) 6687—6694.
374 J. Finne, T. Krusius and H. Rauvala, Biochem. Biophys. Res. Commun., 74 (1977) 405—410.
375 J. Finne, T. Krusius, H. Rauvala and K. Hemminki, Eur. J. Biochem., 77 (1977) 319—323.
376 R. Kornfeld, Biochemistry, 17 (1978) 1415—1423.
377 D. Carlson, J. Biol. Chem., 243 (1968) 616—626.
378 D. Aminoff, W. D. Gathmann and M. Baig, J. Biol. Chem., 254 (1979) 8909—8913.
379 H. Van Halbeck, L. Dorland, J. Haverkamp, G. A. Veldink, J. F. G. Vliegenthart, B. Fournet, G. Ricart, J. Montreuil, W. D. Gathmann and D. Aminoff, Eur. J. Biochem., 118 (1981) 487—495.
380 J. Baenziger and S. Kornfeld, Fed. Proc., 31 (1972) 466; J. Biol. Chem., 249 (1974) 7260—7269; 7270—7281.
381 W. Dahr, G. Uhlenbruck, H. H. Gunson and M. Van der Hart, Vox Sang., 28 (1975) 249—252.
382 D. H. van den Eijnden, N. A. Evans, J. F. Codington, V. Reinhold, C. Silber and R. W. Jeanloz, J. Biol. Chem., 254 (1979) 12153—12159.
383 A. Gottschalk and E. R. B. Graham, Biochim. Biophys. Acta, 34 (1959) 380—391.
384 E. R. B. Graham and A. Gottschalk, Biochim. Biophys. Acta, 38 (1960) 513—534.
385 C. G. Lombart and R. J. Winzler, Eur. J. Biochem., 49 (1974) 77—86.
386 W. Dahr, G. Uhlenbruck and G. W. G. Bird, Vox Sang., 28 (1975) 133—148.
387 Nasir-ud-Din, R. W. Jeanloz, V. Reinhold, J. D. Moore and J. W. McArthur, in [75], pp. 241—244.
388 G. Lamblin, M. Lhermitte, P. Humbert, M. C. Tirlemont, P. Roussel and V. Reinhold, in [76], pp. 32—33; G. Lamblin, M. Lhermitte, A. Boersma, P. Roussel and V. Reinhold, J. Biol. Chem., 255 (1980) 4595—4598.
389 B. A. Bray, R. Lieberman and K. Meyer, J. Biol. Chem., 242 (1967) 3373—3380.
390 A. L. DeVries, J. Vandenheede and R. E. Feeney, J. Biol. Chem., 246 (1971) 305—308; W. T. Shier, Y. Lin and A. L. DeVries, FEBS Lett., 54 (1975) 135—138.

391 M. M. Baig and D. Aminoff, J. Biol. Chem., 247 (1972) 6111—6118.
Om P. Bahl, R. B. Carlson, R. Bellisario and N. Swaminathan, Biochem. Biophys. Res. Commun., 48 (1972) 416—422.
393 L. Rovis, B. Anderson, E. A. Kabat, F. Gruezo and J. Liao, Biochemistry, 12 (1973) 5340—5354.
394 J. F. Codington, K. B. Linsley, R. W. Jeanloz, T. Irimura and T. Osawa, Carbohydr. Res., 40 (1975) 171—182.
395 W. M. Glöckner, R. A. Newman and G. Uhlenbruck, Biochim. Biophys. Acta, 443 (1976) 402—413.
396 J. Finne, Biochim. Biophys. Acta, 412 (1975) 317—325.
397 S. Inoue, M. Iwasaki and G. Matsumura, in [76], pp. 40—41.
398 J. K. Wold, B. Smestad and G. Uhlenbruck, Acta Chem. Scand., B 29 (1975) 703—709.
399 R. A. Newman, R. Harrison and G. Uhlenbruck, Biochim. Biophys. Acta, 63 (1976) 344—356.
400 A. Crétel, M. Pamblanco, H. Egge, G. Strecker, J. Montreuil and G. Spik, in [76], pp. 26—27.
401 A. Pierce-Crétel, M. Pamblanco, G. Strecker, J. Montreuil and G. Spik, Eur. J. Biochem., 114 (1981) 169—178.
402 H. G. Garg, D. A. Swann and L. R. Glasgow, Carbohydr. Res., 78 (1979) 79—88.
403 D. B. Thomas and R. J. Winzler, J. Biol. Chem., 244 (1969) 5943—5946.
404 A-M. Fiat, C. Alais and P. Jollès, Eur. J. Biochem., 27 (1972) 408—412.
405 B. Fournet, A-M. Fiat, J. Montreuil and P. Jollès, Biochimie, 57 (1975) 161—165.
406 B. Fournet, A-M. Fiat, C. Alais and P. Jollès, Biochim. Biophys. Acta, 576 (1979) 339—346.
407 B. Nilsson, N. E. Nordén and S. Svensson, J. Biol. Chem., 254 (1979) 4545—4553.
408 Y. Endo, K. Yamashita, Y. N. Han, S. Iwanaga and A. Kobata, J. Biochem. (Tokyo), 82 (1977) 545—550.
409 M. Fukuda and T. Osawa, J. Biol. Chem., 248 (1973) 5100—5105.
410 G. F. Springer and P. R. Desai, Biochem. Biophys. Res. Commun., 61 (1974) 470—475; Carbohydr. Res., 40 (1975) 183—192.
411 H. J. Yang and G. F. Springer, in [75], pp. 563—564.
412 M. J. Kessler, T. Mise, R. D. Ghai and Om P. Bahl, J. Biol. Chem., 254 (1979) 7909—7914.
413 T. Krusius and J. Finne, Eur. J. Biochem., 78 (1977) 369—379.
414 R. A. Newman, W. M. Glöckner and G. Uhlenbruck, Eur. J. Biochem., 64 (1976) 373—380.
415 M. L. Hayes and F. J. Castellino, J. Biol. Chem., 254 (1979) 8768—8711; 8772—8776; 8777—8780.
416 A. Kato, S. Hiraba and K. Kobayashi, Agric. Biol. Chem., 42 (1978) 1025—1029.
417 S. Takasaki, K. Yamashita and A. Kobata, J. Biol. Chem., 253 (1978) 6086—6091.
418 G. Aspinall, Personal communication

419 N. K. Kotchetkov, V. A. Derevitskaya and N. P. Arbatsky, Eur. J. Biochem., 67 (1976) 129—136.
420 T. Feizi, E. A. Kabat, G. Vicari, B. Anderson and W. L. Marsh, J. Immunol., 106 (1971) 1578—1592.
421 M. D. Oates, A. C. Rosbottom and J. Schrager, Carbohydr. Res., 34 (1974) 115—137.
422 V. A. Derevitskaya and N. P. Arbatsky, in [76], pp. 72—73.
423 V. A. Derevitskaya, N. P. Arbatsky and N. K. Kotchetkov, Eur. J. Biochem., 86 (1978) 423—437.
424 K. O. Lloyd and E. A. Kabat, Proc. Natl. Acad. Sci. USA, 61 (1968) 1470—1477.
425 A. Slomiany and B. L. Slomiany, J. Biol. Chem., 253 (1978) 7301—7306.
426 J. R. Baker, J. A. Cifonelli, M. B. Mathews and L. Rodén, Fed. Proc., 28 (1969) 605.
427 H. W. Stuhlsatz, R. Kisters, A. Wollmer and H. Greiling, Z. Physiol. Chem., 352 (1971) 289.
428 R. W. Jeanloz, in M. Florkin and E. H. Stotz (Eds.), Comprehensive Biochemistry, Vol. 5, Elsevier, Amsterdam, 1963, pp. 262—296.
429 A. C. Stoolmiller and A. Dorfman, in M. Florkin and E. H. Stotz (Eds.), Comprehensive Biochemistry, Vol. 17, Elsevier, Amsterdam, 1969, pp. 241—275.
430 N. E. Artz and E. M. Osman (Eds.), Biochemistry of Glycuronic Acid, Academic Press, New York, 1950.
431 K. Meyer, Faraday Soc. Disc., 3 (1953) 271.
432 R. L. Whistler and D. I. McGilvray, Annu. Rev. Biochem., 23 (1954) 79.
433 R. W. Jeanloz, Proc. 3rd Int. Congress Biochemistry, Academic Press, New York, 1956, p. 65.
434 A. Dorfman and M. B. Mathews, Annu. Rev. Physiol., 18 (1956) 69.
435 K. Meyer, Harvey Lectures, 51 (1957) 88.
436 W. Pigman and D. Platt, in W. Pigman (Ed.), The Carbohydrates, Academic Press, New York, 1957, p. 465.
437 H. Gibian (Ed.), Mucopolysaccharide und Mucopolysaccharidasen, Deuticke, Wien, 1959.
438 R. W. Jeanloz, Bull. Soc. Chim. Biol., 42 (1960) 303.
439 F. Clark and J. K. Grant (Eds.), The Biochemistry of Mucopolysaccharides of Connective Tissue, Biochem. Soc. Symposium, Cambridge University Press, Cambridge, 1961.
440 R. W. Jeanloz, Adv. Enzymol., 25 (1963) 433—456.
441 H. Muir, Int. Rev. Connective Tissue Res., 2 (1964) 101—154.
442 M. R. J. Salton, Annu. Rev. Biochem., 34 (1965) 143—174.
443 G. Quintarelli (Ed.), The Chemical Physiology of Mucopolysaccharides, Little and Brown, Boston, 1968.
444 K. Meyer, Am. J. Med., 47 (1969) 664—672.
445 H. Muir, Am. J. Med., 47 (1969) 673—690.
446 R. W. Jeanloz, in [41], pp. 590—625.
447 L. Rodén, in [79], pp. 267—371.

448 L. Rodén, J. R. Baker, N. B. Schwartz, A. C. Stoolmiller, S. Yamagata and T. Yamagata, in [49], pp. 345—385.
449 H. Muir, in H. G. Hers and F. Van Hoof (Eds.), Lysosomes and Storage Diseases, Academic Press, New York, 1973, pp. 79—104.
450 A. Dorfman, Mol. Cell. Biochem., 4 (1974) 45—65.
451 M. B. Mathews (Ed.), Connective tissue. Macromolecular Structure and Evolution, Springer, Berlin, 1975.
452 H. Muir and T. E. Hardingham, MTP Int. Rev. Sci. Biochem., Ser. 1, 5 (1975) 153—222.
453 L. Rodén and N. B. Schwartz, MTP Int. Rev. Sci. Biochem., Ser. 1, 5 (1975) 96—152.
454 U. Lindahl, MTP Int. Rev. Sci. Org. Chem., Ser. 2, 7 (1976) 283—312.
455 U. Lindahl and M. Höök, Annu. Rev. Biochem., 47 (1978) 385—417.
456 S. Ito, K. Yamashita, R. G. Spiro and A. Kobata, J. Biochem. (Tokyo), 81 (1977) 1621—1631.
457 A. Chapman and R. Kornfeld, J. Biol. Chem., 254 (1979) 824—828.
458 E. Li and S. Kornfeld, J. Biol. Chem., 254 (1979) 1600—1605.
459 A. Chéron and G. Spik (unpublished results).
460 A. Chapman and R. Kornfeld, J. Biol. Chem., 254 (1979) 816—823.
461 D. Burke, Ph. D. Thesis, SUNY (Stony Brook), 1976; D. Burke and K. Keegstra, J. Virol., 20 (1976) 676; 29 (1979) 546—554.
462 T. Tai, K. Yamashita, A. M. Ogata, N. Koide, T. Muramatsu, S. Iwashita, Y. Inoue and A. Kobata, J. Biol. Chem., 250 (1975) 8569—8575.
463 J. Conchie and I. Strachan, Carbohydr. Res., 63 (1978) 193—213.
464 T. Tai, K. Yamashita and A. Kobata, Biochem. Biophys. Res. Commun., 78 (1977) 434—440.
465 S. Minobe, H. Nakajima, N. Itoh, I. Funakoshi and I. Yamashina, J. Biochem. (Tokyo), 86 (1979) 1851—1854.
466 H. Lis and N. Sharon, J. Biol. Chem., 253 (1978) 3468—3476.
467 T. Nakajima and C. E. Ballou, J. Biol. Chem., 249 (1974) 7685—7694; Proc. Natl. Acad. Sci. USA, 72 (1975) 3912—3916.
468 L. Lehle, R. E. Cohen and C. E. Ballou, J. Biol. Chem., 254 (1979) 12209—12218.
469 C. E. Ballou, Adv. Microb. Physiol., 14 (1976) 93—158.
470 A. Chéron, B. Fournet, G. Spik and J. Montreuil, Biochimie, 58 (1976) 927—942.
471 J. U. Baenziger, J. Biol. Chem., 254 (1979) 4063—4071.
472 T. Mizuochi, K. Yonemasu, K. Yamashita and A. Kobata, J. Biol. Chem., 253 (1978) 7404—7409.
473 S. Takasaki, K. Yamashita, K. Suzuki, S. Iwanaga and A. Kobata, J. Biol. Chem., 254 (1979) 8548—8553.
474 A. Kobata, K. Yamashita, S. Takahasi, T. Mizuochi and Y. Tachibana, in [76], pp. 6—7.
475 B. Ekström, A. Lundblad and S. Svensson, in [76], pp. 19—20.
476 B. Fournet, Unpublished results.
477 A. B. Zinn, J. S. Marshall and D. M. Carlson, J. Biol. Chem., 253 (1978) 6768—6774.
478 T. Mizuochi, K. Yamashita, K. Fujikawa, W. Kisiel and A. Kobata, J.

Biol. Chem., 254 (1979) 6419—6425.
479 R. Prasad, B. G. Hudson, D. K. Strickland and K. E. Ebner, J. Biol. Chem., 255 (1980) 1248—1251.
480 Y. Endo, K. Yamashita, Y. Tachibana, S. Tojo and A. Kobata, J. Biochem. (Tokyo), 85 (1979) 669—679.
481 Om P. Bahl, L. März and M. J. Kessler, Biochem. Biophys. Res. Commun., 84 (1978) 667—676; M. J. Kessler, M. S. Reddy, R. H. Shah and Om P. Bahl, J. Biol. Chem., 254 (1979) 7901—7908.
482 T. Tai, S. Ito, K. Yamashita, T. Muramatsu and A. Kobata, Biochem. Biophys. Res. Commun., 65 (1975) 968—974.
483 W. G. Carter and S. I. Hakomori, Biochemistry, 18 (1979) 730—738; M. Fukuda and S. I. Hakomori, J. Biol. Chem., 254 (1979) 5451—5457.
484 A. Crétel, M. Pamblanco, G. Strecker, L. Dorland, J. F. G. Vliegenthart, J. Montreuil and G. Spik, Eur. J. Biochem. (1981) in press.
485 G. Spik, G. Strecker, B. Fournet, S. Bouquelet, L. Dorland, J. F. G. Vliegenthart and J. Montreuil, Eur. J. Biochem. (1981) in press.
486 G. Spik, B. Fournet, A. Chéron, G. Strecker, J. Montreuil, L. Dorland and J. F. G. Vliegenthart, in [76], pp. 21—22.
487 C. François-Gérard, J. Brocteur, A. André, C. Gerday, A. Pierce-Crétel, J. Montreuil and G. Spik, Blood Transf. Immunohaematol., 23 (198) 579—588; C. François-Gérard, A. Pierce-Crétel, H. van Halbeek, L. Dorland, J. F. G. Vliegenthart, J. Brocteur, A. André, C. Gerday, J. Montreuil and G. Spik, Eur. J. Biochem. (1982) in press.
488 T. Kondo, M. Fukuda and T. Osawa, Carbohydr. Res., 58 (1977) 405—414.
489 T. Krusius and J. Finne, Carbohydr. Res., 90 (1981) 203—214.
490 C. L. Reading, E. E. Penhoet and C. E. Ballou, J. Biol. Chem., 253 (1978) 5600—5612.
491 G. Spik, B. Fournet and J. Montreuil, C.R. Acad. Sci. Paris, 288D (1979) 967—970.
492 S. Inoue and M. Iwasaki, Biochem. Biophys. Res. Commun., 83 (1978) 1018—1023.
493 B. L. Slomiany, A. Slomiany and A. Herp, Eur. J. Biochem., 90 (1978) 255—260.
494 J. Finne, T. Krusius, H. Rauvala, R. Kekomäki and G. Myllylä, FEBS Lett., 89 (1978) 111—115.
495 J. Järnefelt, J. Finne, T. Krusius and H. Rauvala, Trends Biochem. Sci., 3 (1978) 111—113; J. Järnefelt, J. Rush, Y-T. Li and R. A. Laine, J. Biol. Chem., 253 (1978) 8006—8009.
496 T. Krusius, J. Finne and H. Rauvala, Eur. J. Biochem., 92 (1978) 289—300.
497 E. Li, R. Gibson and S. Kornfeld, Arch. Biochem. Biophys., 199 (1980) 393—399.
498 M. Fukuda, M. N. Fukuda and S. I. Hakomori, J. Biol. Chem., 254 (1979) 3700—3703.
499 T. Tsuji, T. Irimura and T. Osawa, in [76], p. 39.
500 C. J. Liang, K. Yamashita, C. G. Muellenberg, H. Shichi and A. Kobata,

J. Biol. Chem., 254 (1979) 6414—6418.
501 K. Yamashita, Y. Tachibana and A. Kobata, J. Biol. Chem., 253 (1978) 3862—3869.
502 F. Miller, Immunochemistry, 9 (1972) 217—228.
503 A. Misaki and I. J. Goldstein, J. Biol. Chem., 252 (1977) 6995—6999.
504 H. Ishibara, N. Takahashi, S. Oguri and S. Tejima, J. Biol. Chem., 254 (1979) 10715—10719.
505 P. Prehm, A. Scheid and P. W. Choppin, J. Biol. Chem., 254 (1979) 9669—9677.
506 A. Lundblad, B. Nilsson, N. E. Nordén, S. Svensson, P-A. Öckerman and R. D. Jolly, Eur. J. Biochem., 59 (1975) 601—605.
507 L. W. Cunningham, in [36], pp. 141—160; in [45], pp. 16—34.
508 R. Montgomery, in [46], pp. 518—528.
509 K. Schmid, J-P. Binette, K. Tokita, L. Moroz and H. Yoshizaki, J. Clin. Invest., 43 (1964) 2347—2352; K. Schmid, K. Tokita and H. Yoshizaki, J. Clin. Invest., 44 (1965) 1394—1401.
510 T. Yamauchi and I. Yamashina, J. Biochem. (Tokyo), 66 (1969) 213—223.
511 L. W. Cunningham, B. J. Nuenke and R. H. Nuenke, Biochim. Biophys. Acta, 26 (1957) 660; L. W. Cunningham, J. D. Ford and J. M. Rainey, Biochim. Biophys. Acta, 101 (1965) 233—235.
512 C. C. Huang, H. E. Mayer and R. Montgomery, Carbohydr. Res., 13 (1970) 127—137.
513 V. Warin, F. Baert et R. Fouret; G. Strecker, G. Spik, B. Fournet and J. Montreuil, in [75], pp. 317—320; V. Warin, F. Baert et R. Fouret; G. Strecker, G. Spik, B. Fournet and J. Montreuil, Carbohydr. Res., 76 (1979) 11—22.
514 J. Montreuil, FEBS Lett., in press.
515 R. Huber, Klin. Wochenschr., 58 (1980) 1217—1231.
516 A. Douy and B. Gallot, Biopolymers, 19 (1980) 493—507.
517 J. Davoust, V. Michel, G. Spik, J. Montreuil and P. Devaux, FEBS Lett., 125 (1981) 271—275.
518 C. H. Li (Ed.), Hormonal Proteins and Peptides, Academic Press, New York, 1973.
519 V. T. Marchesi, P. W. Robbins, V. Ginsburg and C. F. Fox (Eds.), Cell Surface Carbohydrates and Biological Recognition, Progress in Clinical and Biological Research, Vol. 23, Liss, New York, 1978.
520 R. T. A. MacGillivray, E. Mendez and K. Brew, in E. B. Brown, P. Aisen, J. Fielding and R. R. Crichton (Eds.), Proteins of Iron Metabolism, Grune and Stratton, New York, 1977, pp. 133—141.
521 D. A. Sly, D. Grohlich and A. Berkorovainy, in R. E. Harmon (Ed.), Cell Surface Carbohydrate Chemistry, Academic Press, New York, 1978, pp. 255—268; H. Y. Yang Hu and P. Aisen, J. Supramol. Struct., 8 (1978) 349—360.
522 R. W. Jeanloz, in [46], pp. 565—611.
523 K. Schmid, in F. W. Putnam (Ed.), The Plasma Proteins: Structure, Function and Genetic Control, Academic Press, New York, 1975, pp. 183—228.

# REFERENCES

524 R. G. Spiro, in [36], pp. 59—78.
525 M. T. McQuillan and V. M. Trikojus, in [46], pp. 926—963.
526 R. G. Spiro and M. J. Spiro, J. Biol. Chem., 240 (1965) 997—1011; R. G. Spiro, J. Biol. Chem., 240 (1965) 1603—1610.
527 T. Arima, M. J. Spiro and R. G. Spiro, J. Biol. Chem., 2477 (1972) 1825—1835; 1836—1848.
528 V. C. Hascall and S. W. Sajdera, J. Biol. Chem., 245 (1970) 4920—4930; V. C. Hascall and R. L. Riolo, J. Biol. Chem., 247 (1972) 4529—4538.
529 D. Heinegård and I. Axelsson, J. Biol. Chem., 252 (1977) 1971—1979.
530 H. Muir and T. E. Hardingham, in [75], pp. 375—391.
531 D. Heinegård and V. C. Hascall, J. Biol. Chem., 249 (1974) 4250—4256.
532 L. Rosenberg, H. Choi, S. Pal and L. Tang, in [197], pp. 186—216.
533 M. J. Tanner, in F. Bronner and A. Kleinzeller (Eds.), Current Topics in Membranes and Transport, Vol. 11, Academic Press, New York, 1978, pp. 279—325.
534 V. T. Marchesi, T. W. Tillack, R. L. Jackson, J. P. Segrest and R. E. Scott, Proc. Natl. Acad. Sci. USA, 69 (1972) 1445.
535 R. J. Winzler, in G. A. Jamieson and T. J. Greenwalt (Eds.), Red Cell Membrane, Structure and Function, Lippincott, Philadelphia, 1969, pp. 157—171.
436 M. Tomita, H. Furthmayr and V. T. Marchesi, Biochemistry, 17 (1978) 4756—4770.
437 M. Tomita and V. T. Marchesi, Proc. Natl. Acad. Sci. USA, 72 (1975) 2964—2968.
538 R. J. Winzler, in [35], pp. 226—243.
539 H. Clauser, G. Herman, B. Rossignol and S. Harbon, in [46], pp. 1151—1169.
540 P. Louisot, in [51], pp. 73—79.
541 P. J. O'Brien and E. F. Neufeld, in [46], pp. 1170—1186.
542 H. Schachter and L. Rodén, in W. H. Fischman (Ed.), Metabolic Conjugation and Metabolic Hydrolysis, Vol. 3, Academic Press, New York, 1973, pp. 1—149.
543 H. B. Bosmann, in [61], pp. 893—920.
544 R. G. Spiro, New England J. Med., 288 (1973) 1337—1342.
545 J. Molnar, in [61], pp. 921—935; Mol. Cell Biochem., 6 (1975) 3—14.
546 H. Schachter, in [61], pp. 937—955.
547 H. Schachter, Biochem. Soc. Symp., 40 (1974) 57—71.
548 H. Schachter, Adv. Cytopharmacol., 2 (1974) 207—218.
549 H. Schachter, in M. Elstein and D. V. Parke (Eds.), Mucus in Health and Disease, Plenum, New York, 1977, pp. 103—129.
550 H. Schachter, in [73], pp. 87—181.
551 R. J. Ivatt, Biosystems, 7 (1975) 154—159.
552 H. Schachter and C. A. Tilley, Int. Rev. Biochem., 16 (1978) 209—246.
553 H. Schachter, S. Narasimhan and J. R. Wilson, in [74], pp. 21—46.
554 H. Schachter, S. Narasimhan and J. R. Wilson, in [75], pp. 575—596.
555 J. Sturgess, M. Moscarello and H. Schachter, in F. Bronner and A. Kleinzeller (Eds.), Current Topics in Membranes and Transport, Vol. 11,

Academic Press, New York, 1978, pp. 16—105.
556 R. U. Margolis and R. K. Margolis (Eds.), Complex Carbohydrates of Nervous Tissues, Plenum, New York, 1979.
557 A. P. Corfield and R. Schauer, Biol. Cell., 36 (1979) 213—226.
558 H. Schachter and S. Roseman, in [79], pp. 85—160.
559 R. Gibson, S. Kornfeld and S. Schlesinger, Trends Biochem. Sci., 5 (1980) 290—293.
560 H. Schachter, in J. W. Callahan and J. A. Lowden (Eds.), Lysosomes and Lysosomal Storage Diseases, Raven Press, New York, 1981, pp. 73—93.
561 E. F. Neufeld and W. Z. Hassid, Adv. Carbohydr. Chem., 18 (1963) 309; V. Ginsburg, Adv. Enzymol., 26 (1964) 35—88; S. Kornfeld, R. Kornfeld, E. F. Neufeld and P. J. O'Brien, Proc. Natl. Acad. Sci. USA, 52 (1964) 371—379; L. F. Leloir, Biochem. J., 91 (1964) 1; C. F. Phelps, T. E. Hardingham and P. J. Winterburn, in P. Boulanger, M. F. Jayle and J. Roche (Eds.), Exp. Ann. Biochim. Méd., Vol. 30, Masson, Paris, 1970, pp. 79—95; D. S. Feingold, in [49], pp. 79—112; L. Warren, in [46], pp. 1097—1126; J. L. Strominger, in [36], pp. 375—393.
562 R. Schauer, H. P. Buscher and J. Casals-Stenzel, Biochem. Soc. Symp., 40 (1974) 87—116.
563 N. H. Behrens and L. F. Leloir, Proc. Natl. Acad. Sci. USA, 66 (1970) 153—159.
564 A. Herscovics, S. W. Rostad and R. W. Jeanloz, in [75], pp. 691—700.
565 W. J. Lennarz and M. G. Scher, Bioenergetics, 4 (1972) 441—453; W. J. Lennarz and M. G. Scher, Biochim. Biophys. Acta, 265 (1972) 417—441.
566 F. W. Hemming, Biochem. Soc. Trans., 1 (1973) 1029—1033; 5 (1977) 1223—1231.
567 W. J. Lennarz, Science, 188 (1975) 986—991.
568 C. J. Waechter and W. J. Lennarz, Annu. Rev. Biochem., 45 (1976) 95—112.
569 J. J. Lucas and C. J. Waechter, Mol. Cell. Biochem., 11 (1976) 67—79.
570 A. P. Parodi and L. F. Leloir, Biochim. Biophys. Acta, 1979 (559) 1—37.
571 C. J. Waechter and M. G. Scher, in [556], pp. 75—102.
572 N. H. Behrens, in E. Y. C. Lee and E. E. Smith (Eds.), Biology and Chemistry of Eucaryotic Cell Surfaces, Academic Press, New York, pp. 159—180.
573 R. G. Spiro and M. J. Spiro, in [75], pp. 613—635.
574 D. K. Struck and W. J. Lennarz, in [79], pp. 35—83.
575 N. H. Behrens, A. J. Parodi and L. F. Leloir, Proc. Natl. Acad. Sci. USA, 68 (1971) 2857—2860.
576 F. W. Hemming, Phil. Trans. Roy. Soc. Lond., B, 284 (1978) 559—568.
577 J. Chambers, W. T. Farsee and A. D. Elbein, J. Biol. Chem., 252 (1977) 2498—2506; M. S. Kang, J. P. Spencer and A. D. Elbein, J. Biol. Chem., 253 (1978) 8860—8866.
578 J. S. Schutzbach, J. D. Springfield and J. W. Jensen, J. Biol. Chem.,

255 (1980) 4170—4175.
579 A. Chapman, O. S. Trowbridge, R. Hyman and S. Kornfeld, Cell, 17 (1979) 509—515; A. Chapman, K. Fuzimoto and S. Kornfeld, J. Biol. Chem., 255 (1980) 4442—4446.
580 J. A. Hanover and W. J. Lennartz, J. Biol. Chem., 255 (1980) 3600—3604.
581 R. J. Staneloni, R. A. Ugalde and L. F. Leloir, Eur. J. Biochem., 105 (1980) 275—278.
582 G. C. Rosso, L. DeLuca, C. D. Warren and G. Wolf, J. Lipid Res., 16 (1975) 235—243; W. Sasak and L. DeLuca, in [75], pp. 767—770.
583 I. P. Trayer and R. L. Hill, J. Biol. Chem., 246 (1971) 6666—6675.
584 J. C. Paulson, J. I. Rearick and R. L. Hill, J. Biol. Chem., 252 (1977) 2362—2371.
585 B. Kaufman, S. Basu and S. Roseman, in S. M. Aron and B. W. Volk (Eds.), Proc. 3rd Int. Symp. Cerebral Sphingolipidosis, Pergamon, Oxford, 1966, pp. 193—213.
586 D. M. Carlson, G. W. Jourdian and S. Roseman, J. Biol. Chem., 248 (1973) 5742—5750.
587 B. A. Bartholomew, G. W. Jourdian and S. Roseman, J. Biol. Chem., 248 (1973) 5751—5762.
588 D. M. Carlson, E. J. McGuire, G. W. Jourdian and S. Roseman, J. Biol. Chem., 248 (1973) 5763—5773.
589 J. C. Paulson, W. E. Beranek and R. L. Hill, J. Biol. Chem., 252 (1977) 2356—2362.
590 J. C. Paulson, J. I. Rearick and R. L. Hill, J. Biol. Chem., 252 (1977) 2363—2371.
591 J. E. Dadler, J. I. Rearick, J. C. Paulson and R. C. Hill, J. Biol. Chem., 254 (1979) 4434—4443; 4444—4451; J. E. Sadler, J. I. Rearick and R. L. Hill, J. Biol. Chem., 254 (1979) 5934—5941.
592 R. Morelis and P. Louisot, Biochimie, 55 (1973) 671—677; R. Morelis, P. Broquet and P. Louisot, Biochim. Biophys. Acta, 373 (1974) 10—17.
593 P. Broquet, R. Morelis and P. Louisot, J. Neurochem., 24 (1975) 989—995.
594 P. Louisot, M. Richard and O. Gateau, Ann. Biol. Clin., 34 (1976) 243—253.
595 O. Gateau, R. Morelis and P. Louisot, in [75], pp. 663—668; Eur. J. Biochem., 88 (1978) 613—622; in [76], pp. 285—286.
596 O. Gateau, M. Rocha de Morillo, P. Louisot and R. Morelis, Biochim. Biophys. Acta, 595 (1980) 157—160.
597 M. Richard, A. Martin and P. Louisot, Biochem. Biophys. Res. Commun., 64 (1975) 108—114; G. Berthillier; J. P. Benedetto and R. Got, Biochim Biophys. Acta, 603 (1980) 245—254.
598 B. D. Shur and S. Roth, Biochim. Biophys. Acta, 415 (1975) 473—512.
599 M. Pierce, E. A. Turley and S. Roth, Int. Rev. Cytol., 65 (1980) (in press).
600 A. Verbert, R. Cacan, B. Hoflack and J. Montreuil, in [75], pp. 1091—1094.
601 T. W. Keenan and D. J. Morre, FEBS Lett., 55 (1975) 8—13.

602  R. Cacan, A. Verbert and J. Montreuil, FEBS Lett., 63 (1976) 102—106.
603  A. Verbert, R. Cacan and J. Montreuil, Eur. J. Biochem., 70 (1976) 49—53.
604  B. Hoflack, R. Cacan, J. Montreuil and A. Verbert, Biochim. Biophys. Acta, 568 (1979) 348—356.
605  E. Morel, G. Spik and J. Montreuil, C.R. Acad. Sci. Paris, 282D (1976) 317—320; E. Morel, N. Achy-Sachot, G. Spik and J. Montreuil, FEBS Lett., 69 (1976) 171—174.
606  P. W. Robbins, S. C. Hubbard, S. J. Turco and D. F. Wirth, Cell, 12 (1977) 893—900.
607  A. M. Adamany and R. G. Spiro, J. Biol. Chem., 250 (1975) 2852—2854.
608  D. K. Struck and W. J. Lennarz, J. Biol. Chem., 252 (1977) 1007—1013.
609  J. A. Hanover and W. J. Lennarz, J. Biol. Chem., 254 (1979) 9237—9246.
610  B. Sela, A. Lis and L. Sachs, J. Biol. Chem., 247 (1972) 7575—7590.
611  W. M. Evans, Nature, 250 (1974) 391—395.
612  G. Spik, P. Six and J. Montreuil, Biochim. Biophys. Acta, 584 (1979) 203—215; G. Spik, P. Six, S. Bouquelet, T. Sawicka and J. Montreuil, in [75], pp. 933—936.
613  J. T. Mills and A. M. Adamany, J. Biol. Chem., 253 (1978) 5270—5273.
614  M. M. Carson and W. J. Lennartz, Proc. Natl. Acad. Sci. USA, 76 (1979) 5709—5713.
615  F. W. Hemming, in [76], pp. 232—233.
616  R. Cacan, B. Hoflack and A. Verbert, Eur. J. Biochem., 106 (1980) 473—479.
617  J. F. Wedgwood and J. L. Strominger, J. Biol. Chem., 255 (1980) 1120—1123.
618  H. Eylar, J. Theoret. Biol., 10 (1965) 89—112.
619  H. Sinohara and T. Maruyama, J. Mol. Evol., 2 (1973) 117—122.
620  R. Marshall, Biochem. Soc. Symp., 40 (1974) 17—26.
621  L. T. Hunt and M. O. Dayhoff, Biochim. Biophys. Res. Commun., 39 (1970) 757—765.
622  C. Ronin, S. Bouchilloux, C. Granier and J. Van Rietschoten, FEBS Lett., 96 (1978) 179—182.
623  E. Bause, FEBS Lett., 103 (1979) 296—299; E. Bause and H. Hettkamp, FEBS Lett., 108 (1979) 341—344.
624  P. H. Morgan, H. G. Jacobs, J. P. Segrest and L. W. Cunningham, J. Biol. Chem., 245 (1970) 5042—5048.
625  M. Isemura, T. Ikenaka, T. Mega and Y. Matsushima, Biochem. Biophys. Res. Commun., 57 (1974) 751.
626  J. D. Young, D. Tsuchiya, D. E. Sandlin and M. J. Holroyde, Biochemistry, 17 (1979) 4414—4445.
627  G. W. Hart, K. Brew, G. A. Grant, R. A. Bradshaw and W. J. Lennarz, J. Biol. Chem., 254 (1979) 9747—9753.
628  I. D. Kuntz, J. Amer. Chem. Soc., 94 (1972) 4009—4012.

# REFERENCES

629 V. I. Lin, J. Mol. Biol., 88 (1974) 873—894.
630 P. Y. Chou and G. D. Fasman, Biochemistry, 13 (1974) 211—245; J. Mol. Biol., 115 (1977) 135—175.
631 M-H. Loucheux-Lefebvre and J-P. Aubert, C.R. Acad. Sci. Paris, 282D (1976) 585—587.
632 J-P. Aubert and M-H. Loucheux-Lefebvre, Arch. Biochem. Biophys., 175 (1976) 400—409.
633 J-P. Aubert, G. Biserte and M-H. Loucheux-Lefebvre, Arch. Biochem. Biophys., 175 (1976) 410—418.
634 J. G. Beeley, Biochem. J., 159 (1976) 335—345; Biochem. Biophys. Res. Commun., 76 (1977) 1051—1055.
635 M. H. Loucheux-Lefebvre, L'Actualité Chimique, (1980) 12—27.
636 R. Prasad, B. G. Hudson, R. Butkovski, J. W. Hamilton and K. E. Ebner, J. Biol. Chem., 254 (1979) 10607—10614; [75], pp. 257—259.
637 F. Melchers and P. M. Knopf, Cold Spring Harbor Symp. Quant. Biol., 32 (1967) 255.
638 G. B. Robinson, Biochem. J., 115 (1969) 1077.
639 J. W. Uhr, Cell Immunol., 1 (1970) 228.
640 M. L. Kiely, S. McKnight and R. T. Schimke, J. Biol. Chem., 251 (1976) 5490—5495.
641 L. W. Bergman and W. M. Kuehl, Biochemistry, 16 (1977) 4490—4497.
642 J. E. Rothman and H. F. Lodish, Nature, 269 (1977) 775—780.
643 V. R. Lingappa, J. R. Lingappa, R. Prassad, K. F. Ebner and F. Blobel, Proc. Natl. Acad. Sci. USA, 75 (1978) 2338—2342.
644 J. Kruppa, Biochem. J., 181 (1979) 295—300.
645 C. G. Glabe, J. A. Hanover and W. J. Lennartz, J. Biol. Chem., 255 (1980) 9236—9242.
646 S. Weitzman, M. Grennon and K. Keegstra, J. Biol. Chem., 254 (1979) 5377—5382.
647 I. Jabbal and H. Schachter, J. Biol. Chem., 246 (1971) 5154—5161; J. R. Munro and H. Schachter, Arch. Biochem. Biophys., 156 (1973) 534—542; J. R. Munro, S. Narasimhan, S. Wetmore, J. R. Riordan and H. Schachter, Arch. Biochem. Biophys., 169 (1975) 269—277.
648 J. R. Wilson, D. Williams and H. Schachter, Biochem. Biophys. Res. Commun., 72 (1976) 909—916.
649 P. Stanley, S. Narasimhan, L. Siminovitch and H. Schachter, Proc. Natl. Acad. Sci. USA, 72 (1975) 3323—3327; S. Narasimhan, P. Stanley and H. Schachter, Fed. Proc., 34 (1975) 679; 35 (1976) 1441; J. Biol. Chem., 252 (1977) 3926—3933; N. Harpaz and H. Schachter, J. Biol. Chem., 255 (1980) 4885—4893.
650 N. Harpaz and H. Schachter, J. Biol. Chem., 255 (1980) 4894—4902.
651 A. K. Rao and J. Mendicino, Biochemistry, 17 (1978) 5632—5638.
652 D. H. van den Eijnden, D. H. Joziasse, L. Dorland, H. van Halbeek, J. F. G. Vliegenthart and K. Schmid, Biochem. Biophys. Res. Commun., 92 (1980) 839—845.
653 J. C. Paulson, J-P. Prieels, L. R. Glasgow and R. L. Hill, J. Biol. Chem., 253 (1978) 5617—5624.
654 L. A. Hunt, J. R. Etchison and D. F. Summers, Proc. Natl. Acid. Sci.

USA, 75 (1978) 754—758.
655 I. Tabas, S. Schlesinger and S. Kornfeld, J. Biol. Chem., 253 (1978) 716—722.
656 E. Li, I. Tabas and S. Kornfeld, J. Biol. Chem., 253 (1978) 7762—7770; S. Kornfeld, E. Li and I. Tabas, J. Biol. Chem., 253 (1978) 7771—7778.
657 I. Tabas and S. Kornfeld, J. Biol. Chem., 253 (1978) 7779—7786.
658 E. Li and S. Kornfeld, J. Biol. Chem., 254 (1979) 2754—2758.
659 A. Chapman, E. Li and S. Kornfeld, J. Biol. Chem., 254 (1979) 10243—10249.
660 S. Kornfeld, W. Gregory and A. Chapman, J. Biol. Chem., 254 (1979) 11649—11654.
661 W. W. Chen and W. J. Lennarz, J. Biol. Chem., 253 (1978) 5780—5785; L. S. Grinna and P. W. Robbins, J. Biol. Chem., 254 (1979) 8814—8818; R. A. Ugalde, R. J. Staneloni and L. F. Leloir, Eur. J. Biochem., 113 (1980) 97—103.
662 I. Tabas and S. Kornfeld, J. Biol. Chem., 254 (1979) 11655—11663.
663 H. Schachter, N. Harpaz, S. Narasimhan, D. Williams and G. Longmore, in [76], pp. 305—306.
664 N. Harpaz and H. Schachter, in [76], pp. 307—309.
665 T. A. Beyer, J. I. Rearick, J. C. Paulson, J. P. Prieels, J. E. Sadler and R. L. Hill, J. Biol. Chem., 254 (1979) 12531—12541.
666 J. E. Sadler, J. I. Rearick, J. C. Paulson and R. L. Hill, J. Biol. Chem., 254 (1979) 4434—4443; J. I. Rearick, J. E. Sadler, J. C. Paulson and R. L. Hill, J. Biol. Chem., 254 (1979) 4444—4451.
667 H. Boström and L. Rodén, in [29], pp. 45—80.
668 K. S. Dodgson and A. G. Lloyd, in F. Dickens, P. J. Randle and W. T. Wheelan (Eds.), Carbohydrate Metabolism and its Disorders, Academic Press, New York, 1968, pp. 169—212.
669 L. Rodén, in W. H. Fishman (Ed.), Metabolic Conjugation and Metabolic Hydrolysis, Vol. 2, Academic Press, New York, 1970, pp. 345—442.
670 L. Rodén and N. B. Schwartz, in W. J. Whelan (Ed.), Biochemistry of Carbohydrates, MTP Int. Rev. Sci., Vol. 5, Butterworth, London, 1975, p. 95.
671 L. Rodén and M. I. Horowitz, in [73], pp. 3—71.
672 L. Rodén and N. B. Schwartz, in [60], pp. 73—84.
673 G. W. Jourdain, in [556], pp. 103—125.
674 G. Hart and W. J. Lennarz, J. Biol. Chem., 253 (1978) 5795—5801.
675 N. B. Schwartz and L. Rodén, Carbohydr. Res., 37 (1974) 167—180.
676 M. E. Richmond, S. DeLuca and J. F. Silbert, Biochemistry, 12 (1973) 3898—3903.
677 M. E. Richmond, S. DeLuca and J. E. Silbert, Biochemistry, 12 (1973) 3904—3910.
678 S. DeLuca, M. E. Richmond and J. E. Silbert, Biochemistry, 12 (1973) 3911—3915.
679 C. J. Waechter, J. J. Lucas and W. J. Lennartz, Biochem. Biophys. Res. Commun., 56 (1974) 343—349.
680 S. J. Turco and E. C. Heath, J. Biol. Chem., 252 (1977) 918—923.

# REFERENCES

681 W. T. Butler, in [73], pp. 79—85.
682 K. Kivirikko and L. Ristelli, Med. Biol., 54 (1976) 159.
683 V. Patel and A. L. Tappel, in [45], pp. 133—163.
684 N. N. Aronson, in [46], pp. 1211—1227.
685 V. Patel, in [73], pp. 185—234.
686 A. Gottschalk and E. Buddecke, in [46], pp. 1201—1210.
687 G. Spik, in [716], pp. 31—41.
688 G. Tettamanti and A. Pretti, in [76], pp. 342—343.
689 R. J. Pierce, G. Spik and J. Montreuil, Biochem. J., 180 (1979) 673—676.
690 R. J. Pierce, G. Spik and J. Montreuil, Biochem. J., 185 (1980) 261—264.
691 J. Montreuil, G. Spik, G. Strecker and A. Verbert, unpublished results.
692 E. F. Neufeld, in A. G. Steinberg and A. B. Bearn (Eds), Progress in Medical Genetics. Vol. 10, Grune and Stratton, New York, 1974, pp. 81—101.
693 G. Ashwell and A. G. Morell, in [73], pp. 231—234; Adv. Enzymol., 41 (1974) 99—128.
694 W. S. Sly, in L. Svennerholm, P. Mandel, H. Dreyfus and P. F. Urban (Eds.), Structure and Function of Gangliosides, Plenum, New York, 1980, 433—451.
695 E. F. Neufeld and G. Ashwell, in [79], pp. 241—266.
696 A. G. Morell, R. A. Irvine, I. Sternlieb, I. H. Scheinberg and G. Ashwell, J. Biol. Chem., 243 (1968) 155—159.
697 R. L. Hudgin, W. E. Pricer, G. Ashwell, R. J. Stockert and A. G. Morell, J. Biol. Chem., 249 (1974) 5536—5543; T. Kawasaki and G. Ashwell, J. Biol. Chem., 251 (1976) 5292—5299.
698 C. P. Stowell and Y. C. Lee, Adv. Carbohyd. Chem. Biochem., 37 (1980) 225—281.
699 J. G. Leroy, M. W. Ho, M. C. McBrinn, K. Zielke, J. Jacob and J. S. O'Brien, Pediatr. Res., 6 (1972) 752—757.
700 G. Strecker, J-C. Michalski, J. Montreuil and J-P. Farriaux, Biomedicine Exp., 25 (1976) 238—240; G. Strecker and J-C. Michalski, FEBS Lett., 85 (1978) 20—24; G. Strecker, in L. Svennerholm, P. Mandel, H. Dreyfus and P. F. Urban (Eds.), Structure and Function of Gangliosides, Plenum, New York, 1980, pp. 371—384.
701 G. H. Thomas, G. E. Tiller, L. W. Reynolds, C. S. Miller and J. W. Bace, Biochem. Biophys. Res. Commun., 71 (1976) 188—195.
702 S. Hickman and E. F. Neufeld, Biochem. Biophys. Res. Commun., 49 (1972) 992—999.
703 P. G. Pentchev, J. A. Barranger, A. E. Gal, F. S. Furbish and R. O. Brady, in [74], pp. 150—159.
704 L. H. Rome, B. Weissmann and E. F. Neufeld, Proc. Natl. Acad. Sci. USA, 76 (1979) 2331—2334.
705 S. Hickman, L. J. Shapiro and E. F. Neufeld, Biochem. Biophys. Res. Commun., 57 (1974) 55—61.
706 A. Kaplan, D. T. Achard and W. S. Sly, Proc. Natl. Acad. Sci. USA, 74 (1977) 2026—2030; J. Clin. Invest., 60 (1977) 1088—1093; W. S. Sly

and P. Stahl, Life Sci. Res. Rep., 11 (1978) 229—245; A. Gonzalez-Noriega, J. H. Grubb, V. Talkad and W. S. Sly, J. Cell. Biol., 85 (1980) 839—852; H. D. Fischer, A. Gonzalez-Noriega, W. S. Sly and D. J. Morré, J. Biol. Chem., 255 (1980) 9608—9615.
707 J. Diestler, V. Hieber, G. Sahagian, R. Schmickel and G. W. Jourdain, Proc. Natl. Acad. Sci. USA, 76 (1979) 4235—4239; M. R. Natowicz, N. M. Y. Chi, O. H. Lowry and W. S. Sly, Proc. Natl. Acad. Sci. USA, 76 (1979) 4322—4326; A. Hasilik and E. F. Neufeld, J. Biol. Chem., 255 (1980) 4946—4950; K. von Figura and U. Klein, Eur. J. Biochem., 94 (1979) 347—354; G. Bach, R. Bargal and M. Cantz, Biochem. Biophys. Res. Commun., 91 (1979) 976—981.
708 A. Hasilik, U Klein, G. Strecker and K. von Figura, Proc. Natl. Acad. Sci. USA, 77 (1980) 7074—7078; A. Hasilik, Trends Biochem. Sci., 5 (1980) 237—240; A. Varki and S. Kornfeld, J. Biol. Chem., 255 (1980) 10847—10858.
709 I. Tabas and S. Kornfeld, J. Biol. Chem., 255 (1980) 6633—6639; A. Varki and S. Kornfeld, J. Biol. Chem., 255 (1980) 8398—8401.
710 K. von Figura and E. Weber, Biochem. J., 176 (1978) 943—950; G. D. Vladutiu and M. C. Rattazzi, J. Clin. Invest., 63 (1979) 595—601; W. S. Sly and P. Stahl, in S. Silverstein (Ed.), Transport of Macromolecules in Cellular Systems, Dahlem Konferenzen, Berlin, 1979, pp. 229—244.
711 G. N. Sando and E. F. Neufeld, Cell, 12 (1977) 619—627; K. Ullrich, G. Mersmann, E. Weber and K. von Figura, Biochem. J., 170 (1978) 643—650; E. F. Neufeld, in J. W. Callahan and J. A. Lowden (Eds.), Lysosomes and Lysosomal Storage, Raven Press, New York, 1981, pp. 115—129.
712 H. D. Fischer, A. Gonzalez-Noriega and W. S. Sly, J. Biol. Chem., 255 (1980) 5069—5074.
713 P. Stahl, P. Schlesinger, J. S. Rodman and T. Doebber, Nature, 264 (1977) 86—88; Proc. Natl. Acad. Sci. USA, 73 (1976) 4045—4049; Arch. Biochem. Biophys., 177 (1976) 594—605.
714 P. H. Schlesinger, T. W. Doebber, B. F. Mandell, R. White, C. DeSchryver, J. S. Rodman, M. J. Miller and P. Stahl, Biochem. J., 176 (1978) 103—109.
715 J. W. Spranger and H. Wiedeman, Humangenetik, 9 (1970) 113—139; A. Dorfman and R. Matalon, in J. B. Stanbury, J. B. Wyngaarden and D. S. Fredrickson (Eds.), The Metabolic Basis of Inherited Disease, McGraw-Hill, New York, 1972, pp. 1218—1272; Proc. Natl. Acad. Sci. USA, 73 (1976) 603—637; V. A. McKusick, Heritable Disorders of Connective Tissue, Mosby, St. Louis, 1972, pp. 521—686; C. I. Scott Jr., Prog. Med. Genet., 8 (1972) 243—297; E. F. Neufeld and R. W. Barton, in G. E. Gaull (Ed.), Biology of Brain Dysfunction, Plenum, New York, 1972, pp. 1—30; R. Hirschhorn and G. Weissman, Progr. Med. Genet., 1 (1976) 49—102; J. W. Callahan and J. A. Lowden, Lysosomes and Lysosomal Storage Diseases, Raven, New York, 1981.
716 J-P. Farriaux (Ed.), Les Oligosaccharidoses, Cronan et Roques, Lille, 1977.
717 J. N. Isenberg, in [74], pp. 123—134.

# REFERENCES

718 G. Dawson, in [556], pp. 347—375.
719 N. E. Nordén, A. Lundblad, S. Svensson, P. A. Öckerman and S. Autio, J. Biol. Chem., 248 (1973) 6210—6215.
720 N. E. Nordén, A. Lundblad, S. Svensson and S. Autio, Biochemistry, 13 (1974) 871—874.
721 G. Strecker, B. Fournet, S. Bouquelet, J. Montreuil, J. L. Dhondt and J-P. Farriaux, Biochimie, 58 (1976) 579—586.
722 G. Strecker, T. Hondi-Assah, B. Fournet, G. Spik, J. Montreuil, P. Maroteaux and J-P. Farriaux, C.R. Acad. Sci. Paris, 282D (1976) 671—673.
723 G. Strecker, T. Hondi-Assah, B. Fournet, G. Spik, J. Montreuil, P. Maroteaux, P. Durand and J-P. Farriaux, Biochim. Biophys. Acta, 444 (1976) 349—358.
724 J-C. Michalski, G. Strecker, B. Fournet, M. Cantz and J. Spranger, FEBS-Letters, 79 (1977) 101—104.
725 A. Federico, A. Cecio, G. Apponi Battini, J-C. Michalski, G. Strecker and G. C. Guazzi, J. Neurol. Sci., 48 (1980) 157—169.
726 G. Strecker, J-C. Michalski, M-C. Herlant-Peers, B. Fournet and J. Montreuil, in [75], pp. 945—948.
727 G. Strecker, M-C. Peers, J-C. Michalski, B. Fournet, G. Spik, J. Montreuil, J-P. Farriaux, P. Maroteaux and P. Durand, Eur. J. Biochem., 75 (1977) 391—403.
728 G. C. Tsay, G. Dawson and S. S. Sung, J. Biol. Chem., 251 (1976) 5852—5859.
729 G. Strecker, B. Fournet, G. Spik, J. Montreuil, P. Durand and M. Tondeur, C.R. Acad. Sci. Paris, 284D (1977) 85—88.
730 A. Lundblad, J. Lundsten, N. E. Nordén, S. Sjöblad, S. Svensson, P. A. Öckerman and M. Gehlhoff, Eur. J. Biochem., 83 (1978) 513—521.
731 A. Kobata, K. Yamashita, Y. Tachibana, M. Nishigaki, I. Matsuda and S. Arashima, in [76], pp. 398—399.
732 A. Lundblad, S. Sjöblad and S. Svensson, Arch. Biochem. Biophys., 188 (1978) 130—136.
733 L. S. Wolfe, R. G. Senior and N. M. K. Ng Ying Kin, Fed. Proc., 32 (1973) 484; J. Biol. Chem., 249 (1974) 1828—1838.
734 G. Strecker, unpublished results.
735 G. Strecker and J. Montreuil, Clin. Chim. Acta, 33 (1971) 395—401.
736 N. M. K. Ng Ying Kin and L. S. Wolfe, Biochem. Biophys. Res. Commun., 59 (1974) 837—844.
737 R. J. Pollit and K. M. Pretty, Biochem. J., 141 (1974) 141—146.
738 A. Lundblad, P. Masson, N. E. Nordén, S. Svensson, P. A. Öckerman and J. Palo, Eur. J. Biochem., 67 (1976) 209—214.
739 G. Strecker, unpublished results.
740 M. Nishigaki, T. Muramatsu and A. Kobata, Biochem. Biophys. Res. Commun., 59 (1974) 638—645.
741 G. Strecker, Glycoprotéines et Glycoprotéinoses, in J-P. Farriaux (Ed.), Les Oligosaccharides, Cronan et Roques, Lille, 1977, pp. 13—30; oligosaccharides in Lysosomal Diseases, in J. W. Callahan and J. A. Lowden (Eds.), Lysosomes and Lysosomal Storage Diseases, Raven, New York,

1981, pp. 95—113.
742 P. J. Winterburn and C. F. Phelps, Nature, 236 (1972) 147—151.
743 H. Faillard and R. Schauer, in [46], pp. 1246—1267.
744 A. Rosenberg and C. L. Schengrund (Eds.), Biological Roles of Sialic Acids, Plenum, New York, 1976.
745 A. Meager and R. C. Hughes, Virus Receptors, in P. Cuatrecasas and M. F. Greaves (Eds.), Receptors and Recognition, Ser. A, Vol. 4, Chapman and Hall, London, 1977, pp. 141—196.
746 E. Köttgen, C. Bauer, W. Reutter and W. Gerok, Klin. Wochenschr., 57 (1979) 151—179; 199—214.
747 R. D. Marshall, Biochem. Soc. Trans., 7 (1979) 800—805.
748 A. Gottschalk, Nature, 222 (1969) 452—454.
749 R. T. Schwartz and R. Datema, Trends Biochem. Sci., 5 (1980) 65—67; D. A. White (Ed.), Uses of Tunicamycin, Biochem. Soc. Trans., 8 (1980) 163—171.
750 A. Takatsuki, K. Arima and G. Tamura, J. Antibiotics, 24 (1971) 215—223.
751 J. S. Tkacz and J. O. Lampen, Biochem. Biophys. Res. Commun., 65 (1975) 248—257.
752 A. Heifetz, R. W. Keenan and A. D. Elbein, Biochemistry, 18 (1979) 2186—2192.
753 R. K. Keller, D. Y. Boon and F. C. Crum, Biochemistry, 18 (1979) 3946—3952.
754 E. W. Silverton, M. A. Navia and D. R. Davies, Proc. Natl. Acad. Sci. USA, 74 (1977) 5140—5144.
755 J. Deisenhofer, P. M. Colman, O. Epp and R. Huber, Z. Physiol Chem., 357 (1975) 1421—1434; R. Huber, Trends Biochem. Sci., 1 (1976) 174—178; R. Huber, J. Deisenhofer, P. M. Colman, M. Matsushima and W. Palm, Nature, 264 (1976) 415.
756 N. Korde, M. Nose and T. Muramatsu, Biochem. Biophys. Res. Commun., 75 (1977) 838.
757 S. C. Kuo and J. O. Lampen, Biochem. Biophys. Res. Commun., 58 (1974) 287—295; H. R. Onishi, J. S. Tkacz and J. O. Lampen, J. Biol. Chem., 254 (1979) 11943—11952.
758 A. Hasilik and W. Tanner, Antimicrob. Agents Chemother., 253 (1976) 402—410.
759 H. Neumann, U. Zehavi and T. D. Tanksley, Biochem. Biophys. Res. Commun., 36 (1969) 151.
760 P. Jollès, in [46], pp. 782—809.
761 R. Gibson, S. Schlesinger and S. Kornfeld, J. Biol. Chem., 254 (1979) 3600—3607.
762 K. Meyer, in E. A. Balazs (Ed.), Chemistry and Molecular Biology of the Intercellular Matrix, Vol. 1, Academic Press, New York, 1970, pp. 5—24.
763 A. Gottschalk and S. Fazekas de St. Groth, Biochim. Biophys. Acta, 43 (1960) 513; Y. Hashimoto, S. Tsuiki, K. Nisizawa and W. Pigman, Ann. N.Y. Acad. Sci., 106 (1963) 233.
764 H. Faillard and W. Pribilla, Lin. Wochenschr., 42 (1964) 686; H. Faillard,

Blut, 19 (1969) 238.
765 Y. P. Loh and H. Gainer, FEBS Lett., 96 (1978) 269—272.
766 J. R. Vandenheede, A. I. Ahmed and R. E. Feeney, J. Biol. Chem., 247 (1972) 7885—7889.
767 K. Olden, P. M. Pratt, C. Jaworski and K. M. Yamada, Proc. Natl. Acad. Sci. USA, 76 (1979) 791—795.
768 M. Peterson and C. P. Leblond, J. Biophys. Biochem. Cytol., 21 (1964) 143.
769 D. K. Struck, P. B. Siuta, M. D. Lane and W. J. Lennartz, J. Biol. Chem., 253 (1978) 5332—5337.
770 R. K. Keller and G. D. Swank, Biochem. Biophys. Res. Commun., 85 (1978) 762—768.
771 A. Mizraki, J. A. O'Malley, W. A. Carter, A. Takatsuki. G. Tamura and E. Sulkowski, J. Biol. Chem., 253 (1978) 7612—7615.
772 M. L. Tanzer, F. N. Rowland, L. W. Murray and J. Kaplan, in [75], pp. 817—818; T. J. Housley, F. N. Rowland, P. W. Ledger, J. Kaplan and M. L. Tanzer, J. Biol. Chem., 255 (1980) 121—128.
773 M. Monsigny, C. Kieda and A. C. Roche, Biol. Cell., 36 (1979) 239—300.
774 L. Glaser, Trends Biochem. Sci., 1 (1976) 84—86.
775 R. J. Winzler, Intern. Rev. Cytol., 23 (1970) 77—125.
776 R. J. Winzler, in [46], pp. 1268—1293.
777 G. M. W. Cook and R. W. Stoddart (Eds.), Surface Carbohydrates of the Eukaryotic Cell, Academic Press, New York, 1973.
778 E. Y. C. Lee and E. E. Smith (Eds.), Biology and Chemistry of Eucaryotic Cell Surfaces, Academic Press, New York, 1974.
779 A. Rosenberg and C. L. Schengrund (Eds.), Biological Role of Sialic Acid, Plenum, New York, 1976.
780 R. W. Jeanloz and J. F. Codington, in [779], pp. 201—238.
781 B. R. Brinkley and K. P. Porter (Eds.), Intern. Cell. Biol., Rockefeller University Press, 1977.
782 H. Popper, L. Bianchi and W. Reutter (Eds.), Membrane Alterations as Basis of Liver Injury, MTP Press, Lancaster, 1977.
783 F. Bronner and A. Kleinzeller (Eds.), Current Topics in Membranes and Transport, Vol. 11, Academic Press, New York, 1978.
784 M. C. Glick and H. Flowers, in [73], pp. 337—384.
785 R. E. Harmon (Ed.), Cell Surface Carbohydrate Chemistry, Academic Press, New York, 1978.
786 M. Monsigny and J. Schrével (Eds.), Membrane Glycoconjugates, Biol. Cell., 36 (1979) 209—330.
787 L. Warren, C. A. Buck and G. P. Tuszinski, Biochim. Biophys. Acta, 516 (1978) 97—127.
788 P. H. Atkinson and J. Akimi, in [79], pp. 191—239.
789 P. Stanley, in [79], pp. 161—189.
790 K. D. Noonan, in [783], pp. 397—461.
791 J. Pouysségur, M. Willingham and I. Pastan, Proc. Natl. Acad. Sci. USA, 74 (1977) 243—247.
782 S. Roth, E. J. McGuire and S. Roseman, J. Cell Biol., 51 (1971) 525—

535; 536—547.
793 E. J. McGuire, in C. F. Fox (Ed.), First ICN-UCLA Symposium on Molecular Biology, Membrane Research, Academic Press, New York, 1972, pp. 347—368.
794 S. Roth, Quart. Rev. Biol., 48 (1973) 541—563.
795 B. D. Shur and S. Roth, Biochim. Biophys. Acta, 415 (1975) 473—512.
796 R. C. Greig and M. N. Jones, Biosystems, 9 (1977) 43—55.
797 W. Frazier and L. Glaser, Annu. Rev. Biochem., 48 (1979) 491—523.
798 S. Roseman, in [778], pp. 317—354.
799 L. A. Culp, in [783], pp. 327—396.
800 J. M. Jancik and R. Schauer, Z. Physiol. Chem., 355 (1974) 395—400.
801 J. M. Jancik, R. Schauer, K. H. Andres and M. von Düring, Cell Tiss. Res., 186 (1978) 209—226.
802 T. Bachi, J. E. Deas and C. Howe, in G. Poste and G. L. Nicolson (Eds.), Cell Surface Reviews, Virus Infection and the Cell Surface, Vol. 2, North-Holland, Amsterdam, 1977, pp. 83—127.
803 J. C. Paulson, J. E. Sadler and R. L. Hill, J. Biol. Chem., 254 (1979) 2120—2124.
804 J. R. Bartles and W. A. Frazier, J. Biol. Chem., 255 (1980) 20—38.
805 R. E. Hausman and A. A. Moscona, Proc. Natl. Acad. Sci. USA, 72 (1975) 916—920.
806 R. O. Hynes, Biochim. Biophys. Acta, 458 (1976) 73—107.
807 K. M. Yamada and I. Pastan, Trends Biochem. Sci., 1 (1976) 222—224.
808 M. Mosesson, Thrombos. Haemostas., 38 (1977) 742—750.
809 F. Grinnell, Int. Rev. Cytol., 53 (1978) 65—144.
810 A. Vaheri and D. F. Mosher, Biochem. Biophys. Acta, 516 (1978) 1—25.
811 K. M. Yamada and K. Olden, Nature, 275 (1978) 179—184.
812 R. C. Hughes, G. Mills and Y. Courtois, Biol. Cell., 36 (1979) 321—330.
813 R. O. Hynes, A. T. Destree, M. E. Perkins and D. D. Wagner, J. Supramol. Struct., 11 (1979) 95—104.
814 H. K. Kleinman, A. T. Hewitt, J. C. Murray, L. A. Liotta, S. I. Rennard, J. P. Pennypacker, E. B. McGoodwin, G. R. Martin and P. H. Fishman, J. Supramol. Struct., 11 (1979) 69—78.
815 K. Olden, R. M. Pratt and K. M. Yamada, Cell, 13 (1978) 461—473; Proc. Natl. Acad. Sci. USA, 76 (1979) 3343—3347.
816 S. D. J. Pena and R. C. Hughes, Nature, 276 (1978) 80—83.
817 T. D. Butters, V. Devalia, J. D. Aplin and R. C. Hughes, J. Cell Sci., 44 (1980) 33—58.
818 J. W. Baynes and F. Wold, J. Biol. Chem., 251 (1976) 6016—6024; J. F. Day, R. W. Thornburg, S. R. Thorpe and J. W. Baynes, J. Biol. Chem., 255 (1980) 2360—2365.
819 R. W. Thornburg, J. F. Day, J. W. Baynes and S. R. Thorpe, J. Biol. Chem., 255 (1980) 6820—6825.
820 M. C. Glick, in [74], pp. 404—411.

Chapter 2

# Immunoglobins and Histocompatibility Antigens
# Their Structure, Function and Metabolism

DAVID I. STOTT and ALAN R. WILLIAMSON*

*Department of Bacteriology and Immunology, and the Department of Biochemistry, University of Glasgow, Glasgow, Scotland (U.K.)*

## Immunoglobulins and Histocompatibility Antigens

Higher organisms possess a complex, highly efficient, specific defence system against invading organisms such as bacteria, viruses, acellular and multicellular parasites. Immunology is the study of this defence mechanism and may be arbitrarily divided into two categories: (1) cellular immunology, viz. the study of the response of the cells of the immune system (lymphocytes, monocytes, macrophages and granulocytes), (2) humoral immunology, viz. the study of the production of antibodies (immunoglobulins) in response to an antigenic stimulus. Immunochemistry is the study of the immune system by biochemical methods, in particular the biochemistry of the proteins of the immune system, viz. immunoglobulins, histocompatibility antigens and the proteins of the complement system. The latter has been dealt with separately (Ch. 6, Vol. 19B/I) and this chapter will consider only the properties of the immunoglobulins and products of the major histocompatibility complex (MHC), a set of genes with many control functions in immune response.

---

* Present address: Glaxo Group Research Ltd., Greenford Road, Greenford, Middlesex UB6 0HE (U.K.).

## TABLE I
### Physicochemical properties of human immunoglobulins

| | IgG | IgM | IgA | IgD | IgE |
|---|---|---|---|---|---|
| Subclasses | 1–4 | 1, 2? | 1, 2 | — | — |
| Heavy chain | $\gamma 1$–$\gamma 4$ | $\mu 1, \mu 2$ | $\alpha 1, \alpha 2$ | | |
| Light chain | $\kappa$ or $\lambda$ | $\kappa$ or $\lambda$ | $\kappa$ or $\lambda$ | $\kappa$ or $\lambda$ | $\kappa$ or $\lambda$ |
| Other chains | — | $J^a$ | J, S.P.[a] | — | — |
| Formula | $\gamma_2 \kappa_2$ or $\gamma_2 \lambda_2$ | $(\mu_2 \kappa_2)_5$ J or $(\mu_2 \lambda_2)_5$ J | $\alpha_2 \kappa_2, \alpha_2 \lambda_2,$ $(\alpha_2 \kappa_2)_2$ J, S.P.[a] $(\alpha_2 \lambda_2)_2$ J, S.P.[a] | $\delta_2 \kappa_2$ or $\delta_2 \lambda_2$ | $\epsilon_2 \kappa_2$ or $\epsilon_2 \lambda_2$ |
| Sedimentation coefficient (S) | 7 | 19 | 7, 11 | 7 | 8 |
| $M_r$ | (1, 2, 4) 146 000 (IgG3) 170 000 | 970 000 | 160 000 or 400 000 | 184 000 | 188 000 |
| $M_r$ of heavy chain | ($\gamma 1$, 2, 4) 51 000 ($\gamma 3$) 60 000 | 65 000 | ($\alpha 1$) 56 000 ($\alpha 2$) 52 000 | 60–70 000 | 72 500 |
| $M_r$ of light chain | 23 000 | 23 000 | 23 000 | 23 000 | 23 000 |
| $M_r$ of S.P. | — | — | 60 000 | — | — |
| $M_r$ of J chain | — | 15 000 | 15 000 | — | — |
| No. of heavy chain domains | 4 | 5 | 4 | 4? | 5 |
| Carbohydrate (%) | 2–3 | 12 | 7–11 | 9–14 | 12 |
| Serum concentration (mg/ml) | 8.0–16 | 0.5–2.0 | 1.5–4.0 | 0.003–0.4 | $2 \cdot 10^{-5}$–$5 \cdot 10^{-4}$ |

[a] J, J chain; S.P., Secretory Piece (not present in myeloma IgA).

# INTRODUCTION

## 1. Immunoglobulins

### 1.1. Introduction

The immunoglobulins were first recognised as a group of serum proteins, though they also occur in all internally and externally secreted body fluids. Their importance is that they function as antibodies and as such are secreted in large amounts in response to stimulation by macromolecular antigens (immunogens) and bind directly to the inducing antigen. There are many immunoglobulins, both normal and pathological e.g. myeloma proteins, for which the stimulating immunogen and hence the antibody activity is not known, but it is assumed that every immunoglobulin has antibody activity. Immunoglobulins are found in the serum and secretions of all groups of vertebrates from the Agnatha (hagfish and lampreys) to mammals. They are divided into five classes (Table I) according to their molecular structure and serological reactions. IgG and IgA are further divided into subclasses. The distribution and biological properties of the immunoglobulins vary with class. Thus, IgA is found predominantly in the external secretions of the alimentary, respiratory and reproductive tracts and mammary, lachrymal and salivary glands. IgG is the predominant immunoglobulin in the blood and tissue fluids and is able to cross the placenta in man. Both IgG and IgA are transmitted to the newborn in colostrum and milk. This is the sole mode of transfer of antibodies in ruminants, since the placenta is impermeable to immunoglobulins of all classes. There are 10 classes or subclasses of Ig in man. The number of subclasses varies and some classes are absent or as yet unidentified in other species. IgM is largely confined to the blood stream and is extremely efficient in fixing complement, which plays an important role in phagocytosis and lysis of bacteria. Certain subclasses of IgG fix complement though less efficiently and IgA only fixes complement by the alternate pathway.

Little is known about the function of IgD but it has recently been discovered on the surface of lymphocytes in conjunction with IgM. Serum levels of IgD are extremely low. IgE is also present at very low concentrations in serum and is responsible for allergic reactions such as hay fever and asthma. The benefits conferred by IgE are not clear but IgE antibodies are thought to be

*References p. 315*

involved in rejection of parasitic worms, particularly helminths.

One of the major problems of immunology has been to understand how the immunoglobulins, having such an apparently limited heterogeneity defined by the classes and subclasses, can bind specifically to such an enormous range of antigens. An antigenic determinant, or epitope, is defined as that part of an antigen molecule that occupies the binding site of the antibody and it has been estimated that an individual mammal (e.g. mouse) can respond to as many as $10^6$—$10^7$ epitopes [1]. As a result of the work carried out by many groups over the past 15 years, an understanding of the diverse specificity and genetics of antibodies in terms of their molecular structure, has emerged. Other properties, such as complement fixation and binding to cell surfaces are beginning to yield to the onslaught though it may be some time before these properties are fully explainable in molecular terms. The results of these studies have shown that immunoglobulins are very curious molecules, uniquely adapted to their function. The majority of this work would have been impossible without the discovery that the paraprotein found in the sera of patients with multiple myelomatosis (myeloma) is identical with antibody except that, in most cases, its antibody activity is unknown. The myeloma proteins are secreted by plasma cells which have resulted from the neoplastic clonal proliferation of a single plasma cell. Since these proteins are homogeneous clonal products, as opposed to the extremely heterogeneous population of normal immunoglobulins, they have greatly facilitated studies on the structure, genetics and biosynthesis of immunoglobulins. A further boost to the study of immunoglobulins was provided by the discovery that it is possible to induce plasmacytomas, analogous with the human myelomas in certain inbred strains of mice [3]. These tumours can be maintained indefinitely by serial transplantation between syngeneic inbred mice and have provided the means whereby most of the studies of immunoglobulin biosynthesis and, recently, direct analysis of gene structure have been carried out.

*1.2. Structure of immunoglobulins*

*1.2.1. Polypeptide chain structure.* Immunoglobulins, in order to function as antibodies, exhibit extensive diversity in their specific binding function. This variation in specificity is based entirely on the primary structure of the immunoglobulins (see below). However, within each class (or subclass) of immunoglobulin all antibodies are sufficiently similar to be regarded in gross structural terms as a single group of proteins.

*1.2.1.1. IgG.* The structure of IgG is considered first, not only because it was the first class of immunoglobulin for which the structure was solved, but also because it provides the basic structure upon which the model for the more complex classes of immunoglobulins is built. The elucidation of the multichain structure of IgG is largely due to the efforts of Porter, using rabbit IgG, and Edelman, using human IgG [3—5]. By digesting IgG with papain, Porter showed that the molecule could be split into three fragments: two identical monovalent antigen-binding fragments (Fab), and one crystallisable fragment (Fc) which did not possess antibody activity. Digestion with pepsin resulted in one bivalent antigen binding fragment F(ab)$_2$. By reducing IgG ($M_r$ = 150 000) with mercaptoethanol followed by gel filtration in dissociating medium (6M urea), Edelman showed that the molecule was composed of two heavy chains ($M_r$ approx. 50 000) and two light chains ($M_r$ approx. 23 000). As a result of these and other studies Porter proposed the four chain structure of IgG (Fig. 1). The proteolytic fragments were shown to be the result of cleavage of the molecule at the positions shown in Fig. 2.

Subsequent studies showed that there are two types of disulphide bridges involved in stabilising the structure of the molecule.

(1) Interchain bridges: these connect the light and heavy or two heavy chains together and vary in both number (inter-heavy chain bridges) and position according to the class and subclass of immunoglobulin (Fig. 3). (2) Intrachain bridges: these connect cysteine residues at different positions on the same polypeptide chain; each intrachain bridge spans an approximately similar-sized stretch of polypeptide (50—70 amino acid residues, Fig. 1). Unlike the interchain bridges, the positions of the intrachain bridges vary remarkably little between different subclasses of IgG; this conservation is due to the preservation of domain structure (see below).

*1.2.1.2. IgA.* Human serum IgA (sedimentation coefficient 7S) has the same basic four-chain structure as IgG except that the heavy

*References p. 315*

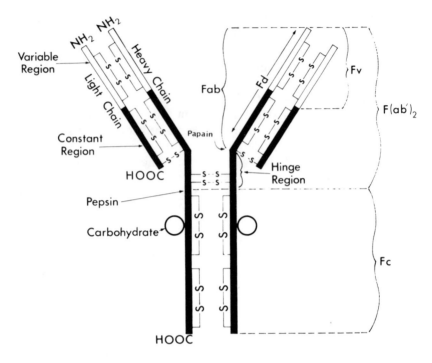

Fig. 1. Four-chain model of human IgG1 showing the arrangement of intra- and inter-chain disulphide bridges, sites of cleavage by proteolytic enzymes and fragments.

($\alpha$) chains differ serologically and in amino acid sequence from those of IgG ($\gamma$ chains). The interchain bridges also differ, particularly in the human IgA$_2$ subclass and Balb/c mouse IgA, in both of which the light chains are disulphide-bonded to each other and only bound to the α chains by noncovalent interaction (Fig. 4). Thus, the L chains are released as a dimer (L$_2$) by dissociating agents. IgA molecules readily form dimers and higher oligomers. In some other species such as the dog and mouse, the serum IgA is predominantly in the oligomeric form, mainly dimer.

Secretory IgA also exists mainly as a dimer (sedimentation coefficient 11S) and to a lesser extent as higher oligomers. The subunits are linked together by the C-termini of the Fc and an extra polypeptide chain ($M_r$ = 15 000) called J chain is attached to the cysteine residue at the penultimate position of the α chain. It is

# STRUCTURE OF IMMUNOGLOBULINS

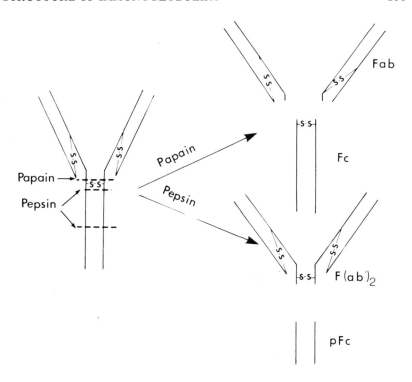

Fig. 2. Cleavage of rabbit IgG by papain and pepsin. The Fc region is normally extensively degraded during pepsin digestion but a pepsin Fc fragment (pFc'), corresponding to the $C_H3$ domain (Section 1.2.4.2) can be isolated under mild conditions.

not clear whether it links the two subunits together, although this is generally assumed (Fig. 4). It has been proposed that J chain may play a role in the intracellular polymerisation of IgA (Section 1.4.6.2.).

A fourth polypeptide, secretory or transport piece, is also bound to dimeric secretory IgA. This has an $M_r$ of 60 000 and is bound by disulphide bridges to the subunits. This interaction possibly involves an extra pair of cysteine residues on the $C_H2$ domain. Approx. 20% of human IgA molecules have noncovalently bound secretory piece [6] whereas this proportion is 60% for rabbit IgA. Unlike J chain, secretory piece is not synthesised by the plasma cells which synthesise IgA, but by epithelial cells present

References p. 315

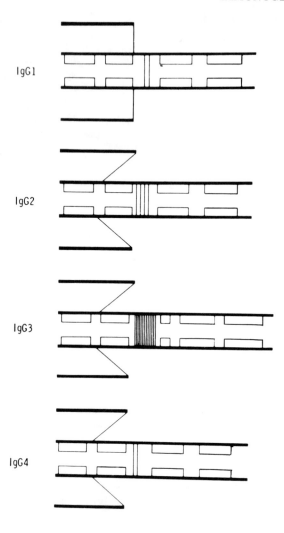

Fig. 3. Four-chain structure of human IgG subclasses showing the arrangement of intra- and inter-chain disulphide bridges. The number of inter-heavy chain bridges in IgG3 is not certain but may be as high as 15. Reproduced from [317] by permission of John Wiley and Sons.

## STRUCTURE OF IMMUNOGLOBULINS

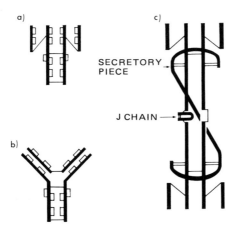

Fig. 4. Structure of human IgA. (a) IgA1 and IgA2.m(2). (b) IgA2.m(1). This molecule has no L-H disulphide bridges. (c) Hypothetical model of secretory IgA. The precise arrangement of disulphide bridges linking J chain and secretory piece to the α chains is not known.

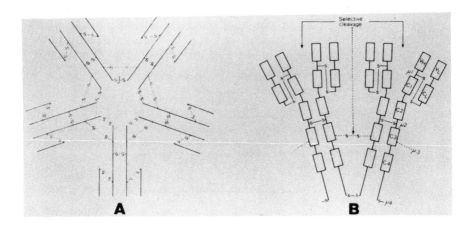

Fig. 5. Diagram of the structure of IgM. (A) Pentameric IgM showing the probable arrangement of intra- and inter-subunit disulphide bridges. (B) Sites of cleavage of disulphide bridges by selective reduction. Reproduced from (9).

References p. 315

in the organ which secretes the IgA. The secretory piece becomes attached to IgA after release from the plasma cells and during the process of external secretion. For a more detailed review of IgA structure and function see [7].

*1.2.1.3. IgM.* Human IgM exists chiefly as a cyclic pentamer of $M_r$ = 970 000 in which one pair of subunits is linked by a single J chain (Fig. 5a). In other species cyclic tetramers or hexamers are known.

The heavy ($\mu$) chain is larger than that of IgG, possessing an extra domain, and is rich in carbohydrate. Selective reduction of human IgM with very low concentrations of dithiothreitol produces a pentameric IgM molecule in which most of the $\mu 2$ and $\mu 3$ disulphide bridges are cleaved but the molecule is held together by noncovalent interaction between the heavy chains. On polyacrylamide gel electrophoresis in sodium dodecyl sulphate (SDS) the molecules dissociate into $H_2L_2$ units plus some HL units and oligomers. Comparison of human IgM with mouse IgM, in which bridge $\mu 3$ appears to be absent, suggests that bridge $\mu 4$ links the subunits and it follows that the $H_2L_2$ units produced by selective reduction are "inside out" compared with the natural subunits which are held together by strong noncovalent interaction [8, 9] (Fig. 5b). Since J chain is bound to $\mu 5$ [10, 11], in this model it links two subunits rather than linking two heavy chains within a single subunit, as in earlier models of IGM in which $\mu 4$ was regarded as an intrasubunit bridge. The J chains of human IgA and IgM are indistinguishable by amino acid composition and antigenic specificity. The arrangement of disulphide bridges, positions of carbohydrate groups (each

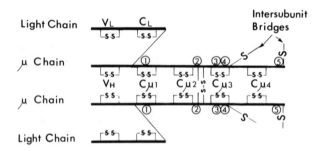

Fig. 6. Four-chain model of a human IgM subunit showing the arrangement of disulphide bridges, numbered positions of carbohydrate groups (O) and homology regions (Section 1.2.2.2).

one linked to an asparagine residue) and the homology regions (section 2(b)(ii)) are shown in Fig. 6.

*1.2.1.4. IgD and IgE.* These classes of immunoglobulin both have a four chain structure of the IgG type. The heavy $\epsilon$ chain of IgE has five domains (one variable and four constant) but that of human IgD ($\delta$) probably has only four, as $\gamma$ and $\alpha$, with an extended hinge region, although this has not yet been confirmed*. IgE has an extra intra-chain bridge in the $C_H 1$ domain, as also found in rabbit IgG, and interchain bridges at each domain (Fig. 7). Both immunoglobulins are rich in carbohydrate.

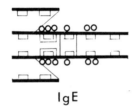

IgE

Fig. 7. Four-chain model of human IgE. O = carbohydrate.

*1.2.2. Primary structure.* The major difficulty in determining the primary structure of immunoglobulins was the extensive heterogeneity of either normal immunoglobulins or even elicited antibodies specific for a single antigen. This problem was overcome by sequencing the monoclonal immunoglobulins present as paraproteins in the sera of patients with multiple myeloma. The normal immunoglobulin levels are suppressed in this disease and large amounts of the myeloma protein, secreted by the malignant plasma cells, appear in the blood giving a characteristic spike on electrophoretic patterns. The myeloma protein can be readily purified for sequencing. Furthermore, many of these patients excrete Bence-Jones protein in their urine. This protein is a dimer of the immunoglobulin light chains, produced in excess and secreted by the plasma cells, and is also easily purified and sequenced.

*1.2.2.1. Variable and constant regions.* When the first Bence-Jones

---

* See Note added in proof, p. 314.

*References p. 315*

proteins from different patients were sequenced by Hilschman and Craig, the remarkable observation was made that, whereas the C-terminal part (107 residues) was identical save for a valine/leucine interchange at position 191, the N-terminal part (105—115 amino acid residues) of the two light chains differed at many positions. As more sequences were completed a pattern of variability in the N-terminal half with a constant C-terminal sequence was established. A similar situation was also found to be true for the heavy chains, the variable region occupying the N-terminal 105—115 residues and the remaining three quarters of the heavy chain being essentially constant for a given class and subclass. In addition to differences in the sequence, variations in the length of the variable region that could be accounted for by deletions or insertions were noted. The observation that a single polypeptide chain possesses a constant and a variable region resulted in the hypothesis that one polypeptide chain could be the product of two genes, posing several interesting genetic problems (Section 1.4.9). There is now good evidence that this is the case for the variable and constant regions of both the light and heavy chains of immunoglobulin.

Our picture of the antibody molecule thus must include the fact that each Fab fragment contains two N-terminal variable (V-) regions, one from the light chain and one from the heavy chain (Fig. 1). This naturally led to the hypothesis that the two variable regions form the antigen binding site.

*1.2.2.2. Variable region families, groups and subgroups.* The sequences of variable regions can be classified into three major families according to their constant region association. Thus we can recognise $\kappa$, $\lambda$ and heavy chain variable region sequence families. This implies that variable region sequences in the heavy chain family are shared between all heavy-chain constant regions. There is extensive evidence consistent with this idea for the $\gamma$, $\mu$ and $\alpha$ constant regions. This sharing of variable regions poses interesting questions concerning gene translocation (Section 1.4.10).

Comparison of the sequence of many variable regions of the same family shows that they may be divided into groups or subgroups. The simplest definition of a group or subgroup is that its members have greater similarity in sequence with each other than with all other variable regions. In this way, human $\kappa$ variable regions ($V_\kappa$) sequences have been divided into three subgroups, human $V_\lambda$ sequences have been divided into five subgroups and $V_H$ sequences into three subgroups. Mouse $\kappa$ chain sequences have been most extensively studied and the number of subgroups necessary to account for the data has increased linearly with accumulated sequences. By contrast the $\lambda$ chains have only two $V_\lambda$ subgroups (possibly sub-families since they apparently correspond to the two $C_\lambda$ types). The number of subgroups has been postulated to represent the minimum number of V-region genes. More recently murine $V_\kappa$ sequences have been arranged in groups on the basis of the sequence of the N-terminal peptide extending to $CySH_{23}$ with approx. 30 groups being recognised (Table II) [12]. Where many sequences were determined in a single group it is possible to subdivide them further into subgroups. The nomenclature is at present confusing since the mouse groups are equivalent to the human subgroups but it is probable that the latter will eventually also be named groups to bring them into line with the mouse system. For one group, $V_{\kappa 21}$, an antiserum has been raised against group-specific determinants and antiserum used to select more examples of $V_{\kappa 21}$ proteins [13]. This group has been divided into subgroups $V_{\kappa 21A}$, $V_{\kappa 21B}$ and $V_{\kappa 21C}$ on the basis of sequence differences.

Mouse $V_H$ sequences have similarly been divided into subgroups, some of which have been further subdivided (Table III).

The genetic origin of immunoglobulin variability is an interesting problem that has caused a great deal of controversy. There are two main opposing theories: the germ-line theory states that all the variable regions are coded by separate genes inherited in the germ line, while the somatic theory states that there are only a limited number of germ-line genes and further variability is generated by somatic mutation. It is not proposed to delve into the arguments for or against the theories here but the interested reader is referred to recent reviews on the subject [1, 14—17].

*1.2.2.3. Hypervariable regions.* Wu and Kabat [18] devised a plot of the degree of variability at each position in the $\kappa$ light chain variable region against distance along the chain. An updated version of their plot using all the available sequence data for human light and heavy chains is shown in Fig. 8. This clearly demonstrates that the variability is not random throughout the varible region sequences but there are three well defined regions which show very pronounced variability; these statistically determined areas which

*References p. 315*

TABLE II

Balb/c Mouse $V_L$ Regions Classified by $NH_2$-$Cys_{23}$ Peptide Sequence[a]

| VK Isotype: | 1 | 2 | 3 | 4 | 5 | 6 | 7 | 8 | 9 | 10 | 11 | 12 | 13 | 14 | 15 | 16 | 17 | 18 | 19 | 20 | 21 | 22 | 23 |
|---|---|---|---|---|---|---|---|---|---|---|---|---|---|---|---|---|---|---|---|---|---|---|---|
| | D | I | V | M | T | Q | S | P | A | S | L | S | V | S | L | G | E | R | V | T | I | T | C |
| VK-1  |   | V |   |   |   |   | T |   | L |   |   | T |   |   |   |   | D |   | A | S |   | S |   |
| VK-26 |   | V | L |   |   |   | T |   | L |   |   | P |   |   |   |   | D | Q | A | S |   | S |   |
| VK-2  |   | V |   |   |   |   | T |   | L |   |   |   |   | T |   |   |   | P | A | S |   | S |   |
| VK-3  |   | V |   | V |   |   | T | G | L |   |   | P |   |   |   |   | D |   |   | S |   | S |   |
| VK-4  | E |   |   | L |   |   |   |   |   | I | T | A | A |   | I |   |   | K |   |   |   |   |   |
| VK-5  | E | N |   | L |   |   |   |   |   | I | M |   | A |   | M |   |   |   |   |   | L |   |   |
| VK-6  | E | V |   | L |   | Z |   |   |   | I | M |   | A |   |   |   | Q |   |   |   |   |   |   |
| VK-7  |   |   |   |   |   |   |   |   | S |   |   | Q | A |   | I |   | L |   |   | S | M | S |   |
| VK-8  |   |   |   |   |   |   |   |   | S |   | M |   |   |   | A |   |   |   |   |   |   | S |   |
| VK-9  |   |   | Q |   |   |   | T | T | S |   |   |   | A |   | A |   | D | K |   | S | M | S |   |
| VK-10 |   |   | Q |   |   |   | T | T | S |   |   |   | A |   | A |   | D | R |   |   | M |   |   |
| VK-11 |   |   | Q |   | I |   |   | T | S |   |   |   | A |   | A |   | D |   |   |   | L | S |   |
| VK-12 |   |   | Q |   |   |   |   |   | S |   |   |   | A |   | V |   |   |   |   |   |   |   |   |
| VK-13 |   |   | Q |   |   |   |   | P | B | Y |   | A | A |   | V |   |   | T |   |   | M |   |   |
| VK-14 | N |   |   |   |   |   |   | T | K | F | M | T |   |   | V |   | ( ) | T |   | S |   |   |   |
| VK-15 |   |   |   |   |   |   |   |   | K |   | M | M | T |   | V |   | G |   |   |   | L |   |   |
| VK-16 |   |   |   |   |   |   |   | S | S | E | L |   |   |   | V |   | G |   |   |   |   |   |   |

STRUCTURE OF IMMUNOGLOBULINS 203

| | | | | | | | | | | | | | | | | | | | | | |
|---|---|---|---|---|---|---|---|---|---|---|---|---|---|---|---|---|---|---|---|---|---|
| VK-17 | | | | | | | Q | S | F | M | T | V | | D | S | S |
| VK-18 | | | | | | | | S | | M | A | I | | D | S | S |
| VK-19 | N | | Q | | | | H | K | F | F | T | V | | B | S | • |
| VK-20 | E | T | T | V | | | | | | M | M | A | | | | S |
| VK-21 | | | | I | | | | | | | A | I | K | | | S |
| VK-22 | | | | | | | | T | | A | T | A | S | Q | A | |
| VK-23 | | | | L | | B | E | L | D | P | T | P | K | K | | S |
| VK-24 | | | | I | | B | E | L | K | P | | S | K | D | | S |
| | | | | I | | A | A | F | N | P | | S | S | | | L |
| VK-25 | E | | | | | | | | | | T | | S | T | A | S |
| Vλ Isotype: | P | A | V | V | T | Q | Q | S | A | • | L | T | S | P | G | E | T | V | T | L | T | C |
| Vλ-1 | | | | | | | | | | | | | | | | | | | | | | |
| Vλ-2 | | | | | E | E | | | | | | | | G | | | | | | | | |

[a] ( ), amino acid not identified; •, no amino acid at this position; VK-isotypes circumscribed by boxes are related. The continuous letter sequences across the table are hypothetical sequences based on the most common amino acid found at that position in mouse kappa and lambda chains. Reproduced from [12] by permission of Academic Press.

*References p. 315*

TABLE III
V_H-domain isotypes, Balb/c mouse[a]

| VH Isotype: | 1 | 2 | 3 | 4 | 5 | 6 | 7 | 8 | 9 | 10 | 11 | 12 | 13 | 14 | 15 | 16 | 17 | 18 | 19 | 20 | 21 | 22 | 23 | 24 | 25 | 26 | 27 |
|---|---|---|---|---|---|---|---|---|---|---|---|---|---|---|---|---|---|---|---|---|---|---|---|---|---|---|---|
|  | E | V | K | L | V | E | S | G | G | P | L | V | Q | P | G | G | S | L | K | L | S | C | A | A | S | G | F |
| VH-1A |  |  |  |  | L |  |  |  |  |  |  |  |  |  |  |  |  |  |  |  |  |  |  |  |  |  |  |
| VH-1B |  |  |  |  | L |  |  |  |  |  |  |  |  |  |  |  |  |  |  |  |  |  |  |  |  |  |  |
| VH-1C |  |  |  |  | L |  |  |  |  |  |  |  |  |  |  |  |  |  |  |  |  |  |  |  |  |  |  |
| VH-1D |  |  |  |  | L |  |  |  |  |  |  |  |  | L |  |  |  |  |  |  |  |  |  |  |  |  |  |
| VH-2 |  |  | Q |  | V |  | T |  |  |  |  |  |  |  |  |  |  |  |  |  |  |  |  |  |  |  |  |
| VH-3 |  |  |  |  | V |  |  |  |  |  |  |  | Z |  | K |  |  |  |  |  |  |  |  |  |  |  | — |
| VH-4A |  |  |  |  | V |  |  |  |  |  |  |  |  |  |  |  |  | R |  |  |  |  |  | T |  | T | — |
| VH-4B |  |  |  |  | V |  |  |  |  |  |  |  |  |  |  |  |  | R |  |  |  |  |  | T |  |  |  |
| VH-4C |  |  |  |  | V |  |  |  |  |  |  |  |  |  |  |  |  | R |  |  |  |  |  | T |  |  |  |
| VH-4D |  |  |  |  | V |  |  |  |  |  |  |  |  |  |  |  |  | R |  |  |  |  |  | T |  |  |  |
| VH-4E |  |  |  | V | E |  |  |  |  |  |  |  |  |  |  |  |  |  |  |  |  |  |  |  |  |  |  |
| VH-5 |  |  |  |  | V |  |  |  |  |  |  |  |  |  |  |  |  | M |  |  |  |  |  |  |  |  |  |
| VH-6 | D | Q |  |  |  |  |  |  |  |  |  |  |  |  |  |  |  | R |  |  |  |  | V |  |  |  |  |
| VH-7 |  | Q | Q | Q | Q |  |  | P | E | P |  | K |  |  | A |  |  | M | M |  |  | K |  | V |  | Y |  |
| VH-8 | D | Q | Q | Q | Q |  |  | P | E | P |  | K | K |  | S | Q | V | S | S | T | T | S | S | V | F | Y | Y |
| VH-9 |  |  |  |  | Q |  |  |  | P | S |  |  |  |  | S | Z | V | S |  |  |  |  |  | V | T |  | S |
| VH-10 |  |  | Q |  | Q |  |  |  | T | V |  |  | A | R |  | S |  |  | M |  |  |  |  |  |  |  |  |

[a] The continuous letter sequence across the top of the table is hypothetical and based on the most common amino acid found at that position. Dash means no sequence determined. Reproduced from [12] by permission of Academic Press.

appear to allow for more variability than elsewhere in the V region are termed the hypervariable regions. Similar comparative analyses of heavy-chain variable region sequences [19, 20] also revealed three (or possibly four) hypervariable regions. Certain of these light and heavy chain hypervariable regions have been shown by affinity labelling and X-ray crystallography to form part of the antigen-binding site (Sections 1.2.3.4 and 1.2.4.2). For this reason hypervariable residues are now sometimes referred to as complementarity-determining residues. The stretches of sequence surrounding the hypervariable regions have been termed framework regions since they contain relatively constant amino acids which are thought to provide a rigid framework within which high variability can occur.

*1.2.2.4. The constant region.* Sequences of IgG heavy and light chains showed that the chains could be divided into homology regions of approx. 110 amino acids, each containing one intrachain disulphide bridge and showing 30% sequence homology between each section with the exception of the variable region. This led to the "domain hypothesis" in which Edelman and Gall [21] proposed that immunoglobulin heavy and light chains are composed of repeating globular units or domains showing structural similarity. This has now been confirmed by X-ray crystallography (Section 1.2.4.2)

Each light-chain type and subtype, and each heavy chain class and subclass, originally defined by their antigenic differences, has its own characteristic constant region present in all normal individuals. These are termed isotypes. In the case of the heavy chains, the constant region determines the class of immunoglobulin and its biological properties.

*1.2.2.5. Light-chain types.* As previously indicated, there are two light-chain types, $\kappa$ and $\lambda$ each having a particular C-region sequence. The homology between human and mouse $\kappa$ C-regions is much greater than between human $\kappa$ and $\lambda$ C-regions, indicating early divergence of the two genes during evolution. The reason for two light-chain types is not clear since there appears to be no difference in the range of specificities or biological properties of the two types. The ratio of $\kappa:\lambda$ varies considerably between species, being 2:1 in man compared with about 95% $\kappa$ in the adult mouse and almost 100% in the horse, with no obvious differences in the immune response.

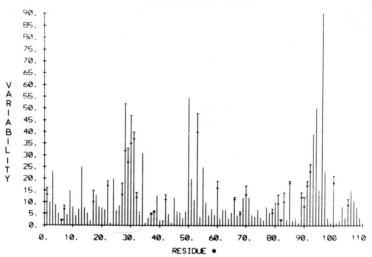

ALL HUMAN LIGHT CHAINS

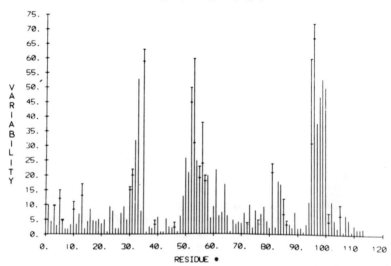

ALL HUMAN HEAVY CHAINS

Three subtypes of human λ chains present in all normal individuals have been detected. These consist of the antigenically defined $Oz^+$ and $Oz^-$ subtypes which have been correlated with lysine ($Oz^+$) or arginine ($Oz^-$) at position 191 [22], and independent variation at position 153 due to glycine ($Kern^+$) or serine ($Kern^-$) [23]. Three of the possible four combinations have been found [24]. No subtypes of human κ chains have been observed.

Studies on mouse myeloma λ chains have revealed the existence of two subtypes $\lambda_1$ and $\lambda_2$, with 27% difference between them in the C-region [25]; by contrast 60% of the amino acid residues of $\lambda_2$ differ from mouse κ chain. $\lambda_1$ is the more common subtype. Each of these light-chain types and subtypes is believed to be the product of separate, non-allelic genes, κ and λ residing on different chromosomes.

*1.2.2.6. Heavy-chain classes and subclasses.* The immunoglobulin heavy chains are divided into classes, determined by the constant region sequence. Human and most mammalian immunoglobulin heavy chains can be classified as α, γ, μ, δ and ε, corresponding to IgA, IgG, IgM, IgD and IgE, respectively. Thus, IgA is either $\alpha_2 \kappa_2$ or $\alpha_2 \lambda_2$, IgG is $\gamma_2 \kappa_2$ or $\gamma_2 \lambda_2$ etc. In addition, most mammals possess isotype immunoglobulins with small differences in the heavy-chain constant region within a given class. These are the subclasses. For example in man IgG has four recognised subclasses, IgG1—4 determined by the heavy chains γ1-γ4. Similarly, the two known subclasses of human IgA, IgA1 and IgA2, are determined by the heavy chains α1 and α2. These subclasses show differences in interchain disulphide bridges (Figs. 3 and 4). An exception to this scheme appears to be the rabbit in which no subclasses of IgG have so far been found. The class and subclass of the heavy chains determine the biological properties of the immunoglobulin (Section 1.3). As a result of the homology observed between the heavy-chain classes and subclasses, and linkage data from studies of allotypic markers (Section 1.2.2.8) the heavy-chain C-region genes

---

Fig. 8. Histogram of variability* of amino acids in the variable region of immunoglobulin (a) light chains, (b) heavy chains. Extra amino acids occurring in some chains have been omitted where indicated by arrows in order to facilitate the comparison.

*Variability = $\dfrac{\text{Number of different amino acids occurring at a given position}}{\text{Frequency of the most common amino acid at that position}}$.

Reproduced from [318] with permission.

are believed to have arisen by duplication and mutation to form a tandem array of genes on the same chromosome as the heavy-chain variable region genes. The $\kappa$ and $\lambda$ light-chain C-region genes are believed to be arranged in a similar manner [15].

*1.2.2.7. Carbohydrate.* All the immunoglobulins are glycoproteins, carbohydrate being covalently bound to the constant region of the heavy chain, but not the light chain. Occasionally, there is glycosylation of the variable region of myeloma heavy or light chains but this appears to be due to the fortuitous occurrence of the recognition sequence for the attachment enzyme. The quantity of bound carbohydrate varies between the classes from 3% (IgG) to 12% (IgM, IgD and IgE) as shown in Table I, and may affect the biological properties of the immunoglobulin. There is also considerable variation in carbohydrate content between individual IgG1 myeloma proteins but no significant difference between subclasses.

The carbohydrate consists of oligosaccharides attached either to asparagine residues, through an $N$-glycosidic linkage with $N$-acetylglucosamine, or to serine or threonine residues through an $O$-glycosidic linkage with $N$-acetylgalactosamine.

The major carbohydrate group of the $\gamma$ chain is attached to an asparagine residue in the $C_H2$ domain [26], a second group being present in the hinge region in some species [27, 28]. The $\mu$ chain of IgM has five carbohydrate groups, also attached to asparagine residues as shown in Fig. 6 [29]. IgA1 has five O-linked carbohydrate groups in the hinge region [30]. Four of the carbohydrate groups are identical and are present in a region consisting of two identical runs of six amino acids, probably the result of a duplication of part of the gene [31]. This region is absent from IgA2. Secretory piece contains 11.6% carbohydrate and the $\alpha$ chains of secretory IgA appear to have a higher carbohydrate content than serum IgA. J chain also contains a small amount of carbohydrate [32].

The recognition sequence for attachment of oligosaccharides via $N$-acetylglucosamine appears to be provided by the simple sequences:

| | |
|---|---|
| Tyr-Asx-Thr-Ser | (myeloma protein Cor) |
| Tyr-Asx-Ser-Thr | (myeloma protein Eu) |
| Phe-Asx-Ser-Thr | (rabbit Fc) |
| Glu-Asx-Ile-Ser | (mouse light chain) |

which bear a resemblance to each other [33]. These sequences are probably necessary but not sufficient for recognition by the enzyme responsible for attachment of the carbohydrate moiety.

Three types of oligosaccharide are found in immunoglobulins [34, 35]: Type 1 is connected to serine or threonine by an O-glycosidic linkage with N-acetylgalactosamine and it contains one N-acetylneuraminic acid residue. This is the type of oligosaccharide found in the hinge region of IgA1. Type 2 and type 3 are linked to asparagine by an N-glycosidic bond with N-acetylglucosamine. Type 2 has a highly branched structure and generally consists of one fucose, three mannose, two galactose, three or four N-acetyl glucosamine residues and 0—2 sialic acid residues. The type 3 oligosaccharide is less branched than the type 2 and is composed of 4—11 mannose residues and 2—3 N-acetylglucosamine residues. Human γ chains contain a single type 2 oligosaccharide, μ chains contain both type 2 and type 3 moieties [36], α and possibly δ chains contain all three types, and ε chains contain type 2 and type 3 moieties [37].

*1.2.2.8. Allotypes.* Allotypes are allelic genetic markers of immunoglobulins inherited in a simple Mendelian manner. The prevailing interpretation is that allotypes are due to structural polymorphism of genes coding for each individual isotype, although there are now reservations concerning this interpretation in view of the deviations from simple allelic behaviour of certain rabbit allotypes that may show that the allelic genes are control elements rather than structural genes (see below). For a more detailed review of allotypes than can be presented here see [38, 39]. Many allotypic specificities have been found to correlate with amino acid sequence differences, in some cases due to a substitution of a single amino acid residue, in immunoglobulins from humans, rabbit and other species. Allotypic specificities are detected by alloantisera (i.e. antisera raised in a different member of the same species. Human anti-allotype sera are obtained from multiparous females with offspring of a different allotype. Leakage of foetal blood across the placenta results in sensitisation of the mother against the foetal allotype. Oudin raised anti-allotype sera in rabbits by injection of immune complexes formed by antibodies from one rabbit into a second rabbit. If the immunoglobulin in the complexes was of a different allotype to that of the second rabbit, anti-allotype antibodies were produced.

Human immunoglobulin allotypes have been detected by the extremely sensitive technique of inhibition of agglutination, although in the rabbit they have been detected by immune precipitation. Precipitation by anti-allotype sera is consistent with the presence of multiple amino acid differences between allotypes, since at least three antigenic determinants are usually required to form large enough immune complexes for precipitation to occur and a single amino acid substitution can result in only two new antigenic determinants (one on each chain). For certain rabbit allotypes amino acid sequence analysis has shown that alleles differ by multiple substitutions.

Human $\kappa$ chains can exist in any one of three allotypic forms: Km1, Km2, and Km3 (InV1, InV2 or InV3 in the old nomenclature). These can be correlated with amino acid substitution at positions 191 and 153, similar positions to the Oz markers of the $\lambda$ chain. Unlike the isotypic Oz markers, the InV markers are inherited as alleles. The residues determining the allotype are exposed at the bends of neighbouring loops in the polypeptide chain, hence their antigenicity, as shown in Fig. 9.

Human $\gamma$ chains exist as a large number of allotypes classified as the Gm system. Most Gm types are restricted to a single heavy chain subclass. Other Gm markers exist in two or more subclasses and have been correlated with amino acid differences at the homologous position(s). For example, G1m(a) is found only on $\gamma_1$ chains, whereas the homoallelic marker (nG1m(a)) is found on $\gamma_1$, $\gamma_2$ and $\gamma_3$ (Table IV). This finding is consistent with $\gamma_1$, $\gamma_2$ and $\gamma_3$ being the result of tandem gene duplications, the G1m(a) mutation having occurred after the duplication events.

Human IgA2 exists in two allotypic forms: A2m(1) and A2m(2). It is the A2m(1) molecules in which the light chains are bound together by a disulphide bridge and are not covalently bound to the heavy chains (Section 1.2.1.2).

Allotypes have been located on the light-chain constant region and both the heavy-chain constant and variable regions of rabbit IgG (Fig. 10). Many of these have been correlated with simple amino acid substitutions, for example, the heavy-chain constant region allotypes d11 and d12 correlate with methionine and threonine, respectively, at position 225 in the hinge region. The heavy-chain variable region allotypes appear to be more complex. There are four of these at the a locus, a1, a2, a3 and $\bar{a}$. These

STRUCTURE OF IMMUNOGLOBULINS

TABLE IV

Probable structural location of some human γ- and κ-chain allotypes [317]

| Allotype | Chain | Homology region | Position | Amino acid(s) |
|---|---|---|---|---|
| G1m(a) | $\gamma_1$ | $C_\gamma 3$ | 355—358 | Arg — *Asp* — Glu — *Leu* |
| nG1m(a) | $\gamma_1, \gamma_2, \gamma_3$ | $C_\gamma 3$ | 355—358 | Arg — Glu — Glu — Met |
| G1m(f) | $\gamma_1$ | $C_\gamma 1$ | 214 | Arg |
| G1m(z) | $\gamma_1$ | $C_\gamma 1$ | 214 | Lys |
| nG4m(a) | $\gamma_1, \gamma_3, \gamma_4$ | $C_\gamma 2$ | 309 | Val — Leu — His |
| nG4m(b) | $\gamma_2, \gamma_4$ | $C_\gamma 2$ | 309 | Val — His |
| G3m(g) | $\gamma_3$ | $C_\gamma 2$ | 296[a] | Tyr |
| nG3m(g) | $\gamma_2, \gamma_3$ | $C_\gamma 2$ | 296[a] | Phe |
| G3m(b°) | $\gamma_3$ | $C_\gamma 3$ | 436[a] | Phe |
| nG3m(b°) | $\gamma_1, \gamma_2, \gamma_3$ | $C_\gamma 3$ | 436[a] | Tyr |
| Km1 (In V1) | | $C_\kappa$ | 153, 191 | Val, Leu |
| Km2 (In V2) | | $C_\kappa$ | 153, 191 | Ala Leu |
| Km3 (In V3) | | $C_\kappa$ | 153, 191 | Ala Val |

[a] Correlative sequence differences observed in incompletely sequenced chains.
Reproduced by permission of John Wiley and Sons.

have been correlated with sequence differences at several positions in the heavy-chain variable region [38]. The most important evidence relating to these allotypes is that they are found on γ, μ, α and ε chains; this was the first indication that the heavy chain variable region genes are shared between the constant region genes [40, 42].

Recent evidence indicates that the a and b locus rabbit allotypes are not true alleles since markers previously thought to be mutually exclusive have been shown to coexist in the same animal, although one allotype is always present at a very much lower concentration than the other. For this reason it has been suggested that these allotypes may represent control genes, the genes for all allotypic variants being present in each individual, their expression being regulated by the control genes.

*1.2.2.9. Idiotypes.* An idiotype is an antigenic specificity which is unique (greek: *idios*) to a particular antibody or immunoglobulin from an individual animal. Idiotypes were first observed and named by Oudin [43]. The experiment was as follows: (1) injection of a

*References p. 315*

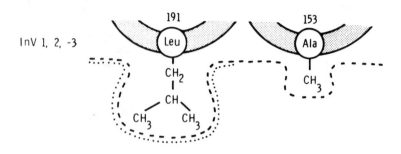

(a)

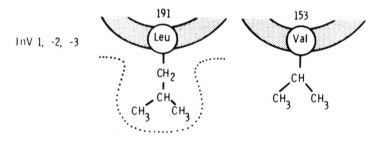

(b)

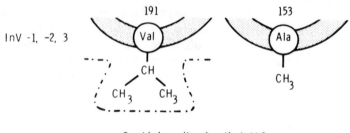

(c)

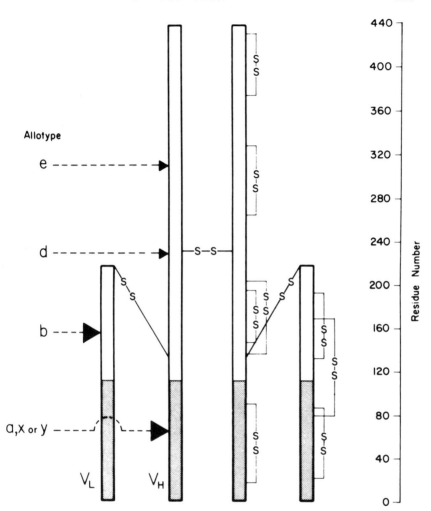

Fig. 10. Four-chain model of rabbit IgG showing the positions of amino acid substitutions that have been correlated with allotypic markers. Note the extra intra-chain disulphide bridges linking $V_L$ and $C_L$, and in the $C_H1$ domain. Reproduced from [39] by permission of John Wiley and Sons.

Fig. 9. In V or Km antigenic determinants. The antigenic determinant In V1 (Km1) is formed by a leucine residue at position 191 (a and b), whereas InV2 is formed by both the leucine at position 191 and an alanine residue at position 153 (a). InV3 (Km3) is due to the substitution of a valine at position 191 and may be independent of residue 153, although this is not certain (c). Reproduced from [317] by permission of John Wiley and Sons.

*References p. 315*

rabbit (A) with bacteria; (2) isolation of the antibodies produced as antibody-bacteria complexes; (3) injection of these complexes into a second rabbit (B) chosen to match the allotype of rabbit A. The antibodies produced by B were found to react only with the original antibacterial antibodies of rabbit A against which they were raised. No reaction was observed with normal immunoglobulin of rabbit A or with antibodies, to identical bacteria, raised in other rabbits. Thus, the antibodies against a given antigen produced by each individual possess a unique antigenic specificity and Oudin postulated that these idiotypic specificities were not inheritable. Slater et al. [44] and later Kunkel et al. [45] demonstrated similar individual antigenic specificities in human myeloma immunoglobulins by extensive cross-absorption of antisera raised in a heterologous species against individual immunoglobulins.

Anti-idiotypic antisera raised against an anti-hapten antibody will inhibit the binding of antibody to the hapten and conversely, hapten inhibits the binding of anti-idiotype to the anti-hapten, demonstrating that the idiotypic specificity resides in the antigen-binding site, probably in the hypervariable regions. Idiotypes have been used extensively as genetic markers for variable region genes, particularly in the mouse [46, 47].

### 1.2.3. The antigen-binding site

*1.2.3.1. Valency.* The valency of an antibody is the number of sites per molecule capable of combining specifically with antigen. The valency of IgG has been determined to be two by measurement of the ratio of bound antigen to antibody in soluble complexes formed in large antigen excess [48], and by measurement of the ratio of bound hapten molecules to antibody by equilibrium dialysis [49–52]. A hapten is a small molecule or group (e.g., dinitrophenyl lysine; DNP-Lys) which occupies only part of a single antigen-binding site. A hapten is chosen to avoid errors in equilibrium dialysis measurements due to steric hindrance which can result with higher $M_r$ antigens and lead to underestimation of valency. It was for this reason that some experiments with IgM gave valencies varying from 10 to as low as 2 depending on the size of antigen used, although, from the pentameric structure, one would expect the valency to be 10. Definitive measurements with anti-hapten IgM antibodies indicated a valency of 10 [53—55]. Evidence for steric hindrance came from studies showing that the valency of

IgM antidextran appears to be inversely related to the $M_r$ of the antigen [56].

The effect of multiple binding sites on antibody molecules is two-fold. (1) It makes possible the cross-linking of antigens, resulting in the immobilisation of bacteria and formation of large complexes which are readily phagocytosed by polymorphonuclear leukocytes and macrophages. (2) It results in much tighter binding of an antibody molecule to a polyvalent antigen, such as a bacterium or virus, with their repeated carbohydrate or capsid proteins, e.g. the association constant of IgG anti-DNP for bacteriophage coupled with multiple DNP groups was $10^4$ times greater than for the free hapten [57]. With IgM antibodies this effect can be even greater.

*1.2.3.2. Location of the antigen-binding site.* The discovery of the variable and hypervariable regions of the light and heavy chains led to the idea that these regions are responsible for the formation of the antigen-binding site, since only these regions of the molecule could produce the enormous number of conformations required to bind the large number (probably at least $10^6 - 10^7$) of different antigenic determinants (epitopes). Proof of the involvement of the variable regions was provided by Inbar et al. [58] who succeeded in isolating, by enzymic digestion, a fragment (Fv) of an anti-DNP myeloma protein which contained the heavy and light chain variable regions but only a few residues from the constant region (Fig. 2). This fragment was shown to have a similar binding capacity to that of Fab.

Further evidence for the location of the antigen-binding site has been provided by affinity labelling experiments, electron microscopy and X-ray crystallography (q.v. Sections 1.2.3.4, 1.2.4.1 and 1.2.4.2).

*1.2.3.3. Size and shape of the antigen-binding site.* Several groups have studied the size of the antigen-binding site by observing the minimum size of a hapten/oligomer that will give optimal inhibition of precipitation of a polymeric antigen. The first experiments of this type were carried out by Kabat [59—61] who showed that maximal inhibition of human antidextran serum was induced by the hexasaccharide isomaltohexaose, isomaltoheptaose showing little, if any, further inhibition. Smaller oligomers were less effective (Fig. 11). From these results it was deduced that the antigen-binding site is in the form of a groove measuring $3.4 \times 1.2$ nm based on the extended dimensions of isomaltohexaose. Similar

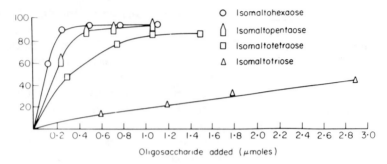

Fig. 11. Inhibition of precipitation of dextran and human antidextran by isomaltose oligosaccharides. Reproduced from Experimental Immunochemistry, Kabat and Mayer, 2nd ed., 1961.

studies with poly-L-lysine [62, 63] and poly-L-alanine [64—66] indicated that the binding site would accommodate the tetra- or pentapeptide, whereas the octa- or dodecapeptide were required for maximal inhibition of anti-silk fibroin [67]. The range of dimensions of the antigen-binding site estimated from these studies is in the region of 2.5—3.6 × 1.0—1.7 × 0.6—0.7 nm. It seems probable that the shape and size of the antigen-binding site will vary with that of the antigen. Cisar et al. [68] have demonstrated the existence of two types of antibodies against $\alpha(1\text{—}6)$ dextrans in rabbit and human antisera and mouse IgA myelomas. One group of antibodies precipitated branched dextrans and bound but did not precipitate a linear dextran. These antibodies exhibited specificity for the terminal non-reducing ends of the dextran molecules. The second group of antibodies precipitated both branched and linear dextrans and bound to non-terminal locations on the $\alpha(1\text{—}6)$dextran chains. It was proposed that the first type of antibody has an antigen-binding site that holds the terminal glucose residue in a fixed location in a partially enclosed space, whereas the antigen-binding site of the second type of antibody is in the form of a shallow groove or depression that is able to bind to the non-terminal regions of the molecule.

The importance of antigen conformation was originally shown by the observation of Landsteiner [69] that antibodies against native proteins do not react with the same protein in its denatured state and v.v. This was elegantly confirmed by Crumpton [70] in an experiment in which antiserum against apomyoglobin was added

STRUCTURE OF IMMUNOGLOBULINS 217

to a solution of metmyoglobin which exists in the equilibrium:
Metmyoglobin ⇌ Apomyoglobin + haem.

The antibody precipitated out the apomyoglobin resulting in a white precipitate, leaving free haem in solution. Thus the antibody was able to recognise the apomyoglobin conformation, shifting the equilibrium to the right by removal of apomyoglobin from the solution. An alternative explanation is that the antibody stabilised a conformational change in the metmyoglobin, allowing the haem to escape.

*1.2.3.4. Affinity labelling.* Attempts have been made by means of affinity-labelling reagents to identify the amino acid residues forming the antigen-binding site. The general approach is to immunise with a hapten coupled to a protein. The anti-hapten antibody is purified by affinity chromatography and allowed to react with a radiolabelled derivative of the hapten containing a potentially reactive group. The latter is then activated, ideally under mild conditions, resulting in a covalent bond between the bound hapten and a neighbouring amino acid (Fig. 12). The heavy and light chains are then prepared, split into peptides by chemical or enzymic action and the radiolabelled peptide is identified. One of the first classes of affinity labelling reagent to be used was based on aromatic diazonium derivatives; examples are *m*- and *p*-nitrobenzene diazonium chloride which bind to anti-*m*-nitrophenyl and anti-dinitrophenyl (DNP) antibodies respectively, and *p*-diazobenzene arsonate, which binds to anti-azobenzene arsonate

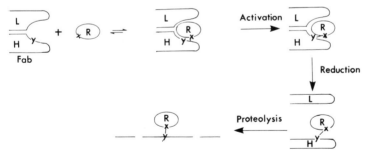

Fig. 12. Diagrammatic representation of the method of affinity labelling of the antigen-binding site. A radioactive reagent R with an active group x binds reversibly to the antigen-binding site. Upon activation a covalent bond is formed between x and a neighbouring amino acid residue y. After reduction and separation of heavy and light chains the amount of reagent R bound to each chain may be determined and the peptide to which R is bound identified following proteolysis and separation of the peptides.

*References p. 315*

## TABLE V
### Affinity-labelling systems [71]

| Species of antibody | Antigenic determinant | Labelling reagent | Molar ratio of label (H:L chain) |
|---|---|---|---|
| Rabbit | HO₃As—⟨⟩—N=N—(Tyr)(His)(Lys) | HO₃As—⟨⟩—N₂⁺ | 2.1 |
| Rabbit | (CH₃)₃N⁺—⟨⟩—N=N—(Tyr)(His)(Lys) | (CH₃)₃N⁺—⟨⟩—N₂⁺ | 1.5 |
| Rabbit | O₂N—⟨⟩—NH(CH₂)₄—CH— (NO₂) | O₂N—⟨⟩—N₂⁺ | 1.3 |
| Rabbit | ⟨⟩(NO₂)—N=N—(Tyr)(His)(Lys) | ⟨⟩(NO₂)—N₂⁺ | 2.4 |
| Sheep | ⟨⟩(NO₂)—N=N—(Tyr)(His)(Lys) | ⟨⟩(NO₂)—N₂⁺ | 4.5 |
| Guinea pig IgG1 | ⟨⟩(NO₂)—N=N—(Tyr)(His)(Lys) | ⟨⟩(NO₂)—N₂⁺ | 2.5 |

| | | | |
|---|---|---|---|
| Guinea pig IgG2 | [structure: nitrobenzene-N=N-(Tyr)(His)(Lys)] | [structure: nitro-aminobenzene $-N^+_2$] | 3 |
| Mouse | [structure: nitrobenzene-N=N-(Tyr)(His)(Lys)] | [structure: nitro-aminobenzene $-N^+_2$] | 2 |
| Pig | [structure: nitrobenzene-N=N-(Tyr)(His)(Lys)] | [structure: nitro-aminobenzene $-N^+_2$] | 2 |

*References p. 315*

(ABA) antibodies. These reagents generally label tyrosine in or adjacent to the hypervariable regions of heavy and light chains [71], the heavy chains being labelled to higher specific activity (Table V).

Bromoacetyl derivatives have also been used [72], e.g. bromo-acetyl-DNP-ethylene diamine (BADE) and bromoacetyl-DNP-lysine (BADL). These reagents react with a wider variety of amino acid side chains than the diazonium compounds. Another advantage of these reagents is that the distance of the reactive groups from the haptenic groups can be varied. When these were reacted with a mouse IgA myeloma protein (MOPC 315) with anti-DNP activity, it was found that BADL labelled a lysine residue in the heavy chain whereas BADE labelled a tyrosine in the first hypervariable region of the light chain [72]. Since MOPC 135 is a homogeneous protein, unlike antibodies induced by antigen, this showed that both heavy and light chain participate in the formation of the antigen-binding site on the same molecule. Using a series of such reagents it was found that the ratio of label in lysine to that in tyrosine increased with the length of the reagent suggesting that the distance between the two labelled residues might be equal to the difference in length of the side chains of BADE and BADL, viz. 0.5 nm. Givol et al. [73], therefore, synthesised a bifunctional reagent DIBAB: DNP $NH(CH_2)_2CH(-NHCOCH_2Br)CO(NH)_2COCH_2Br$, in which the reactive groups were separated by 0.5 nm. This reagent cross-linked the heavy and light chains via the lysine and tyrosine, confirming the previous observation. It is necessary, however, to make the reservation that the conformation adopted by these reagents when in the combining site is not known, so the distances are only approximate.

Problems with such reagents are that they only label tyrosine, lysine or histidine residues, the activation conditions might alter the conformation of the binding site and, since the reactive group is not part of the haptenic group, they do not label the residues in contact with the hapten. Also, being continuously active, they may label sites close to but not in the combining site during entry or exit from the site. Fleet et al. [74] devised a reagent which overcomes many of these problems. This reagent is the photo-activatable hapten 2-nitro-azidophenyl (NAP) lysine. Coupled to a protein NAP may be used as an antigen. The active group of NAP is part of the hapten and occupies the binding site. On activation by light NAP is converted to a nitrene which can react with any C—H group and therefore with any amino acid residue. Converse and Richards [75, 76] have synthesised a similar reagent, a DNP-diazoketone that can be photoactivated to a carbene or ketene. With such reagents it is generally found that the labelled residues were in or near the hypervariable regions, confirming that these take part in the formation of the antigen-binding site. However, the high reactivity of these photoactivatable affinity reagents result in heterogeneity of substitution products and difficulty in identification of contact amino acids. For this reason, Smith and Knowles [77] synthesised an azulene reagent that is photoactivated to a reagent with a much shorter half-life than NAP and the DNP-diazoketone. This has the advantage that there is less possibility of residues in the vicinity of the antigen-binding site becoming labelled by activated molecules in the vicinity of, or diffusing out of the site.

More detailed information on the structure of the antigen-binding site has been obtained for a small number of myeloma proteins by X-ray crystallography (Section 1.2.4.2) and model building (Section 1.2.4.3).

## 1.2.4. Tertiary structure

*1.2.4.1. Electron microscopy.* On the basis of hydrodynamic studies, Noelken et al. [78] predicted that the IgG molecule is composed of three compact subunits linked by flexible peptide bonds and arranged in a Y shape. This was elegantly confirmed by electron microscopy of antigen-antibody complexes [79, 80]. In particular Valentine and Green [81], using the bifunctional hapten bis-DNP-octamethylenediamine, showed that anti-DNP antibody forms a series of cyclic complexes of various shapes and sizes in which the hapten is too small to be visible (Fig. 13a). Digestion with pepsin removed the protrusions from the corners showing that they were the Fc region of the molecule (Fig. 13b). This demonstrated that antigen is bound at the tip of the Fab; it also showed the remarkable variability in the angle between the Fab arms from as little as 10° to as much as 180°. These pictures also made it possible to measure directly the dimensions of the molecule (Fig. 14). IgA anti-DNP bound to DNP-ferritin and free IgA molecules were revealed as two Y-shaped subunits linked

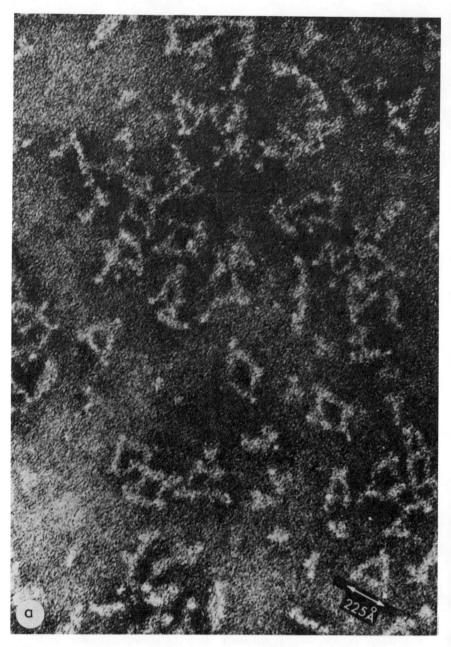

Fig. 13. (a) Electron micrograph of complexes of IgG anti-DNP with a divalent hapten. The hapten is too small to be detected and oligomeric complexes of IgG molecules bound together at the tips of their Fab arms are visible. The

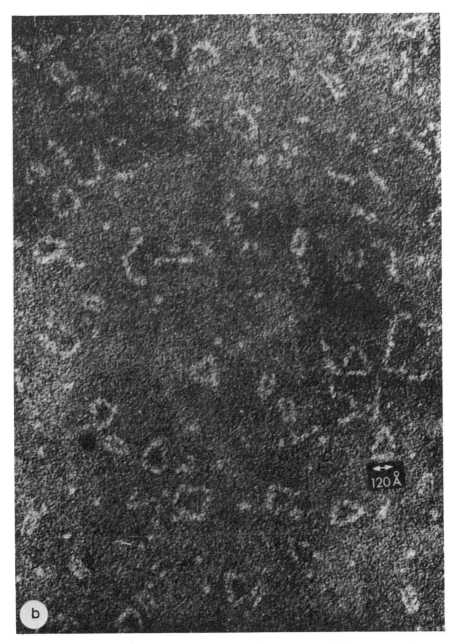

angles between the Fab arms vary. The Fc regions are visible as protrusions at the corners. (b) As (a) after digestion with pepsin. Reproduced from [81].

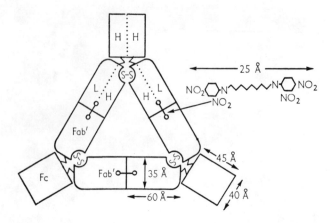

Fig. 14. Diagrammatic representation of a trimer of IgG anti-DNP from Fig. 13(a) showing the position of the bivalent hapten and the dimensions of the molecule. Reproduced from [81].

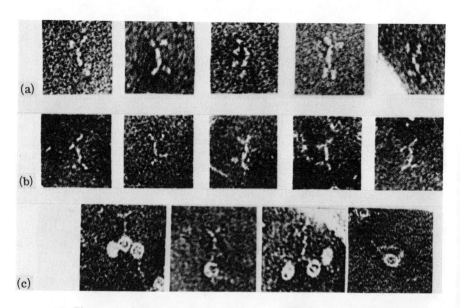

Fig. 15. Electron micrographs of IgA. (a) Free human IgA dimer molecules. (b) Mouse IgA anti-DNP dimer molecules produced by plasmacytoma MOPC 315. (c) Mouse IgA anti-DNP dimer molecules attached to DNP-ferritin. The electron micrographs were kindly provided by Dr. A. Feinstein.

# STRUCTURE OF IMMUNOGLOBULINS

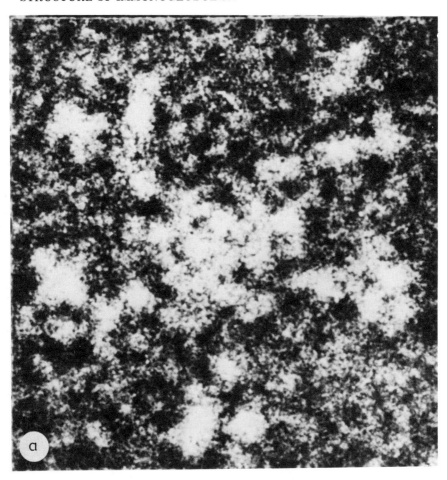

Fig. 16. Electron micrographs of (a) human IgM, (b) dogfish IgM, showing the central disc with five protruding arms which, in some cases are visibly bifurcated. The photographs were kindly provided by Dr. A. Feinstein. (c) A hexameric IgM-like protein from the toad *Xenopus laevis*, (d) a tetrameric IgM-like protein from the carp, reproduced from [319].

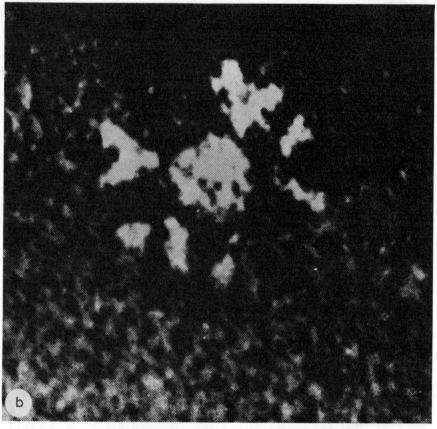

Fig. 16 (Legend on p. 225)

end to end by their Fc regions (Fig. 15) [80]. The appearance of mammalian and dogfish IgM under the electron microscope resembles a starfish in which the five $F(ab')_2$ arms are arranged around a central disc of about 10 nm diameter (Fig. 16a and b), although IgM-like proteins from the toad *Xenopus laevis* (Fig. 16c) and the carp (Fig. 16d) are hexameric and tetrameric, respectively. When viewed bound to a polymeric antigen, such as *Salmonella* flagella, mammalian IgM is able to take up a variety of conformations in which two or more of the Fab arms may crosslink two neighbouring flagella, or several Fab arms may be bound to the same flagellum with the molecule assuming a crab-like appearance

References p. 315

STRUCTURE OF IMMUNOGLOBULINS 227

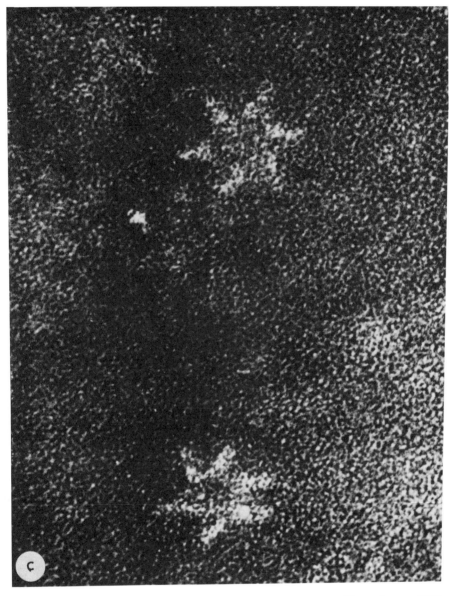

Fig. 16 (Legend on p. 225)

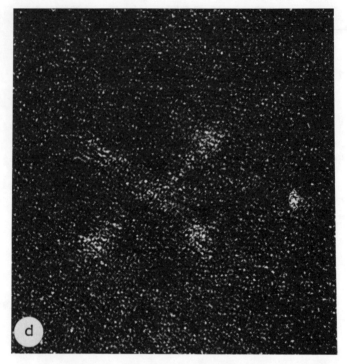

Fig. 16 (Legend on p. 225)

(Figs. 17, 18). Thus the F(ab')$_2$ arms are able to move in a plane at right angles to that of the central disc but the angles between the two Fab arms in any pair do not appear to vary, unlike IgG and IgA.

*1.2.4.2. X-Ray crystallography.* Much detailed information on the tertiary structure of immunoglobulins is now available as a result of X-ray crystallographic studies by several groups. For a more detailed discussion of the subject than can be presented here, the reader is referred to the excellent reviews of Poljak [82, 83] and others [84, 85].

A problem with such studies has been the difficulty of crystallising immunoglobulin. Normal immunoglobulin is much too heterogeneous and even the homogeneous myeloma proteins have proved difficult to crystallise. Early attempts gave poor resolution but more recently IgG 1 (Kol) has been analysed at 40 nm [86, 87]. This gave the structure of the α-carbon backbone of the F(ab')$_2$

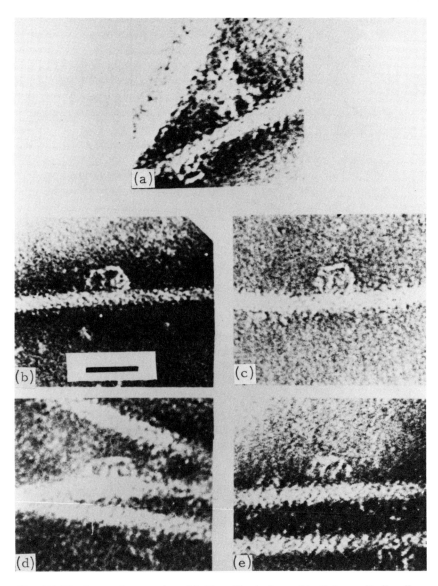

Fig. 17. Electron micrographs of IgM antibody bound to *Salmonella* flagella. The IgM was obtained from (a), (b) and (c) dogfish, (d) sheep and (e) rabbit. The bar represents 25 nm. The photographs were kindly provided by Dr. A. Feinstein.

*References p. 315*

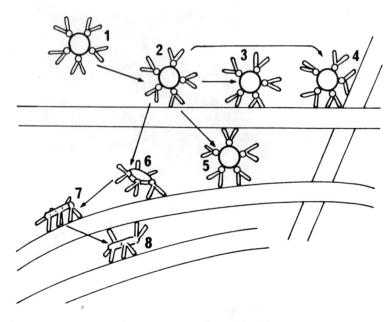

Fig. 18. Diagrammatic representation of the interaction between anti-flagella IgM antibody and the antigen, based on electron micrographs. It is proposed by Feinstein et al. [80] that the free molecule (1) first binds to a flagellum by one Fab (2); subsequently, more Fab arms bind to the same flagellum (3) or, if another flagellum is within range, cross-linking occurs (4) and (5). The authors envisage that the crab-like form (7) is derived via a transient intermediate (6). The crab-like form may also be able to change to a cross-linking form (8). Other sequences are possible and the process is reversible. Reproduced from [80].

region but the Fc region could not be resolved. The structure showed no contact between the Fab arms, and the angle between the pseudo two-fold axes of rotation of the variable and constant region domain pairs was 170°, unlike the angle of 130° observed in Fab' crystals.

Because of the difficulties of analysing whole IgG molecules, most of the investigations have been carried out on either Bence-Jones protein (light-chain dimer) [88, 89] or fragments of the immunoglobulin molecule which are more easily crystallised. Most data have come from four systems: (1) the Fab' of a human IgG (New) which binds vitamin K [90]; (2) a mouse IgA Fab which binds phosphorylcholine [91]; (3) the variable region fragments ($V_\kappa$ dimers) of two Bence-Jones proteins [92, 93]; and the Fc fragment of human IgG [87, 94].

These studies have confirmed the domain theory of Edelman and Gall [21], originally based on amino acid sequence homology. Thus it was found that the L chain dimer, Fab' and Fab fragments are each composed of four globular regions (domains) i.e. 2 $V_L$ and 2 $C_L$ domains in the L chain dimer, $V_L V_H C_L$ and $C_H$ in the Fab' and Fab. The V region dimer is composed of two domains $V_L$ and $V_H$. It should be emphasised at this point that we are using the term "domain" in the sense originally proposed by Edelman and Gall, i.e., the globular structure formed by a single homology region and not, as it is sometimes used, to describe the globular structure formed by two homology regions from two polypeptide chains in close association.

The angle between the major axes of the $V_L$ and $C_L$ domains of the light chain of Fab' New is 100—110° whereas that between the $V_H$ and $C_H 1$ domains of the heavy chain is 80—85°. It is interesting that one of the light chains of the dimeric Bence-Jones protein Mcg [88] assumes an angle between the domains similar to that of the heavy-chain domains of Fab' New, while the other takes up the characteristic light-chain angle. In this instance the dictates of quaternary structure supersede those of primary sequence in determining protein conformation.

Each domain, whether from the light or heavy chain, possesses a similar basic conformation, the "immunoglobulin fold", being formed from two roughly parallel β-pleated sheets one being composed of four and the other of three anti-parallel (approximately linear) extended polypeptide chains connected by loops (Figs. 19 and 20). In the variable region domain there is an extra loop from the sheet of three chains and this sheet is in contact with the corresponding sheet of the other variable region domain, whereas the four chain sheets are in contact in the unit formed by the $C_H 1$ and $C_L$ domains. This allows the formation of the disulphide bridge between the C-terminus of the light chain and the $C_H 1$ domain.

The Fc fragment has been obtained by cleavage of pooled human serum IgG with plasmin and analysed to 35 nm resolution [87, 94]. The fragment has been described as being in the form of a "Mickey Mouse head", the two $C_H 3$ domains forming the globular

*References p. 315*

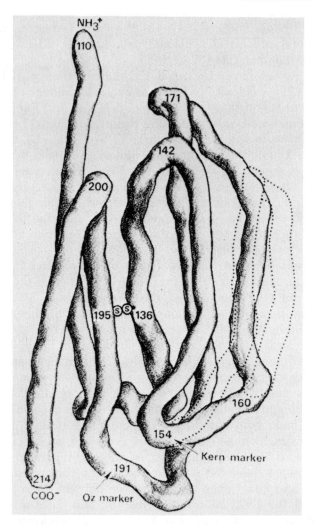

Fig. 19. Model of the polypeptide chain of a single immunoglobulin domain, reproduced from [90]. Solid lines show the folding of the polypeptide chain in the $C_L$ and $C_H1$ domains of Fab. Numbers indicate the positions of λ chain amino acid residues. Broken lines indicate the position of the additional loop of polypeptide chain found in the $V_L$ and $V_H$ domains.

# STRUCTURE OF IMMUNOGLOBULINS

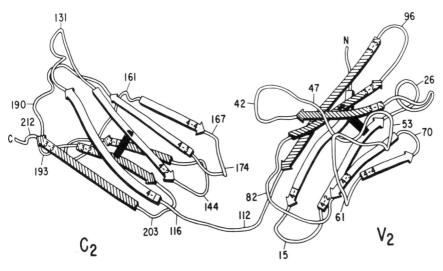

Fig. 20. Diagrammatic representation of the constant and variable domains of one light chain in a light chain dimer showing the two β-pleated sheets of four (white arrows) and three (striated arrows) anti-parallel polypeptide chains. The approximate right angle between the major axes of the two domains is also evident. Reproduced from [320].

head and the $C_H 2$ domains forming the ears (Fig. 21a). The conformation of polypeptide chains of both the $C_H 2$ and $C_H 3$ domains is in the form of the immunoglobulin fold. The two $C_H 3$ domains interact closely in a similar manner to the $C_H 1$-$C_L$ domain interaction of Fab' New, whereas the $C_H 2$ domains do not interact. Part of the hinge region is at the N-terminus of the Fc fragment and this region appears to be disordered. The carbohydrate moiety attached to $C_H 2$ lies between the domains.

Using the known domain coordinates of human IgG Fc from the above work and of mouse Fab (McPC 603), Silverton et al. [94a] determined the best fit to their 0.6 nm electron density map of human IgG 1 (Dob). The Fc region of the resulting model (Fig. 21b) was similar to the isolated Fc but the angle between the variable and constant region domain pairs of the Fab was 147°, intermediate between that of IgG (Kol) (170°) and the isolated Fab of IgG (New) and McPC 603 (approx. 130°), indicating the flexibility of the switch region between the variable and constant regions.

*References p. 315*

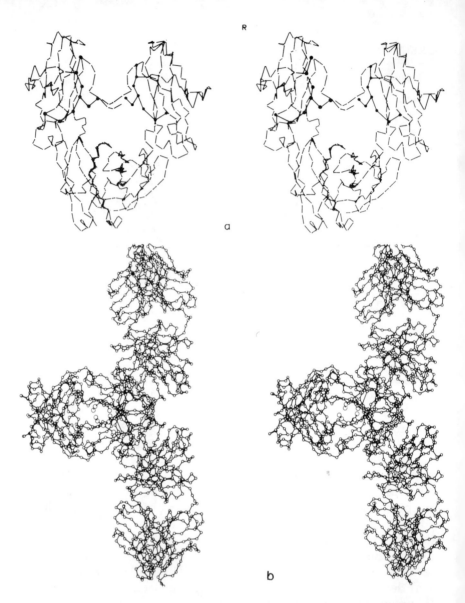

Fig. 21. (a) Stereo-pair drawing of the α-carbon backbone of human Fc and the carbohydrate hexo units (●). Reproduced from [102] with the permission of Springer Verlag. (b) Stereo view of the α-carbon (○) backbone of IgG1 (Dob) and carbohydrate hexose units (O). The Fab arms of the molecule are aligned vertically, and a horizontal 2-fold axis of symmetry bisects the molecule through the Fc. The light chain is in the foreground of the upper Fab and the heavy chain is in the foreground of the lower Fab. Reproduced from [94a].

Screening revealed that IgG New binds several haptens such as uridine, orceine, menadione etc. with low affinity ($K \simeq 10^3$ l/mol) and a γ-hydroxyl derivative of vitamin K (vitamin $K_1OH$) with relatively high affinity ($K$=1. 7 × $10^5$ l/mol). This has made it possible to study the binding of a hapten to the antigen-binding site [95] (Fig. 22). Both the H and L chain contribute to the shape of the site, the hypervariable regions being clustered around the site and exposed to the solvent at bends in the polypeptide chains, in agreement with the affinity labelling data (Section 1.2.3.4). The site is relatively flat, occupying an area of about 2.5 × 2.0 nm with a shallow groove 1.6 × 0.7 × 0.6 nm, in approximate agreement with estimates from hapten binding studies (Section 1.2.1.3). Vitamin K is in close contact with the amino acids of the hypervariable regions, bound mainly by Van der Waal's forces.

The results of Padlan et al. [96] on a mouse myeloma IgA (McPC 603), which binds phosphorylcholine, were in general agreement with those of Poljak et al. [95]. Phosphorylcholine binds to a cleft between $V_H$ and $V_L$ which is deeper than the site of IgG New, occupying a space 1.2 nm deep by 1.5 nm by 2.0 nm. It is probable that the shape of the antigen-binding site will differ for each antibody. No major conformational change was observed in the Fab' of IgG New or McPC 603 on binding to hapten.

*1.2.4.3. Model building and spectroscopic methods.* Following a preliminary description by Poljak [82], Padlan et al. [97] have built a model of the antigen-binding site of the mouse IgA myeloma protein MOPC 315, which binds the DNP group. This model combines the known amino acid sequences of MOPC 315 H and L chains with the X-ray crystallography data from McPC 603 Fab', IgG New Fab and the Bence-Jones protein Mcg. The tertiary structure of the light- and heavy-chain framework regions (those parts of the variable region outside the hypervariable regions) is very similar in these different molecules but the hypervariable region loops differ. Thus it was possible to build the framework of the variable regions based on the known structure of McPC 603 Fab'. The hypervariable loops were then built, using the known amino acid sequence, by comparison with the tertiary structure of the hypervariable loops of the other proteins and in such a way as to maximise structural stability with hydrogen bonds, minimise steric hindrance and maintain the $\phi$ and $\psi$ angles within

*References p. 315*

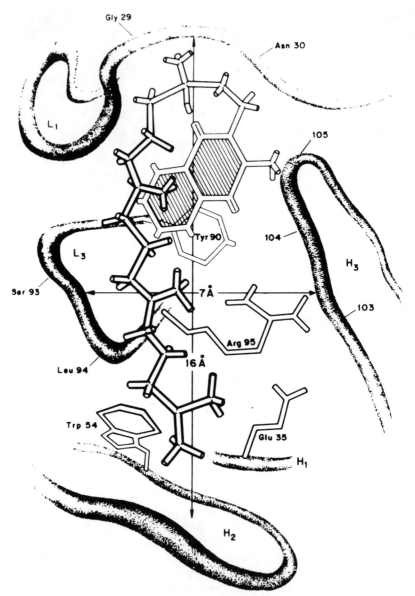

Fig. 22. Diagram showing γ-hydroxy vitamin K, bound to the antigen binding site of human myeloma IgG NEW. $L_1$ and $L_2$ are the first and third light chain hypervariable regions ($L_2$ is deleted in this molecule). $H_1$, $H_2$ and $H_3$ are the heavy chain hypervariable regions. The dimensions of the cleft are approx. 0.6 nm deep, 0.7 nm wide and 1.6 nm long. Reproduced from [95].

# STRUCTURE OF IMMUNOGLOBULINS

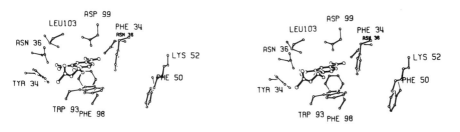

Fig. 23. Stereoscopic drawing of the hypervariable residues projecting into the antigen-binding site of MOPC 315. BADE (without the bromine atom) is shown in the cavity to demonstrate the interactions of the hapten with amino acid residues and the possibility of affinity labelling Tyr 34 (L). Reproduced from [97].

reasonable limits. The resulting structure is shown in Fig. 23. The antigen-binding site consists of a cavity into which the hapten fits, surrounded by the hypervariable loops which contain several aromatic residues. The aromatic ring of a tryptophan is only 0.35—0.4 nm from the dinitrophenyl ring and parallel to it. The nitro groups of the hapten can form hydrogen bonds with two asparagine residues in the light and heavy chain. In Fig. 23 it is seen that BADE and BADL can react with tyrosine 34 in the light chain and lysine 52 in the heavy chain, and the bifunctional reagent DIBAB can cross-link both residues, as observed in affinity labelling studies (Section 1.2.3.4), although this agreement is derived from coarse rather than fine details of the model.

That the structure of MOPC 315 Fab is described by the model has been confirmed in outline and the model further refined by Dweck et al. [98, 99] using electron spin resonance and nuclear magnetic resonance studies. Advantages of these methods are that they can be used to study proteins that cannot be crystallised and they provide information about the structure of the molecule in its natural state, in free solution.

Using spin-labelled derivatives of dinitrophenol bound to the Fv fragment of MOPC 315, which contains the antigen-binding site, Dwek et al. obtained electron spin resonance spectra which they interpreted as indicating a combining site with a cleft of overall dimensions 1.1—1.2 nm × 0.9 nm × 0.6 nm having consider-

*References p. 315*

able structural rigidity. Data for the DNP-binding proteins XRPC 25 and MOPC 460 gave somewhat different dimensions, compatible with their different affinities for related haptens. Analysis of nuclear magnetic resonance difference spectra in the presence and absence of a spin-labelled hapten indicated about 30 aliphatic and 30 aromatic protons to be affected by the proximity of the hapten. The resonance of two histidine residues was no longer observable in the presence of the spin-labelled hapten indicating that they are in the region of the binding site. The hapten is surrounded by a "box" of aromatic residues, as predicted by Padlan's model and the nuclear magnetic resonance data confirm the belief that the DNP group interacts with a tryptophan residue in a similar orientation to that in the model.

No large conformational change in the Fv fragment was detectable outside the antigen-binding site on binding hapten, confirming the inability to detect such a change in IgG (New) by X-ray crystallography. In contrast to this it has been reported that antibodies (and their Fab fragments) to ribonuclease and the "loop" peptide of lysozyme show alterations in circular polarisation of luminescence on binding antigen [100]. This was interpreted as being due to conformational changes. For intact antibodies alterations occur in regions of the spectrum in which no change is observed for the Fab fragment, so it was concluded that conformational change is transmitted from Fab to the Fc region on antigen binding. The phenomenon appears to be independent of antigen:antibody ratio or the nature of the antigen. No such alterations in circular polarisation occur on binding the haptens phosphoryl choline or tetra-alanine, in agreement with the observation by X-ray crystallography and nuclear magnetic resonance.

X-ray crystallographic data have also been used by Feinstein et al. [9, 85, 101] to build low-resolution models of entire immunoglobulin molecules. The possibility of doing this rests on the assumption, to which there is at present no exception, that all constant region domains have the same basic structure (Section 1.2.4.2). Feinstein used the Fab' (New) X-ray crystallography data of Poljak to build plastercast models of the $C_\gamma 1$ domain, the $C_L$ $C_\gamma 1$ domain pair and the $V_L$ $V_\gamma$ domain pair (Fig. 24). The domains were arranged according to the following rules: (1) The known positions of the interchain disulphide bridges in the amino acid sequence were used to pinpoint their positions on the domains,

# STRUCTURE OF IMMUNOGLOBULINS

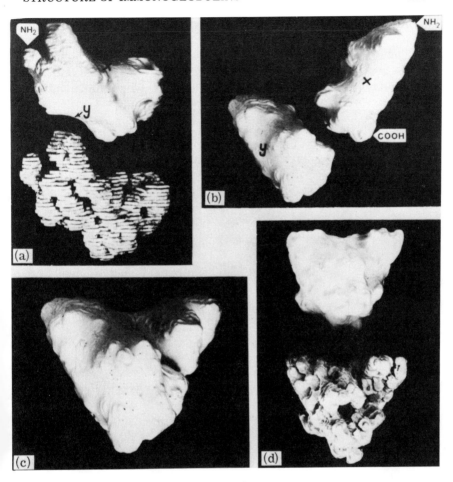

Fig. 24. (a) Models of $C_\gamma 1$ domains of Fab' (NEW). Bottom: glued wooden sections representing density at 0.1-nm intervals. Top: plaster of Paris cast of the wooden model. (b) Two $C_\gamma 1$ domain casts oriented to give an "exploded view" of a Fab-C-like pair. (c) Model showing Fab-C-like pairing of two $C_\gamma 1$ domains. (d) Models of $C_L$-$C_\gamma 1$ pairs of Fab' (NEW). Bottom: glued wooden sections representing density at 0.1-nm intervals. Top: plaster of Paris cast of the wooden model (seen from a slightly different aspect to that in (c). Photographs kindly provided by Dr. A. Feinstein.

*References p. 315*

defining minimal areas of contact between pairs of domains from adjacent chains. (2) It was assumed that the longitudinal interaction between $V_H$ and $C_H 1$ in Fab' (New) occurs along the length of the heavy chain, resulting in a zig-zag pattern. This assumption has been confirmed by Deisenhofer et al. [102] who found a similar

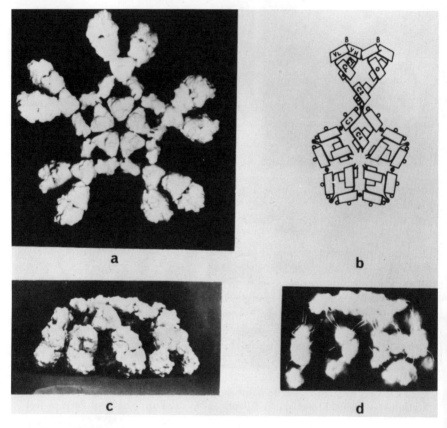

Fig. 25. IgM models. (a) Model made from cast domains showing all five $F(ab)_2'$ arms; (b) diagram showing only one of the five $F(ab')_2$ arms. ●, interchain disulphide bridge; ○, site of attachment of oligosaccharide; □, regions of polypeptide chains folded into domains. $C\mu 2$ and $C\mu 4$ each form Fab-C-like pairs. The $C\mu 3$ domains have been arranged to resemble the spatial relationship of $C_\gamma 2$ domains of he Fc of IgG. (c) "Crab" form of the IgM model corresponding to that seen in electron micrographs (see Fig. 17). (d) X-ray photograph of the "crab" form to simulate that seen in electron micrographs. Photograph kindly provided by Dr. A. Feinstein.

relationship between $C_\gamma 2$ and $C_\gamma 3$ in Fc crystals. (3) Paired domains in two adjacent heavy chains make contact in the same manner as $C_L$ and $C_\gamma 1$ in Fab' (New) (Fab C-like pairing) where the known contact residues permit. This type of pairing occurs between the $C_\gamma 3$ domains in Fc crystals but cannot occur sequentially along the heavy chain since it would disrupt the longitudinal interactions.

Applying these rules, Feinstein built a model of IgM which conforms to the shape and dimensions of the molecule from electron microscopy (Fig. 25a). In this model the $C\mu 4$ domains form a compact, symmetrical ring. The $C\mu 3$ domain is probably equivalent to $C_\gamma 2$ since IgG has no equivalent of $C\mu 2$, the latter being replaced by the hinge region. $C\mu 3$ cannot form Fab C-like pairs since the N-termini make contact to form an interchain disulphide bridge within the subunit (Fig. 5). In human IgM, the $C\mu 3$ domains also form an intersubunit disulphide bridge at their C-termini, although this bridge appears to be absent in mouse IgM. The arrangement of $C\mu 3$ and $C\mu 4$ corresponds to the structure of IgG Fc (Fig. 26). The $C\mu 2$ domains are linked by a disulphide bridge at the C-terminus and therefore probably pair in a Fab C-like manner as shown.

If the $F(ab)_2$ arms of this model are bent downwards it assumes a crab-like form and an X-ray photograph of this bears a striking resemblance to electron micrographs of IgM bound to the surface of an antigen (Figs. 17 and 25c and d).

In a similar manner a model of human IgG1 has been built based on the X-ray crystallographic data of Colman et al. [86] and Deisenhofer et al. [102] (Fig. 26). In this model the pseudo twofold axes of rotation of the V and C domain pairs are aligned at about 180° as in the IgG crystal, unlike the Fab' crystal in which the angle between the axes is about 130°.

### 1.3. Function

*1.3.1. Introduction.* It has long been known that immunoglobulins exhibit several biological properties, even within a single molecule. The most obvious of these is the ability to bind to antigen and form insoluble complexes or, in the case of particulate antigens such as erythroctes or bacteria, to cause them to agglutinate. Other properties of immunoglobulins are: (1) the ability of some but not

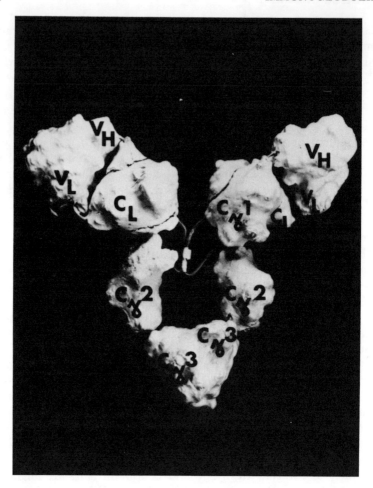

Fig. 26. Plaster of Paris model of human IgG 1, based on the data of Colman et al. [86] and Deisenhofer et al. [102]. In the right-hand Fab the pseudo twofold axes of rotation for the variable domains and constant domains have been aligned in the manner seen by Colman et al. [86] in an IgG (Kol) crystal. In crystals of Fab fragments, the pseudo axes are at approx. 130° to each other so that the longitudinal contact between $V_H$ and $C_H1$ is different from that between $V_L$ and $C_L$. Photograph kindly provided by Dr. A. Feinstein.

all classes to be actively transported across membranes; (2) the ability to bind to cell membranes where they are able to exert various physiological effects depending upon the cell to which they are bound; (3) the ability of immune complexes, but not free antibody to fix the components of the complement system resulting in the death of a bacterial, viral or cellular antigen by enzymatic lysis or phagocytosis. A summary of the biological properties of the immunoglobulins is presented in Table VI. Knowledge of the domain structure of immunoglobulin molecules suggested a means whereby this multiplicity of functions within a single molecule could be rationalised if each domain could be shown to exert a particular function.

*1.3.2. Antigen binding.* The ability of antibodies to form insoluble complexes with soluble antigens and to agglutinate particulate antigens follows from their overall shape, the flexibility of the molecule and the possession of two or more antigen-binding sites at the tips of the Fab arms. The greater the valency of the anti-

TABLE VI

Biological properties of human immunoglobulins

| Immuno-globulin | Complement fixation | | Placental transfer | Binding to cell membranes | |
|---|---|---|---|---|---|
| | Classical pathway | Alternative pathway | | Mono-nuclear cells | Mast cells and basophils |
| IgG1 | ++ | — | + | + | — |
| IgG2 | + | — | + | — | — |
| IgG3 | +++ | —[a] | + | + | — |
| IgG4 | — | —[a] | + | — | (?) |
| IgM | +++ | — | — | — | — |
| IgA1 | — | + | — | — | — |
| IgA2 | — | + | — | — | — |
| sIgA | — | — | — | — | — |
| IgD | — | — | — | — | — |
| IgE | — | —[a] | — | — | +++ |

[a] Aggregated molecules may activate complement by the alternative pathway.
Adapted from [317] and reproduced by permission of John Wiley and Sons.

body the more avidly will it bind and cross-link polymeric antigens. Thus IgM, with its ten antigen-binding sites, is the most efficient class of antibody for agglutinating bacteria and erythrocytes.

The formation of the antigen-binding site by the three-dimensional arrangement of the polypeptide chains of the light- and heavy-chain domains and the importance of the hypervariable regions in determining specificity have already been discussed in some detail. It has been postulated that each antibody molecule need not necessarily bind only one antigen and may in fact bind many different antigens, perhaps in the region of 100 [103, 104]. Consistent with these proposals there is evidence that at least some antibody molecules are able to bind several different antigenic determinants. Several myeloma proteins have been found to be capable of binding more than one hapten e.g. human IgG (New) binds vitamin K, uridine, MSH and menadione [105], and mouse IgA 460 binds both menadione and DNP. In the latter case it has been shown that the two haptens occupy different positions, separated by 1.2—1.4 nm within the antigen-binding site [106, 107]. Further evidence is provided by the observations of Varga et al. [108] that a single band of immunoglobulin on an isoelectric focusing gel binds two dissimilar haptens and is stimulated by either. It is not known how many different antigenic determinants a single immunoglobulin is capable of binding but, if multiple binding is a physiologically significant phenomenon, it would clearly provide an additional means of increasing the total antigenic repertoire of the individual [1]. Specificity of the response is not lost since, as discussed by Talmage [103] all clones of lymphocytes stimulated by antigen A would produce antibodies with specificities for A but their other specificities would not be the same and therefore would be diluted out provided sufficient clones respond. It is now clear that antibody elicited in response to a single determinant can be very heterogeneous. The binding and cross-reactive properties of antibodies have been determined using such heterogeneous populations of antibody molecules. With the advent of hybridoma technology (Section 1.4.8) monoclonal antibody is becoming available and specificity may have to be re-evaluated in the light of multiple specificity of individual combining sites.

### 1.3.3. Class of antibody and the immune response.

IgM is the predominant antibody formed during the primary response and in response to blood-borne bacteria and protozoa, e.g. in malaria and trypanosomiasis. IgM is predominantly located in the plasma, being prevented from crossing the capillary wall into the interstitial fluid, and its efficiency in agglutination and complement fixation (Section 1.3.6) make it eminently suitable for the removal of such organisms.

IgG is much more widely spread, being found in both the blood and interstitial fluid, it is the predominant antibody of the secondary immune response, exhibiting much higher affinity than IgM. Some carbohydrate antigens show subclass restriction, e.g. human antibody to dextrans and levans is mainly restricted to IgG 2 and some rabbits produce extremely restricted responses to polysaccharides in which only one or a few clones of lymphocytes are stimulated.

IgD is found in only trace amounts in the plasma but is present on the cell membranes of a large proportion of lymphocytes, mainly in conjunction with IgM. The function of IgD is not known but its presence on the lymphocyte membrane suggests that it may play a role in the susceptibility of the cell to either induction of proliferation and differentiation leading to memory cell and antibody production or to the induction of tolerance [109].

IgE in the body is almost entirely bound to the membranes of mast cells (Section 1.3.5.4) and is involved in immediate hypersensitivity reactions, e.g. hay fever, asthma etc. The unpleasant effects of these reactions are presumably outweighed by the beneficial effects of IgE or it would have been selected against long ago. The function of IgE is not clear but a role in immunity against intestinal parasites has been postulated.

### 1.3.4. Membrane transmission.

This subject has been reviewed previously by Waldmann and Strober [110] and Waldmann and Hemmings [111]. The secretions of the alimentary, respiratory and genitourinary tracts and the exocrine glands (lachrymal, sudorific, salivary and mammary glands) contain secretory IgA which is not derived from serum but is synthesised locally. This provides an important defence against infection at external surfaces and large amounts of secretory IgA are transmitted to the young in colostrum and, to a lesser extent, in milk. The mechanism of transfer of secretory IgA across the epithelium of these glands is not understood, nor is the function of secretory piece, though it is thought that it may protect the molecule against degradation by proteolytic enzymes.

Transfer of IgG antibodies from mother to offspring take place in either of two ways: prenatally, by transfer of immunoglobulin across the placenta as in humans, or postnatally, via colostrum and milk, across the wall of the intestine as in the sheep, goat, cow, horse and pig. Both mechanisms occur in the rabbit, rat and mouse. In man all four subclasses of IgG cross the placenta and this appears to be a property of the Fc piece since Gitlin et al. [112] showed that the Fc fragment is transferred to the foetus from the maternal circulation more rapidly than the Fab, whole IgG or other proteins. Transport of IgG across the placenta can cause haemolytic disease of the newborn where there is Rhesus incompatibility between mother and foetus. Rhesus-positive (D) foetal erythrocytes, leaked into the maternal circulation during the first parturition, stimulate production of anti-D IgG which then crosses the placenta to the foetus during subsequent pregnancies, resulting in lysis of the foetal erythrocytes. Since antibodies against the major (A, B, O) blood group antigens are normally restricted to IgM, which cannot cross the placenta, haemolytic disease is not usually a problem with the much more common A, B, O incompatibilities. If IgG anti-A or anti-B antibodes are formed they are usually neutralised by A or B antigens on other foetal tissues such as the placental endothelium, or in soluble form in the foetal plasma, before haemolysis occurs. Erythroblastosis, due to A, B, O incompatibility, does occur but only rarely. The colostrum of ruminants and the pig is rich in IgG1 derived from plasma and this is absorbed through the intestinal wall of the suckling offspring, whose placenta is impermeable to immunoglobulin. The intestine of the newborn ruminant is specially adapted to uptake of protein and the mechanism is non-specific but ceases 24—48 h after birth. Uptake of immunoglobulin across the intestinal wall appears to be more specific in rats and mice. Brambell et al. [113] have proposed a mechanism for transport of IgG across epithelial membranes involving pinocytosis and protection of IgG from degradation by binding to Fc receptors within the vacuoles.

# FUNCTION

*1.3.5. Binding to cell membranes.* Immunoglobulins bind to the cell membranes of lymphocytes, monocytes, macrophages, neutrophils, platelets, basophils and mast cells, through Fc receptors on the membranes of these cells. These Fc receptors are readily detected by the formation of "rosettes" with erythrocytes to which anti-erythrocyte antibody is attached. When the cells to be tested are mixed with the sensitised erythrocytes the latter adhere to any cell with Fc receptors to form a rosette of erythrocytes around the central Fc-bearing cell. The nature of the cell to which immunoglobulin binds determines the physiological effect.

*1.3.5.1. Lymphocytes.* The membrane of the B lymphocyte contains bound immunoglobulin, mainly 8S IgM and IgD, synthesised by the cell that bears it. This surface (receptor) immunoglobulin is capable of binding antigen leading to a stimulus that induces the B lymphocyte to divide and differentiate resulting in the synthesis and secretion of more immunoglobulin of the same specificity, the central feature of the humoral immune response. T lymphocytes also possess antigen receptors which are believed by some workers in this field to be immunoglobulin-like, possibly a separate class of immunoglobulin (IgT). This T-cell receptor is presumably involved in antigen-specific T-cell effector functions viz. stimulation and suppression of B lymphocytes and cell-mediated cytotoxicity. For a more detailed discussion of lymphocyte surface immunoglobulin see Section 1.4.10.

In addition, B lymphocytes and activated T lymphocytes have receptors for the Fc region of IgG on their surface. These Fc receptors are specific for IgG1 and IgG3; IgG2 and IgG4 are only bound weakly and binding of other classes is not usually observed [114, 115]. However, two subsets of T cells have been defined by virtue of Fc receptors specific for IgG and IgM respectively. Rosette formation with IgG antibody-coated erythrocytes has been shown to be inhibited by mouse plasmacytoma IgG but not by a mutant lacking the $C_\gamma 3$ domain [116, 117], indicating that it is this domain that binds to the Fc receptor. In support of this, the Facb (fragment, antigen and complement binding) fragment of IgG, which also lacks the $C_\gamma 3$ domain, does not induce K cells to lyse sensitised target cells, unlike complete antibody [118].

*1.3.5.2. Monocytes and macrophages.* Binding of immune complexes to these cells promotes phagocytosis. As in the case of lymphocytes, binding of free serum immunoglobulin to monocytes and macrophages is restricted to IgG1 and IgG2 and takes place via the $C_\gamma 3$ domain [119—121]. Capron et al. [122] found that complexes of rat IgE antibody bound to the schistosomulae of *Schistosoma mansoni* also adhere to macrophages, although it was not established whether binding was through the Fc region. This could point towards a possible role for IgE antibody in combating helminthic parasites.

*1.3.5.3. Neutrophils.* Binding of immune complexes to neutrophils promotes phagocytosis and also occurs through the $C_\gamma 3$ domain since, unlike complete antibody, the Facb fragment does not activate neutrophils to phagocytosis of bacteria to which it is bound [118]. The Fc receptors of neutrophils bind unaggregated IgG1, IgG3, IgG4, IgA1, IgA2 and secretory IgA [115]. No other cells are known to bind IgA. This suggests either that the Fc receptor of neutrophils has a different structure from that of lymphocytes, monocytes and macrophages, or that it has two types of Fc receptors, one for IgG and one for IgA. Upon aggregation all classes of immunoglobulin bind to neutrophils.

*1.3.5.4. Basophils and mast cells.* As previously stated, IgE binds very strongly to the membranes of basophils and mast cells. Guinea pig IgG1 and IgG2 also bind but for much shorter periods (1—2 days) compared with IgE (4 weeks) as measured by persistence of skin sensitisation [123]. As might be expected, IgG binding is through the $C_\gamma 3$ domain as shown by the inhibition of passive cutaneous anaphylaxis in guinea pigs by a $C_\gamma 3$ fragment [124]. IgE binding has been localised to the $C_\epsilon 3$-$C_\epsilon 4$ region and it has been proposed that both domains may be involved in a cooperative binding mechanism [125].

Antigen binding to antibody on the membrane of mast cells results in an explosive degranulation with release of histamine and serotonin, which are stored in the mast cell granules, resulting in an immediate hypersensitivity reaction which, if generalised, can be so severe as to cause death. Examples of immediate hypersensitivity reactions are asthma, hay fever and allergies to various foodstuffs and drugs.

*1.3.6. Complement fixation.* The interacting system of plasma proteins known as the complement system has been stydied in great detail [126, 127] and has been reviewed in Chapter 6, Vol. 19B/I, so it will be dealt with only briefly in this section, discussion being restricted mainly to the mechanism of activation of complement by the classical pathway. There is also an alternative pathway which does not require the mediation of immunoglobulin, although aggregates of immunoglobulin (and Fab')$_2$ fragments but not Fab fragments) will activate the alternative pathway. Activation of complement by either pathway results in lysis or phagocytosis of bacteria, viruses or eukaryotic cells.

IgM is an extremely efficient activator of complement by the classical pathway. It has been estimated that one molecule of IgM antibody is sufficient for lysis of an erythrocyte. IgG1 and IgG3 are also able to activate complement, though not as efficiently as IgM; IgG2 is a poor activator and IgG4 and other classes of immunoglobulin are unable to bind complement. Activation of complement by IgG requires at least two IgG molecules bound to antigen in close proximity and the pentameric subunit structure of IgM is presumably the reason for its greater efficiency, providing five binding sites on the one molecule. For IgG-antigen complexes the initial step in the classical pathway is the binding of the first complement component (C1q) to the $C_\gamma 2$ domain. This was demonstrated by Ellerson et al. [128, 129] who isolated, by trypsin digestion, a fragment of human IgG1 containing the paired $C_\gamma 2$ domains and hinge region. This fragment bound C1q with the same affinity as Fc. Also, Connell and Porter [130] removed the $C_\gamma 3$ domain from rabbit IgG by plasmin digestion to produce Facb consisting of F(ab)$_2$ linked to $C_\gamma 2$ by the hinge region. This fragment bound C1 as efficiently as whole IgG [131]. On the basis of evidence from amino acid residue accessibility, sequence conservation analyses and inhibition of C1q binding by competitive inhibitors and chemical modification of amino acid residues, Burton et al. [424] have proposed that the C1q binding site is formed by the two C-terminal $\beta$-strands of the $C_\gamma 2$ domain.

The complement binding site of IgM has not been identified with certainty with a single domain. An (Fc$\mu$)$_5$ fragment, containing the C$\mu$3 and C$\mu$4 domains, was found to fix complement with 20-fold greater efficiency than intact IgM in the absence of antigen, restricting the site to C$\mu$3 or C$\mu$4 and suggesting that the site is not available for complement binding in free IgM but becomes accessible due to movement of the Fab arms of the molecule [132]. Hurst et al. [133] have detected weak binding of C1 to a fragment from part of the $C\mu 4$ domain but the significance of this is not certain. The $C\mu 3$ domain is perhaps the more likely candidate since this shows homology with $C_\gamma 2$.

Since complement does not bind to free IgG or IgM, some change must take place when antibody binds to antigen which enables C1q to bind to the immunoglobulin molecule. The change could be due to aggregation resulting in multiple binding sites, so enabling C1q to bind much more strongly, as proposed by Metzger [126]. This explanation is ruled out for IgM, however, since one molecule is sufficient for complement activation. An alternative explanation is that antigen induces some type of conformational change, resulting in the formation of a C1q binding site [80]. In support of this theory Brown and Koshland [134] demonstrated that non-aggregated complexes of mono-Lac*—RNase with anti-Lac IgM fixed complement although the free hapten (Lac) did not. By contrast no conformational change has been detected by X-ray crystallography of hapten-antibody complexes (Section 1.2.4.2) although alterations in circular polarisation of luminescence have been detected in antibodies and Fab fragments binding ribonuclease and lysozyme. These phenomena were attributed to conformational changes in the Fab and Fc regions (Section 1.2.4.2).

An alternative explanation is a gross conformational change caused by movement of the Fab arms resulting in a change from a Y- to a T-shaped molecule or v.v. [135]. In support of this, Hyslop et al. [136], studying complexes of IgG and a bivalent hapten, found that only tetramers and larger complexes fixed complement. The angles between the Fab arms of the complement-fixing complexes were 90—180° whereas the angles for the smaller complexes were less than 60°. These results could also be used as evidence for the importance of the size of complex and it is difficult to distinguish between the two factors. An interesting observation of Isenman et al. [137] is that, although human IgG4 is unable to fix complement, the isolated Fc piece is active, suggesting that removal of the Fab arms either exposes an otherwise masked site or causes a conformational change in the Fc. In the case of IgM it is possible that the bending of the $F(ab')_2$ arms rel-

---

*Lac, $p$-aminophenyl-$\beta$-lactoside.

ative to the central disc, observed by electron microscopy (Section 1.2.2.1) may cause similar changes in the Fc without the need for more than one antibody molecule to be involved.

## 1.4. Biosynthesis of immunoglobulins

*1.4.1. Introduction — Cells involved in antibody formation.* Immunoglobulins serve both as receptor antibody and as effector antibody. All antibody is synthesised by cells of the B lymphocyte lineage. Receptor antibody is synthesised by B lymphocytes in the absence of antigenic stimulation. Interaction of antigen with receptor antibody on a B lymphocyte can, under the appropriate conditions (Fig. 27), induce cell proliferation and differentiation. Both macrophages and antigen-specific helper T lymphocytes are usually involved in antigenic stimulation of a B lymphocyte (Fig. 27). Cell-surface differentiation antigens coded at the major histocompatibility locus are involved in these cell—cell interactions (see below) but biochemical mechanisms have not been worked out.

The cellular basis of antibody diversity is one cell-one antibody. This fact is fundamental to the clonal selection hypothesis [138—

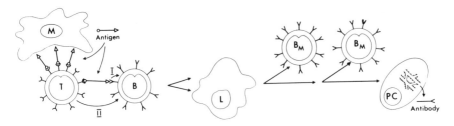

Fig. 27. Antigenic stimulation of a B lymphocyte resulting in proliferation and differentiation. Antigen molecules bind non-specifically to a macrophage (M) which then activates a T or possibly a B lymphocyte. Signals I and II from the T lymphocyte are thought to result in the activation of the B lymphocyte leading to proliferation and differentiation into lymphoblasts (L), B memory cells ($B_m$) and plasma cells (PC). B memory cells are long-lived lymphocytes of similar morphology to and the same specifity as the original B lymphocyte from which the clone was derived. The plasma cells actively secrete antibody of the same specificity as the receptors of the original B lymphocytes.

*References p. 315*

140]. Each cell in the population of B lymphocytes elaborates a receptor antibody of single specificity. Each antigen interacts with the cell, or cells, making an antibody complementary to the antigen. The interaction initiates clonal cell proliferation leading to production of (1) a population of memory B lymphocytes essentially similar to the original B lymphocyte with which the antigen interacted; and (2) a population of plasma cells, each secreting antibody at a high rate. The specificity of antibody is constant for cells of a given clone though the class of antibody produced can vary during clonal expansion.

The amount of antibody synthesised varies greatly during clonal expansion. The B lymphocyte displays about $10^4-10^5$ molecules of antibody on the surface membrane where the half life is about a day. The plasma cell synthesises and secretes about 2000 molecules of antibody per second.

The details of clonal selection and of the cell—cell interactions involved in B-cell differentiation have been reviewed elsewhere [141—150] and are beyond our present scope. In this section we will concentrate on the biochemistry of biosynthesis, assembly, membrane deposition and secretion of immunoglobulin. Biosynthesis of secreted immunoglobulin has been studied extensively and is described in the following sections, terminating with a section on the biosynthesis of membrane immunoglobulin. Murine plasmacytoma cells tranplanted in Balb/c mice or grown in culture have provided the system for most of the studies on biosynthesis and secretion of immunoglobulin. Some information has also been obtained with cultured mouse lymphocytes stimulated by the polyclonal B lymphocyte activator lipopolysaccharide from gram negative bacteria. For studies on the biosynthesis of membrane immunoglobulin, lymphoma or lymphoblastoid cell lines have been used.

*1.4.2. Polyribosomes synthesising immunoglobulin chains.* The heavy and light chains of immunoglobulin have been shown to be synthesised on separate polyribosomes in lymphoid tissue from immunised animals and plasmacytomas (reviewed in [151]). This was shown by separation of the polyribosomes according to size on sucrose gradients and precipitation with antisera against heavy and light chains. The size of polyribosomes was as expected for synthesis of the complete chain, providing evidence that the H and L chains are each synthesised as a single unit and not as separate V and C regions. The size of H and L polyribosomes has also been measured directly by electron microscopy [152], giving a value of 11—18 ribosomes for H chain and 4—5 ribosomes for L chain, in agreement with the sucrose gradient data, though the size of the L chain polyribosomes may have been underestimated. Reports of immunoglobulin chains on smaller polyribosomes or monoribosomes are probably due to degradation or association of released immunoglobulin during extraction. Confirmation of these results has been obtained by analysis of the nascent chains on polyacrylamide gels, showing H and L chains to be associated with the appropriate size of polyribosomes [153].

Synthesis of immunoglobulin occurs mainly on the polyribosomes of the rough endoplasmic reticulum [154] and the nascent chains are vectorially released by extrusion through the membrane of the endoplasmic reticulum into the cysternae where assembly takes place [155, 157]. Segregation of immunoglobulin-synthesising polyribosomes to the endoplasmic reticulum is postulated to be effected by a hydrophobic precursor sequence (Section 1.4.3) at the N-termini of nascent light and heavy chains. This precursor peptide (about 20 amino acid residues long) is also thought to be responsible for leading nascent light and heavy chains into the cysternae, thus vectorially predetermining that Ig is either attached to the outside of the plasma membrane (Section 1.4.11) or is secreted from the cell.

*1.4.3. Cell-free synthesis.* Cell-free synthesis of H and L chains has been demonstrated in a variety of systems. Cell-free extracts of lymphocytes or myeloma cells containing microsomes or polyribosomes will incorporate radioactive amino acids into H and L chains and (where microsomes were used) assembled IgG molecules. Total poly(A)-containing RNA or partially purified mRNA from myeloma cells can be translated in heterologous cell-free translation systems. The H and L chains synthesised are identified by incorporation of radioactive amino acids, precipitation with specific antibody and analysis by polyacrylamide gel electrophoresis. Synthesis of H and L chains has been confirmed by identification of the peptides resulting from enzymic digestion of the polypeptides produced [158]. That the heavy and light chains synthesised in such a cell-free system are complete from N- to C-terminus, and even assemble within microsomes to form active antibody, has been demonstrated [155, 156]. Comparison of biosynthesis by membrane-bound and by free ribosomes has shown that the majority of immunoglobulin synthesis is carrried out by the former [159].

Cell free translation of L chain mRNA in systems lacking membranes has shown that the L chain is synthesised as a precursor molecule which has an extra peptide composed of 15—20, predominantly hydrophobic, amino acid residues at the N-terminus [160—165]. The heavy chain is also synthesised in a similar precursor form but the product of a cell-free translation system is slightly smaller than the secreted heavy chain due to the absence of carbohydrate [165—167]. These precursor peptides are rapidly cleaved in the intact cells and are believed to determine the vectorial release of the chains across the membrane of the endoplasmic reticulum, resulting in commitment of the molecules for membrane insertion or secretion as previously mentioned. Sequence studies in which labelled amino acids were used have shown the precursor peptides to be highly variable in sequence. For the light and heavy chain of a given immunoglobulin, the N-terminal precursor peptides show no significant sequence homology with one another [167a—169].

*1.4.4. Addition of carbohydrate.* Studies of attachment of sugar residues, both in plasmacytomas [170—175], normal lymphocytes [176, 177] and mitogen stimulated lymphocytes [178], were originally interpreted to mean that the oligosaccharide grows sequentially, *N*-acetylglucosamine, the sugar that is directly attached to the polypeptide chain, being added to IgG at about the time of completion of the polypeptide chains, possibly on the polyribosomes [179, 180]. Mannose is also added at an early stage. From the rough endoplasmic reticulum, immunoglobulin passes to the Golgi zone where glucosamine and galactose are also added as shown by cell fractionation and carbohydrate analysis [173, 181—184], and electron microscopy with autoradiography of cells labelled with radioactive amino acids or sugars [185]. Fucose and sialic acid are added shortly before secretion, as is galactose in the case of IgM synthesis [174, 186]. However, Behrens et al. [187], using liver microsomes, have demonstrated that the core sugars N-acetylglucosamine, glucose and mannose are added to glycoproteins as a single oligosaccharide unit attached to the lipid carrier dolichol phosphate. It seems very probable that this mechanism of glycosylation of proteins also applies to immunoglobulins.

It has been suggested that carbohydrate addition may be a necessary prerequisite for secretion of immunoglobulin, but the secretion of free L chains and other proteins, which do not normally contain carbohydrate, weighs against this hypothesis and there is no firm evidence for it.

*1.4.5. Assembly of four chain structures.* Light and heavy chains appear to be synthesised mainly at a balanced rate in normal lymphoid tissue and some plasmacytomas [188—190]. Many plasmacytomas secrete an excess of light chains, however, which may be a feature of malignant cells which readily lose the ability to produce heavy chain and ultimately light chain. An intracellular pool of light chains is, however, a feature of all immunoglobulin-secreting cells, including those in which heavy- and light-chain synthesis are balanced and it has been shown that light chains from this pool are drawn upon for assembly of immunoglobulin, even when synthesis is imbalanced [188—192].

*1.4.5.1. Non-covalent assembly.* It is difficult to determine the order of non-covalent assembly after release of H and L chains. Knowledge of the physical properties of heavy and light chains suggests that the heavy chains will associate rapidly to form dimers, and possibly polymers, provided the concentration is sufficiently high. Free light chains would then associate with the heavy chain dimers if they have not already done so on the polyribosome [151].

Analysis by sucrose gradient centrifugation of heavy- and light-chain synthesising polyribosomes has revealed the presence of completed light chains non-covalently associated with heavy-chain polyribosomes [193, 194] and of binding sites for free light chains [195]. Caution should be exercised in interpreting these results, however, owing to the difficulty of eliminating association of light chains with heavy-chain polyribosomes after disruption of the cell.

*1.4.5.2. Covalent assembly.* The order of covalent assembly, through the formation of disulphide bridges, is much more easily ascertained by means of pulse-labelling experiments with radioactive amino acids and analysis of the products, after precipitation of cell lysates with the appropriate antisera, by polyacrylamide gel electrophoresis under dissociating conditions (SDS and urea). By chasing with unlabelled amino acids, the flow of radioactivity from one intermediate to another may be determined. Experiments of this type have revealed two pathways of assembly:

(1) $2H \to H_2 \xrightarrow{+L} H_2L \xrightarrow{+L} H_2L_2$

(2) $H + L \to HL \xrightarrow{+HL} H_2L_2$

The mode of assembly varies with class, subclass and species but is consistent within those limitations. Patterns of assembly of various classes and subclasses of immunoglobulin are shown in Table VII.

It is interesting to note that selective reduction of immunoglobulin at increasingly low concentrations of reducing agent (mercaptoethanol, dithiothreitol or dithioerythritol) causes release of the same intermediates but in reverse order to that of assembly, e.g. reduction of mouse $IgG_2a$, human IgG or human IgA results in the release of free light chain, $H_2L$ and $H_2$, free heavy and light chains being released at high concentrations of reducing agent

TABLE VII

Intermediates in the covalent assembly of immunoglobulins

| Species | Immunoglobulin | Intermediate | | |
|---|---|---|---|---|
| | | $H_2$ | $H_2L$ | HL |
| Mouse | IgG2a | + | + | − |
| Mouse | IgG2b | + | + | + |
| Mouse | IgG1 | + | + | − |
| Mouse | IgM | − | − | + |
| Mouse | IgA | + | − | − |
| Rabbit | IgG | − | − | + |
| Human | IgG | + | + | − |
| Human | IgM | − | − | + |

Adapted from [151] by permission of Pergamon Press Ltd.

[196, 197]. Rabbit IgG and mouse IgM, on the other hand, are reduced to HL [198, 199]. Thus the pathway of assembly is determined by the oxidation-reduction potential of the cysteine residues involved. This suggests that the reducing conditions in the microenvironment of the cisternae where assembly takes place are intermediate between those required for formation of L—H and H—H disulphide bonds. An exception is mouse IgG2b in which both pathways occur (Table VII). This may be explained in either of two ways: (1) the L—H and H—H disulphide bonds have similar oxidation-reduction potentials; or (2) the reducing conditions of the microenvironment do not lie between those required for the formation of L—H and H—H disulphide bonds.

Although the intracellular pool of free light chain normally exists in the monomeric form with the thiol group that forms the interchain bridge being free [200—202], dimeric light chain ($L_2$) has been found intracellularly and recently it has been shown that $L_2$ could also serve as an intermediate in assembly, undergoing disulphide interchange in the process [202a]. In mouse IgA and human IgA2 the light chains exist in the dimeric form, associated with the heavy chains by noncovalent interaction only. Analysis of an IgA-secreting mouse plasmacytoma [204] has shown that dimerisation of the light chains is a slow process occurring mainly after secretion.

*1.4.6. Assembly of polymeric immunoglobulin*

*1.4.6.1. IgM.* The IgM antigen receptor on the surface of B lymphocytes exists in the form of an 8S four-chain molecule (Section 1.4.10). On the other hand IgM is secreted by plasma cells as a 19S pentamer (Section 1.2.1.3). The mechanism whereby polymerisation of the 8S subunit (IgMs) is controlled is controversial. Most human myelomas [204] and at least one mouse plasmacytoma [205, 206] contain both IgMs and a large pool of intracellular 19S IgM, whereas the majority of mouse plasmacytomas and mitogen-stimulated mouse B lymphocytes contain predominantly IgMs and only a small intracellular pool of 19S IgM, the 8S subunits polymerising shortly before secretion [191, 205, 207, 208]. Early studies on a mouse plasmacytoma showed that intracellular IgMs could not be induced to polymerise unless it was first reduced and reoxidised. Askonas and Parkhouse [209] believed that the intersubunit cysteine residues of the intracellular IgMs were reversibly blocked, but Melchers [210] showed that the polymerisation was only induced by reduction in the presence of a large excess of secreted IgM and concluded that intracellular IgMs is in a different physical state from the IgMs released by reduction of secreted IgM. Despite this inability to polymerise, the intracellular IgMs of all mouse plasmacytomas examined, including one containing a large intracellular pool of 19S IgM, were shown by the charge-shift technique [200, 201, 211] to possess at least two and possibly four free thiol groups per subunit. One of these pairs of cysteines was shown to be the penultimate residue of the $\mu$ chain which forms the intersubunit bridge and is therefore available for polymerisation [9, 212]. IgMs of slow turnover, which is inserted into the membrane of a mouse lymphoid tumour, was also found to have a similar structure [211]. Since the terminal sugar residues and J chain are added to IgM just before secretion it has been thought possible that either or both of these events may initiate the polymerisation process. Chapuis and Koshland [213] proposed a mechanism involving initiation of polymerisation by J chain based on the assumption, now shown to be false, that the intersubunit cysteine residues of intracellular IgMs are blocked. It has been shown that IgMs can be induced to polymerise in the presence of J chain and a disulphide interchange enzyme [214]. Unfortunately, it was not unequivocally demonstrated that the polymeric product formed was pentameric IgM. Evidence against the model was provided by the discovery that some types of human monoclonal IgM [215, 216] and the serum IgM of three species of fish [217] contain no J chain. Furthermore, several authors [218–220] have shown that IgMs obtained by selective reduction of serum IgM (which has the full complement of carbohydrate) will polymerise in the absence of J chain. Although a small proportion of the polymers obtained were not pentameric, Feinstein [221] using milder conditions, obtained pentameric IgM indistinguishable from serum IgM as judged by electron microscopy. Furthermore, Beale [222] was able to polymerise Fc$\mu$ to normal 12S Fc pentamers in the absence of J chain. Stott [206] has shown that 70% of the intracellular 19S IgM molecules in a mouse plasmacytoma (Y5781) have no J chain. This demonstrates that addition of J chain to IgMs is probably not the initiating event of polymerisation, since in that case it should be present on all 19S IgM molecules as in serum IgM, but is more probably a terminal event. This is consistent with the fact that the J—H disulphide bridge is the most sensitive to reduction and therefore would be expected to be the last to be formed by analogy with the patterns of assembly of heavy and light chains of all classes of immunoglobulin (Section 1.4.5). J chain was also detected on 3% of the intracellular IgM molecules in plasmacytoma Y5781 but not in another plasmacytoma (MOPC 104E), or mitogen-stimulated mouse B lymphocytes, which both contain only a very small pool of intracellular 19S IgM. Consequently, there appear to be two pathways for addition of J chain; (1) previously bound to IgMs or (2) as a free chain (Fig. 28). Both pathways occur in Y5781 but the second appears to predominate. Mitogen-stimulated lymphocytes may assemble IgM exclusively by pathway (2) since no J chain was detected on IgMs. Mitogen stimulation of B-lymphocytes has been shown to induce synthesis of J chain [223].

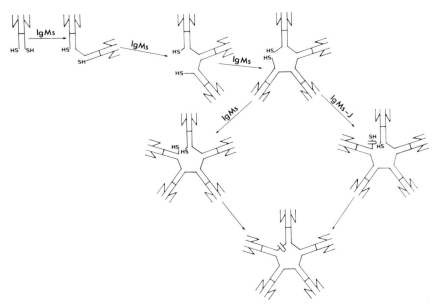

Fig. 28. Possible pathways of assembly of IgM. Polymerisation of IgM is envisaged as being initiated by a conformational change in the IgMs subunit enabling disulphide bridge formation to occur between two subunits, further subunits being added sequentially. The final step involves sealing the pentameric ring by insertion of J chain either free or bound to IgMs.

*References p. 315*

The initiation of polymerisation is, therefore, probably triggered by some structural change, e.g. addition of carbohydrate, enabling association of the subunits to take place followed by formation of disulphide bridges under appropriate oxidising conditions, possibly in the presence of a disulphide interchange enzyme. The pentameric ring structure is probably determined by noncovalent interactions between the Fc pieces and is sealed by the addition of J chain, linking together the first and last subunits.

*1.4.6.2. IgA.* As in the case of the majority of mouse IgM, both mouse and human IgA are polymerised from four-chain subunits just before secretion, since predominantly 7S IgA is found inside the cells. Unlike IgM, however, IgA is secreted as a heterogeneous mixture of monomer and dimer with small amounts of trimer and tetramer, the relative proportions varying from one plasmacytoma to another. Della Corte and Parkhouse [214] have shown that 7S IgA can be polymerised to the 11S dimer in the presence of J chain and a disulphide interchange enzyme, and have suggested that the availability of J chain within the cell may determine the degree of polymerisation. Measurement of the rates of synthesis of J chain and IgA (224) revealed a deficiency of J chain, which thus would favour secretion of monomeric IgA, which has no J chain. Further polymerisation beyond the dimer is probably not favoured due to the weak interactions between IgA monomer, as demonstrated by the fact that IgA subunits do not associate, whereas IgM subunits remain associated as 19S pentamers in non-dissociating solutions after selective reduction at concentrations of reducing agent that cleave the intersubunit disulphide bridges only [8, 225, 226].

Secretory IgA, but not serum IgA, contains a fourth polypeptide, secretory piece or secretory component (Section 2.1.2). Unlike heavy, light and J chain, secretory piece is not synthesised by the plasma cell [227] but by columnar epithelial cells lining the ducts of the secretory gland. Localisation of secretory piece and IgA by immunofluorescence suggests that assembly may take place in the apical cytoplasm of the epithelial cells, presumably after entry of IgA by pinocytosis [228].

*1.4.7. Synthesis of immunoglobulin during the cell cycle.* Studies on synchronised human lymphoid and mouse plasmacytoma cell lines have shown that synthesis of IgG and IgM reaches a peak during late G1 and early S phase, diminishing markedly during late S phase and G2 with minimal synthesis during mitosis [229—232].

*1.4.8. Hybrid cells.* Fusion of immunoglobulin-producing cells to form hybrid cell lines containing the chromosomes of both parent cells is a technique which is useful both for the analysis of immunoglobulin synthesis and for the production of monoclonal antibodies. Cell fusion involving two myeloma cell lines has been used to demonstrate that V-C integration does not occur after translation. Two plasmacytoma lines producing immunoglobulins with different variable and constant regions were fused and no scrambling of V and C regions could be detected [233, 234] (see also Section 1.4.9).

It is also possible to produce hybrid cell lines ("hybridomas") making antibody of any desired specificity. Spleen cells from an animal previously injected with the antigen are fused with cells from a plasmacytoma line. The resulting hybrids are cloned and clones secreting antibody of the desired specificity grown up in bulk culture. A plasmacytoma which does not synthesise immunoglobulin can be used for the fusion so that the monoclonal antibody produced by the resulting hybridoma is not admixed with myeloma protein and mixed products. Large amounts of antibody may be obtained by growing the cells as tumours in mice, provided syngeneic mouse cells have been used for the fusion. The technique was originally demonstrated by Köhler and Milstein [234] who obtained hybridomas secreting monoclonal antibody against sheep red blood cells, but there is no theoretical reason why hybridomas cannot be made with specificity for any antigen, provided it is capable of inducing an immune response in a suitable animal. Hybridomas have already been produced with specificity for a variety of antigens, especially cell surface antigens [235—239]. Advantages of the technique are the ability to isolate pure antibody with specificity for one antigen in a complex mixture or to study a particular antigenic determinant on a protein molecule. Since the tumours can be maintained, probably indefinitely, by serial transplantation they provide a continuous source of large amounts of antibody of defined specificity and affinity.

*1.4.9. Variable and constant region genes.* Since the variable and constant regions appear to be the products of separate genes it is important to establish whether V-C joining occurs at the level of gene integration, transcription or translation. The early evidence which was mainly against integration at the level of translation and to some extent, transcription has been reviewed in detail by others [151, 241] and is considered briefly here. The size of H and L chain mRNAs can be calculated approximately from the size of the polyribosomes and this was found to be adequate for the synthesis of complete polypeptide chains, indicating, but not proving, that joining does not take place during or after translation. More definitive experiments [242—245] involving pulse-labelling of immunoglobulin-producing cells for various short periods of time, showed that a gradient of radioactivity from the —COOH to the —$NH_2$ terminus of the heavy or light chain resulted. The single gradient showed that the chains grew from a single point at the $NH_2$-terminus. If the V and C regions were synthesised separately and then joined, two gradients would have resulted. Cell-free synthesis of L chain and H chain from their respective mRNAs using non-lymphoid protein synthesising systems such as wheat-germ and reticulocyte preparations, presumably lacking any hypothetical V—C joining enzymes, have confirmed the above conclusions. Furthermore, L chain mRNA has been purified sufficiently to allow sequencing, which has shown it to contain contiguous V and C regions with an untranslated sequence of nucleotides at each end [246], ruling out V—C joining at the translational but not at the transcriptional level.

Fusion of plasmacytoma cells synthesising different immunoglobulins has also shown no evidence of hybrid products formed from the variable region of one chain and the constant region of the other [233, 234]. Only mixing of complete chains occurred, i.e. the light chain from one cell was capable of combining with the heavy chain from the other cell in any of the possible combinations, thus arguing against the possibility of V—C joining after synthesis of separate V and C mRNA.

Recently, experiments involving nucleic acid hybridisation and gene cloning techniques have provided direct evidence that the V and C genes are separate in embryonic DNA and DNA from non-lymphoid organs but are joined in immunoglobulin-secreting cells, indicating a somatic rearrangement of V and C genes during maturation of lymphoid cells. Hozumi and Tonegawa [247] and Rabbitts and Forster [248] demonstrated this by digesting DNA from mouse embryo, adult liver or kidney (germ-line DNA) or plasmacytoma DNA with restriction enzymes, which cut the DNA at a small number of sites only, resulting in large fragments. These fragments were fractionated by preparative agarose gel electrophoresis and the fractions hybridised with $\kappa$ chain mRNA or complementary DNA (cDNA) prepared by using $\kappa$mRNA as a template for reverse transcriptase. The cDNA was synthesised under conditions such that it corresponded either to the variable plus constant regions ((V+C) cDNA) or to the constant region alone (CcDNA). Using these probes, DNA from mouse embryo or adult liver or kidney gave two peaks, one containing the C region gene and the other the $V_K$ gene. Mouse plasmacytoma DNA, however, gave only one peak containing both $V_K$ and $C_K$ genes. These results have been confirmed and extended to other $\kappa$-producing plasmacytomas by Seidman and Leder [249] using cloned DNA fragments. The fragments were characterised by agarose gel electrophoresis, hybridisation analysis and R-loop mapping. By these means it was shown that each plasmacytoma contains both the rearranged V+C genes and the embryonic pattern, indicating that the V and C genes are rearranged on only one chromosome of the pair.

Brack et al. [250] have demonstrated a similar rearrangement of mouse $\lambda$ V and C genes. Fractionation of restriction enzyme fragments of mouse embryo DNA produced three fragments corresponding to $V_{\lambda 1}$, $V_{\lambda 2}$ and $C_{\lambda 1}$. These were analysed by R-loop and heteroduplex mapping (Fig. 29). A $\lambda_1$-secreting plasmacytoma produced similar fragments plus a fourth containing both a $V_{\lambda 1}$ and a $C_{\lambda 1}$ sequence separated by a 1.2 kilobase (kb) intervening sequence or "intron". An intervening sequence separating the $V_K$ and $C_K$ genes of plasmacytoma DNA was demonstrated by analysis of the size of DNA or cDNA protected from S1 nuclease digestion by hybridisation of restriction enzyme fragments with $\kappa$mRNA or $\kappa$cDNA [248, 251] and by analysis of cloned immunoglobulin genes from plasmacytomas using the technique of R-loop mapping [249]. Nucleotide sequence analysis of cloned embryonic and plasmacytoma $\lambda$ genes has confirmed these results [257—259] and has revealed that the hydrophobic leader sequence of the precursor light chain (Section 1.4.3) is separated from the rest

*References p. 315*

of the variable region by an untranslated region of 93 base pairs ($I_1$, Fig. 30). Sequence analysis of cloned DNA fragments containing the $C_{\gamma 1}$, $C_{\gamma 2b}$, $C_\alpha$ and $C_\mu$ genes demonstrated that the sequences coding for the hinge region and constant region domains of the heavy chains are also separated by short intervening sequences [256, 409—413]. Intervening sequences are not unique to immunoglobulin genes as they have been found to interrupt the coding sequences of other eukaryotic structural genes, e.g the gene coding for ovalbumin is composed of seven coding segments and six intervening sequences or "introns" [252, 253]; the chick ovomu-

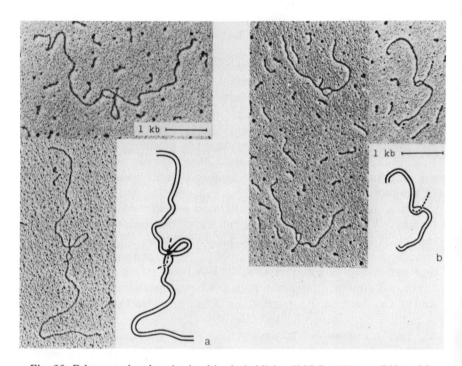

Fig. 29. R loop molecules obtained by hybridising HOPC 2020 $\lambda_1$ mRNA with the EcoRI fragments of the embryo DNA clones. (a) Ig25λ DNA displays one R loop corresponding to the constant-region gene, the double-stranded DNA loop and a long RNA tail corresponding to variable region sequences. The short tail observed in some molecules is the 3' poly(A) tail. (b) Ig99λ DNA has one R loop corresponding to variable region sequences and a long RNA tail that is composed of the constant region gene sequences plus poly(A) tail.
Reproduced from [250] with the permission of the MIT Press.

coid gene also contains at least six intervening sequences [254], and the globin gene contains two [255].
In addition, Brack et al. [250] and Bernard et al. [414] found a 40-bp segment "J" for joining (not to be confused with the J chain of IgA and IgM, Sections 1.2.1.2. and 1.2.1.3) 1.2 kb upstream from the 5' end of the $C_{\lambda 1}$ gene in embryonic DNA and this was homologous with an amino acid sequence at the V—C junction of the light chain. J was also found at the boundary of V and the 1.2 kb untranslated DNA in the integrated gene of the plasmacytoma.

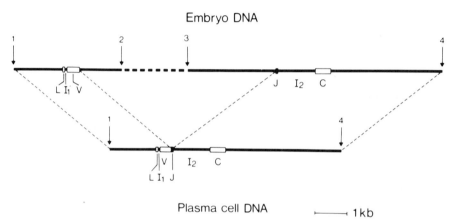

Fig. 30. Arrangement of mouse $\lambda_1$ gene sequences in embryos and $\lambda_1$ chain-producing plasma cells. In embryo DNA, a full $\lambda_1$ gene sequence consists of two parts that lie on two separate EcoRI fragments. On one of these fragments, the coding sequence is further split into two parts, one for most of the leader peptides (L) and the other for the rest of the leader peptides plus the variable region peptides (V). The two coding sequences are separated by a 93-nucleotide intervening sequence ($I_1$). On the second EcoRI fragment, the coding sequence is also split into two parts by a 1250-nucleotide intervening sequence ($I_2$). The two parts code for the constant region (C) and approx. 13 residues near the junction of the variable and constant regions (J). The relative orientation of and the distance between the two EcoRI fragments are unknown. In the DNA of a $\lambda_1$ chain-producing myeloma (H2020), the $\lambda_1$ gene sequence is rearranged as a result of one (or more) recombination(s) that involve(s) sequences in the two embryonic EcoRI fragments. One recombination takes place at the ends of the V and J sequences and brings the two sequences into direct contact. The limits of the corresponding sequences in the embryo and the myeloma DNAs are indicated by thin dotted lines. Arrows with numbers indicate EcoRI sites. Reproduced from [250], with the permission of the MIT Press.

References p. 315

Thus the V and J-C genes are separated in the germ-line DNA and undergo a somatic rearrangement which joins them together with the "intron" ($I_2$) between them in the plasma cell (Fig. 30). This event presumably occurs during the maturation of a lymphoid stem cell with the sequence upstream from J acting as a recognition signal for joining of the V gene.

It has since been shown that five different J sequences are associated with the $C_K$ gene and four with the $C_\mu$ gene [415—419]. Since recombination of V can occur with any one of four J sequences (one of the $J_K$ sequences appears to be non-functional) and at alternative sites within the last V codon and the first or second J codon, this allows considerable diversity at the third hypervariable region. Furthermore, part of the third heavy-chain hypervariable region appears to be encoded by neither J nor the $V_H$ gene, so Hood et al. [417, 420] have postulated an additional set of gene segments "D" (for diversity) which make up the missing part of the third hypervariable region. This would allow even further generation of diversity at this site if there should be similar mechanisms operating for recombination of $V_H$ with multiple D sequences, followed by recombination of $V_H$-D with $J_H$ to form a $V_H$-D-$J_H$ sequence.

By means of S1 nuclease protection experiments similar to those described above, Rabbitts [260] has shown that the high molecular weight nuclear RNA from a κ chain-producing plasmacytoma also contains an insertion between V and C, whereas cytoplasmic κ mRNA does not, indicating that the light-chain gene is transcribed faithfully, including the intervening sequences, and the latter are then removed by excision and ligase enzymes to form cytoplasmic mRNA. Pulse-chase labelling and hybridisation analysis with cDNA have established this precursor—product relationship for κ mRNA and $\gamma_{2b}$ mRNA [261, 262]. A similar sequence of events appears to take place following transcription of the β-globin gene [262a]. An interesting relationship between a gene and its product has been revealed by the sequence of the $C_\mu$ gene [412, 413] since, in addition to the coding sequences for the four constant region domains and the C-terminal tail of the μ chain of secreted IgM ($\mu_s$), two sequences were found 1.6 kb downstream from the 3' end of the tail sequence which together code for the hydrophobic tail of the μ chain of membrane IgM ($\mu_{mem}$) (Sections 1.4.10.2 and 1.4.10.3). Thus the same gene codes for two products, $\mu_{mem}$ apparently being formed by transcription of the entire gene followed by excision of the sequence coding for the tailpiece of $\mu_s$ and the intron between each of the coding sequences from the nuclear RNA. It may be presumed that $\mu_s$ is formed either by termination of transcription at the 3' end of the gene or by transcription of the whole gene followed by excision of the $\mu_{mem}$ tail and introns.

For further discussion of the structure, rearrangement and expression of immunoglobulin genes and the effect of work in this field on our understanding of the generation of diversity and the mechanism of immunoglobulin class switching during the clonal development of B-lymphocytes, the reader is referred to recent reviews [421–423].

*1.4.10. Cell membrane immunoglobulin*

*1.4.10.1. General properties.* The ability of certain classes of immunoglobulin to bind to the membranes of various types of cells has been discussed in relation to function in Section 1.3.5. In the case of macrophages, monocytes, neutrophils, basophils and mast cells, the membrane immunoglobulin is extraneous, having been picked up from the surrounding medium and not synthesized by the cell that bears it. It is not, therefore, monospecific. In contrast, lymphocyte membrane immunoglobulin is synthesized by the cell that bears it and, since one lymphocyte can only synthesise immunoglobulin of one specificity, all the molecules on the surface of the same cell bear the same set of variable regions. This ensures that a B lymphocyte binding antigen to its membrane immunoglobulin, which acts as a receptor for that antigen, is stimulated to divide and differentiate into plasma cells that synthesise and secrete antibody of the same specificity. Since this hypothesis is central to an understanding of the immune response, a great deal of work has been carried out on the nature of the lymphocyte antigen receptor.

Cell-surface immunoglobulin has been detected by a variety of methods, e.g. fluorescent anti-immunoglobulin, radio-labelled anti-immunoglobulin, the cytotoxic effect of anti-immunoglobulin, lactoperoxidase-catalysed iodination of the membrane proteins with $^{125}I$ or $^{131}I$, and pulse-chase labelling with radioactive amino acids. Fluorescent antibody has been used widely to study the distribution of the different classes of immunoglobulin on lymphocytes in normal individuals and in various diseases. A difficulty encountered here is that lymphocytes also have Fc receptors (Section 1.3.5.1) so it is essential to ensure that the surface immunoglobulin is not derived from the serum as a result of binding by these receptors. Taking this into account, the general picture is that the majority of B lymphocytes in a healthy individual (human or mouse) bear *both* IgM and IgD with small numbers bearing either IgM, IgD or IgG alone. Very few lymphocytes bear surface IgA or IgE. This is in marked contrast to the distribution of immunoglobulins in the serum where IgG is the predominant class, IgM being present in relatively small amounts and only trace amounts of IgD are normally present. Moreover, whereas serum IgM is exclusively in the 19S pentameric form, membrane IgM has been shown to be in the form of the 8S monomeric subunit (q.v.).

Another important observation was that of Pernis et al. [263] who showed, using fluorescent antibody specific for a single allotype, that each B-lymphocyte in a heterozygous rabbit expresses only one allotype, showing that the gene from only one chromosome is expressed in each cell. The choice of expression of maternal or paternal chromosome is apparently non-random. Allelic exclusion was subsequently confirmed for human lymphocytes with the Gm markers [264]. Reports to the contrary [265] could be due to the uptake of serum immunoglobulin by Fc receptors, or due to the markers being studied not being true alleles.

An interesting phenomenon exhibited by membrane immunoglobulin and other membrane components is that of "capping". This may be demonstrated using fluorescein or radio-labelled anti-immunoglobulin [266, 267] or by electron microscopy [268]. If fluorescein-labelled anti-immunoglobulin is added to B lymphocytes at $4°C$ it uniformly coats the surface of the cell, appearing as a ring of fluorescence under the light microscope (Fig. 31a), showing the uniform distribution of the membrane immunoglobulin. Upon warming to room temperature or $37°C$ this ring of fluorescence breaks up to form patches which accumulate at one pole of the cell to form a "cap", which is then ingested by the cell and disappears (Fig. 31b and c). If the cells are incubated for 12—24 h resynthesis of immunoglobulin occurs and it can again be detected on the cell surface [266, 267].

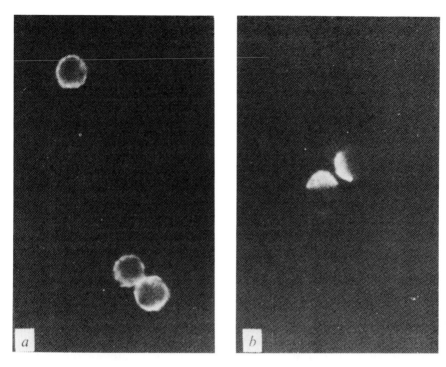

Fig. 31. Pattern of immunofluorescence of mouse splenic lymphocytes incubated in fluorescein-conjugated rabbit anti-mouse Ig for 30 min, (a) at 0°C ("ring" pattern), (b) at room temperature ("cap" pattern). Reproduced from [266] with the permission of Macmillan Journals Ltd.

The interpretation of this phenomenon is that the immunoglobulin is not rigidly fixed in the membrane but is floating freely in the lipid bilayer. At room temperature or above, a two-dimensional aggregation occurs analagous with the formation of immune precipitates, resulting in the formation of patches. These are then transported to one pole of the cell where they are ingested due to the active flow of the membrane. Cap formation is an active process and is inhibited by metabolic inhibitors such as sodium azide, dinitrophenol, oligomycin etc., whereas patch formation is not energy dependent. Capping of membrane molecules provides evidence for the fluid mosaic model of the cell membrane postulated by Singer [269].

*References p. 315*

The amount of membrane-bound immunoglobulin on lymphocytes has been estimated by measuring the degree of inhibition of binding of anti-Ig to radio-labelled mouse Ig, Fab or Fc by mouse lymphocytes [270, 271]. A calibration curve is plotted from measurements with unlabelled mouse immunoglobulin and the amount of immunoglobulin giving the same degree of inhibition as a known number of lymphocytes is estimated. This gives a figure of $5 \times 10^4 - 1.5 \times 10^5$ molecules per B lymphocyte.

*1.4.10.2. Structure.* Iodination of mouse splenic lymphocyte membrane proteins by the lactoperoxidase technique followed by lysis in detergent, precipitation with class-specific anti-immunoglobulin and analysis of the precipitates by polyacrylamide gel electrophoresis with and without reduction has revealed the major membrane immunoglobulin to be an 8S monomeric IgM composed of $\mu$ and light chains [272—274]. An immunoglobulin with properties similar to human IgD is also labelled by this method [275]. IgD was not detected in the earlier studies since it comigrates with 8S IgM on 5% polyacrylamide gels with and without reduction, and is rapidly degraded after cell lysis. On higher percentage gels the $\mu$ and $\delta$ chain are readily distinguishable. IgD does not appear on the membranes of lymphocytes from young mice until they are about 2 weeks old, whereas membrane IgM is detectable on the splenocytes of newborn mice [276]. Immunofluorescence studies of human peripheral blood lymphocytes [277] have shown that single cells bear both IgM and IgD on their surface and that IgD is resynthesised after capping, showing that it is not passively acquired from the plasma. The IgM and IgD present on the lymphocytes of patients with chronic lymphocytic leukaemia or Waldenström's macroglobulinaemia, or on normal lymphocytes, share the same idiotype [278] and antigenic specificity [279, 280], and therefore presumably the same variable regions. Therefore, unless the mRNA for one of these immunoglobulins is extremely long lived, it appears to be possible for one variable region gene to be shared between two constant region genes and both forms expressed simultaneously.

How lymphocyte immunoglobulin is bound to the membrane has been a problem since its discovery. It is assumed that it is bound by the Fc region with the $F(ab)_2$ exposed to the medium, since membrane immunoglobulin binds both antigen, anti-Fab and some, but not all, anti-Fc. Fu and Kunkel [281] have shown that antibody directed against the C-terminal 50 amino acids of the $\mu$ chain does not bind to the surface IgM of human leukaemia cells, suggesting that it is masked by the plasma membrane. There are, therefore, two possible modes of binding to the membrane: (1) The molecule may bind directly to the membrane via a hydrophobic region of the Fc or (2) it may bind to another membrane protein, such as the Fc receptor, which has a hydrophobic region inserted into the lipid bilayer and a hydrophilic region with affinity for the Fc. The problem with the first hypothesis is that serum IgM has no contiguous hydrophobic region in the Fc that could be used for binding to the lipid bilayer. There are, however, differences in properties between secreted IgM subunits and membrane IgM, e.g. $\mu$ chains of membrane IgM bind more detergent than secreted $\mu$ chains [282], membrane IgM is less dense than intracellular or secreted IgM [283] and is less soluble in the absence of detergent than secreted IgM [283a], suggesting that membrane-bound IgM has hydrophobic regions not present on secreted IgM. Studies on a mouse lymphoblastoma which has surface IgM and secretes both 19S and 8S IgM showed that pulse-chase labelled membrane IgM had the same number of free thiol groups (two or four) and the same isoelectric point as the cytoplasmic IgM, whereas the secreted 8S IgM had a lower isoelectric point, possibly due to the addition of sialic acid residues, and its thiol groups were blocked, probably due to the formation of internal disulphide bridges [284]. There are also several reports of size differences between membrane-bound $\mu$ chains and secreted $\mu$ chains. One report [285] suggested that membrane-bound $\mu$ chains are heterogeneous in size and similar to or smaller than secreted $\mu$ chains, but this could have been the result of degradation. Other workers [286, 287] have found that $\mu$ and $\delta$ chains from murine spleen cell membrane immunoglobulin have a higher molecular weight than the $\mu$ and $\delta$ chains of secreted immunoglobulin. In none of these studies could size differences be clearly ascribed to peptide or carbohydrate differences.

However, recent studies have defined three different size classes of $\mu$ chain [288, 289]. Membrane-bound $\mu$ chains have the highest apparent $M_r$, about 2000 greater than secreted $\mu$ chain which in turn is about 2000 greater than the intracellular $\mu$ chain which is the precursor of secreted $\mu$ chain. The difference in size between intracellular and secreted $\mu$ chains is believed to be due to addition of terminal carbohydrate. It has been shown that cell lines producing both membrane-bound and secreted immunoglobulin synthesise $\mu$ chains of all three sizes. $\gamma$ Chains from the membrane of a cell line which produces membrane IgG and secreted IgG were also larger than the secreted $\gamma$ chains [288].

Analysis of the $\mu$ chains synthesised in the presence of the drug tunicamycin [290], which blocks $N$-glycosylation, allows comparison of the sizes of the polypeptide chains without the complication of carbohydrate groups. Two intracellular $\mu$ chains have been defined, differing in apparent $M_r$ by about 2000. The smaller of the two $\mu$ chains appears to be the precursor of the secreted $\mu$ chain and secretion does not require glycosylation. The larger $\mu$ chain is thought to be the precursor of membrane bound $\mu$ chain but deposition of $\mu$ chain on the cell surface is blocked in the absence of glycosylation. Analysis of the C-terminal residues of these $\mu$ chains by carboxypeptidase digestion showed that the smaller $\mu$ chain has C-terminal tyrosine, as does secreted $\mu$ chain, whereas the larger chain has no C-terminal tyrosine, suggesting that it has an extra C-terminal peptide.

McIlhinney et al. [291] have reported that they were unable to detect any difference in the C-termini of $\mu$ chains from membrane and secreted IgM. By contrast Williams et al. [291] reported finding extra hydrophobic residues at the C-terminus of membrane $\mu$ chain relative to the C-terminal tyrosine found on secreted $\mu$ chain. This discrepancy could be due to a difference in the method of isolation or of carboxypeptidase digestion. In conclusion the data suggest that membrane IgM is an integral membrane protein held by the insertion of a C-terminal hydrophobic tail peptide of about 20 amino acid residues in length.

The above studies have all been carried out on either B-lymphocytes or total populations of lymphocytes in which the majority of the membrane-bound immunoglobulin comes from B-lymphocytes. There has been a great deal of controversy for several years concerning the nature of the T-cell antigen receptor as it has been extremely difficult to demonstrate the presence of immunoglobulin on the surface of T-lymphocytes by techniques that have been successful for B-lymphocytes, although immunoglobulin-like molecules have been detected by very sensitive immunological techniques [293, 294]. Marchalonis and Cone [295, 296] have detected an 8S IgM-like molecule by iodination of T-lymphocytes but other workers [297—299] have been unable to reproduce these experiments. Feldmann et al. [300] have shown that this putative T-cell immunoglobulin is capable of binding to macrophages and initiating a response to antigen by B-lymphocytes. It has been suggested that the T-cell antigen-receptor may be a new class of immunoglobulin (IgT) that cross-reacts with IgM; alternatively it may be a molecule unrelated to immunoglobulin. The controversy concerning T-lymphocyte immunoglobulin has been extensively reviewed by others [301—304].

*1.4.10.3. Biosynthesis and turnover.* Studies on murine spleen cells and human lymphoma cell lines [178, 275, 299, 305—307] suggest that membrane-bound immunoglobulin is synthesised in the rough endoplasmic reticulum and transported to the Golgi apparatus in a similar manner to secreted immunoglobulin. Small lymphocytes contain very little endoplasmic reticulum but this may be sufficient for synthesis of the small amount of surface immunoglobulin present on these cells. The transit time from the site of synthesis to the cell membrane is 1—2 h, as judged by the time taken for label to appear in surface IgM [308]. This suggests that membrane immunoglobulin is not synthesised close to the plasma membrane.

Pulse-chase experiments [178] have shown that murine splenocytes synthesis immunoglobulin of two types, one with a half-life of about 4 h and the other with a half-life of 20 h or more. Iodination of cell surface immunoglobulin gives similar results [309, 310]. It is possible that this is due to the presence of two populations of cells, the IgM of high turnover rate being produced by lymphoblasts and that of slow turnover rate produced by small lymphocytes. In support of this hypothesis, it was found that the IgM synthesised by mouse plasmacytomas has a high turnover rate while that synthesised by tumours resembling early stimulated lymphocytes is predominantly of the type with a long half-life [311]. The IgM of slow turnover rate is released into the medium as an 8S molecule and is removed by capping with anti-IgM, showing that it is on the cell surface [178, 308].

Vitetta and Uhr [308] claim to have detected incorporation of galactose and fucose, as well as glucosamine into the slow-turnover IgM. Melchers and Andersson [178], however, showed incorporation of mannose, but were unable to detect incorporation of galactose or fucose. The reason for this discrepancy is not clear, but it may just reflect heterogeneity of cell types.

mRNA from cell lines synthesising both surface and secreted immunoglobulin has been translated in a cell-free system under conditions in which carbohydrate addition and post translational processing do not occur [425]. Two sizes of $\mu$ chain were produced, suggesting the existence of two species of $\mu$ chain mRNA; one coding for $\mu$ chain destined for secretion as 19S IgM, the other coding for $\mu$ chain destined for incorporation into the cell membrane as 8S receptor IgMs.

In summary, lymphocyte membrane immunoglobulin is probably synthesised by polyribosomes on the rough endoplasmic reticulum and released vectorially into the cysternae in the same way as secreted immunoglobulin, probably coded by a separate mRNA with a coding sequence for an extra C-terminal hydrophobic peptide (Fig. 32). Assembly of heavy and light chains occurs in the cisternae of the endoplasmic reticulum and anchoring to the membrane by the hydrophobic tail of the Fc may occur at this point. Addition of carbohydrate probably commences on the nascent chains and continues in the Golgi apparatus. Vesicles containing membrane-bound immunoglobulin migrate to and fuse with the cell membrane (reverse pinocytosis), the bound IgM thus becoming exposed to the environment as a result of eversion of the vesicle. IgM lacking the anchoring site at its C-terminus is not anchored to the membrane of the vesicle and is able to polymerise to 19S IgM, possibly as a result of the addition of the terminal sugar residues, the pentameric structure being sealed by J chain. On fusion of the vesicle with the cell membrane the free 19S IgM is released into the environment.

### 1.5. Catabolism

*1.5.1. Definitions and methodology.* Before considering the catabolism of immunoglobulins it is necessary to define the terms commonly used. These are: (i) Plasma pool: The quantity of immunoglobulin in the plasma per unit body weight, normally expressed as mg of immunoglobulin/kg body weight. (ii) Distribution ratio: The ratio of total immunoglobulin in the plasma to that in the whole body. (iii) Plasma half-life ($T_{\frac{1}{2}}$): The time taken for half of

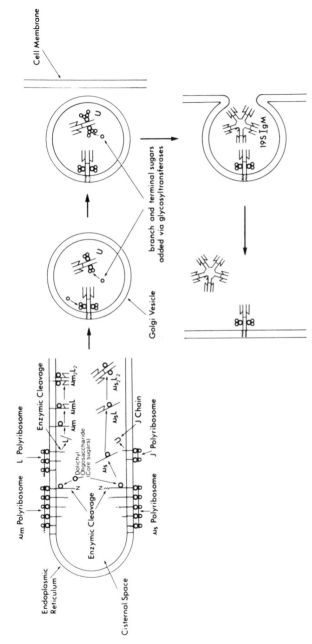

Fig. 32. A possible mechanism of synthesis, transport and release of membrane-bound and secreted IgM. μ chains destined for secretion ($\mu_s$), μ chains destined for insertion into the cell membrane ($\mu_m$), light chains (L) and J chains (U) are coded for by separate mRNAs and synthesised on separate polyribosomes on the rough endoplasmic reticulum. The nascent polypeptide chains with their hydrophobic ends (∿∿∿) are released vectorially into the cisternae where addition of the core oligosaccharide as a single unit occurs by transfer from dolichyl oligosaccharide before or possibly after release of the nascent chains. The assembled IgM is then transported to the Golgi apparatus where addition of branch sugars takes place. Membrane IgM binds to the membrane through its hydrophobic tail sequences at the C-termini of the μ chains, whereas IgM destined for secretion polymerises to 19S IgM, possibly as a result of a conformational change induced by addition of the terminal sugar residues. J chain is probably added at the final stage, free or bound to an IgMs subunit (Fig. 28) to complete the ring closure.

*References p. 315*

the immunoglobulin in the plasma to be catabolised. This is measured from the linear part of the plasma immunoglobulin decay curve with the use of a semi-logarithmic plot. (iv) Fractional catabolic rate: The percentage of total plasma immunoglobulin catabolised per day. This is estimated from either the plasma half-life by the formula: Fractional catabolic rate $= \frac{\ln 2}{T_{\frac{1}{2}}}$, or from the rate of excretion of products of catabolised immunoglobulin in the urine.

Methods used for studying immunoglobulin catabolism involve injection of immunoglobulin in a form that can readily be measured, estimation of the plasma pool by dilution analysis and plotting the rate of loss of immunoglobulin from the plasma or the rate of appearance of catabolic products in the urine. The injected immunoglobulin may be normal immunoglobulin in immunodeficient patients, antibody which is then measured by its antigen specificity, or isotopically labelled immunoglobulin. The latter may be labelled internally with radioactive amino acids or externally by means of $^{125}I$ or $^{131}I$. Errors encountered in such studies may result from reutilisation of radioactive amino acids, altered catabolism of the tracer immunoglobulin due to denaturation or its effect on the total immunoglobulin concentration, or contamination of the tracer with other classes and subclasses of immunoglobulin having a different catabolic rate. Errors may also arise as a result of incorrect assumptions about equilibration between intravascular pools, the existence of a steady state of balanced synthesis and catabolism and other causes.

*1.5.2. Catabolic rates of immunoglobulin classes and subclasses.*
Both synthetic and catabolic rates vary with the class and subclass of immunoglobulin (Table VIII). For example, IgG1, 2 and 4 have similar distribution ratios, plasma half-lives and fractional catabolic rates, whereas a higher proportion of IgG3 is found in the plasma; it also has a shorter half-life and higher fractional catabolic rate, presumably due to structural differences in the γ3 chain.

A much higher proportion (74%) of IgM is found in the plasma and it is catabolised much more rapidly than IgG, having a shorter

## TABLE VIII
### Data on immunoglobulin turnover in man

| Immuno-globulin | Plasma pool (mg/kg) | Distribution ratio | Total body pool (mg/kg) | Plasma $T_{\frac{1}{2}}$ (days) | Fractional catabolic rate (% per day) | Synthetic rate (mg/kg/day) |
|---|---|---|---|---|---|---|
| IgG | 280–820 (500) | 0.32–0.64 (0.52) | 570–2050 (1030) | 14–28 (21) | 4.3–9.8 (6.9) | 20–60 (36) |
| IgG 1 | 298 | 0.51 | 589 | 21 | 8.0 | 25.4 |
| IgG 2 | — | 0.53 | — | 20 | 6.9 | — |
| IgG 3 | 20 | 0.64 | 31 | 7 | 16.8 | 3.4 |
| IgG 4 | — | 0.54 | — | 21 | 6.9 | — |
| IgA 1 | 101 | 0.55 | 185 | 5.9 | 24.0 | 24 |
| IgA 2 | 14.0 | 0.55 | 24.5 | 4.5 | 34.0 | 4.3 |
| IgM | 23 | 0.74 | 36 | 5.1 | 10.6 | 2.2 |
| IgD | 0.83 | 0.75 | 1.1 | 2.8 | 37.0 | 0.4 |
| IgE | 0.004 | 0.41 | 0.01 | 2.7 | 94.3 | 0.004 |

Numbers in parentheses are mean values. Reproduced from [312], by permission of Lange Medical Publications.

half-life and higher fractional catabolic rate in addition to a much lower synthetic rate. IgA2 has a higher catabolic rate than IgA1 which in turn has a much higher turnover than IgG. As expected, their half-lives show similar differences in reverse order.

IgD also has a high fractional catabolic rate and short half-life with a low synthetic rate. The concentration and rate of synthesis of IgD vary widely. IgE has by far the highest fractional catabolic rate and lowest synthetic rate of all the immunoglobulins.

Free light chains are rapidly eliminated due to both a high fractional catabolic rate and excretion in urine without degradation. Impaired renal function results in reduced light chain catabolism indicating that this occurs at least partly in the kidney.

*1.5.3. Control of immunoglobulin catabolism.* The fractional catabolic rate of IgG is directly proportional to its serum concentration and this phenomenon is known as the "concentration-catabolism effect". It is observed in diseases involving both hypo- and hypergammaglobulinaemia, e.g. in sex-linked hypogammaglobulinaemia the fractional catabolic rate is reduced, although this is not true for all diseases with low serum immunoglobulin. Conversely, a high immunoglobulin turnover is associated with the high concentrations of IgG in multiple myeloma of the IgG type, with the result that the normal serum IgG level is depressed resulting in susceptibility to infection. Catabolism of IgM and IgA is, on the other hand, independent of serum concentration, while the fractional catabolic rate of IgE, and possibly IgD, is inversely proportional to the serum concentration. The site and mechanism of control of catabolism are not known, although the liver, gastrointestinal tract, kidney and the whole reticuloendothelial system are all possible sites.

The catabolic rate of IgG is not altered by removal of sialic acid residues (by enzymic digestion) or by reduction. Fab and $F(ab')_2$ fragments have very short half-lives (3—5 h), as does the pFc' fragment, which corresponds to the $C_\gamma 3$ domain, whereas the Fc fragment has a much longer half-life, approaching that of intact IgG which suggests that the catabolic rate is controlled by the $C_\gamma 2$ domain [313, 314]. This is supported by the results of Dorrington and Painter [315] who showed that the isolated $C_\gamma 2$ domain of human IgG has a similar half-life to Fc and IgG (60—

# INTRODUCTION

70 h) in rabbits, whereas the $C_H3$ domain and Fab have half-lives of only 16—17 h. Brambell et al. [316] have proposed a mechanism for the control of IgG catabolism similar to their transport mechanism (Section 1.3.4). In this model IgG becomes attached to Fc receptors in pinocytotic vesicles, possibly on the microvilli of the intestinal mucosa. This leads to protection of the molecules from catabolism whereas those molecules which are free in solution in the vesicles are degraded by proteolytic enzymes. The attached molecules are then returned to the circulation by some unexplained process.

## 2. Histocompatibility antigens

### 2.1. Introduction

All immune responses are regulated by antigens. The histocompatibility antigens encoded by a gene complex which is now referred to as the major histocompatibility complex (MHC) play a very special role in immune responsiveness. These antigens were initially recognised by the most pronounced effect which they exert on control of allograft rejection. It has been shown in a variety of species that alloantigenic differences at the MHC predetermine rapid graft rejection. The historical link between antibodies and the antigens involved in rejection of grafts was provided by Landsteiner [321] who, on the basis of his important work on ABO erythrocyte antigens predicted the existence and encouraged the search, for other blood group antigens important in matching for transplantation. Evidence that sensitivity to transplanted tumours in mice shows strain dependencies was being obtained at about the same time as Landsteiner's work defining the ABO system in man. In 1914 Little [322], using classical genetic techniques, showed that many genes were involved in controlling susceptibility to a tumour graft. Gorer [323] described what we now know to be the major histocompatibility antigens of the mouse. He showed that these antigens which were crucial in determining susceptibility to tumour transplantation were also present on red blood cells. Snell [324] described his development of congenic lines of mice selected for resistance to tumour grafts. The term histocompatibility antigen was introduced by Snell and the H2 system described previously by Gorer became recognised as the major histocompatibility system.

Parallel with the work on mice, Medawar was studying the rejection of skin grafts [325], having been led to this work by the problems encountered by surgeons involved in carrying out major skin grafts on military personnel during the second world war. Medawar used the rabbit as his model system and noticed that the relevant histocompatibility antigens were present on leukocytes but not on red blood cells, as judged by presensitisation for rejection of skin grafts. In man leukocyte antigens were described in the early sixties. As in the rabbit, human histocompatibility antigens are not detectable on red blood cells, in contrast with the mouse and chicken. Independently, van Rood [326] and Payne [327] showed that leukocyte agglutinins, first noted by Dausset [328] in the serum of patients who had received multiple blood transfusions were also present in multiparous women as a result of foetal-maternal immunisation. These workers used such alloantisera to define two series of antigens termed "Group four" and LA. These two antigens are encoded by two separate, but closely linked genes now known as HLA-A (LA) and HLA-B (four). Each individual has two alleles at each genetic locus. There is extensive polymorphism with more than twenty alleles being recognised at each locus. With the subsequent discovery of the polymorphic loci HLA-C and HLA-D also closely linked to HLA-A and -B it has been calculated that the known alleles of these four loci can generate more than $3 \times 10^8$ genetically different individuals [329]. Clearly the MHC is a major factor determining individuality in the immune system. However, this fact alone would not necessarily warrant special consideration of the biochemical properties of the products of the MHC in this chapter. The MHC also has major control functions in the immune system.

Products of the MHC play a role in regulating immune responses, both cellular and humoral, to the vast majority of antigens. These controls manifest themselves in a variety of ways each apparently reflecting involvement of a different MHC product operating at a different level in the immune response. The discovery that genetic control of the production of antibodies to certain antigens is linked to the MHC was described by McDevitt and Chinitz [330]. The genes controlling such antibody specificities are termed the immune response (Ir) genes. They have been extensively mapped and studied in mice and are located in the I region of the MHC. Analagous genes have been identified in a variety of other species [331]. The identification of individual Ir genes by studies of defects in immune responsiveness was followed by a series of experiments which indicated that products of the I region (immune associated, Ia, antigens) are involved in co-operation between T lymphocytes and B lymphocytes necessary for antibody production [332]. The Ia antigens also regulate the presentation of antigen by macrophages to T lymphocytes. Most recently the studies of Zinkernagel and Doherty [333] on the specificity of cytotoxic T cells directed against virus infected target cells demonstrated that killing was restricted to target cells sharing major histocompatibility antigens (K or D antigens of the mouse H2 system) with the cytotoxic T cells. It now appears that this MHC restriction for cell mediated immunity is a general phenomenon [334—336] with the specificity of restriction being dictated by the thymus during differentiation of T lymphocytes [337].

Mechanisms for the highly specific and important role of MHC products in immune responsiveness have not been established, although interesting hypotheses abound. It seems likely that proper understanding of the mode of action of each of the MHC products will require detailed knowledge of their biochemistry. Current information concerning the products of the MHC, although less extensive than our knowledge of antibodies, is presented in the following sections.

*2.2. Genetics and serology of the major histocompatibility complex*

*2.2.1. HLA and H-2.* Genetic maps of the MHC of mouse (H2) and man (HLA) are shown in Fig. 34. Genetic analysis of the mouse MHC is more extensive than in other species because of the availability of inbred stains and congenic strains [338]. The latter differ, at a single selected locus, from the congenic strain from which they are derived. The construction of congenic lines is illustrated in Fig. 33. However, clear recognition that the polymorphisms of the major antigens of H2 define two genetic loci, K and D, dates only from 1971 and resulted from independent analysis of the data by Snell [339], Shreffler [340] and Thorsby [341]. A comprehensive and accessible account of H2 genetics has been

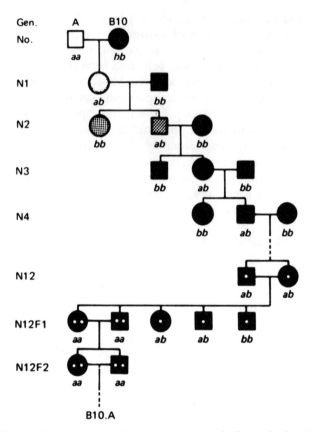

Fig. 33. The backcross system of producing congenic lines of mice. Gen. No., generation of backcrossing; A, donor strain; B10, background strain; a and b are the alleles of the gene being selected from the donor strain. The final congenic strain B10A is identical with B10 mice at all loci except at that locus selected for the *a* allele. From J. Klein, *Biology of the Mouse Histocompatibility-2 Complex*, 1975, with the permission of Springer Verlag, Berlin.

given by Klein [338]. The detailed serology of MHC antigens describes the extensive polymorphism of that locus. Some order has been brought to the complexity by the recognition of private, rarely expressed, specificities and public, frequently expressed specificities. From the biochemical point of view, the complexity should resolve itself when the complete primary structures of a sufficient number of MHC products have been determined.

The size of the MHC, shown in Fig. 34 in recombination units, indicates a stretch of DNA which has been calculated to be sufficient to code for the order of $10^3$ polypeptide chains [342]. The presently determined number of genes within the MHC is about two orders of magnitude less than the calculated coding potential. Further genes continue to be mapped in the MHC as more products and properties of the MHC are recognised and mapped by new recombinants. However, it seems safe to predict that molecular genetics will overtake classical genetics and that the true coding capacity of the MHC will be defined by the application of recombinant DNA technology to the cloning of MHC DNA.

There is a basic similarity between the H2 and the HLA genetic maps. Division of the maps into regions (Fig. 34) helps to reveal the similarities. The products of each of these regions, K, G and D in the mouse and A, B and C in man, have been detected by serology. In the mouse alloantisera are raised by immunisation between histoincompatible inbred strains [338]. In man study of histocompatibility antigens has relied upon fortuitous antisera obtained from multiparous women, immunised by natural leakage into their system of foetal cells, or from patients who have received multiple blood transfusions. In both species, but especially in man, serology should now undergo a dramatic clarification with the availability of monoclonal antibodies directed against products of the MHC. In both mouse and man the major histocompatibility antigens are expressed on all adult cells with the exception of sperm and mature human red blood cells. The K and D regions

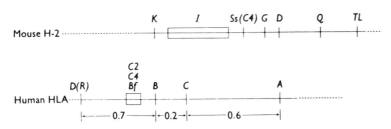

Fig. 34. The HLA-A and B regions are aligned with the analogous K and D regions of the H-2 complex. The D (R) region codes for antigens analogous to the immune associated (Ia) antigens coded for by the I region of H-2. Ss (C4) and C2, C4 and Bf are genes coding for complement components. HLA-C codes for a product homologous to those of the A and B loci. H-2G codes for an erythrocyte antigen. Q and TL code for lymphocyte antigens. From Bodmer.

*References p. 315*

of the H2 are analagous to the A and B regions of the HLA complex. These regions code for proteins which were first recognised as major histocompatibility antigens and which are now known to be very similar in their biochemical properties and in their role in restriction of immune recognition. These properties are discussed in more detail in subsequent sections. In man the HLA-C region codes for a third protein, similar to the A and B products carrying histocompatibility antigens. No direct counterpart of the C region has yet been recognised in the H2 complex. The G region of the H2 complex codes for alloantigenic molecules present on red blood cells.

The I region of H2 and the D region of HLA appear to be homologous in function. The I region has been divided into five subregions A, B, J, E, C on the basis of recombinant events [343]. Individual Ir genes controlling the production of antibody to specific antigens have been mapped to the A, B, E and C subregions. In certain cases complementation between genes in two of these regions has been demonstrated indicating the need for simultaneous function for two of these regions in the control of antibody formation. Immune-associated, Ia, antigens defined serologically have been mapped to the A, J, E and C subregions of the I region [344]. So far there is no serological product identified for the B subregion. Prior to serological characterisation, antigenic differences controlled by the I region had been recognised by the mixed lymphocyte reaction (MLR) [345]. In this reaction antigenic differences on cells of a given strain of mice or individual are recognised not by antibody but by T lymphocytes from a different strain or individual. In a positive MLR the T cells respond by proliferation which can be measured by the incorporation of radioactive thymidine into DNA [346, 347]. Present evidence suggests that the MLR is recognising alloantigenic differences on the same Ia molecule recognised serologically.

*2.2.2. Polymorphism.* The MHC shows greater polymorphism than any other genetic system. This statement can be put in perspective by considering the frequency of heterozygosity which is often taken as a convenient measure of the degree of polymorphism at a locus. Polymorphic gene loci are certainly not unusual. More than a third of all gene loci studied have been shown to be polymorphic. For this broad class of polymorphisms, covering many different enzymes and the blood group system the average frequency of heterozygosity is approx. 15%. By contrast over 80% of individuals are heterozygous at the HLA-A, -B, and -D loci. This extensive heterozygosity has led to a search for explanations for why and how such balanced polymorphism can be maintained. What is the selective advantage of heterozygotes over homozygotes? Bodmer and Bodmer [329] have suggested that natural selection might act through the association of certain HLA genes with susceptibility to specific infectious diseases. Those authors pointed out that the well documented associations between a disease and an HLA type involve syndromes which do not show any obvious selective significance. However, the presently existing polymorphisms could well have been selected by the major infectious diseases that have mostly now been eradicated.

In addition to natural selection favouring particular alleles, there may also be an accumulation of neutral polymorphic differences at the MHC locus. Accumulation of neutral differences could reflect a high mutation rate at the MHC or a novel mechanism for establishing the extensive polymorphism. The finding (see below) that the allelic MHC products show multiple amino acid sequence differences has provided a stimulus for the elaboration of two novel models [348] which might explain the evolution and control of the extensive balanced polymorphism at the MHC. The term complex allotypes has been proposed for allelic products which differ by multiple amino acid residues despite the fact that they segregate in a Mendelian manner. The two models suggested to explain the existence of complex alleles at the MHC are (1) the duplication-deletion model and (2) the control gene model. These models are shown schematically, and compared with the conventional model, in Fig. 35.

The duplication-deletion model proposes that the MHC genes underwent duplication during evolution and that many sequence differences became fixed in these duplicated genes. As evolution proceeded the number of copies of each gene at the MHC was again reduced to one. This model requires that expansion and contraction of the number of genes at the MHC must have occurred subsequent to individual speciation events.

The control gene model also invokes the idea that genes at the MHC have undergone multiple duplication events. In this model

*References p. 315*

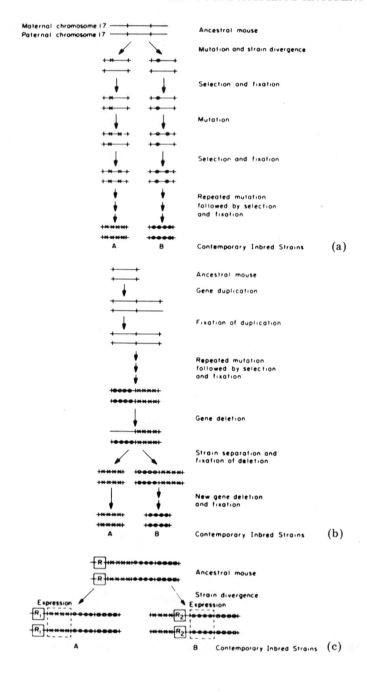

it is assumed that the present day MHC of each species consists of an array of multiple copies of each gene, with one copy for each of the alleles that the species can express. The behaviour of individual duplicated genes as alleles with expression following an apparent Mendelian inheritance pattern is postulated to be due to a control mechanism operating cis on the MHC to allow only one gene of each duplicated set to be expressed per chromosome. In order to explain the allelic behaviour of transplantation genes the proposed control gene must be inherited in a Mendelian manner.

Ultimately, detailed biochemical and molecular genetic studies should differentiate between the various hypotheses invoked to explain the extensive polymorphism and complex allelic nature of the products of the MHC.

*2.2.3. Linkage disequilibrium.* Linkage disequilibrium describes the tendency for alleles at different loci to be paired more often on the same chromosome than would be expected on a random basis calculated from the frequency of each allele in the population. Linkage disequilibrium is found for a number of combinations of alleles at individual loci of the MHC. It is also noteworthy that linkage disequilibrium occurs for allelic forms of immunoglobulin genes. The functional significance of linkage disequilibrium is unclear. However, it is most likely that most examples of linkage disequilibrium at MHC loci are due in some way to natural selection. In this regard, and in consideration of all functions of the MHC, it should be borne in mind that an effect or function may be the result of an allele at a locus closely linked to the identified locus or loci.

---

Fig. 35. Three genetic models to account for evolution of the allotypes of the MHC. (a) The conventional mutation and selection model. (b) A gene duplication involving expansion and contraction of the number of genes at each locus. (c) A gene duplication model postulating retention of multiple genes at each locus with expression limited by a control mechanism. From Silver et al. [358] with the permission of the Cold Spring Harbor Laboratory.

*References p. 315*

## 2.3. Structure of the products of the major histocompatibility locus

2.3.1. *Major transplantation antigens.* The structures of the major transplantation antigens in mouse and man (and presumably other species) are analogous. The products of the K and D genes in the mouse and the A, B and C genes in man are each glycosylated polypeptides with $M_r$s of about 45 000 (Table IX). In each case the $M_r$ = 45 000 product is found on the cell surface in non-covalent association with an $M_r$ = 12 000 polypeptide identified by serology and complete amino acid sequence (see below) as $\beta_2$ microglobulin. A diagrammatic representation of the structure of a cell surface major transplantation antigen is shown in Fig. 36.

The human major transplantation antigens have been isolated by biochemical fractionation of the membrane proteins of human lymphoblastoid cells, which are a rich source of HLA antigens [349]. The antigenic molecules can be released intact by lysis of the cell, or disruption of isolated cell membranes, with appropriate detergents. The antigenic molecules obtained by this method form complexes with detergent. The amount of detergent bound can be calculated from the physical properties of the HLA-A and -B molecules, solubilised by papain treatment as shown in Table X. The enzymatically released products do not bind detergent, have a lower $M_r$ and are more symmetrical than the intact antigens measured as complexes with detergent. Treatment of detergent-solubilised HLA-A or -B molecules with papain cleaves the $M_r$ = 45 000 protein chain at two sites (Fig. 36) removing a C-terminal hydrophilic domain and a penultimate hydrophobic domain [350]; each of these domains comprises approx. 10% of the molecule. Cleavage at site 1 results in an $M_r$ = 39 000 polypeptide chain while cleavage at site 2 leaves a 34 000 dalton chain. The papain treatment of intact cells results in an $M_r$ = 34 000 HLA heavy chain. The resultant soluble molecule lacks the hydrophobic domain and therefore does not bind detergent. The function of the hydrophobic domain is to bind the heavy chain of HLA antigen into the cell membrane (Fig. 36). The amino-terminal—carboxy-terminal orientation of the HLA heavy chain as shown in Fig. 36 was indicated by the fact that papain and detergent-released HLA heavy chains have identical amino-terminal amino acid sequences.

## TABLE IX

A summary of the major regions of the H-2 complex and their products and functions
$\beta_2 M = \beta_2$ microglobulin.

| | K | I | S | G | D | TL |
|---|---|---|---|---|---|---|
| Region marker locus | H-2K | Ir-1 | Ss | H-2G | H-2D | T1a |
| Products | Cell membrane glycoproteins | Cell membrane glycoproteins; mediators | Serum proteins | Erythrocyte membrane molecules | Cell membrane glycoproteins | Cell membrane glycoproteins |
| Product size | 45K | $\alpha$ 33K $\beta$ 25K | 200K | Not known | 45K | 45K |
| Associated chains | $\beta_2 M$ 12K | | | | $\beta_2 M$ 12K | $\beta_2 M$ 12K |
| Histocompatibility role | Cytotoxic targets | MLR-GVHR stimulation | None defined | None defined | Cytotoxic targets | Transplantation antigens |
| Function | Marker for cytotoxicity vs. deviant cells | Immune response; cell–cell interactions | Complement components ($C'_4$) | Unknown | Marker for cytotoxicity vs. deviant cells | Unknown |

*References p. 315*

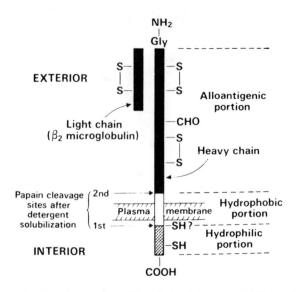

Fig. 36. A model for the polypeptide chain structure of HLA antigens and their insertion in the plasma membrane. From Strominger et al. [350], with the permission of the Cold Spring Harbor Laboratory.

The papain-released molecules retain their alloantigenicity [351].

The finding of a hydrophilic domain carboxy-terminal to the hydrophobic domain strongly suggested a transmembrane orientation for the HLA heavy chain. Direct evidence for such a transmembrane structure came from vectorial catalytic radioiodination with lactoperoxidase [352]. The HLA-A, -B and -C heavy chains were shown to be exposed on both sides of the plasma membrane by labelling either intact cells or isolated inside-out membrane vesicles. The extracellular portion of the heavy chain is organised into two intrachain disulphide bonded loops (Fig. 36). Each loop spans a length of polypeptide chain similar to that found in immunoglobulin intrachain disulphide loops (see above) suggesting a two domain structure for the extracellular region of HLA heavy chain. The heavy chain is glycosylated at a single site; an asparagine residue located approx. 100 residues from the amino terminus carries an oligosaccharide of approx. $M_r$ = 3 000 [353].

The light chain of the major transplantation antigen is a species-specific $\beta_2$ microglobulin [354]. All of the major transplantation

## TABLE X

Molecular size and shape of HLA antigens in the presence of sodium deoxycholate

| Antigen | Antigenic activity | $S_{20,w}$ | Stokes' radius (nm) | $M_r$ DOC[a] | $M_r$ SDS[b] | Frictional ratio | Deoxycholate-bound (g/g protein) |
|---|---|---|---|---|---|---|---|
| Deoxycholate-solubilised antigens | A | 5.15 | 4.4 | 88 000 | | 1.49 | 0.60 |
| | B | 4.55 | 4.4 | 78 000 | 55 000 | 1.55 | 0.42 |
| | C | 5.15 | 4.4 | 88 000 | | 1.49 | 0.60 |
| | Drw | 4.48 | 4.4 | 77 000 | 61 000 | 1.57 | 0.26 |
| Papain-solubilised antigens | A | 3.68 | 2.98 | 46 000 | 45 000 | 1.26 | 0 |
| | B | | | | | | |

[a] Calculated from $S_{20,w}$ and Stokes radius measured in presence of sodium deoxycholate (DOC).
[b] Calculated from measurements made by polyacrylamide gel electrophoresis in the presence of sodium dodecyl sulphate (SDS).

Reproduced from [426].

antigens of a given species share identical light chains. No alloantigenic forms of $\beta_2$ microglobulin have been found in any species examined so far, except the mouse. The $M_r = 12\,000$ $\beta_2$ microglobulin has a single intrachain disulphide bond in a position analogous to that in an immunoglobulin domain [355]. The similarity to an immunoglobulin domain also extends to sequence homology (see below). There is no glycosylation of $\beta_2$ microglobulin either in its free state or when it functions as the light chain of transplantation antigens.

The structure of mouse transplantation antigens has been largely determined by immunochemical methods [348, 356]. Mouse spleen cells have been catalytically radioiodinated or labelled biosynthetically with radioactive amino acids, lysed with appropriate detergents and transplantation antigens precipitated with alloantisera. The radioactive transplantation antigens are then characterised by polyacrylamide gel electrophoresis. This approach, in which biosynthetic radiolabelling is used, has permitted a micro-amino acid sequence determination (see below). The ubiquitous presence of $\beta_2$ microglobulin on major transplantation antigens allows the use of antibodies directed against $\beta_2$ microglobulin in the isolation of transplantation antigens from any species.

*2.3.2. Immune associated (Ia) antigens.* The Ia alloantigens are products of the I region of the murine MHC and analogous regions of the MHC in other species. In the mouse the products of I region genes have been defined by using antisera raised in congenic mice to precipitate radiolabelled cell surface molecules [345]. Available alloantisera may be directed against the products of several I subregions. The Ia molecules of a single subregion can be selected even with such antisera by choosing to radiolabel splenocytes from appropriate recombinant mouse strains having only a single alloantigenic specificity in common with the set against which the antiserum is directed. Using this approach the molecules isolated by antibodies directed against I-A subregion differences consist of two glycosylated polypeptide chains of approx. $M_r = 34\,000$ and $M_r = 28\,000$ [357]. However, molecular weight heterogeneity of the heavier of the two polypeptide chains has been reported [358]. The two polypeptide chains appear to be associated noncovalently [359]. It is important to establish whether the heavy chain, or the light chain, or both chains are coded for by genes in the I region of the MHC. The distribution of Ia specificities across the I subregions leads us to expect different Ia molecules from each subregion. Evidence for antigenic specificities from different subregions being on separate molecules stems from antibody induced redistribution of antigens on the surface [360] ("capping" — Section 1.4.10.1) and from the immunoprecipitation of distinct molecules [361].

Guinea pig antigens have been characterised by similar radiochemical and serological methodology [362]. They too consist of two glycosylated polypeptide chains, one of $M_r$ = 33 000 and the other of $M_r$ = 25 000. Antiserum directed against one particular specificity, Ia.7, precipitates an $M_r$ 58 000 component which on reduction and alkylation yields the $M_r$ = 33 000 and an $M_r$ = 25 000 polypeptide chain. Thus the molecule carrying Ia.7 appears to consist of two different polypeptide chains linked together by disulphide bonding. The other specificities investigated were carried on molecules made up of two different polypeptide chains noncovalently linked.

The human analogues of mouse Ia antigens have been isolated by biochemical fractionation of the membrane proteins of lymphoblastoid cells [349, 363]. The progress of purification procedures is estimated quantitatively in Table XI. The degree of purification achieved is illustrated in Fig. 37. The glycoprotein fraction isolated from lectin columns comprises the Ia, HLA-A and -B antigens and is resolved by polyacrylamide gel electrophoresis into 5 bands, $M_r$ = 43 000, 39 000, 33 000, 28 000, and 12 000. The $M_r$ 43 000 and 12 000 components are the heavy and light chains of the tranplantation antigens and the complex between them can be removed on an anti-$\beta_2$ microglobulin immunoadsorbent column. The nature of the $M_r$ 39 000 component is presently unclear. The $M_r$ 28 000 and $M_r$ 33 000 polypeptides appear to comprise the Ia antigen determined by the HLA-Drw locus [364]. Gel electrophoresis of samples not exposed to heat or urea reveals a component of apparent $M_r$ = 55 000 which is a complex of the two Ia polypeptide chains. This complex can be stabilised by chemical cross-linking with the reagent dimethyl-3,3'-dithiobispropionimidate dihydrochloride. This reagent contains a disulphide bond susceptible to reduction. The cross-linked molecule characterised under gel electrophoretic conditions com-

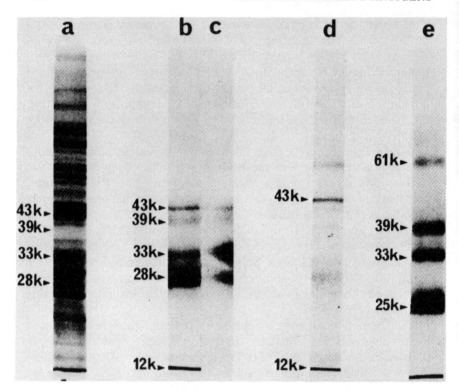

Fig. 37. Purification from human lymphoblastoid cells BRI-8 of HLA antigens demonstrated by polyacrylamide gel electrophoresis in SDS. (a) BRI-8 plasma membrane; (b) The glycoprotein fraction (see Table XI); (c) as (b) but stained for glycoprotein; (d) Fraction selected by anti-$\beta$2 microglobulin antibody; (e) The glycoprotein fraction after cross-linking with dimethyl-3,3'-dithiobispropionimidate. Staining was with Coomassie blue except (c). The 12K band is $\beta$2 microglobulin. From Snary et al. [349], with the permission of the Cold Spring Harbor Laboratory.

parable to those used for analysis of the dissociated polypeptide chain had an apparent $M_r$ = 61 000 prior to reduction. The molecules of apparent $M_r$s = 55 000 and 61 000 are assumed to be two different conformational forms of the two chain Ia molecule. Similarly the Ia light chain appears to have two conformations with apparent $M_r$s = 25 000 and 28 000 according to reduction state and yielded two polypeptides of $M_r$ = 28 000 and 33 000 after

TABLE XI

Purification of Ia and A2 antigens from BRI 8 cells

| Fraction | Protein (% recovery) | A2 activity | | Ia activity | |
|---|---|---|---|---|---|
| | | Recovery (%) | Degree of purification | Recovery (%) | Degree of purification |
| BRI 8 cells | 100 | 100 | 1 | 100 | 1 |
| Plasma membrane | 1.0 | 46 | 45 | 46 | 45 |
| Sodium deoxycholate-soluble | 0.95 | 39 | 41 | 48 | 51 |
| Gel filtration | 0.26 | 36 | 136 | 41 | 158 |
| Glycoprotein | 0.026 | 33 | 1240 | 32 | 1390 |

reduction [349]. The suggested model for the complex of two glycosylated polypeptide chains is shown in Fig. 38; it bears a striking resemblance to the proposed structure of the Ia antigens of mouse and guinea pig.

Evidence correlating the human Ia antigens with products of the HLA-Drw locus has come from the use of xenogeneic antisera prepared against the isolated glycoprotein complex [364]. Rabbit antisera against this complex behaved similarly to alloantisera specific for HLA-Drw in the following criteria: (1) they show specificity for B lymphocytes and monocytes; (2) Fab fragments of these antibodies block cytotoxic-lysis of B lymphocytes by allogeneic HLA-Drw antisera, and (3) the antisera are potent inhibitors of the mixed lymphocyte reaction which is driven by HLA-Drw antigenic differences. The rabbit antisera lyse B cells from most individuals and so do not recognise the alloantigenic determinants seen by human anti-HLA-Drw antiserum. However, the Ia glycoprotein complex of apparent $M_r$ = 55 000 specifically inhibits the cytotoxicity of an allogeneic HLA-Drw antiserum. The rabbit antiserum will precipitate the complex of $M_r$ 33 000 and 28 000 polypeptide chains but prior dissociation of the complex allows the antiserum to precipitate the $M_r$ 28 000 polypeptide chain preferentially [349].

*References p. 315*

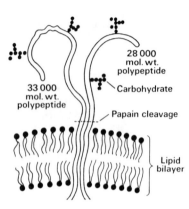

Fig. 38. A model for the polypeptide chain structure of Ia antigens and their insertion in the plasma membrane. From Snary et al. [349], with the permission of the Cold Spring Harbor Laboratory.

It is important to know whether only one or both of the polypeptide chains of human Ia antigens are coded for at the HLA-Drw locus. This question has been explored by somatic cell genetic techniques [364]. Hybrid cells have been produced by fusion of Daudi, a human Burkitt lymphoma line, with A9, a metabolic mutant of mouse L-cells. Analysis of the hybrid lines shows that expression of HLA-Drw alloantigens correlates completely with the presence of chromosome 6. However, none of the hybrids showed reactivity with the xenogeneic antiserum directed against the $M_r$ 28 000 polypeptide chain of Ia. This evidence is consistent with the proposal that the alloantigenic determinants are present only on the $M_r$ 33 000 polypeptide and that it is this chain that is encoded on chromosome 6.

*2.3.3. Amino acid sequences.* Further understanding of the nature and function of the products of the major histocompatibility complex requires knowledge of their amino acid sequences. Such knowledge should allow definition of (a) the nature of the extensive polymorphism of transplantation antigens and Ia antigens; (b) the evolutionary relationship between the products of different genes within the locus; (c) homologies between the products of

STRUCTURE OF THE PRODUCTS    297

analogous genes clustered in the MHC of different species; (d) possible homologies between products of the MHC and other proteins.

Amino acid sequence analysis of proteins found only on the cell

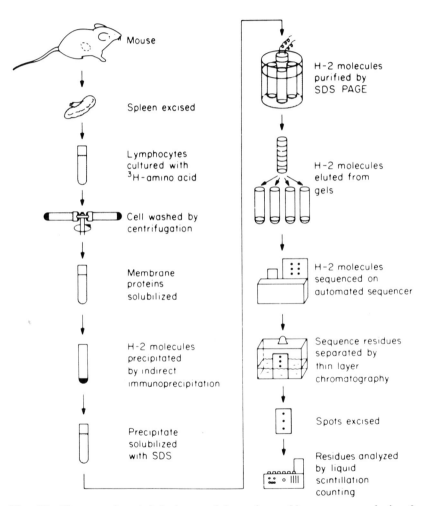

Fig. 39. The experimental design used in amino acid sequence analysis of murine MHC products. From Silver et al. [358], with the permission of the Cold Spring Harbor Laboratory.

*References p. 315*

surface and therefore available only in small amounts has necessitated the development of microsequencing techniques. The procedures used are shown schematically in Fig. 39. The proteins are biosynthetically labelled with one radioactive amino acid per culture. Individual products of the MHC are isolated by immunoprecipitation with specific alloantisera. The individual polypeptide chains of each antigen are separated by polyacrylamide gel electrophoresis, eluted from the gel and subjected to successive cycles of automated sequence analyses in the presence of a carrier protein. Identification of the phenylthiohydantoin amino acids is made by thin-layer chromatography or high performance liquid chromatography. The presence of the incorporated radioactive amino acid at a given position is determined by liquid scintillation counting. This approach has been applied by several groups and has yielded partial amino-terminal sequence data for several MHC products of mouse, man and guinea pig (Fig. 40). These data also include sequences determined by conventional sequencing applied to purified membrane proteins, determinations being made on greater than 10 nmol of protein [365]. Other sequence data were obtained by a solid phase, microsequencing method requiring about 1 nmol of polypeptide purified by polyacrylamide gel electrophoresis [366]. This latter method requires only about one order of magnitude more protein than microsequencing of biosynthetically labelled protein. The results of the various methods are internally consistent and the micromethods have been checked by sequencing the $M_r$ = 12 000 light chain of the transplantation

Fig. 40. Comparison of the amino acid sequences of the major histocompatibility antigens of mouse, human, guinea pig and chicken. From Vitetta and Capra [356] with the permission of Academic Press.

antigens ensuring the amino-terminal sequence to be identical with that of $\beta_2$ microglobulin.

Comparisons between the amino acid sequences of the major transplantation antigens lead to the following initial conclusions.

(a) Allelic products, e.g. $K^b$ and $K^k$, are only about 85% homologous, i.e. show three differences in the first twenty residues. Extrapolating this difference to the complete sequences of allelic products would imply more than sixty residue differences, presupposing that such differences are randomly spread throughout the molecule. Certainly, peptide mapping of allelic products is consistent with multiple sequence differences. This extent of difference is greater than that known for any other allelic proteins and casts some doubt upon whether the polymorphism reflects allelic structural genes or an inheritable control process (see above).

(b) The products of K and D genes show a similar extent of homology with each other as they do with their allelic variants. A similar relationship exists between the HLA-A and -B products. These data support the idea that the genes coding for the heavy chains of the major transplantation antigens arose by gene duplication.

(c) Comparison of the partial sequences of H2-K and -D products with HLA-A and -B products shows almost 70% sequence homology. This finding is consistent with a postulated common ancestral gene from which the genes coding for major transplantation antigens have evolved.

(d) In searching for homology between the sequences of transplantation antigens and those of other proteins, the immunoglobulins are the most intriguing candidates. Impetus for comparison of transplantation antigen sequences with immunoglobulin sequences came from the identification of $\beta_2$ microglobulin as the light chain of HLA-A and -B products [354] (and their analogues in other species). The amino acid sequence of $\beta_2$ microglobulin has been determined on material isolated from urine [367]. The sequence of $\beta_2$ microglobulin shows striking homology with a single domain of an immunoglobulin molecule (Fig. 41). However, within the limited amino-terminal sequence information available for the transplantation antigens no obvious homology with any immunoglobulin sequences can be inferred. Partial internal sequence data around the glycosylation site and around one of the half cystine residues of HLA-A2 and -B7 heavy chains does show

*References p. 315*

## HISTOCOMPATIBILITY ANTIGENS

```
                                            1                                    10
β₂-MICROGLOBULIN                  ILE GLN |ARG| THR |PRO| LYS ILE |GLN VAL| TYR| SER
EU C_L  (RESIDUES 109-214)        THR VAL ALA ALA |PRO|  -   -   SER VAL  PHE  ILE
EU C_H1 (RESIDUES 119-220)        SER THR LYS GLY |PRO|  -   -   SER VAL  PHE  PRO
EU C_H2 (RESIDUES 234-341)        LEU LEU GLY GLY |PRO|  -   -   SER VAL  PHE  LEU
EU C_H3 (RESIDUES 342-446)        GLN PRO |ARG| GLU |PRO| -   -  |GLN VAL| TYR| THR


                                                         20
ARG HIS |PRO| ALA  -  |GLU|  -  -  -  -  ASX |GLY| LYS SER ASX PHE |LEU| ASN |CYS| TYR |VAL|
PHE PRO |PRO| SER ASP|GLU| GLN  -  -  LEU LYS SER GLY THR ALA SER VAL VAL |CYS| LEU  LEU
LEU ALA |PRO| SER SER LYS SER  -  -  THR SER |GLY| GLY THR ALA ALA |LEU| GLY |CYS| LEU  VAL
PHE PRO |PRO| LYS PRO LYS ASP THR LEU MET ILE SER ARG THR PRO GLU VAL THR |CYS| VAL  VAL
LEU PRO |PRO| SER ARG|GLU| GLU  -  -  MET THR LYS ASN GLN VAL SER |LEU| THR |CYS| LEU  VAL


              30                                     40
SER |GLY| PHE  HIS |PRO| SER| ASP| ILE| GLU |VAL|  -  -  ASP LEU LEU LYS |ASP| GLY| GLU ARG ILE
ASN ASN |PHE| TYR  PRO  ARG GLU ALA LYS VAL  -  -   GLN TRP LYS VAL |ASP| ASN  -  ALA LEU
LYS ASP TYR PHE |PRO| GLU PRO VAL THR VAL  -  -  SER TRP ASN SER  -  |GLY|  -  ALA LEU
VAL ASP VAL SER HIS GLU |ASP| PRO GLN VAL LYS PHE ASN TRP TYR VAL |ASP| GLY  -   -  VAL
LYS |GLY| PHE  TYR |PRO| SER| ASP| ILE| ALA |VAL|  -  -  GLU TRP GLU SER ASN ASP  -  -  GLY


              50                                     60
|GLX| LYS VAL |ASX|  -  |HIS| SER GLX LEU SER PHE SER LYS ASN  -  |SER| TRP PHE |TYR| LEU |LEU|
|GLN| SER GLY |ASN| SER GLN GLU SER VAL THR GLU GLN ASP SER LYS ASP SER THR |TYR| SER  LEU
THR SER GLY  -  VAL |HIS| THR PHE PRO ALA VAL LEU GLN SER  -  |SER| GLY LEU |TYR| SER  LEU
|GLN| VAL HIS |ASN| ALA LYS THR LYS PRO ARG GLU GLN GLN TYR  -  ASP SER THR |TYR| ARG  VAL
|GLU| PRO GLU |ASN| TYR LYS THR THR PRO PRO VAL LEU ASP SER  -  ASP GLY SER PHE PHE |LEU|


              70                                     80
|TYR| SER| TYR  -  |THR| GLU PHE THR PRO THR  -  |GLU| LYS|  -  ASP |GLU| TYR| ALA |CYS| ARG |VAL|
 SER  SER  THR LEU  THR  LEU SER LYS ALA ASP TYR |GLU| LYS| HIS LYS VAL  TYR  ALA |CYS| GLU  VAL
 SER  SER  VAL VAL  THR  VAL PRO SER SER SER LEU GLY THR GLN  -  THR  TYR  ILE |CYS| ASN  VAL
 VAL  SER  VAL LEU  THR  VAL LEU HIS GLN ASN TRP LEU ASP GLY LYS |GLU| TYR  LYS |CYS| LYS  VAL
|TYR| SER| LYS LEU |THR| VAL ASP LYS SER ARG TRP GLN GLN GLY ASN VAL PHE SER |CYS| SER  VAL


              90                                     100
ASX |HIS| VAL THR |LEU| SER |GLX| PRO  -  -  -  LYS  ILE |VAL|  -  |LYS| TRP ASP ARG ASP MET
THR |HIS| GLN GLY |LEU| SER  SER |PRO| VAL THR  -  LYS  SER PHE  -   -  ASN ARG GLY GLU CYS
ASN |HIS| LYS PRO SER ASN THR LYS VAL  -  ASP LYS ARG |VAL|  -   -  GLU PRO LYS SER CYS
SER ASN LYS ALA |LEU| PRO ALA |PRO| ILE  -  GLU LYS THR ILE SER |LYS| ALA LYS GLY
MET |HIS| GLU ALA |LEU| HIS ASN HIS TYR THR GLN |LYS| SER LEU SER LEU SER PRO GLY
```

Fig. 41. Comparison of the amino acid sequence of β2-microglobulin with the homology regions C_L, C_H1, C_H2 and C_H3 of the γG1 immunoglobulin Eu. Deletions, indicated by dashes, have been inserted to maximise homologies. Identical residues are enclosed in boxes. Numbering is for β2-microglobulin.

*References p. 315*

some limited homology with sequences at similar sites in immunoglobulin molecules. The functional and evolutionary meaning of this finding remains unclear.

Internal sequence data around the glycosylation site of H2 and HLA heavy chains extends the known homology between these mouse and human transplantation antigens. With the limited amino acid sequence data available no correlations can yet be made between sequence differences and alloantigenic determinants. However, it is known that a single amino acid difference is sufficient to give rise to a different alloantigenic determinant [368]. The complete amino acid sequence of HLA-B7 has recently been reported [368a]. The most interesting feature of this sequence is that it reveals 35% homology with $\beta_2$ microglobulin and with certain immunoglobulin domains.

## 2.4. Functions of MHC products

*2.4.1. Functions as histocompatibility antigens.* The most detailed functional studies have been carried out in mice. It has been shown that alloreactive cytotoxic T cells are mainly directed against K and D antigenic differences [369]. Cytotoxic T cells are generated by immunisation of a mouse with allogeneic cells. Cytotoxicity is then assayed in vitro by lysis of $^{51}$Cr-labelled target cells. With a large range of congenic mouse strains the exact nature of the target can be manipulated and identified.

When allogeneic lymphocytes are allowed to interact in culture either mutual or one way activation occurs. This activation, termed a mixed lymphocyte reaction (MLR) is essentially a proliferative response of T cells and is generally measured by incorporation of radioactive thymidine into DNA [346, 347]. Analysis of the genetics of mixed lymphocyte reactivity suggests that it is mainly triggered by Ia antigenic differences. Since Ia antigens are mainly expressed on B-lymphocytes, the mixed lymphocyte reaction is largely directed against B-lymphocytes [343]. However, I region differences are not essential for MLR activation of T cells [370]. The MLR is taken to be the correlation in vitro of a graft-vs.-host (GVH) reaction in vivo.

In addition to the finding that K and D antigenic differences are the major allogeneic targets for cytotoxic T cells it has been found that there is a role for these antigens in the targeting of cytotoxic T cells directed against other cellular antigens [333, 334, 336, 371]. From a physiological point of view the role of MHC antigens in syngeneic responses to foreign antigens expressed on syngeneic cells is probably the main function of these antigens. Their role in graft rejection, by which they were first recognised and named, thus appears to be consequential to their principal function. It is clear that MHC antigens are qualitatively, as well as quantitatively distinct from minor histocompatibility antigens.

*2.4.2. MHC restriction.* The initial demonstration of the role of MHC products in cytotoxic T cell function came from studies on virus infected target cells. Zinkernagel and Doherty [333, 371] showed that cytotoxic T cells generated against virus-infected

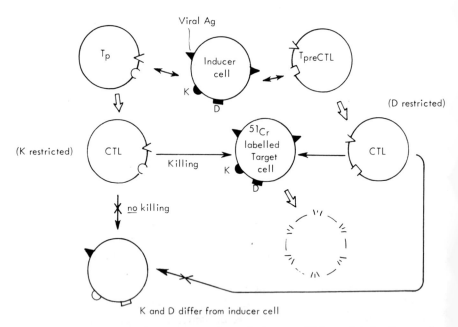

Fig. 42. Diagrammatic representation of the specificity of induction and effector function for cytotoxic T lymphocytes (CTL). Precursors of CTL (Tp or T preCTL) recognise both the foreign viral antigen and a self MHC antigen (K or D) and the CTL shows specificity for both antigens in killing the target cell. Killing is assayed by lytic release of $^{51}Cr$ from prepared target cells.

## FUNCTIONS OF MHC PRODUCTS

target cells show specificity of killing for both the correct viral antigen and the K or D antigens of the target cell. Generation and specificity of cytotoxic lymphocytes is illustrated in Fig. 42. This phenomenon has been termed MHC restriction; for the mouse the term H-2 restriction is used. The original finding has been extended to many different systems [334, 336] and it is now accepted as a general phenomenon that lytic killing by cytotoxic T cells requires the joint presence on the surface of the target cell of the appropriate foreign antigen and the appropriate histocompatibility K or D-antigen (or the analogous product in other species) originally present on the cell inducing the killer T cell.

Two types of hypothesis have been proposed to explain MHC restriction [372]. One hypothesis is referred to as the dual recognition hypothesis and the other as the altered self hypothesis. The two hypotheses are summarised in Fig. 43. In the dual recognition hypothesis it is assumed that the T cell has two different receptor molecules; one specific for MHC antigens and the other specific for non-MHC antigens. It is generally assumed that the second receptor, which recognises foreign antigens, uses the antibody gene repertoire. The receptor specific for MHC antigens is more enigmatic although in one form of the hypothesis the two receptors are assumed to be identical at an early stage of development with mutation and selection acting on one of the receptors to generate the full repertoire of specificities for non-MHC antigens [373]. The altered self hypothesis invokes a single receptor on the T cell. It is postulated that this receptor can only recognise foreign antigens when they are in association with MHC antigens on the cell surface. It is

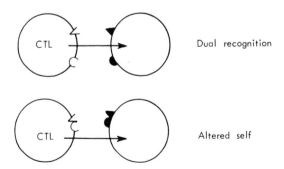

Fig. 43. Diagrammatic representation of two types of models proposed to explain MHC restriction.

*References p. 315*

assumed that each foreign antigen interacts with an MHC antigen to form a new antigenic determinant; in this hypothesis it is generally assumed that the antibody V genes code for the T cell receptor [374].

*2.4.3. Recognition of self.* In the initial demonstration of MHC restriction the evidence was consistent with the idea that cells presenting the foreign antigen, either as immunising cell, or as target cell needed to share histocompatibility antigens with the cytotoxic T cell. Thus MHC restriction was seen as being synonymous with the recognition of self antigens. Subsequently it has been demonstrated that identity of histocompatibility between the immunising or the target cell and the cytotoxic T cell is not necessary for recognition of self. It has been demonstrated that the T cell "learns" to recognise as self the histocompatibility antigens displayed on the epithelial cells of the thymus during the processing of T cell precursors through that organ. This finding stems from experiments in which an animal termed a bone-marrow chimaera is produced and to which grafts of irradiated thymus tissue are donated. A bone-marrow chimaera is constructed by lethally irradiating an animal and reconstituting its haemopoietic system with a graft of bone-marrow cells from an allogeneic animal. Using type $(A \times B)F_1$ chimaeric mouse bone-marrow injected into irradiated parent A animals, Bevan [375] showed that cytotoxic T cells, generated against minor histocompatibility antigens, recognise these antigens preferentially when they are on target cells in association with the H-2 haplotype of parent A, i.e. the haplotype of the thymus in which the bone-marrow cells mature to T lymphocytes. In similar chimaeric mice Zinkernagel et al. [376] showed exclusive restriction of cytotoxic T cell killing for parent A H-2 type of either virally infected or hapten modified target cells. In a further experiment chimaeric mice of the type $(A \times B)F_1 \rightarrow (A \times C)F_1$ were constructed and the lymphocytes sensitised against hapten modified $(B \times C)F_1$ cells with the result that cytotoxicity was found to be restricted to hapten-modified cells of the C haplotype. This result dramatically demonstrates that the genotype of the T cell does not determine which MHC type the T cell will recognise as self. In the latter experiment T cells heterozygous for the haplotypes of A and B parents have learnt to recognise the H-2 type of the C mouse as self but fail to recognise the H-2 type of the B mouse (genotypically a self haplotype) as self. The MHC type recognised as self by the T cell has thus been determined by the phenotype of the irradiated host used in constructing the chimaeric mouse.

The thymus is the crucial organ whose phenotype determines which MHC antigens the mature T cell will recognise as self. The evidence of this statement comes from experiments in which irradiated thymus grafts were placed under the kidney capsule of adult thymectomised irradiated recipients of bone-marrow cells [337]. In this system the H-2 type of the thymus epithelium determines what the newly generated T cell learns to recognise as the self H-2 type. Since T cells of one H-2 type can, by thymus processing learn to recognise a different H-2 type as self it is clear that self recognition is not determined by a reaction of identity between the H-2 haplotypes of cytotoxic T cells and target cells. Learning of self recognition in the thymus is strong evidence in favour of the dual recognition hypothesis. Zinkernagel et al. [376] pointed out that a single-receptor altered-self hypothesis would require the special rule that the specificities generated by modified self in thymus A do not overlap with the specificities generated in thymus B. However, those workers made no clear prediction as to whether the two recognition structures are separate receptor molecules or two parts of a single receptor molecule.

*2.4.4. Prevalence of allo-reactive T cells.* The high frequency of T cells showing specificity for allo-histocompatibility antigens is a phenomenon which must be explained by any hypothesis relating to T cell receptors and MHC restriction [377, 379]. The prevalence of allo-aggressive cells was first observed in graft versus host and mixed lymphocyte reactions (MLR) when MHC antigen differences were involved. The frequency of allo-aggressive cells has been measured most exactly in limiting dilution experiments [380, 382]. The problem can be stated simply in the following way. The frequency of cells exhibiting specific allo-aggression for any given non-self MHC type can be shown to be of the order of, or greater than, 1%. However, in the mouse, probably the most intensively studied species, more than 100 foreign H-2 haplotypes are known. Simultaneous immunisation with two different H-2 antigens shows that distinct populations of cytotoxic T cells are directed against two different antigens. There is currently no generally accepted explanation for the high frequency of allo-aggressive T cells. Bevan has suggested that the problem might be explained on the basis of cross-reactions between HMC haplotypes [374]. This hypothesis has the advantage that it retains the idea that cytotoxic T cells respond in a clonal manner. Another explanation also retaining clonal selection for T cell responses stems from a version of dual recognition that invokes three receptors on each T cell [383]. In this model one type of T cell receptor is encoded by immunoglobulin V genes and therefore exhibits the same repertoire and similar rules of expression as B cell receptor antibodies. A second set of T cell receptors distinct from immunoglobulins, is postulated to be encoded in a multi-gene family and to cover a specificity repertoire of receptors each specific for an allelic form of one of the MHC antigens of the species. Inheritance of a repertoire of receptor genes reflecting the extent of polymorphism of the MHC antigens of the species would explain why pre-T cells of one genotype can learn to recognise the phenotype of an allogeneic thymus as self by expressing an appropriate non-antibody receptor. The model postulates that the expression of MHC receptor genes is limited to a single gene per haplotype in each cell with no necessity for allelic exclusion. Selection, taking place in the thymus, from pre-T cells exhibiting at random two anti-MHC receptor molecules will result in a population of T cells each of which exhibits at least one receptor molecule specific for one of the MHC antigens displayed by the thymus epithelium. If a family of $n$ genes is expressed in a random diploid manner T cells will be divided clonally into approximately $n$ equal proportions with each clone bearing one anti-self-MHC receptor together with either a second anti-self-MHC receptor or a receptor for an allo-MHC specificity. Superimposed on this division of T cells will be the expression by each T cell type of the repertoire of antibody specificities. MHC restriction in this model occurs by dual recognition involving an anti-self-MHC receptor and a V gene encoded receptor. Allogeneic cellular immune reactions are postulated to involve that population of T cells expressing a specific anti-allo-MHC receptor together with V gene encoded receptors for allo-antigens. In this way all antigenic differences between the cytotoxic T cell and the target cell will be seen in an MHC restricted manner.

*2.4.5. Immune response (Ir) genes.* An important class of genes that control immune responsiveness (Ir genes) has been found to map at the MHC [331]. These specific Ir genes have been extensively studied and mapped in inbred strains of mice and guinea pigs. However, it is probable that MHC linked Ir genes exist in all immunologically competent species.

Defined antigens such as immunoglobulin allo-antigens [384] or synthetic polypeptides of limited structural heterogeneity [385] have provided the most useful system for demonstrating Ir genes. In each case the ability to mount an effective antibody response can be shown to be controlled by an autosomal dominant gene linked to the MHC [331, 384].

This is illustrated by the studies (reviewed in [331] and [386]) on the responsiveness of guinea pigs to poly(L)-lysine (PLL) summarised in Table XII. An antibody response to PLL is evoked in guinea pigs of strain 2 but not in those of strain 13. Responder strain 2 guinea pigs, but not strain 13 guinea pigs exhibit delayed-type hypersensitivity (DTH) in response to PLL. A hapten (DNP) presented on PLL induces anti-DNP antibodies in strain 2 but not in strain 13 guinea pigs. If, however, DNP-PLL is presented as a conjugate with bovine serum albumin the non-responder strain-13 guinea pigs are able to make anti-DNP antibody but the strain-13 animal remains non-responsive with respect to DTH when challenged with PLL. In testing back crosses of $(2 \times 13)F_1 \times 13$ guinea pigs responsiveness is found to segregate with the strain-2 histocompatibility type. It is now known that strain-2 and strain-13 guinea pigs are indistinguishable at the loci controlling their major histocompatibility antigens but differ at a region of the MHC analogous to the I region in mice.

TABLE XII

Immune responses to poly-(L)-lysine (PLL) in guinea pigs

| Antigen | Antibody against DNP | | | Delayed-type hypersensitivity to PLL | | |
|---|---|---|---|---|---|---|
| Strain | 2 | 13 | $(2 \times 13)F_1$ | 2 | 13 | $(2 \times 13)F_1$ |
| PLL |   |   |   | + | − | + |
| DNP-PLL | + | − | + | + | − | + |
| DNP-PLL-BSA | + | + | + | + | − | + |

*References p. 315*

TABLE XIII

Production of antibody in response to synthetic polypeptides: relationship of response to H-2 type in mice

| Antigen | H-2 haplotype | |
|---|---|---|
| | High responders | Low responders |
| (TG) AL | H-2$^b$ | H-2$^a$, H-2$^k$, H-2$^q$, H-2$^s$ |
| (HG) AL | H-2$^a$, H-2$^k$ | H-2$^b$, H-2$^q$, H-2$^s$ |
| (Phe G) AL | H-2$^a$, H-2$^b$, H-2$^k$, H-2$^q$ | H-2$^s$ |

The most extensive genetic mapping of Ir genes has been carried out in mice. The family of branched-chain synthetic polypeptides (TG)-AL, (HG)-Al and (pheG)-AL have been used to map the responder status of several different strains of mice [387, 388] (Table XIII). These synthetic polypeptides consist of a PLL backbone with side chains of poly-D,L-alanine terminating in each case with a small number of glutamic acid residues and either tyrosine, histidine or phenylalanine residues. In high-responder strains these antigens induce both specific antibody of the IgG class and immunological memory. Strains of mice that are low responders to these antigens produce only IgM antibody and exhibit no immunological memory [389]. Thymectomy of responder-strain mice converts them to a typical non-responder pattern [390]. Thus genetic non-responders appear to be lacking in T cells of required specificity. This lack of specific T cells in non-responder strains can be circumvented by an allogeneic effect. This was demonstrated in an experiment utilising an $F_1$ cross between two non-responder strains of haplotype H-2$^k$ H-2$^q$. These $F_1$ mice make only IgM antibody in response to (TG)-AL but injection of parental H-2$^k$ cells together with antigen induces the non-responder $F_1$ to make IgG antibody specific for (TG)-AL [391].

Measurement of the responder status of congenic and recombinant inbred strains of mice allowed the Ir genes to be mapped to the I region of the MHC and that I region can now be sub-divided into five sub-regions on the basis of recombinational events [343]. In some instances the $F_1$ animal constructed by crossing two non-responder strains has been found to have responder status [392].

This finding has been interpreted in terms of complementation between two genes mapping in different I sub-regions [393].

Two controversial questions concerning Ir genes are (1) the cellular basis of Ir function and (2) the correlation between Ir gene function and Ia antigens coded for by the I sub-regions.

(a) The evidence quoted above clearly points to a defect at the T cell level in animals expressing the low responder Ir phenotype. However expression of Ir genes in both B cells and macrophages is indicated by an experiment in which helper T cells from an $F_1$ cross between high- and low-responder mice were shown to co-operate with B cells and macrophages of high-responder origin but not those of low-responder origin [394]. A crucial role for the I region phenotype of B cells and macrophages and of the environment in which T cells mature has been shown by the construction of appropriate chimaeric mice. T cells of low responder MHC genotype when allowed to mature in an irradiated $F_1$ produced by crossing a high responder with a low responder were found to be capable of co-operating with B cells and macrophages of high-responder MHC type to generate characteristic high responsiveness [395]. When $F_1$ interparent chimaeric mice were constructed only $F_1$ T cells that have differentiated in a high-responder environment, and not those which have differentiated in a low-responder environment, are able to yield high responses in co-operation with B cells and macrophages of the high-responder type [395]. The crucial role of the B cell and macrophage phenotype is underlined by the fact that when these two cell types are of the low-responder phenotype only low responses can be obtained irrespective of the source of antigen specific helper T cells. These results also have an important bearing on the correlation between Ir gene function and Ia antigens and also complement the results pointing to a role for Ia antigens in cell—cell interaction.

(b) The search for the products of I region genes was initially pursued by allo-immunisation of congenic strains of mice. The Ia antigens map genetically at the I sub-regions as do the Ir genes [344]. An extensive strain survey showed a very high correlation for the association of Ia antigen specificity with Ir gene responder status but the correlation was not absolute [396]. Ir gene status and Ia antigen phenotype are contrasted even more starkly in the experiments with chimaeric mice described above. In those experiments the Ir status of B cells and macrophages invariably correlated with the Ia phenotype (and genotype). By contrast the Ir status of helper T cells was shown to be dictated by the environment in which those cells have differentiated and not by the MHC genotype of the T cells. Moreover, helper T cells do not express Ia antigens so no correlation can be drawn between Ia phenotype and Ir gene status. These findings are best interpreted in terms of the role of Ia antigens in cell—cell interaction (discussed below) with a postulate that helper T cells are specific for both Ia antigens and a foreign antigenic determinant.

*2.4.6. Ia antigen functions.* Ia antigens can be classified as differentiation antigens and as such serve the immunologist as markers of B lymphocytes and T lymphocyte sub-sets [344]. Ia antigens are also expressed on some lymphocyte precursors and on a proportion of macrophages.

In addition to the detection of Ia antigens by serological methods they can also be detected by the fact that they provide a particularly strong proliferative stimulus for T cells in a mixed lymphocyte reaction (MLR) [345]. Indeed the D locus of the human MHC was first detected and is still conventionally assayed by using the MLR [397]. Evidence for the involvement of murine Ia antigens in stimulating the MLR comes from experiments showing that anti-Ia antisera directed against the target cells will block their ability to stimulate the MLR. Similar evidence has been obtained in the human system with antisera directed to the human analogue of Ia antigens to block the MLR [399]. Despite the apparently special role of Ia antigens in the MLR there is no absolute necessity for I region differences between stimulator and proliferating cells for the MLR to occur [400].

The major physiological function of Ia antigens appears to be their involvement in cell—cell interactions during immune responses. An involvement of Ia antigens in functional interactions of cells of the immune system was first deduced from the observed need for compatibility at the I region between the interacting cell [401—403]. The need for H-2 compatibility between B cells and T cells was initially demonstrated by reconstitution of athymic T cell deficient nude mice with T cells of various H-2 types; only H-2 compatible T cells fully restored competence to respond to T dependent antigen [401]. In the course of the production of

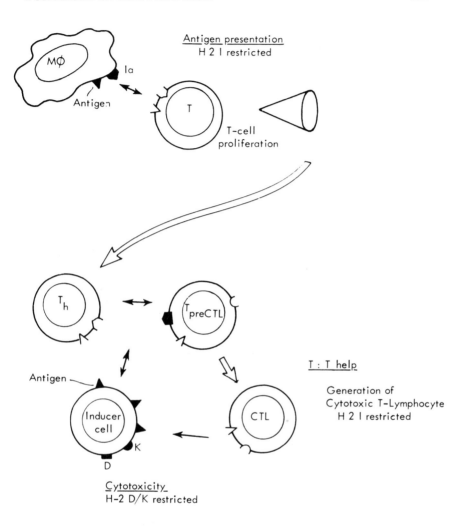

Fig. 44. Schematic illustration of the involvement of Ia antigens in cell—cell interactions necessary for helper T cell ($T_h$) generation by antigen presented on macrophages (M$\phi$) and for cytotoxic T lymphocyte (CTL) generation involving cell surface antigens presented with H-2 K or D histocompatibility antigens.

antibody to a T-dependent antigen the T cell must interact with
a macrophage presenting the antigen [403] and with the B cell that
is to produce the antibody (Fig. 27). Helper T cells generated in
response to antigen presented on macrophages also function in
the generation of cytotoxic T cells (Fig. 44). It would appear that
the requirement for histocompatibility between T cells and
macrophages [403] or between T cells and B cells [401, 402] to
allow effective interactions to occur is similar in nature to the
MHC restriction of T cell-mediated cytotoxicity.

The genes controlling the histocompatibility requirement for
cell co-operation have been mapped to the I region of the H-2
by using an adoptive cell transfer system to study antibody formation [404]. In this system T cells from a donor P are able to co-operate with B cells and macrophages from either P or (P × Q)$F_1$
mice but not with those from Q mice. Consistent with the demonstrated requirement for I region compatibility between co-operating
cells was the finding that T cells from $F_1$ mice with a high responder
Ir phenotype can successfully co-operate with B cells and macrophages of the high-responder parent, or with $F_1$ cells, but not with
cells of the low-responder parent [395].

The requirement for histocompatibility between co-operating
T and B cells has not been accepted without controversy. It was
validly argued that the requirement for I region identity could
result from the generation of adverse allogeneic reactions when
non-histocompatible cells are mixed. In support of this notion is
the evidence that T cells tolerant of allogeneic B cells can effectively
co-operate in the production of antibody by those B cells. Tolerant
T cell populations have been obtained by depletion of allo-reactive
cells, by construction of bone marrow chimaeric mice or in tetra-parental mice. In both chimaeric and tetra-parental mice lymphocytes of two different genotypes can develop simultaneously and
acquire mutual tolerance. In these situations T cells also learn
to see both phenotypes as self and this, rather than the mutual
tolerance, may be the most important factor in the generation of
co-operation between histo-incompatible cells.

The genetics of T cell—macrophage interaction has been studied
in a system involving the stimulation in vitro of T cell proliferation
by antigen presented on macrophages [403]. Rosenthal and Shevach
[403] showed that lymphocytes from immunised strain-2 guinea
pigs would respond to antigen presented on either strain 2 or (2 ×

13)$F_1$ macrophages but not on strain-13 macrophages. Reciprocal results were obtained with immune strain-13 lymphocytes but immune $F_1$ T cells proliferated in response to antigen presented by either 2, 13 or $F_1$ macrophages. Further investigation of T cell—macrophage interaction both in the guinea pig system and using congenic mice led to the conclusion that T cells see both the foreign antigen and syngeneic Ia antigens during the priming process and can then be restimulated only when the same foreign antigen is seen together with the syngeneic Ia antigen [405]. Moreover, $F_1$ T cells have been shown to comprise two different populations responding respectively to the foreign antigen seen together with one or other of the parental Ia phenotypes on macrophages. Thus, functionally, the T cell receptor for Ia antigen appears to exhibit allelic exclusion [405]. This evidence is most easily compatible with a dual recognition model for macrophage presentation of antigen to T cells.

Recent experiments with helper T cells primed in irradiated, bone marrow-reconstituted mice indicate that the I region phenotype of the host animal restricts the range of B cell and macrophage phenotypes with which the mature T cell can cooperate. In such a system T cells with a low responder Ir genotype can develop a high responder Ir phenotype when assayed in cooperation with high responder B cells and macrophages [395].

Present evidence indicates that cell co-operation involves recognition of I region antigens on B cells and macrophages by a receptor on T cells. Several models postulate somatic mutation as a key event in the generation of the T cell receptor. It has been postulated that Ir controlled non-responsiveness is due to the inability of a given anti-I receptor to mutate to a receptor specific for the foreign antigen [406]. Conversely, it has also been postulated that non-responsiveness could be due to the receptor for non-responder Ia type being the only one capable of mutating to give a receptor for the foreign antigen, therefore prohibiting Ia restricted recognition of that foreign antigen [407].

## 3. Conclusions

The understanding of immunity at a molecular level has been advanced rapidly by studies of the antibody molecule, its biosynthesis

*References p. 315*

and now its structural genes. The progression from protein to RNA to DNA in the study of antibodies is now being followed for products of the MHC that play an important role in the control of immune responses. Progress at the DNA level promises to be the most rapid due to the techniques of genetic manipulation. However, the nature of the recognition processes involving MHC products must be elucidated at the cellular and protein levels. The biochemist will have a crucial role in these studies.

## Note added in proof

Recently [427], a mouse plasmacytoma $C\delta$ gene has been sequenced and codes for only two constant domains plus a hinge region. No sequence equivalent to human $C\delta 2$ was found. The IgD secreted by a second mouse plasmacytoma and a rat plasmacytoma also appear to have the same structure [428].

# REFERENCES

1. A. R. Williamson, Annu. Rev. Biochem., 45 (1976) 467.
2. M. Potter, Physiol. Rev., 52 (1972) 631.
3. R. R. Porter, Science, 180 (1973) 713.
4. G. M. Edelman, Science, 180 (1973) 830.
5. R. R. Porter, in A. Gelhorn and E. Hirschberg (Eds.), Symposium on Basic Problems in Neoplastic Disease, Columbia University Press, New York, 1962, p. 177.
6. T. B. Tomasi and N. Calvanico, Fed. Proc., 27 (1968) 617.
7. T. B. Tomasi and H. M. Grey, Prog. Allergy, 16 (1972) 81.
8. C. P. Milstein, N. E. Richardson, E. V. Deverson and A. Feinstein, Biochem. J., 151 (1975) 615.
9. A. Feinstein, N. E. Richardson and E. A. Munn, in R. Markham and R. W. Horne (Eds.), Proc. Third John Innes Symp., Structure-Function Relationships of Proteins, North-Holland, Amsterdam, 1976, p. 111.
10. F. P. Inman and M. J. Ricardo, J. Immunol., 112 (1974) 229.
11. J. Mestecky, R. E. Schrohenloher, R. Kulhavy, G. P. Wright and M. Tomana, Proc. Natl. Acad. Sci. USA, 71 (1974) 544.
12. M. Potter, Adv. Immunol., 25 (1977) 141.
13. M. Weigert and R. Riblet, Cold Spring Harbor Symp. Quant. Biol., 41 (1976) 837.
14. J. A. Gally and G. M. Edelman, Annu. Rev. Genet., 6 (1972) 1.
15. C. Milstein and A. J. Munro, in R. R. Porter (Ed.), Biochemistry Series One, Defence and Recognition, Vol. 10, Butterworth, London, 1973, p. 199.
16. A. R. Williamson, in L. E. Glynn and M. W. Steward (Eds.), Immunochemistry: An Advanced Textbook, Wiley, Chichester, 1977, p. 141.
17. I. Schechter, Y. Burstein and R. Zemell, Immunol. Rev., 36 (1977) 3.
18. T. T. Wu and E. A. Kabat, J. Exp. Med., 132 (1970) 211.
19. E. A. Kabat and T. T. Wu, Ann. N.Y. Acad. Sci., 190 (1971) 382.
20. J. M. Kehoe and J. D. Capra, Proc. Natl. Acad. Sci. USA, 68 (1971) 2019.
21. G. M. Edelman and W. E. Gall, Annu. Rev. Biochem., 38 (1969) 415.
22. D. Ein, Proc. Natl. Acad. Sci. USA, 60 (1968) 982.
23. D. Gibson, M. Levanon and O. Smithies, Biochem (Am. Chem. Soc.), 10 (1971) 3114.
24. M. Hess, N. Hilschmann, L. Rivat, C. Rivat and C. Ropartz, Nature New Biol., 234 (1971) 58.
25. E. P. Schulenberg, E. S. Simms, R. G. Lynch, R. A. Bradshaw and H. N. Eisen, Proc. Natl. Acad. Sci. USA, 68 (1971) 2623.
26. G. M. Edelman, B. A. Cunningham, W. E. Gall, P. D. Gottlieb, U. Rutishauser and M. Waxdal, Proc. Natl. Acad. Sci. USA, 63 (1969) 78.
27. D. S. Smyth and S. Utsumi, Nature, 216 (1967) 332.
28. R. B. Payne, Biochem. J., 11 (1969) 473.
29. F. W. Putnam, A. Shimizu, C. Paul and T. Shinoda, Fed. Proc., 31 (1972) 193.
30. J. Baenziger and S. Kornfeld, J. Biol. Chem., 249 (1974) 7270.

31  B. Frangione and C. Wolfenstein-Todel, Proc. Natl. Acad. Sci. USA, 69 (1972) 3673.
32  B. O. Barger and F. P. Inman, Immunochemistry, 13 (1976) 165.
33  E. M. Press and N. M. Hogg, Biochem. J., 117 (1970) 641.
34  J. R. Clamp and I. Johnson, in A. Gottschalk (Ed.), Glycoproteins, 2nd ed., Vol. 5A, Elsevier, Amsterdam, 1972, 612.
35  R. L. Wasserman and J. D. Capra, in M. I. Horowitz and W. Pigman (Eds.), The Glycoconjugates, Vol. 1, Mammalian Glycoproteins and Glycolipids, Academic Press, New York, 1977, p. 323.
36  A. Shimizu, F. W. Putnam, C. Paul, J. R. Clamp and I. Johnson, Nature New Biol., 231 (1971) 73.
37  J. Baenziger, S. Kornfeld and S. Kochwa, J. Biol. Chem., 249 (1974) 1889.
38  T. J. Kindt, Adv. Immunol., 21 (1975) 35.
39  J. A. Sogn and T. J. Kindt, in L. E. Glynn and M. W. Steward (Eds.), Immunochemistry: An Advanced Textbook, Wiley, Chichester, 1977, p. 113.
40  C. W. Todd, Biochem. Biophys. Res. Commun., 11 (1963) 170.
41  A. Feinstein, Nature (Lond.), 199 (1963) 1197.
42  T. J. Kindt and C. W. Todd, J. Exp. Med., 130 (1969) 859.
43  J. Oudin, Proc. Roy. Soc. Ser. B., 166 (1966) 207.
44  R. J. Slater, S. M. Ward and H. G. Kunkel, J. Exp. Med., 101 (1955) 85.
45  H. G. Kunkel, M. Mannick and R. C. Williams, Science, 140 (1963) 1218.
46  J. D. Capra and J. M. Kehoe, Adv. Immunol., 20 (1975) 1.
47  K. Eichmann, Immunogenetics, 2 (1975) 491.
48  S. J. Singer and D. H. Campbell, J. Am. Chem. Soc., 74 (1952) 1794.
49  H. Eisen and F. Karush, J. Am. Chem. Soc., 71 (1949) 363.
50  F. Karush, J. Am. Chem. Soc., 78 (1956) 5519.
51  F. Karush, Adv. Immunol., 2 (1962) 1.
52  S. Velick, C. Parker and H. N. Eisen, Proc. Natl. Acad. Sci. USA, 46 (1960) 1470.
53  K. Onoue, A. L. Grossberg, Y. Yagi and D. Pressman, Science, 162 (1968) 574.
54  R. Oriol, R. Binaghi and E. Coltorti, J. Immunol., 106 (1971) 932.
55  N. M. Young, I. B. Jocius and M. A. Leon, Biochem. (Am. Chem. Soc.), 10 (1971) 3457.
56  S. C. Edberg, P. M. Bronson and C. J. Van Oss, Immunochemistry, 9 (1972) 273.
57  F. Karush, Ann. N.Y. Acad. Sci., 169 (1970) 56.
58  D. Inbar, J. Hochman and D. Givol, Proc. Natl. Acad. Sci. USA, 69 (1972) 2659.
59  E. A. Kabat, J. Am. Chem. Soc., 76 (1954) 3709.
60  E. A. Kabat, J. Immunol., 84 (1960) 82.
61  E. A. Kabat, Structural Concepts in Immunology and Immunochemistry, 2nd ed., Holt, Rinehart and Winston, New York, 1976.
62  R. Arnon, M. Sela, A. Yaron and H. A. Sober, Biochem. Am. Chem. Soc., 4 (1965) 948.

63  H. Van Vunakis, J. Kaplan, H. Lehrer and L. Levine, Immunochemistry, 3 (1964) 393.
64  H. J. Sage, G. F. Deutsch, G. D. Fasman and L. Levine, Immunochemistry, 1 (1964) 133.
65  I. Schechter, B. Schechter and M. Sela, Biochim. Biophys. Acta, 127 (1966) 438.
66  M. Sela, Ann. N.Y. Acad. Sci., 169 (1970) 23.
67  J. J. Cebra, J. Immunol., 86 (1961) 205.
68  J. Cisar, E. A. Kabat, M. M. Dorner and J. Liao, J. Exp. Med., 142 (1975) 435.
69  K. Landsteiner, The Specificity of Serological Reactions, 2nd ed., Harvard University Press, Cambridge, MA, 1945.
70  M. J. Crumpton, in R. Jaenicke and E. Helmreich (Eds.), Protein—Protein Interactions, Springer, Heidelberg, 1972, p. 395.
71  S. J. Singer, N. Martin and N. O. Thorpe, Ann. N.Y. Acad. Sci., 190 (1971) 342.
72  J. Haimovich, D. Givol and H. N. Eisen, Proc. Natl. Acad. Sci. USA, 67 (1970) 1656.
73  D. Givol, P. H. Strausbach, E. Hurwitz, M. Wilchek, J. Haimovich and H. N. Eisen, Biochem. (Am. Chem. Soc.), 10 (1971) 3461.
74  G. W. J. Fleet, J. R. Knowles and R. R. Porter, Nature (Lond.), 224 (1969) 511.
75  C. A. Converse and F. F. Richards, Fed. Proc., 27 (1968) 683.
76  C. A. Converse and F. F. Richards, Biochem. (Am. Chem. Soc.), 8 (1969) 4431.
77  R. A. G. Smith and J. R. Knowles, J. Am. Chem. Soc., 95 (1973) 5072.
78  M. E. Noelken, C. A. Nelson, C. E. Buckley III and C. Tanford, J. Biol. Chem., 240 (1965) 218.
79  N. M. Green, Adv. Immunol., 11 (1969) 1.
80  A. Feinstein, E. A. Munn and N. E. Richardson, Ann. N.Y. Acad. Sci., 190 (1971) 104.
81  R. C. Valentine and N. M. Green, J. Mol. Biol., 27 (1967) 615.
82  R. J. Poljak, Adv. Immunol., 21 (1975) 1.
83  R. J. Poljak, CRC Crit. Rev. Biochem., 5 (1978) 45.
84  D. R. Davies, E. A. Padlan and D. M. Segal, Annu. Rev. Biochem., 44 (1975) 639.
85  D. Beale and A. Feinstein, Quart. Rev. Biophys., 9 (1976) 135.
86  P. M. Colman, J. Deisenhofer, R. Huber and W. Palm, J. Mol. Biol., 100 (1976) 257.
87  R. Huber, J. Deisenhofer, P. M. Colman, M. Matsushima and W. Palm, in F. Melchers and K. Rajewsky (Eds.), Das Immunsystem, Springer, Berlin, 1976, p. 26.
88  M. Schiffer, R. L. Girling, K. R. Ely and A. B. Edmundson, Biochemistry, 12 (1973) 4620.
89  A. B. Edmundson, K. R. Ely, R. L. Girling, E. E. Abola, M. Schiffer and F. A. Westholm, in L. Brent and J. Holborow (Eds.), Progress in Immunology II, Vol. 1, North-Holland, Amsterdam, 1974, p. 103.

90  R. J. Poljak, L. M. Amzel, H. P. Avey, B. L. Chen, R. P. Phizackerly and F. Saul, Proc. Natl. Acad. Sci. USA, 70 (1973) 3305.
91  D. M. Segal, E. A. Padlan, G. H. Cohen, S. Rudikoff, M. Potter and D. R. Davies, Proc. Natl. Acad. Sci. USA, 71 (1974) 4298.
92  O. Epp, E. E. Lattman, M. Schiffer, K. Huber and W. Palm, Biochemistry, 14 (1975) 4943.
93  H. Fehlhammer, M. Schiffer, O. Epp, P. M. Colman, E. E. Lattman, P. Schwager and W. Steigemann, Biophys. Struct. Mech., 1 (1975) 139.
94  J. Deisenhofer, P. M. Colman, O. Epps and R. Huber, Z. Physiol. Chem., 357 (1976) 1421.
94(a)  E. W. Silverton, M. A. Navia and D. R. Davies, Proc. Natl. Acad. Sci. USA, 74 (1977) 5140.
95  L. M. Amzel, R. J. Poljak, F. Saul, J. M. Varga and F. F. Richards, Proc. Natl. Acad. Sci. USA, 71 (1974) 1427.
96  E. A. Padlan, D. M. Segal, T. F. Spande, D. R. Davies, S. Rudikoff and M. Potter, Nature New Biol., 245 (1973) 165.
97  E. A. Padlan, D. R. Davies, I. Pecht, D. Givol and C. Wright, Cold Spring Harbor Symp. Quant. Biol., 41 (1976) 627.
98  R. A. Dwek, J. C. A. Knott, D. Marsh, A. C. McLaughlin, E. M. Press, N. C. Price and A. I. White, Eur. J. Biochem., 53 (1975) 25.
99  R. A. Dwek, Contemp. Topics Mol. Immunol., 6 (1977) 1.
100  D. Givol, J. Sharon, J. Hochman, M. Gavish, D. Inbar, I. Pecht, I-Z. Steinberg and J. Schlessinger, Cold Spring Harbor Symp. Quant. Biol., 41 (1976) 667.
101  A. Feinstein, J. Physiol., 242 (1974) 32P.
102  J. Deisenhofer, P. M. Colman, R. Huber, H. Haupt and G. Schwick, Z. Physiol. Chem., 357 (1976) 435.
103  D. W. Talmage, Science, 129 (1959) 1643.
104  F. F. Richards, W. H. Konigsberg and R. W. Rosenstein, Science, 187 (1975) 130.
105  J. M. Varga, S. Lande and F. F. Richards, J. Immunol., 112 (1974) 1565.
106  B. N. Manjula, F. F. Richards and R. W. Rosenstein, Immunochemistry, 13 (1976) 929.
107  R. W. Rosenstein and F. F. Richards, Immunochemistry, 13 (1976) 939.
108  J. M. Varga, W. H. Konigsberg and F. F. Richards, Proc. Natl. Acad. Sci. USA, 70 (1973) 3269.
109  D. S. Rowe, K. Hug, L. Forni and B. Pernis, J. Exp. Med., 138 (1973) 965.
110  R. A. Waldmann and W. Strober, Prog. Allergy, 13 (1969) 1.
111  T. A. Waldmann and W. A. Hemmings, in L. Brent and J. Holborow (Eds.), Progress in Immunology II, Vol. 1, North-Holland, Amsterdam, 1974, p. 230.
112  D. Gitlin, J. Kumate, J. Urrusti and C. Morales, J. Clin. Invest., 43 (1964) 1938.
113  F. W. R. Brambell, W. A. Hemmings, C. L. Oakley and R. R. Porter, Proc. Roy. Soc. Ser. B., 151 (1960) 478.
114  S. S. Froland, T. E. Michaelson, F. Wishoff and J. B. Natvig, Scand. J. Immunol., 3 (1974) 509.

# REFERENCES

115 D. A. Lawrence, W. O. Weigle and H. L. Spiegelberg, J. Clin. Invest., 55 (1975) 368.
116 R. Ramasamy, D. S. Secher and K. Adetugbo, Nature (Lond.), 253 (1975) 656.
117 R. Ramasamy, N. E. Richardson and A. Feinstein, Immunology, 30 (1976) 851.
118 I. C. M. MacLennan, G. E. Connell and F. M. Gotch, Immunology, 26 (1974) 303.
119 D. Yasmeen, J. R. Ellerson, K. J. Dorrington and R. H. Painter, J. Immunol., 110 (1973) 1706.
120 G. O. Okafor, M. W. Turner and F. C. Hay, Nature (Lond.), 248 (1974) 228.
121 F. Ciccimara, F. S. Rosen and E. Merler, Proc. Natl. Acad. Sci. USA, 72 (1975) 2081.
122 A. Capron, J.-P. Dessaint, M. Capron and H. Bazin, Nature (Lond.), 253 (1975) 474.
123 D. R. Stanworth, Clin. Exp. Immunol., 6 (1970) 1.
124 J. O. Minta and R. H. Painter, Immunochemistry, 9 (1972) 1041.
125 H. Bennich and H. von Bahr-Lindstrom, in L. Brent and J. Holborow (Eds.), Progress in Immunology II, Vol. 1, North-Holland, Amsterdam, 1974, p. 49.
126 H. Metzger, Adv. Immunol., 18 (1974) 169.
127 K. B. M. Reid and R. R. Porter, Contemp. Topics Mol. Immunol., 4 (1975) 1.
128 J. R. Ellerson, D. Yasmeen, R. H. Painter and K. J. Dorrington, FEBS Lett., 24 (1972) 318.
129 K. J. Dorrington and R. H. Painter, in L. Brent and J. Holborow (Eds.), Progress in Immunology II, Vol. 1, North-Holland, Amsterdam, 1974, p. 75.
130 G. E. Connell and R. R. Porter, Biochem. J., 124 (1971) 53P.
131 M. Colomb and R. R. Porter, Biochem. J., 145 (1975) 177.
132 C. Wolfenstein-Todel, F. Prelli, B. Frangione and E. C. Franklin, Biochemistry, 12 (1973) 5195.
133 M. M. Hurst, J. E. Volanakis, R. B. Hester, R. M. Stroud and J. C. Bennet, J. Exp. Med., 140 (1974) 1117.
134 J. C. Brown and M. E. Koshland, Proc. Natl. Acad. Sci. USA, 72 (1975) 5111.
135 A. Feinstein and A. J. Rowe, Nature (Lond.), 205 (1965) 147.
136 N. E. Hyslop, R. R. Dourmashkin, N. M. Green and R. R. Porter, J. Exp. Med., 131 (1970) 783.
137 D. E. Isenman, K. J. Dorrington and R. H. Painter, J. Immunol., 114 (1975) 1726.
138 F. M. Burnet, Aust. J. Sci., 20 (1957) 67.
139 F. M. Burnet, The Clonal Selection Theory of Acquired Immunity, Cambridge University Press, Cambridge, 1959.
140 J. Lederberg, Science, 129 (1959) 1649.
141 P. J. Lachmann, in M. J. Hobart and I. McConnell (Eds.), The Immune System, Blackwell, Oxford, 1975, p. 120.

142 I. McConnell, in M. J. Hobart and I. McConnell (Eds.), The Immune System, Blackwell, Oxford, 1975, p. 130.
143 A. Globerson, Current Topics Microbiol. Immunol., 75 (1976) 1.
144 D. Katz and B. Benacerraf, Adv. Immunol., 15 (1972) 1.
145 J. F. A. P. Miller, Int. Rev. Cytol., 33 (1972) 77.
146 M. Feldmann and G. J. V. Nossal, Transplant. Rev., 13 (1972) 3.
147 E. R. Unanue, Adv. Immunol., 15 (1972) 95.
148 G. Mitchell, Contemp. Topics Immunobiol., 3 (1974) 97.
149 G. Möller (Ed.), Immunology Reviews, Munksgaard, Copenhagen, Vol. 40, 1978.
150 G. Möller (Ed.), Transplantation Reviews, Munksgaard, Copenhagen, Vol. 23, 1975.
151 M. J. Bevan, R. M. E. Parkhouse, A. R. Williamson and B. A. Askonas, Prog. Biophys. Mol. Biol., 25 (1972) 131.
152 D. de Petris, Biochem. J., 118 (1970) 385.
153 D. Schubert, Proc. Natl. Acad. Sci. USA, 60 (1968) 683.
154 D. Cioli and E. S. Lennox, Biochemistry, 12 (1973) 3211.
155 P. Vassalli, B. Lisowska-Bernstein, M. E. Lamm and B. Benacerraf, Proc. Natl. Acad. Sci. USA, 58 (1967) 2422.
156 P. Vassalli, B. Lisowska-Bernstein and M. Lamm, J. Mol. Biol., 56 (1971) 1.
157 M. J. Bevan, Biochem. J., 122 (1971) 5.
158 B. Mach, H. Koblet and D. Gros, Proc. Natl. Acad. Sci. USA, 59 (1968) 445.
159 B. Lisowska-Bernstein, M. E. Lamm and P. Vassalli, Proc. Natl. Acad. Sci. USA, 66 (1970) 425.
160 C. Milstein, G. G. Brownlee, T. M. Harrison and M. B. Mathews, Nature New Biol., 239 (1972) 117.
161 D. Swan, H. Aviv and P. Leder, Proc. Natl. Acad. Sci. USA, 69 (1972) 1967.
162 B. Mach, C. Faust and P. Vassalli, Proc. Natl. Acad. Sci. USA, 70 (1973) 451.
163 S. Tonegawa and I. Baldi, Biochem. Biophys. Res. Commun., 51 (1973) 81.
164 B. J. Schmeckpeper, J. Cory and J. M. Adams, Mol. Biol. Rept., 1 (1974) 355.
165 M. Green, P. N. Graves, T. Zehavi-Willner, J. McInnes and S. Pestka, Proc. Natl. Acad. Sci. USA, 72 (1975) 224.
166 N. J. Cowan and C. Milstein, Eur. J. Biochem., 36 (1973) 1.
167 N. J. Cowan, T. M. Harrison, G. G. Brownlee and C. Milstein, Biochem. Soc. Trans., 1 (1973) 1247.
167(a) I. Schechter, D. J. McKean, R. Guyer and W. Terry, Science, 188 (1975) 160.
168 R. L. Jilka and S. Pestka, Proc. Natl. Acad. Sci. USA, 74 (1977) 5692.
169 H. H. Singer, F. T. Gates, T. J. Kindt and A. R. Williamson, unpublished work.
170 F. Melchers and P. Knopf, Cold Spring Harbor Symp. Quant. Biol., 32 (1967) 255.

# REFERENCES

171 I. Schenkein and J. W. Uhr, J. Cell Biol., 46 (1970) 42.
172 F. Melchers, Biochem. J., 119 (1970) 765.
173 F. Melchers, Biochemistry, 10 (1971) 653.
174 R. M. E. Parkhouse and F. Melchers, Biochem. J., 125 (1971) 235.
175 F. Melchers, Biochem. J., 125 (1971) 241.
176 R. M. Swenson and M. Kerr, Proc. Natl. Acad. Sci. USA, 59 (1968) 546.
177 H. J. Cohen and M. Kern, Biochim. Biophys. Acta, 188 (1969) 255.
178 F. Melchers and J. Andersson, Transplant. Rev., 14 (1973) 76.
179 C. Moroz and J. W. Uhr, Cold Spring Harbor Symp. Quant. Biol., 32 (1967) 263.
180 C. J. Sherr and J. W. Uhr, Proc. Natl. Acad. Sci. USA, 64 (1969) 381.
181 F. Melchers, Biochemistry, 8 (1969) 938.
182 J. W. Uhr and L. Schenkein, Proc. Natl. Acad. Sci. USA, 66 (1970) 952.
183 Y. S. Choi, P. M. Knopf and E. S. Lennox, Biochemistry, 10 (1971) 659.
184 Y. S. Choi, P. M. Knopf and E. S. Lennox, Biochemistry, 10 (1971) 668.
185 D. Zagury, J. W. Uhr, J. D. Jamieson and G. E. Palade, J. Cell Biol., 46 (1970) 52.
186 J. Andersson, J. Buxbaum, R. Citronbaum, S. Douglas, L. Forni, F. Melchers, B. Pernis and D. I. Stott, J. Exp. Med., 140 (1974) 742.
187 N. H. Behrens, H. Carminatti, R. J. Staneloni, L. F. Leloir and A. I. Cantarella, Proc. Natl. Acad. Sci. USA, 70 (1973) 3390.
188 B. A. Askonas and A. R. Williamson, Nature (Lond.), 216 (1967) 264.
189 B. A. Askonas and A. R. Williamson, in J. Killander (Ed.), Gamma Globulins, Proc. Third Nobel Symp., Wiley, London, 1967, p. 369.
190 R. Baumal and M. D. Scharff, Transplant. Rev., 14 (1973) 163.
191 R. M. E. Parkhouse, Biochem. J., 123 (1971) 635.
192 D. I. Stott, Biochem. J., 130 (1972) 1151.
193 A. L. Shapiro, M. D. Scharff, J. V. Maizel and J. W. Uhr, Proc. Natl. Acad. Sci. USA, 56 (1966) 216.
194 D. Schubert, Proc. Natl. Acad. Sci. USA, 60 (1968) 683.
195 B. A. Askonas, A. R. Williamson and Z. L. Awdeh, FEBS Symp., 15 (1969) 105.
196 A. R. Williamson and B. A. Askonas, Biochem. J., 107 (1968) 823.
197 R. M. E. Parkhouse, G. Virella and R. R. Dourmashkin, Clin. Exp. Immunol., 8 (1971) 581.
198 R. Hong and A. Nisonoff, J. Biol. Chem., 240 (1965) 3883.
199 B. A. Askonas and R. M. E. Parkhouse, Biochem. J., 123 (1971) 629.
200 A. Feinstein and D. I. Stott, J. Physiol., 226 (1972) 34P.
201 D. I. Stott and A. Feinstein, Eur. J. Immunol., 3 (1973) 229.
202 A. Cooke, and A. Feinstein, Immunochemistry, 14 (1977) 627.
202(a) A. R. Kazin and S. Beychock, Science, 199 (1978) 688.
203 M. J. Bevan, Eur. J. Immunol., 1 (1971) 133.
204 J. N. Buxbaum, S. Zolla, M. D. Scharff and E. C. Franklin, J. Exp. Med., 133 (1971) 1118.
205 J. N. Buxbaum and M. D. Scharff, J. Exp. Med., 138 (1973) 278.
206 D. I. Stott, Immunochemistry, 13 (1976) 157.

207  R. M. E. Parkhouse and B. A. Askonas, Biochem. J., 115 (1969) 163.
208  R. M. E. Parkhouse, Transplant. Rev., 14 (1975) 131.
209  B. A. Askonas and R. M. E. Parkhouse, Biochem. J., 123 (1971) 629.
210  F. Melchers, Biochemistry, 11 (1972) 2204.
211  D. I. Stott, Immunol. Commun., 7 (1978) 519.
212  N. E. Richardson and A. Feinstein, Biochem. J., 175 (1978) 959.
213  R. M. Chapuis and M. E. Koshland, Proc. Natl. Acad. Sci. USA, 71 (1974) 657.
214  E. Della Corte and R. M. E. Parkhouse, Biochem. J., 136 (1973) 596.
215  T. Eskeland and P. Brandtzaeg, Immunochemistry, 11 (1974) 161.
216  D. M. Parr, G. E. Connell, A. J. Powell and W. Pruzanski, J. Immunol., 113 (1974) 2020.
217  P. F. Weinheimer, J. Mestecky and R. G. Acton, J. Immunol., 107 (1971) 1211.
218  T. Eskeland and M. Harboe, Scand. J. Immunol., 2 (1973) 511.
219  T. Eskeland, Scand. J. Immunol., 3 (1974) 757.
220  E. Kownatzki, Immunol. Commun., 2 (1973) 105.
221  A. Feinstein, IUB Abstracts 9th Int. Congress of Biochem., Stockholm, 1973, p. 300.
222  D. Beale, Biochim. Biophys. Acta., 351 (1974) 13.
223  E. L. Mather and M. E. Koshland, in E. E. Sercarz, L. A. Herzenberg and C. F. Fox (Eds.), The Immune System: Genetics and Regulation, Academic Press, New York, 1978, p. 727.
224  R. M. E. Parkhouse and E. Della Corte, Biochem. J., 136 (1973) 607.
225  T. B. Tomasi, Jr., Proc. Natl. Acad. Sci. USA, 70 (1973) 3410.
226  R. M. E. Parkhouse, Immunology, 27 (1975) 1063.
227  A. B. Lawton, R. Asofsky and R. Mage, J. Immunol., 104 (1970) 397.
228  M. E. Poger and M. E. Lamm, J. Exp. Med., 139 (1974) 629.
229  D. N. Buell and J. L. Fahey, Science, 164 (1969) 1524.
230  M. Takahashi, Y. Yagi, G. E. Moore and D. Pressman, J. Immunol., 103 (1969) 834.
231  R. A. Lerner and L. D. Hodge, J. Cell. Physiol., 77 (1971) 265.
232  N. Byars and C. Kidson, Nature (Lond.), 226 (1970) 648.
233  R. G. H. Cotton and C. Milstein, Nature (Lond.), 244 (1973) 42.
234  G. Köhler and C. Milstein, Nature (Lond.), 256 (1975) 495.
235  G. Köhler and C. Milstein, Eur. J. Immunol., 6 (1976) 511.
236  G. Galfre, S. C. Howe, C. Milstein, G. W. Butcher and J. C. Howard, Nature (Lond.), 266 (1977) 550.
237  T. Springer, G. Galfré, D. S. Secher and C. Milstein, Eur. J. Immunol., 8 (1978) 539.
238  I. S. Trowbridge, J. Exp. Med., 148 (1978) 313.
239  F. Melchers, M. Potter and N. L. Warner (Eds.), Current Topics Microbiol. Immunol., Vol. 81, Springer, Berlin, 1978.
240  G. Galfre, C. Milstein and B. Wright, Nature (Lond.), 277 (1979) 131.
241  R. S. Nezlin, Structure and Biosynthesis of Antibodies, Consultants Bureau, New York, 1977.
242  J. B. Fleischman, J. Immunol., 91 (1963) 163.
243  J. B. Fleischman, Biochemistry, 6 (1967) 1311.

# REFERENCES

244  P. M. Knopf, R. M. E. Parkhouse and E. S. Lennox, Proc. Natl. Acad. Sci. USA, 58 (1967) 2288.
245  E. S. Lennox, P. M. Knopf, A. J. Munro and R. M. E. Parkhouse, Cold Spring Harbor Symp. Quant. Biol., 32 (1967) 249.
246  C. Milstein, G. G. Brownlee, E. M. Cartwright, J. M. Jarvis and N. J. Proudfoot, Nature (Lond.), 252 (1974) 354.
247  N. Hozumi and S. Tonegawa, Proc. Natl. Acad. Sci. USA, 73 (1976) 3628.
248  T. H. Rabbitts and A. Forster, Cell, 13 (1978) 319.
249  J. G. Seidman and P. Leder, Nature (Lond.), 276 (1978) 790.
250  C. Brack, M. Hirama, R. Lenhard-Schuller and S. Tonegawa, Cell, 15 (1978) 1.
251  G. Matthyssens and S. Tonegawa, Nature (Lond.), 273 (1978) 763.
252  R. Breathnach, J. L. Mandel and P. Chambon, Nature (Lond.), 270 (1977) 314.
253  M. T. Doel, M. Haughton, E. A. Cook and N. H. Carey, Nucl. Acids Res., 4 (1977) 3701.
254  J. F. Catterall, J. P. Stein, E. C. Lai, S. L. C. Woo, A. Dugaiczyk, M. L. Mace, A. R. Means and B. W. O'Malley, Nature (Lond.), 278 (1979) 323.
255  S. M. Tilghman, D. C. Tiemeier, J. G. Seidman, B. M. Peterlin, M. Sullivan, J. V. Maizel and P. Leder, Proc. Natl. Acad. Sci. USA, 75 (1978) 725.
255(a)  D. C. Tiemeier, S. M. Tilghman, F. I. Polsky, J. G. Seidman, A. Leder, M. H. Edgell and P. Leder, Cell, 14 (1978) 237.
256  H. Sakano, J. Rogers, K. Hüppi, C. Brack, A. Traunecker, R. Maki, R. Wall and S. Tonegawa, Nature (Lond.), 277 (1979) 627.
257  S. Tonegawa, C. Brack, N. Hozumi and R. L. Schuller, Proc. Natl. Acad. Sci. USA, 74 (1977) 3518.
258  N. Hozumi, C. Brack, V. Pirrotta, R. Lenhard-Schuller and S. Tonegawa, Nucl. Acids Res., 5 (1978) 1485.
259  S. Tonegawa, A. M. Maxam, R. Tizard, O. Bernard and W. Giltbert. Proc. Natl. Acad. Sci. USA, 75 (1978) 1485.
260  T. H. Rabbitts, Nature (Lond.), 275 (1978) 291.
261  M. Gilmore-Hebert and R. Wall, Proc. Natl. Acad. Sci. USA, 75 (1978) 342.
262  U. Schibler, K. B. Marcu and R. P. Perry, Cell, 15 (1978) 1495.
262(a)  S. M. Tilghman, P. J. Curtis, D. C. Tiemeier, P. Leder and C. Weissmann, Proc. Natl. Acad. Sci. USA, 75 (1978) 1309.
263  B. Pernis, L. Forni and L. Amante, J. Exp. Med., 132 (1970) 1001.
264  S. S. Froland and J. Natvig, J. Exp. Med., 136 (1972) 409.
265  B. Wolf, J. C. A. Janeway, R. R. A. Coombs, D. Catty, P. G. H. Gell and A. S. Kelus, Immunology, 20 (1971) 931.
266  R. B. Taylor, P. H. Duffus, M. C. Raff and S. de Petris, Nature New Biol., 233 (1971) 225.
267  E. R. Unanue, W. D. Perkins and M. J. Karnovsky, J. Exp. Med., 136 (1972) 885.
268  E. R. Unanue and M. J. Karnovsky, Transplant. Rev., 14 (1978) 184.
269  S. J. Singer, Adv. Immunol., 19 (1974) 1.

270 R. S. Smith, R. L. Longmire, R. T. Reid and R. S. Farr, J. Immunol., 104 (1970) 367.
271 E. Rebellino, S. Colon, H. M. Grey and E. R. Unanue, J. Exp. Med., 133 (1971) 156.
272 J. J. Marchalonis and R. E. Cone, Transplant. Rev., 14 (1973) 3.
273 E. S. Vitetta and J. W. Uhr, Transplant. Rev., 14 (1973) 50.
274 J. J. Marchalonis, Contemp. Topics Mol. Immunol., 5 (1976) 125.
275 E. S. Vitetta and J. W. Uhr, Biochim. Biophys. Acta, 415 (1975) 253.
276 E. S. Vitetta, U. Melcher, M. McWilliams, J. Phillips-Quagliata, M. Lamm and J. W. Uhr, J. Exp. Med., 141 (1975) 206.
277 D. S. Rowe, K. Hug, L. Forni and B. Pernis, J. Exp. Med., 138 (1973) 965.
278 S. M. Fu, R. J. Winchester and H. G. Kunkel, J. Immunol., 114 (1975) 250.
279 B. Pernis, J. C. Brouet and M. Seligmann, Eur. J. Immunol., 4 (1974) 776.
280 C. Stern and I. McConnell, Eur. J. Immunol., 6 (1976) 225.
281 S. M. Fu and H. G. Kunkel, J. Exp. Med., 140 (1974) 895.
282 H. L. Spiegelberg, Immunol. Rev., 37 (1977) 3.
283 U. Melcher and J. W. Uhr, Biochemistry, 16 (1977) 145.
283(a) U. Melcher, L. Eidels and J. W. Uhr, Nature (Lond.), 158 (1975) 434.
284 D. I. Stott, Immunol. Commun., 7 (1978) 519.
285 U. Melcher and J. W. Uhr, J. Exp. Med., 138 (1973) 1282.
286 U. Melcher and J. W. Uhr, J. Immunol., 116 (1976) 409.
287 E. S. Vitetta and J. W. Uhr, Immunol. Rev., 37 (1977) 50.
288 P. A. Singer and A. R. Williamson, Eur. J. Immunol., 10 (1980) 180.
289 Y. Bergman and J. Haimovich, Eur. J. Immunol., 8 (1978) 876.
290 P. A. Singer, Ph.D. Thesis, University of Glasgow.
291 R. A. J. McIlhinney, N. E. Richardson and A. Feinstein, Nature (Lond.), 272 (1978) 555.
292 P. B. Williams, R. T. Kubo and H. M. Grey, J. Immunol., 121 (1978) 2435.
293 M. F. Greaves, Transplant. Rev., 5 (1970) 45.
294 M. F. Greaves and N. M. Hogg, in Amos (Ed.), Progress in Immunology, Vol. 1, Academic Press, New York, 1972, p. 111.
295 J. J. Marchalonis and R. E. Cone, Transplant. Rev., 14 (1973) 3.
296 J. J. Marchalonis, Contemp. Topics Mol. Immunol. 5 (1976) 125.
297 E. S. Vitetta, C. Bianco, V. Nussenzweig and J. W. Uhr, J. Exp. Med., 136 (1972) 81.
298 H. Grey, J. Exp. Med., 136 (1972) 1323.
299 E. S. Vitetta and J. W. Uhr, Transplant. Rev., 14 (1973) 50.
300 M. Feldmann, J. Exp. Med., 136 (1972) 737.
301 H. Wigzell, Contemp. Topics Immunobiol., 3 (1974) 77.
302 N. L. Warner, Adv. Immunol., 19 (1974) 67.
303 R. M. E. Parkhouse and E. R. Abney, in F. Loor and G. E. Roelants (Eds.), B and T Cells in Immune Recognition, Wiley, New York, 1977, p. 211.
304 E. R. Unanue and C. F. Schreiner, in G. Poste and G. T. Nicolson

(Eds.), Cell Surface Reviews, Vol. 3, North-Holland, Amsterdam, 1977, p. 619.
305 C. J. Sherr, I. Schenkein and J. W. Uhr, Ann. N.Y. Acad. Sci., 190 (1971) 250.
306 C. J. Sherr, I. Schenkein and J. W. Uhr, Proc. Natl. Acad. Sci. USA, 66 (1970) 1183.
307 D. Wernet, E. S. Vitetta, E. A. Boyse and J. W. Uhr, J. Exp. Med., 138 (1973) 847.
308 E. S. Vitetta and J. W. Uhr, J. Exp. Med., 139 (1974) 1599.
309 R. E. Cone, J. J. Marchalonis and R. T. Rolley, J. Exp. Med., 134 (1971) 1373.
310 E. S. Vitetta and J. W. Uhr, J. Exp. Med., 136 (1972) 676.
311 J. Andersson, J. Buxbaum, R. Citronbaum, S. Douglas, L. Forni, F. Melchers, B. Pernis and D. I. Stott, J. Exp. Med., 140 (1974) 742.
312 J. V. Wells, in H. H. Fudenberg, D. P. Stites, J. L. Caldwell and J. V. Wells (Eds.), Basic and Clinical Immunology, 2nd ed., Lange Medical Publications, Los Altos, CA, 1978, p. 237.
313 T. A. Waldmann, W. Strober and R. M. Blaese, in B. Amos (Ed.), Progress in Immunology, Vol. 1, Academic Press, New York, 1971, p. 891.
314 J. Watkins, M. W. Turner and A. Roberts, in H. Peeters (Ed.), Protides of the Biological Fluids, Pergamon, Oxford, 1971, p. 461.
315 K. J. Dorrington and R. H. Painter, in L. Brent and J. Holborow (Eds.), Progress in Immunogloby II, Vol. 1, North Holland, Amsterdam, 1974, p. 75.
316 F. W. R. Brambell, W. A. Hemmings, C. L. Oakley and R. R. Porter, Proc. Roy. Soc. Ser. B., 151 (1960) 478.
317 M. W. Turner, in L. E. Glynn and M. W. Steward (Eds.), Immunochemistry: An Advanced Textbook, Wiley, New York, 1977, p. 1.
318 E. A. Kabat, T. T. Wu and H. Bilofsky, Variable Regions of Immunoglobulin Chains. Tabulations and Analyses of Amino Acid Sequences, Belt, Bearanek and Newman, Boston, MA, 1976.
319 H. Metzger, Adv. Immunol., 12 (1970) 57.
320 A. B. Edmundson, K. R. Ely, E. E. Abola, M. Schiffer and N. Panagiotopoulos, Biochemistry, 14 (1975) 3953.
321 K. Landsteiner, Science, 73 (1931) 403.
322 C. C. Little, Science, 40 (1914) 904.
323 P. A. Gorer, Br. J. Exp. Pathol., 17 (1936) 42.
324 G. P. Snell, J. Genet., 49 (1948) 87.
325 P. B. Medawar, Br. J. Exp. Pathol., 27 (1946) 15.
326 J. J. Van Rood, J. G. Eernisse and A. Van Leeuwen, Nature (Lond.), 18 (1958) 1735.
327 R. Payne and M. R. Rolfs, J. Clin. Invest., 37 (1958) 1756.
328 J. Dausset and F. T. Rapaport, Ann. N.Y. Acad. Sci., 129 (1966) 408.
329 W. F. Bodmer and J. G. Bodmer, Br. Med. Bull., 34 (1978) 309.
330 H. O. McDevitt and A. Chinitz, Science, 163 (1969) 1207.
331 B. Benacerraf and H. O. McDevitt, Science, 175 (1972) 273.
332 D. H. Katz, Lymphocyte Differentiation, Recognition and Regulation, Academic Press, New York, 1977.

333  R. M. Zinkernagel and P. C. Doherty, Nature (Lond.), 251 (1974) 547.
334  M. J. Bevan, J. Exp. Med., 142 (1975) 1349.
335  G. M. Shearer, T. G. Rhem and C. A. Garbarino, J. Exp. Med., 141 (1975) 1348.
336  R. D. Gordon, E. Simpson and L. E. Samelson, J. Exp. Med., 142 (1975) 1108.
337  R. M. Zinkernagel, G. N. Callahan, A. Althage, S. Cooper, P. A. Klein and J. Klein, J. Exp. Med., 147 (1978) 882.
338  J. Klein, Biology of the Mouse Histocompatibility-2 Complex, Springer, Berlin, 1975.
339  G. D. Snell, M. Cherry and P. Démant, Transplant. Proc., 3 (1971) 183.
340  D. C. Shreffler, C. S. David, H. C. Passmore and J. Klein, Transplant. Proc., 3 (1971) 176.
341  E. Thorsby, Eur. J. Immunol., 1 (1971) 57.
342  W. F. Bodmer, Nature (Lond.), 237 (1972) 139.
343  D. C. Shreffler and C. S. David, Adv. Immunol., 20 (1975) 135.
344  D. B. Murphy, K. Okumura, L. A. Herzenberg and H. O. McDevitt, Cold Spring Harbor Symp. Quant. Biol., 41 (1976) 497.
345  F. H. Bach, M. B. Widmer, M. L. Bach and J. Klein, J. Exp. Med., 136 (1972) 1420.
346  B. Bain, M. R. Vas and L. Lowenstein, Blood, 23 (1964) 108.
347  F. M. Bach and K. Hirsch-horn, Science, 142 (1964) 813.
348  J. Silver and L. Hood, Contemp. Topics Mol. Immunol., 5 (1976) 35.
349  D. Snary, C. Barnstable, W. F. Bodmer, P. Goodfellow and M. J. Crumpton, Cold Spring Harbor Symp. Quant. Biol., 41 (1976) 379.
350  J. L. Strominger, D. L. Mann, P. Parham, R. Robb, T. Springer and C. Terhorst, Cold Spring Harbor Symp. Quant. Biol., 41 (1976) 323.
351  A. Shimada and S. G. Nathenson, Biochemistry, 8 (1969) 4048.
352  F. S. Walsh and M. J. Crumpton, Nature (Lond.), 269 (1977) 307.
353  P. Parham, B. N. Alpert, H. T. Orr and J. L. Strominger, J. Biol. Chem., 252 (1977) 7555.
354  N. Tanigaki and D. Pressman, Transplant Rev., 21 (1974) 15.
355  B. A. Cunningham and I. Berggard, Transplant Rev., 21 (1974) 3.
356  E. S. Vitetta and J. D. Capra, Adv. Immunol., 26 (1978) 147.
357  T. L. Delovitch and H. O. McDevitt, Immunogenetics, (1975)
358  J. Silver, M. J. Cecka, M. McMillan and L. Hood, Cold Spring Harbor Symp. Quant. Biol., 41 (1976) 369.
359  S. E. Cullen, J. H. Freed and S. G. Nathenson, Transplant Rev., 30 (1976) 236.
360  J. Klein, V. Hauptfeld and E. S. Vitetta, in D. H. Katz and B. Benacerraf (Eds.), The Role of Products of the Histocompatibility Gene Complex in Immune Responses, Academic Press, New York, 1976, p. 53.
361  B. D. Schwartz and S. E. Cullen, in D. H. Katz and B. Benacerraf (Eds.), The Role of Products of the Histocompatibility Gene Complex in Immune Responses, Academic Press, New York, 1976, p. 691.
362  B. D. Schwartz, A. M. Kask and E. M. Shevach, Cold Spring Harbor Symp. Quant. Biol., 41 (1976) 397.
363  T. A. Springer, J. F. Kaufman, L. A. Siddoway, M. Giphart, D. L.

# REFERENCES

Mann, C. Terhorst and J. L. Strominger, Cold Spring Harbor Symp. Quant. Biol., 41 (1976) 387.
364 C. J. Barnstable, E. A. Jones, W. F. Bodmer, J. B. Bodmer, B. Arce-Gomez, D. Snary and J. J. Crumpton, Cold Spring Harbor Symp. Quant. Biol., 41 (1976) 443.
365 E. Appella, N. Tanigaki, O. Henriksen, D. Pressman, D. F. Smith and T. Fairwell, Cold Spring Harbor Symp. Quant. Biol., 41 (1976) 341.
366 J. Bridgen, D. Snary, M. J. Crumpton, C. Barnstable, P. Goodfellow and W. F. Bodmer, Nature (Lond.), 261 (1976) 200.
367 P. A. Peterson, B. A. Cunningham, I. Berggård and G. M. Edelman, Proc. Natl. Acad. Sci. USA, 69 (1972) 1697.
368 M. J. Crumpton, in M. Sela (Ed.), The Antigens, Vol. 2, Academic Press, New York, 1974, p. 1.
369 B. J. Alter, D. J. Schendel, M. L. Bach, F. M. Bach, J. Klein and J. H. Stimpfling, J. Exp. Med., 137 (1973) 1303.
370 H. Festenstein, Transplant. Rev., 15 (1973) 62.
371 R. M. Zinkernagel and P. C. Doherty, Nature (Lond.), 248 (1975) 701.
372 P. C. Doherty, Immunogenetics, 3 (1976) 517.
373 H. Von Boehmer, W. Haas and N. K. Jerne, Proc. Natl. Acad. Sci. USA, 75 (1978) 2439.
374 M. J. Bevan, Cold Spring Harbor Symp. Quant. Biol., 41 (1976) 519.
375 M. J. Bevan, Nature (Lond.), 269 (1977) 417.
376 R. M. Zinkernagel, G. N. Callahan, J. Klein and G. Dennert, Nature (Lond.), 271 (1978) 251.
377 M. Simonsen, Cold Spring Harbor Symp. Quant. Biol., 32 (1967) 517.
378 D. B. Wilson, J. Blyth and P. C. Nowell, J. Exp. Med., 128 (1968) 1157.
379 F. M. Bach, H. Bock, K. Graupner, E. Day and H. Klostermann, Proc. Natl. Acad. Sci. USA, 62 (1969) 377.
380 K. F. Lindahl and D. B. Wilson, J. Exp. Med., 145 (1977) 508.
381 M. A. Skinner and J. Marbrook, J. Exp. Med., 143 (1976) 1562.
382 H-S. Teh, E. Harley, R. A. Phillips and R. G. Miller, J. Immunol., 118 (1977) 1049.
383 A. R. Williamson, Nature, 283 (1980) 527.
384 R. Lieberman and W. Humphrey, Proc. Natl. Acad. Sci. USA, 68 (1971) 2510.
385 H. O. McDevitt and M. Sela, J. Exp. Med., 122 (1965) 517.
386 B. Benacerraf, Harvey Lectures, 67 (1973) 109.
387 H. O. McDevitt and M. Sela, J. Exp. Med., 126 (1967) 969.
388 H. O. McDevitt, B. D. Deak, D. C. Shreffler, J. Klein, J. H. Stimpfling and G. D. Snell, J. Exp. Med., 135 (1972) 1259.
389 F. C. Grumet, J. Exp. Med., 135 (1972) 110.
390 G. F. Mitchell, F. C. Grumet and H. O. McDevitt, J. Exp. Med., 135 (1972) 126.
391 J. C. Ordal and F. C. Grumet, J. Exp. Med., 136 (1972) 1195.
392 M. E. Dorf, J. H. Stimpfling and B. Benacerraf, J. Exp. Med., 141 (1975) 1459.
393 M. E. Dorf and B. Benacerraf, Proc. Natl. Acad. Sci. USA, 72 (1975) 3671.

394  P. Marrack and J. W. Kappler, J. Exp. Med., 149 (1978) 1596.
395  J. W. Kappler and P. Marrack, J. Exp. Med., 148 (1978) 1510.
396  M. W. Dorf, M. B. Twigg and B. Benacerraf, Eur. J. Immunol., 6 (1976) 552.
397  B. A. Bradley and H. Festenstein, Br. Med. Bull., 34 (1978) 223.
398  T. Meo, C. S. David and D. C. Shreffler, in D. H. Katz and B. Benacerraf (Eds.), The Role of the Products of the Histocompatibility Gene Complex in Immune Responses, Academic Press, New York, 1976.
399  R. Ceppellini, G. D. Bonnard, F. Coppo, V. C. Miggiano, M. Pospisil, E. S. Curtoni and M. Pellegrino, Transplant. Proc., 3 (1971) 58.
400  M. Rychlikova, O. Demant and I. K. Egorov, Folia Biol. (Praha), 18 (1972) 360.
401  B. Kindred and D. C. Shreffler, J. Immunol., 109 (1972) 940.
402  D. H. Katz, T. Hamaoka and B. Benacerraf, J. Exp. Med., 137 (1973) 1405.
403  A. S. Rosenthal and E. M. Shevach, J. Exp. Med., 138 (1973) 1194.
404  D. H. Katz, J. Exp. Med., 141 (1975) 263.
405  W. E. Paul, E. M. Shevach, D. W. Thomas, S. F. Pickeral and A. S. Rosenthal, Cold Spring Harbor Symp. Quant. Biol., 41 (1976) 571.
406  H. Von Boehmer, W. Haas and N. K. Jerne, Proc. Natl. Acad. Sci. USA, 75 (1978) 2439.
407  M. Cohn and R. Epstein, Cell. Immunol., 39 (1978) 125.
408  R. Lenhard-Schuller, B. Hohn, C. Brack, M. Hirama and S. Tonegawa, Proc. Natl. Acad. Sci. USA, 75 (1978) 4709.
409  T. Honjo, M. Obata, Y. Yamawaki-Kataoka, T. Kataoka, T. Kawakami, N. Takahashi and Y. Mano, Cell, 18 (1979) 559.
410  T. Kataoka, Y. Yamawaki-Kataoka, H. Yamagishi and T. Honjo, Proc. Natl. Acad. Sci. USA, 76 (1979) 4240.
411  P. W. Early, M. M. Davis, D. B. Kaback, N. Davidson and L. Hood, Proc. Natl. Acad. Sci. USA, 76 (1979) 857.
412  J. Rogers, P. Early, C. Carter, K. Calame, M. Bond, L. Hood and R. Wall, Cell, 20 (1980) 303.
413  P. Early, J. Rogers, M. Davis, K. Calame, M. Bond, R. Wall and L. Hood, Cell, 20 (1980) 313.
414  O. Bernard, N. Hozumi and S. Tonegawa, Cell, 15 (1978) 1133.
415  E. E. Max, J. G. Seidman and P. Leder, Proc. Natl. Acad. Sci. USA, 76 (1979) 3450.
416  H. Sakano, K. Hüppi, G. Heinrich and S. Tonegawa, Nature (Lond.), 280 (1979) 288.
417  P. W. Early, H. V. Huang, M. N. Davis, K. Calame and L. Hood, Cell, 19 (1980) 981.
418  O. Bernard and N. M. Gough, Proc. Natl. Acad. Sci. USA, 77 (1980) 3630.
419  S. Cory, J. Adams and D. J. Kemp, Proc. Natl. Acad. Sci. USA, 77 (1980) 4943.
420  M. Weigert, R. Perry, D. Kelly, T. Hunkapiller, J. Schilling and L. Hood, Nature (Lond.), 283 (1980) 497.
421  J. M. Adams, Immunol. Today, 1 (1980) 10.

## REFERENCES

422 H. V. Molgaard, Nature (Lond.), 286 (1980) 657.
423 M. M. Davis, S. K. Kim and L. Hood, Cell, 22 (1980) 1.
424 D. R. Burton, J. Boyd, A. D. Brampton, S. B. Easterbrook-Smith, E. J. Emanuel, J. Novotny, T. W. Rademacher, M. R. van Schravendijk, M. J. E. Sternberg and R. A. Dwek, Nature (Lond.), 288 (1980) 338.
425 P. A. Singer, H. H. Singer and A. R. Williamson, Nature (London), 285 (1980) 294.
426 C. J. Barnstable, E. A. Jones and M. J. Crumpton, Br. Med. Bull., 34 (1978) 241.
427 P. W. Tucker, C.-P. Lui, J. F. Mushinski and F. R. Blattner, Science, 209 (1980) 1353.
428 G. Alcaraz, A. Bourgois, A. Moulin, H. Bazin and M. Fougerau, Ann. Immunol. (Inst. Pasteur), 131C (1980) 363.

Chapter 3

# Regulation of the Messenger RNA for Hepatic Tryptophan 2,3-Dioxygenase

LOIS A. KILLEWICH and PHILIP FEIGELSON

*Department of Biochemistry and the Institute of Cancer Research, Columbia University, New York, NY 10032 (U.S.A.)*

## Abstract

In the adult rat, the activity of hepatic tryptophan 2,3-dioxygenase (EC 1.13.11.11; L-tryptophan 2,3-oxidoreductase (decyclizing)) also known as tryptophan pyrrolase, is induced by glucocorticoid hormones and its substrate, L-tryptophan. Actinomycin D, if given during the period of hormonal induction, will cause a further superinduced elevation in enzyme activity above the hormonally induced level. In the developing rat, tryptophan 2,3-dioxygenase catalytic activity is absent from liver until the fifteenth postnatal day, after which it increases in a linear fashion to the adult level by 22 days. Glucocorticoids will also induce the enzyme during this period, both before and after its normal appearance at 15 days. Tryptophan 2,3-dioxygenase activity is not detectable in the minimal-deviation Morris hepatoma 7793, although the liver of the host animal has normal levels. Furthermore, hydrocortisone fails to induce the appearance of tryptophan 2,3-dioxygenase activity in the hepatoma unlike its effect in the developing rat.

To determine the biochemical processes underlying these various types of enzymic regulation, we have used cell-free protein synthesizing systems to quantitate functional tryptophan 2,3-dioxygenase mRNA after induction of the enzyme with these various regulators. In the adult rat, we have found that glucocorticoid induction of enzyme activity is mediated through the parallel induction of tryptophan 2,3-dioxygenase mRNA. Neither tryptophan nor actinomycin D, however, have any effect on its mRNA. In the postnatal rat, both the normal and glucocorticoid-induced appearance and development of enzyme activity result from the parallel appearance and development of its mRNA. In the untreated and hydrocortisone-treated hepatoma, the absence of tryptophan 2,3-dioxygenase activity result from the concomitant absence

*References p. 344*

of its mRNA. Thus, developmental and hormonal control of the rate of synthesis of hepatic tryptophan 2,3-dioxygenase is through pretranslational, developmental and hormonal control of its specific mRNA level.

A central problem in differentiation and physiological adaptation is to gain understanding of the biochemical mechanisms which regulate the level and catalytic efficiency of specific enzymes in normal and neoplastic tissues. We have been interested in studying the regulation of the rat liver enzyme tryptophan 2,3-dioxygenase [EC 1.13.11.11; L-tryptophan: oxygen 2,3-oxidoreductase (decyclizing)] whose activity is regulated in normal liver by glucocorticoid hormones, L-tryptophan and a variety of allosteric effectors [4, 10, 18, 19, 30, 33]. In young rats, its activity can be induced by glucocorticoids prior to its normal developmental appearance as well as during its developmental period [5, 7, 11, 22, 29, 44, 45]. Furthermore, its activity is absent from several Morris hepatomas although the host livers have normal levels [27].

To further our understanding of the regulation of the tryptophan 2,3-dioxygenase activity in adult and developing rat liver and in neoplastic transformation, we have determined the level at which control of its synthesis occurs. Using cell-free protein synthesizing systems, we have measured the level of tryptophan 2,3-dioxygenase mRNA during induction of the enzyme with various regulators, during development and during neoplastic transformation. Cell-free protein synthesis permits quantitation of the level of a specific messenger RNA (mRNA) by its ability to synthesize the protein for which it codes. We have used cell-free systems to measure tryptophan 2,3-dioxygenase mRNA, the fortified Krebs II ascites system [31, 32, 34] and the wheat-germ system [14, 16]. Poly(A)-containing rat liver mRNA used to program the synthesis was prepared by phenolization of the liver followed by chromatography of the total cellular RNA on cellulose [32, 34] or poly(U)-mica [25]. Tryptophan 2,3-dioxygenase synthesized in response to this mRNA was isolated by immunoprecipitation followed by electrophoresis of the immunoprecipitate on sodium dodecyl sulfate-polyacrylamide gels. The amount of radioactivity in newly synthesized tryptophan 2,3-dioxygenase was taken as a measurement of the level of its mRNA.

Cell-free protein synthesis permits measurement of the level of functional mRNA coding for a particular mRNA, but it does not distinguish between modifications of the synthetic or degradative rate of the mRNA. An increase in the amount of functional tryptophan 2,3-dioxygenase mRNA following induction of enzyme activity indicates that control of the enzyme induction is pretranslational and, in all likelihood, results from an accelerated transcription of the corresponding structural gene.

In the adult rat, a single injection of hydrocortisone causes a 3—5-fold increase in the activity of rat liver tryptophan 2,3-dioxygenase in 4—6-h, with the activity returning to the basal level by 8—10 h [4, 10, 18]. This increase in enzyme activity is due to an increase in the rate of synthesis of the enzyme [30], and thus in the amount present in the liver [4].

In Fig. 1 the effect of a single injection of 2 mg hydrocortisone/ 100 g body weight on tryptophan 2,3-dioxygenase enzyme activity and mRNA level is compared. The enzyme activity increases 3.5-fold (from 4 to 13 $\mu$mol kynurenine/h/g liver) 4—6 h after hormone administration, and drops to 2.5 times the basal level by 8 h after administration. Tryptophan 2,3-dioxygenase mRNA, measured in the manner described, increases 3-fold 4—6 h after hormone administration, and returns to approx. 1.5 times the basal level 8 h after administration. It is evident that the amount of tryptophan 2,3-dioxygenase mRNA increases in parallel with the enzyme activity during the period on induction. It falls slightly faster than the enzyme activity during the deinduction period. We concluded that the induction of tryptophan 2,3-dioxygenase activity and rate of synthesis by glucocorticoids is mediated through an increase in the level of its mRNA. It cannot be attributed to an increased translational efficiency of a fixed level of this mRNA [34].

If the drug actinomycin D, which is known to inhibit RNA synthesis, is given after tryptophan 2,3-dioxygenase activity has been pre-induced with steroids, the enzyme activity is elevated above the hormonally induced level [6]. This phenomenon, termed superinduction, is paradoxical because actinomycin D, by inhibiting RNA synthesis, in most instances leads to a decrease in the rate of protein synthesis. Numerous other proteins, including glutamine synthetase [13, 21, 25], interferon [42], the secretory proteins of the chick oviduct [23, 24], and tyrosine aminotransferase [15, 20, 28, 37—41], have been found to be superinducible; however, a

*References p. 344*

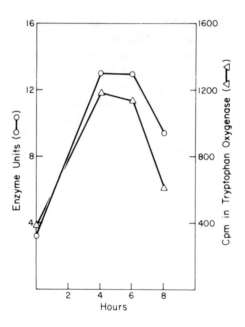

Fig. 1. Comparison of tryptophan 2,3-dioxygenase enzyme activity and mRNA level during induction with hydrocortisone. Rats were injected with 2 mg hydrocortisone acetate/100 g body weight and sacrificed 0, 4, 6, and 8 h later. Enzyme activity was measured as described [34]. Poly(A)-containing mRNA was prepared from each pool of livers by chromatography of total cellular RNA on cellulose [31]. For each time point, 60 µg mRNA was translated in 500 µl of a Krebs II ascites cell-free protein synthesizing system fortified with rabbit reticulocyte initiation factors and rat liver tRNA. [$^3$H]L-Leucine was the amino acid incorporated monitoring protein synthesis. Newly synthesized tryptophan 2,3-dioxygenase was separated from the total released proteins by immunoprecipitation with monospecific anti-tryptophan 2,3-dioxygenase and carrier tryptophan 2,3-dioxygenase, followed by electrophoresis of the dissolved immunoprecipitate on 10% polyacrylamide-0.1% sodium dodecyl sulfate gels. The level of tryptophan 2,3-dioxygenase mRNA was quantitated by the amount of radioactivity incorporated into newly synthesized tryptophan 2,3-dioxygenase [34]. ○——○, Enzyme activity, in µmol kynurenine/h/g liver. △——△, mRNA level, as cpm in tryptophan 2,3-dioxygenase.

clear understanding of the mechanism by which this occurs is lacking. The most extensive studies have been done upon tyrosine aminotransferase, which, like tryptophan 2,3-dioxygenase, is a glucocorticoid-inducible rat liver enzyme. Different laboratories, however, have reported conflicting results on the mechanism of its superinduction. Tomkins' group has suggested that superinduction is the result of an increase in the amount of tyrosine aminotransferase mRNA, based on their finding that actinomycin D increases the rate of tyrosine aminotransferase synthesis [37—41]. Kenney et al. have found that actinomycin D has no effect of the rate of tyrosine aminotransferase synthesis, and have suggested that superinduction of enzyme activity is merely the result of a stabilization of pre-existing enzyme molecules [15, 20, 28]. To elucidate the mechanism of superinduction of enzymes such as tryptophan 2,3-dioxygenase and tyrosine aminotransferase, we measured the level of tryptophan 2,3-dioxygenase mRNA during superinduction of the enzyme with actinomycin D.

Fig. 2 shows an experiment in which three groups of rats were injected with hydrocortisone at 0 h, and a fourth group was sacrificed then for obtaining control values. One of the three groups was sacrificed at 4 h and the second group was injected with 0.4 mg actinomycin D/100 g body weight. The second and third groups were sacrificed at 8 h. Tryptophan 2,3-dioxygenase enzyme activity and mRNA level were measured for each group. The enzyme activity is induced approx. 4-fold by the hormone and is superinduced by actinomycin D. Tryptophan 2,3-dioxygenase mRNA increases and decreases in parallel with the enzyme activity for the hormonally induced animals. However, there is no further increase in the amount of mRNA above the hormonally induced level for the animals which received actinomycin D, although the enzyme activity is almost twice that of the animals which received only hormone for 8 h. We concluded, therefore, that the superinduction of tryptophan 2,3-dioxygenase activity by actinomycin D cannot be explained on the basis of an elevation in the level of its mRNA [17].

Tryptophan 2,3-dioxygenase activity is also induced by its substrate, L-tryptophan [10]. This induction has been attributed to a protection of pre-existing enzyme molecules against degradation, leading to an increase in the amount of enzyme present, without an accompanying increase in its rate of synthesis [4]. To determine the level at which control of this induction occurs, we compared the enzyme activity and mRNA level in control animals and those which received 100 mg L-tryptophan/100 g body weight 4 h before killing. As is shown in Table I, tryptophan induced the enzyme activity 8-fold in 4 h from 1.5—12μmol kynurenine h/g liver. The amount of radioactivity incorporated into total and released proteins by mRNAs extracted from the livers of control and those which received 100 mg L-tryptophan/100 g body weight tophan 2,3-dioxygenase mRNA, measured by the amount of radioactivity incorporated into newly synthesized tryptophan 2,3-dioxygenase, was also the same in the control and tryptophan induced animals. Induction of enzyme activity by tryptophan is not mediated through an increase in the amount of functional tryptophan 2,3-dioxygenase mRNA.

We have also been interested in studying the developmental

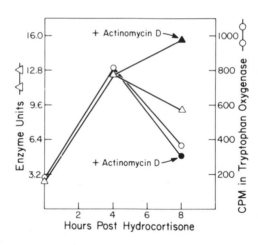

Fig. 2. Comparison of tryptophan 2,3-dioxygenase enzyme activity and mRNA level during superinduction of the enzyme with actinomycin D. 3 groups of rats were injected with 2 mg hydrocortisone acetate/100 g body weight at 0 h and a fourth group was killed for a control. One group was killed at 4 h and the second group was injected with 0.4 mg actinomycin D/100 body weight (solid symbols). The second and third groups were killed at 8 h. Tryptophan 2,3-dioxygenase enzyme activity, expressed as μmol kynurenine/ h/g liver △—△, and mRNA level, expressed as cpm in tryptophan 2,3-dioxygenase ○——○, were determined as in Fig. 1.

TABLE I

Effect of tryptophan on tryptophan 2,3-dioxygenase activity and mRNA

| Animal status | Catalytic activity[a] | | mRNA activity[a] | | |
|---|---|---|---|---|---|
| | Tryptophan 2,3-dioxygenase ($\mu$mol kynurenine/h/g liver) | cpm incorporated into total protein ($\cdot 10^6$) | cpm incorporated into total released proteins ($\cdot 10^6$) | cpm in tryptophan 2,3-dioxygenase | Tryptophan 2,3-dioxygenase as % of total synthesis |
| Control[b] | 1.5 | 6.7 | 2.34 | 606 | 0.026 |
| Tryptophan[b]-induced, 4 h | 12.0 | 5.7 | 1.96 | 572 | 0.029 |

[a] Tryptophan 2,3-dioxygenase enzyme activity and mRNA were determined as described in Fig. 1. Incorporation of [$^3$H]leucine into total released proteins was determined as described [34].
[b] Control rats were killed at 0 h; induced rats were injected with 100 mg L-tryptophan/100 g body weight at 0 h and killed 4 h later.

process in the rat, and the manner in which it is regulated. Many developmental changes are mediated through the programmed appearance and disappearance of specific metabolic enzymes. The appearance of specific enzymes during development probably results from their de novo synthesis; this has been shown for tryptophan 2,3-dioxygenase [29]. It is not known, however, whether the de novo synthesis of specific enzymes is caused by the de novo appearance of their respective mRNAs, or whether the mRNAs are present before enzyme synthesis begins. In at least two systems, the myoblast and the sea urchin embryo, mRNAs have been shown to be present prior to the proteins for which they code. Messenger RNAs coding for histones and microtubule proteins are present in the cytoplasm of the unfertilized sea urchin egg, but are not translated until after fertilization [12, 36]. In the myoblast, myosin is not present although its mRNA is detectable [3].

An additional problem in the study of development is the identification of the 'triggers' which initiate synthesis of specific mRNAs and/or the enzymes for which they code. It has been suggested that these substances may be identical to those which serve as regulators of the adult enzymes, since in certain instances an inducer of a particular enzyme in an adult animal will also induce the enzyme in a developing animal. This is true for glucocorticoid induction of tyrosine aminotransferase [8] and tryptophan 2,3-dioxygenase [5, 9, 29, 44, 45]. In the adult animal, however, changes in enzyme activity are usually reversible, whereas in the developing animal they are most often irreversible.

In the rat, tryptophan 2,3-dioxygenase is one of a cluster of enzymes which appears during the third postnatal week. Its activity is first detectable around the 15th postnatal day, and increases to the adult level by the 22nd postnatal day [5, 7, 9, 11, 22, 29, 44, 45]. The developmental increases in enzyme activity have been attributed to increases in the amount of immunochemically titratable tryptophan 2,3-dioxygenase [9] and in its rate of synthesis [29].

To determine the level at which developmental enzyme synthesis is controlled, we have investigated whether the appearance of tryptophan 2,3-dioxygenase during development is the result of de novo appearance of its mRNA, or whether, as with the sea urchin proteins, the appearance of the mRNA precedes that of the enzyme.

We have quantitated tryptophan 2,3-dioxygenase mRNA using the wheat-germ cell-free protein synthesizing system. Since glucocorticoids regulate tryptophan 2,3-dioxygenase activity in the adult, and since they have been shown to induce its activity precociously in the developing animal, we have also studied their effect on the level of tryptophan 2,3-dioxygenase mRNA.

In Fig. 3, the normal and glucocorticoid-induced development of tryptophan 2,3-dioxygenase enzyme activity is compared with the development of its mRNA. The data in this figure have been expressed as percents of the adult, uninduced levels of enzyme activity and mRNA, which were 3.4 µmol kynurenine/h/g liver

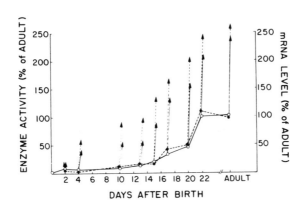

Fig. 3. Comparison of normal and hydrocortisone-induced tryptophan 2,3-dioxygenase activity and mRNA level during postnatal development. Control rats (open circles) were killed on the indicated days after birth; induced rats (solid triangles) received 0.067 mg hydrocortisone acetate/g body weight 9 h prior to killing on the indicated days after birth. Tryptophan 2,3-dioxygenase enzyme activity was measured as described [34]. Poly(A)-containing mRNA was prepared from each pool of livers by chromatography of total cellular RNA on poly(U)-mica [25]. For each time point, 120 µg mRNA was translated in 500 µl of a wheat-germ cell-free protein synthesizing system [14, 16]. [$^3$H]-leucine and [$^3$H]-lysine were used to monitor protein synthesis. Newly synthesized tryptophan 2,3-dioxygenase was isolated, and tryptophan 2,3-dioxygenase mRNA was quantitated as in Fig. 1. In this figure the data are expressed as percents of the adult, uninduced enzyme activity and mRNA level, which were 3.4 µmol kynurenine/h/g liver and 24 000 cpm in tryptophan 2,3-dioxygenase, repectively. ———, tryptophan 2,3-dioxygenase enzyme activity; ------, tryptophan 2,3-dioxygenase mRNA level.

*References p. 344*

and 24 000 counts per minute (cpm) in tryptophan 2,3-dioxygenase, repectively. The enzyme is first detectable around 15 days, increases to one-half the adult level by 17 days, and reaches the adult level by 22 days. Tryptophan 2,3-dioxygenase mRNA is undetectable before 15 days, as indicated by the failure of mRNAs isolated from the livers of rats of this age to stimulate the incorporation of radioactive amino acids into tryptophan 2,3-dioxygenase. The mRNA increases to one-half the adult level by 17 days, and has reached the adult level by 22 days. A single injection of 0.067 mg hydrocortisone/g body weight induces the appearance of tryptophan 2,3-dioxygenase activity at the adult level precociously in 4-, 10-, and 13-day-old rats, but not in 2-day-old rats. It also elevated the enzyme activity above the normal adult level in rats between 17 and 22 days of age, when the enzyme in untreated rat liver is rapidly increasing. Moreover, the precocious induction of enzyme activity at 4, 10 and 13 days as well as the induction during the developmental period are accompanied by parallel inductions of tryptophan 2,3-dioxygenase mRNA. The mRNA is not, however, inducible by hydrocortisone in 2-days-old rats.

From these experiments, we conclude that the appearance and subsequent development of tryptophan 2,3-dioxygenase activity in postnatal rats results from the appearance and development of its mRNA. Tryptophan 2,3-dioxygenase mRNA is not present in rat liver prior to the appearance of the enzyme. Unlike the sea urchin, where control of the synthesis of specific proteins is posttranscriptional, control of tryptophan 2,3-dioxygenase synthesis is pre-translational. In all likelihood, it is the result of de novo synthesis of tryptophan 2,3-dioxygenase mRNA. Furthermore, the premature induction of tryptophan 2,3-dioxygenase activity is the consequence of premature induction of its mRNA [16].

The malignant transformation of normal cells is often accompanied by changes in enzymatic and protein patterns. New proteins, such as the carcinofetal antigens [1], may appear, pre-existing proteins and enzymes may be absent and the relative proportions of isoenzymes may be altered [43]. Moreover, reponsiveness to hormonal regulators may become aberrant. Tryptophan 2,3-dioxygenase activity is absent from several Morris hepatomas, although the livers of the host animals have normal levels. We therefore explored at what level the deletion of enzymatic activity in the hepatoma was controlled. Using the fortified Krebs ascites

TABLE II

Levels of catalytic activity and mRNA for tryptophan 2,3-dioxygenase in host liver and hepatoma

Tryptophan 2,3-dioxygenase activity and mRNA were determined as described in Fig. 1. Incorporation of [$^3$H]leucine into total and released proteins was determined as described [34]. The livers and hepatomas of control rats and those which received 5.0 mg hydrocortisone 100 g body weight 4 h prior to killing were collected and assayed as described.

| Tissue | Treatment | Tryptophan 2,3-dioxygen-catalytic activity ($\mu$mol kynurenine/h/g liver) | Heterogeneous assay of mRNA | | | Tryptophan 2,3-dioxygenase as %of total protein synthesis |
|---|---|---|---|---|---|---|
| | | | Total protein (cpm · $10^{-6}$) | Total released chains (cpm · $10^{-6}$) | cpm in tryptophan 2,3-dioxygenase | |
| Host liver 7793b | None | 3.1 | 5.29 | 1.35 | 337 | 0.025 |
| Host liver 7793 | Hydrocortisone | 10.6 | 5.85 | 1.89 | 931 | 0.069 |
| Hepatoma 7793b | None | Undetectable | 5.42 | 1.90 | Undetectable | |
| Hepatoma 7793 | Hydrocortisone | Undetectable | 5.89 | 1.35 | Undetectable | |

*References p. 344*

protein synthesizing system, we measured the level of tryptophan 2,3-dioxygenase mRNA in the Morris hepatoma 7793 and the liver of its host, to determine whether the absence of tryptophan 2,3-dioxygenase activity in the hepatoma resulted from the concomitant absence of its mRNA. In Table II, tryptophan 2,3-dioxygenase enzymatic activity and mRNA level in hepatoma 7793 and the liver of its host are compared. The liver of the host animal contains a normal level of enzymatic activity which is induced 3—4-fold by hydrocortisone. As was demonstrated previously [34], the hormonal induction of enzymatic activity in the liver is mediated through a parallel increase in the basal level of tryptophan 2,3-dioxygenase mRNA, measured in the mRNA-dependent translational assay as cpm in tryptophan 2,3-dioxygenase. In the hepatoma, tryptophan 2,3-dioxygenase activity is undetectable and, unlike the developing rat, its appearance cannot be induced with hydrocortisone. Although the mRNAs extracted from the hepatoma show the same ability to code for total protein synthesis as those extracted from the host liver, neither the untreated nor hydrocortisone-treated hepatomas contain mRNAs capable of coding for the synthesis of tryptophan 2,3-dioxygenase. Thus, our inability to detect tryptophan 2,3-dioxygenase activity in the untreated and hydrocortisone-treated hepatomas is due to the absence of tryptophan 2,3-dioxygenase mRNA [27].

In the previous studies, we have shown that the induction of tryptophan 2,3-dioxygenase activity by glucocorticoid hormones is accompanied by a parallel increase in the level of tryptophan 2,3-dioxygenase mRNA. These increases in enzyme activity and mRNA are accompanied by increases in the degree of saturation of the nuclear glucocorticoid receptor in vivo [2]. Thus, it is probable that glucocorticoid hormones increase transcription of the genes coding for the $\alpha$ and $\beta$ protomers of tryptophan 2,3-dioxygenase. Neither tryptophan nor actinomycin D have any effect on tryptophan 2,3-dioxygenase mRNA level indicating that these regulators act post-transcriptionally. In the developing rat, the initiation of tryptophan 2,3-dioxygenase synthesis is mediated through the de novo appearance of its mRNA, which in all likelihood, also results from increased transcription of the tryptophan 2,3-dioxygenase genes. In the hepatoma, tryptophan 2,3-dioxygenase activity and mRNA are undetectable, and cannot be induced with hydrocortisone. The level and activity of the glucocorticoid receptor are, however, normal [27], suggesting that the transcription of tryptophan 2,3-dioxygenase genes is suppressed as a consequence of the neoplastic transformation.

*References p. 344*

## REFERENCES

1. G. I. Abelev, Cancer Res., 14 (1971) 295.
2. M. Beato, M. Kalimi, and P. Feigelson, Biochem. Biophys. Res. Commun., 47 (1972) 1464.
3. M. E. Buckingham, D. Caput, A. Cohen, R. G. Whalen and F. Gros, Proc. Natl. Acad. Sci. USA, 71 (1974) 1466.
4. P. Feigelson and O. Greengard, J. Biol. Chem., 237 (1962) 3714.
5. J. M. Franz and W. E. Knox, Biochemistry, 6 (1967) 3464.
6. L. D. Garren, R. R. Howell, G. M. Tomkins and R. M. Crocco, Proc. Natl. Acad. Sci. USA, 52 (1964) 1121.
7. L. Goldstein and W. E. Knox, Ann. N.Y. Acad. Sci., 11 (1963) 227.
8. O. Greengard, Science, 167 (1969) 891.
9. O. Greengard and H. K. Dewey, Proc. Natl. Acad. Sci. USA, 68 (1971) 1698.
10. O. Greengard and P. Feigelson, J. Biol. Chem., 236 (1961) 158.
11. O. Greengard and P. Feigelson, Ann. N.Y. Acad. Sci., 111 (1963) 233.
12. K. Gross, J. Ruderman, M. Jacobs-Lorena, C. Baglioni and P. R. Gross, Nature New Biol., 241 (1972) 272.
13. R. E. Jones, J. H. Moscona and A. A. Moscona, J. Biol. Chem., 249 (1974) 6021.
14. R. E. Jones, P. Pulkrabek and D. Grunberger, Biochem. Biophys. Res. Commun., 74 (1977) 1490.
15. F. T. Kenney, K. L. Lee, C. D. Stiles and J. E. Fritz, Nature New Biol., 246 (1973) 208.
16. L. A. Killewich and P. Feigelson, Proc. Natl. Acad. Sci. USA, (1981) in press.
17. L. Killewich, G. Schutz and P. Feigelson, Proc. Natl. Acad. Sci. USA, 72 (1975) 4285.
18. W. E. Knox and A. H. Mehler, Science, 113 (1951) 237.
19. K. Koike, W. N. Poillon and P. Feigelson, J. Biol. Chem., 244 (1969) 3457.
20. K. L. Lee and F. T. Kenney, Acta Endocrinol. Suppl., 153 (1971) 109.
21. A. A. Moscona, M. H. Moscona and N. Saenz, Proc. Natl. Acad. Sci. USA, 61 (1968) 160.
22. A. M. Nemeth, J. Biol. Chem., 234 (1959) 2921.
23. R. D. Palmiter, T. Oka and R. T. Schimke, J. Biol. Chem., 246 (1971) 724.
24. R. D. Palmiter and R. T. Schimke, J. Biol. Chem., 248 (1973) 1502.
25. P. Pulkrabek, K. Klier and D. Grunberger, Anal. Biochem., 68 (1975) 26.
26. R. A. Raff, H. V. Colot, S. E. Selvig and P. R. Gross, Nature, 235 (1972) 211.
27. L. Ramanarayanan-Murthy, P. D. Colman, H. P. Morris and P. Feigelson, Cancer Res., 36 (1976) 3594.
28. J. R. Reel and F. T. Kenney, Proc. Natl. Acad. Sci. USA, 61 (1968) 200.
29. M. D. Roper and J. M. Franz, J. Biol. Chem., (1977) in press.
30. R. T. Schimke, E. W. Sweeney and C. M. Berlin, J. Biol. Chem., 240 (1965) 322.

31  G. Schutz, M. Beato and P. Feigelson, Biochem. Biophys. Res. Commun., 49 (1972) 680.
32  G. Schutz, M. Beato and P. Feigelson, Proc. Natl. Acad. Sci. USA, 70 (1973) 1218.
33  G. Schutz, E. Chow and P. Feigelson, J. Biol. Chem., 247 (1972) 5333.
34  G. Schutz, L. Killewich, G. Chen and P. Feigelson, Proc. Natl. Acad. Sci. USA, 72 (1975) 1017.
35  R. J. Schwartz, J. Biol. Chem., 248 (1973) 6426.
36  A. Skoultchi and P. R. Gross, Proc. Natl. Acad. Sci. USA, 70 (1973) 2840.
37  R. A. Steinberg, B. B. Levinson and G. M. Tomkins, Cell, 5 (1975) 29.
38  R. A. Steinberg, B. B. Levinson and G. M. Tomkins, Proc. Natl. Acad. Sci. USA, 72 (1975) 2007.
39  E. B. Thompson, D. K. Granner and G. M. Tomkins, J. Mol. Biol., 54 (1970) 159.
40  G. M. Tomkins, T. D. Gelehrter, D. Granner, D. Martin Jr., H. H. Samuels and E. B. Thompson, Science, 166 (1969) 1474.
41  G. M. Tomkins, B. B. Levinson, J. D. Baxter and L. Dethlefsen, Nature New Biol., 239 (1972) 9.
42  J. Vilcek and E. A. Havell, Proc. Natl. Acad. Sci USA 70 (1973) 3909.
43  G. Weinhouse, Cancer Res., 32 (1972) 2007.
44  A. Yuwiler, B. L. Bennett and E. Geller, Neurochem. Res., 1 (1976) 591.
45  A. Yuwiler and E. Geller, Enzyme, 15 (1973) 161.

*References p. 344*

*Chapter 4*

# Human Haemoglobin

H. LEHMANN and R. CASEY*

*University of Cambridge, Department of Biochemistry, Tennis Court Road, Cambridge CB2 1QW, and *John Innes Institute, Colney Lane, Norwich NR4 7UH (U.K.)*

## Haemoglobin

Haemoglobin was described some 50 years ago as the second most interesting substance in the world [1] — the first, presumably, being chlorophyll. The replacement of Mg in chlorophyll by Fe in haem was the key that opened the door to the whole of aerobic life [2]. It is fruitless to argue about the greater or lesser interest of a biological compound, but it might be claimed that from the point of view of basic information provided by a single molecule, haemoglobin now has no rival. There is no other molecule for which it has been possible to so closely relate structure and function, to study in man alone more than 300 variant types and to compare, for some of these latter, their anthropological role and for others, their pathological significance. Globin has also been in the foreground in the study of protein synthesis.

## 1. Distribution of haemoglobin

Erythrocytes can raise the oxygen-combining capacity of plasma from 0.5 ml $O_2$/100 ml to about 25 ml $O_2$/100 ml of mammalian blood. This enables an animal to transfer oxygen rapidly and efficiently to an active tissue far removed from the respiratory surface. For example the red cell supplying the tail of the blue whale completes a round trip of approx. 120 feet in the process.

However, in contrast, if the organism is less than 1 mm in diameter or flat like a tapeworm or eel larva, then simple diffusion is adequate and a respiratory pigment would be valueless. Alternatively, if the basal metabolism is at a low level, which is possible in cold-blooded animals, then again survival without a respiratory pigment would be possible.

Haemoglobin appears in evolution not as a sudden response to increasing size or complexity of the animal, but in a peculiarly haphazard manner.

The occurrence of haemoglobin in such unicellular organisms as yeast and paramecium [3], where diffusion of oxygen is adequate, is at first strange. However, Barcroft [4] realised that although haemoglobin is widespread in nature it occupies only a corner of the field which cytochrome covers. Cytochrome occurs in both plant and animal kingdoms and like haemoglobin it is a protein molecule with haem prosthetic groups. It is absent in absolutely anaerobic organisms such as the anaerobic bacteria. The function of cytochrome is to link cellular metabolites donating hydrogen with molecular oxygen. It appears then possible that haemoglobin might be produced as a by-product of cytochrome formation and requires no new enzyme system in the cell for its manufacture.

In the higher invertebrates the function of a respiratory pigment is not always clear. There is a tendency for the pigment to occur in mud-dwelling animals where there is a potentially anaerobic environment and also in some species in response to comparative oxygen lack. The position of a respiratory pigment has a survival value only in times of stress due to periods of relative or absolute lack of oxygen.

Whether haemoglobin has been arrived at more than once by convergent evolution, or has been transmitted from one species to another by transduction, is a matter for speculation. In the vertebrates, Redfeld [5] considers that haemoglobin appeared de novo and the absence of haemoglobin in present-day primitive chordates suggests that this may be so. Even today some vertebrates are able to survive without haemoglobin. Ice fish living in the Antarctic lack haemoglobin and yet obtain a weight [6] of over 1 kg. With warm-blooded animals, however, the presence of respiratory pigment is obligatory.

# INTRODUCTION

In plants, leghaemoglobin is found in the root nodules of legumes [7]. These are the sites of symbiosis with the soil bacteria of the genus *Rhizobium*. Neither the host legume nor the free-living bacteroid can produce leghaemoglobin by themselves. The globin genes are present in the plant [8] whilst the haem is synthesised within the nodule by the *Rhizobium* [9]. The function of leghaemoglobin seems to be the maintenance of an optimal oxygen tension which permits the simultaneous synthesis of ATP and fixation of nitrogen, the former by oxidative phosphorylation and the latter by an oxygen-sensitive nitrogenase.

## 2. Haemoglobin structure and function

### 2.1. Introduction

Barcroft [10] assessed the physiological importance of haemoglobin by pointing out that it should be
  (a) capable of transporting large quantities of oxygen;
  (b) very soluble;
  (c) capable of taking up oxygen at suitable velocity and in sufficient amounts in the blood and of releasing it to the tissues;
  (d) able to buffer a bicarbonate solution.

This list is still relevant in the face of present knowledge, the requirements being now referred to as *oxygen affinity*, solubility, *co-operativity* and the *Bohr effect*.

The haemoglobins of vertebrates are tetrameric and are composed of two $\alpha$-like and two $\beta$-like chains. At any one time the total circulating haemoglobin may comprise a number of species all based on this $\alpha_2\beta_2$ pattern. The human $\alpha$- and non $\alpha$-chains are different in length and their genes are located on different chromosomes (see below). Such a structure enables the haemoglobin to behave in a co-operative manner; that is, oxygen affinity rises with increasing, and falls with decreasing oxygen saturation so that the oxygen equilibrium curve is sigmoid. Furthermore, the magnitude of the difference between zero and full oxygen saturation, and the oxygen tension at which such a tetrameric haemoglobin is half saturated, will vary with pH, $[CO_2]$, $[Cl^-]$, $[HPO_4^{2-}]$, [2,3-diphosphoglycerate], a change in the activity of these gases or electrolytes

*References p. 411*

affecting the affinity of the haemoglobin for all the others. These are a property of the [$\alpha_2$/non-$\alpha_2$] tetramer and vanish upon the dissociation of the molecule into dimers or free subunits; similarly, the oxygen affinities of single-chain haemoglobins, such as those of the lamprey or leghaemoglobin, or of myoglobin, are generally independent of the ionic composition and exhibit oxygen equilibrium curves with a rectangular hyperbolic rather than sigmoid shape. Thus, haemoglobin can be considered as a reversible oxygen-binding transport protein which will bind or release $O_2$, depending on its physiological environment.

Haemoglobin is unrivalled as a model of protein structure in relation to function and most of our understanding of the structure and function of haemoglobin comes from the X-ray crystallographic studies [11] of Perutz and his colleagues at Cambridge, England, which have defined the atomic structure of the molecule. Although it is highly probable that all tetrameric haemoglobins have similar quaternary structures, it should be borne in mind that the only haemoglobins whose atomic structures have been analysed and refined at resolution better than 0.28 nm are horse met- and human deoxyhaemoglobin and all subsequent discussion is based on these structures.

The arrangement of the $\alpha$ and $\beta$ subunits (the quaternary structure) differs in oxy- and deoxyhaemoglobin; the relative positions of the subunits within the dimers $\alpha 1\beta 1$ and $\alpha 2\beta 2$ (which would be obtained by dividing the tetramer along its true axis of symmetry) are similar in the two quaternary structures, but there is a shift of several nanometres at the interfaces between $\alpha 1$ and $\beta 2$ and $\alpha 2$ and $\beta 1$ on the transition from oxy- to deoxyhaemoglobin. Our present knowledge of haemoglobin atomic structures explains to some extent oxygen affinity and the Bohr effect, while the change in tertiary and quaternary conformation during oxygen binding is the basis for haem—haem interaction, or co-operativity [12]. Before these, or the binding of $CO_2$ the effect of 2,3-diphosphoglycerate (2,3-DPG) on oxygen affinity and the question of allosteric behaviour can be discussed in more detail, it is necessary to describe the structure of haemoglobin.

### 2.2. The structure of haemoglobin

*2.2.1. Haem structure.* As shown in Fig. 1, haem is composed of an iron atom coordinated to four pyrrole rings through their substituent nitrogen atoms to form a so-called metalloporphyrin.

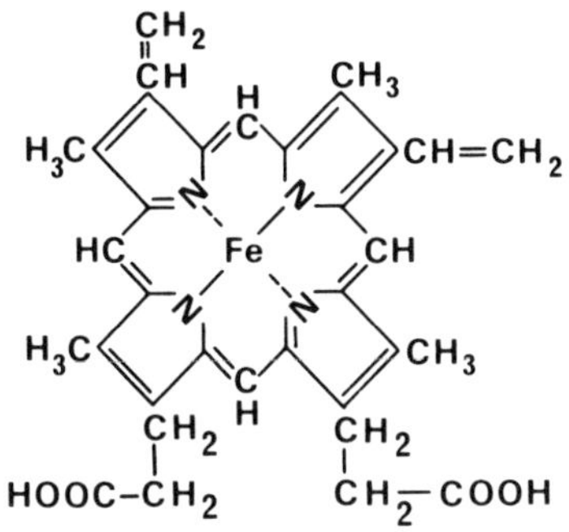

Fig. 1. The structure of haem.

The β positions of the pyrrole rings are fully substituted with four methyl, two propionate and two vinyl groups and the pyrrole rings are joined by methene bridges. These bridges and rings form an extended conjugated ring system which results in a high degree

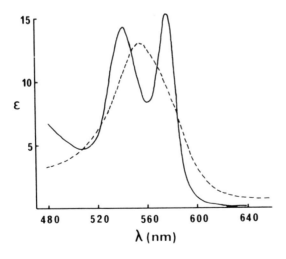

Fig. 2. The visible spectrum of oxy- (-) and deoxy- (----) haemoglobin.

of resonance stabilisation of the molecule. Absorption of visible light by this ring system gives haemoglobin its characteristic red colour and absorption spectrum (see Fig. 2).

In addition to its coordination with the pyrrole rings, the iron atom in haem may bind two other ligands, one above and one below the plane of the porphyrin ring. One of these ligands is a histidine residue of the globin chain to which the haem is bound and which is the only covalent link between haem and globin; the other coordination position is unoccupied in deoxyhaemoglobin and the iron is protected from oxidation by the non-polar nature of the "haem pocket" formed by the globin chain. Oxygen can readily bind to the sixth coordination position, provided the iron is in the ferrous state, to form oxyhaemoglobin.

### 2.2.2. Protein structure

*2.2.2.1. Primary structure.* The primary structures (amino acid sequences) of the $\alpha$-, and non -$\alpha$ chains of human haemoglobin are known as are those of some 30 or more other animal haemoglobins. Fig. 3 shows that all four polypeptide chains have basically similar structures, especially if one introduces hypothetical "gaps" at appropriate places. For instance, if one considers the first seven or eight residues of the human $\alpha$ or $\beta$ sequences

| 1 | 2 | 3 | 4 | 5 | 6 | 7 | 8 |
|---|---|---|---|---|---|---|---|
| *Valyl* | Leucyl | Seryl | Prolyl | Alanyl | Aspartyl | Lysyl | |
| *Valyl* | Histidyl | Leucyl | Threonyl | Prolyl | Glutamyl | Glutamyl | Lysyl |

only one identity is observed. If however, a so-called "Braunitzer gap" is introduced between the first and second residues of the $\alpha$-chain

| 1 | 2 | 3 | 4 | 5 | 6 | 7 | 8 |
|---|---|---|---|---|---|---|---|
| *Valyl* | — | *Leucyl* | Seryl | *Prolyl* | Alanyl | Aspartyl | *Lysyl* |
| *Valyl* | Histidyl | *Leucyl* | Threonyl | *Prolyl* | Glutamyl | Glutamyl | *Lysyl* |

four identities appear, while the residues at positions 4 (serine-threonine) and 7 (aspartic acid-glutamic acid) are extremely similar in both $\alpha$ and $\beta$. Although of course these gaps do not exist in reality, such methods show that more than 42% of the $\alpha$- and $\beta$-chains are identical, while the $\gamma$- and $\beta$-chains differ by only 38

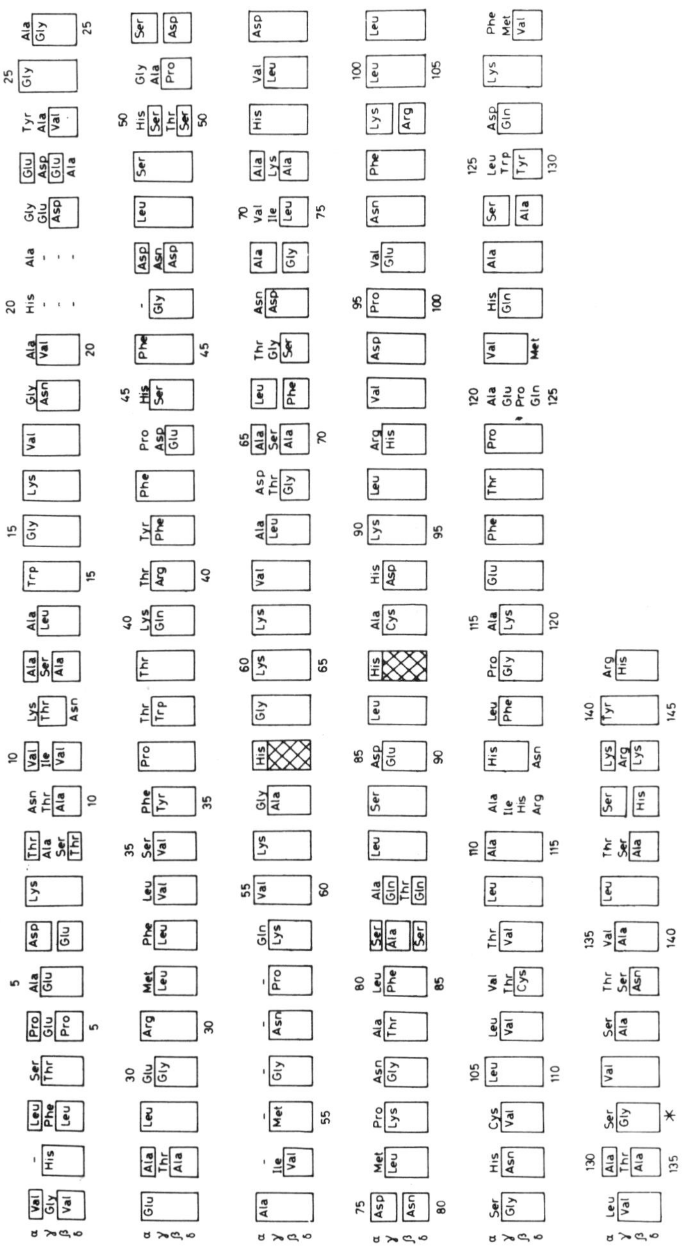

# STRUCTURE OF HAEMOGLOBIN

Fig. 3. Comparison of the amino acid sequences of the α-, β-, γ- and δ-chains of human haemoglobin.

*References p. 411*

354  STRUCTURE AND FUNCTION

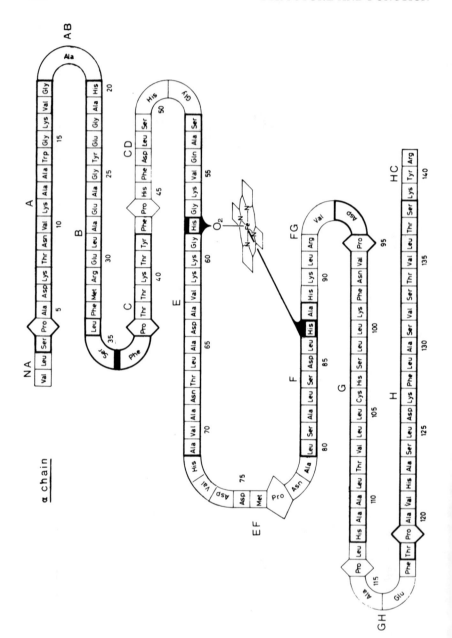

# STRUCTURE OF HAEMOGLOBIN

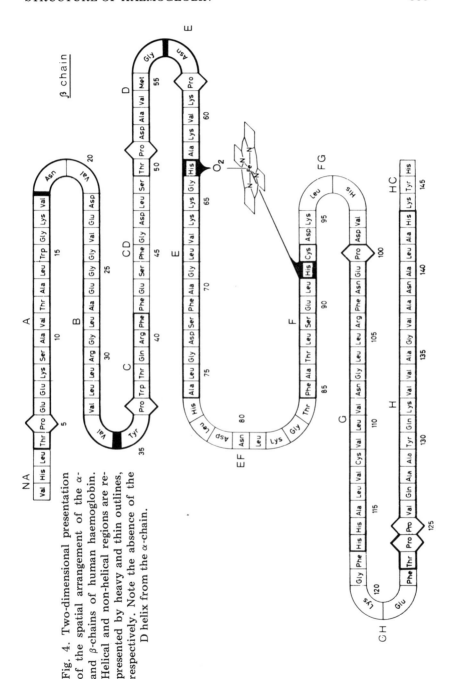

Fig. 4. Two-dimensional presentation of the spatial arrangement of the α- and β-chains of human haemoglobin. Helical and non-helical regions are represented by heavy and thin outlines, respectively. Note the absence of the D helix from the α-chain.

residues out of 146 and the δ- and β-chains by a mere 10 residues. This implies that the four protein chains may well have arisen from a single common ancestor.

There is a polymorphism at position γ136, which can be Gly ($^G\gamma$) or Ala ($^A\gamma$); this a result of gene duplication and mutation.

*2.2.2.2. Secondary structure.* The component chains of haemoglobin can fold into helices which are stabilised by hydrogen bonding between peptide-linked —CO— and —NH— groups of amino acids lying above or below each other in adjacent turns of the helix [13]; between 70 and 80% of the globin chain in fact attain a helical structure, made up of 7 or 8 segments interspersed with non-helical regions (which account for the remaining 20—30% of the amino acids). In considering three-dimensional models of globin chains or haemoglobin it is more convenient to use a numbering system based on these helical units, rather than a system related to the linear sequence of amino acids. Thus the first helical region of the β-chain is designated "A"; it begins at position β3 (A1) and ends at position β18 (A16). Then comes the inter-helical region AB (which in the case of the β-chain is just one residue, β19) followed by helix B, and so on, (see Fig. 4). The α-chain differs from the β-chain in that it has no D-helix. When globin chains are considered from their helical notation, homologies become apparent which were not obvious from the amino acid sequence alone; most strikingly, the haem-linked histidine residue (see below) is always at position F8.

*2.2.2.3. Tertiary structure.* To arrive at the globular shape of the globin subunit (Fig. 5) it is necessary to bend the polypeptide chain wherever the helical portions are connected by non-helical sections. This typical "globin" structure can be observed, by X-ray crystallography, to be present in, for example, myoglobin, the chains of haemoglobin and the monomeric haemoglobin of the marine worm *Glycera dibranchiata* [14] and even leghaemoglobin [7] and thus appears to be a universal model for the haem/globin unit. Since all these molecules have to accommodate a similar insoluble haem group within a globin cleft, it is perhaps not surprising that they have many features in common. The structure is maintained by Van der Waals forces, hydrogen bonds and some salt bridges and is designed to maintain the haem group [Fe(II) protoprophyrin IX] in a non-polar pocket. There are about 60 interactions between the atoms of the globin chains coming to

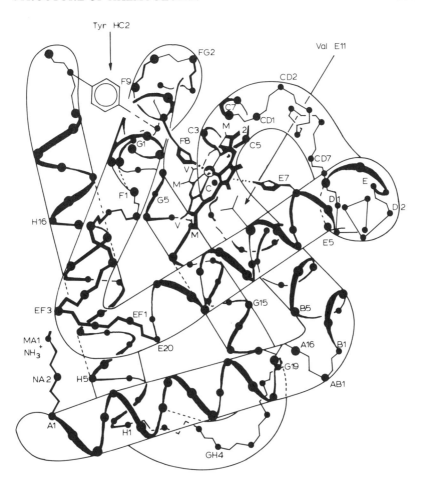

Fig. 5. Diagrammatic sketch of the tertiary structure of the polypeptide chain of the β-subunit of haemoglobin.

within about 0.4 nm of haem atoms (Fig. 6) and virtually all of these interactions are non-polar. Comparison of the amino acid sequences of different mammalian species shows that there is very little variation in the nature of these residues surrounding the haem group. The iron of the haem is linked to the $N_\epsilon$ of the so-called proximal histidine (F8) and the porphyrin is wedged into its pocket by a phenylalanine; apart from these two amino acid residues, it seems that about 35 other specific sites along the poly-

*References p. 411*

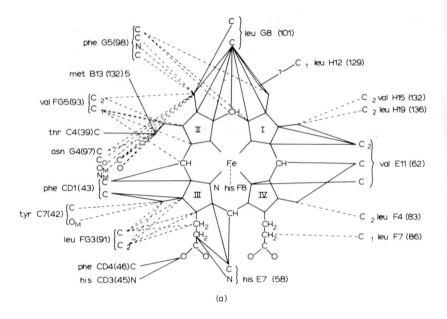
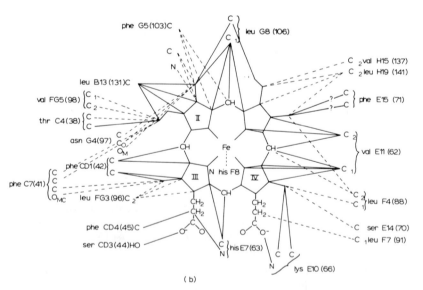

Fig. 6. Contacts between the haem group and residues of the globin in (a) the α-chain and (b) the β-chain of haemoglobin.

peptide chain must be occupied by non-polar residues, but otherwise the rest of the sequence can be variable, the number of differences between any two species being roughly proportional to their phylogenetic separation [15] (see below). Thus it seems that only a small proportion of a unique globin sequence is essential to determine the characteristic fold of the haemoglobin chain. It would, however, be wrong to assume that any substitution of one amino acid for another outside this framework of about 37 is permissible; any changes must be conservative in terms of tertiary structure (a Glu for an Asp, or Ser for Thr, for instance). To introduce a polar residue into a non-polar region of the molecule, for example, would result in an unstable globin. In fact, this does sometimes happen in nature, as the cause of an anaemia due to an abnormal haemoglobin (see below).

*2.2.2.4. Quaternary structure.* The four haem-globin subunits of mammalian haemoglobin interlock to form an ellipsoid with an $M_r$ of about 67 000 and dimensions of about 6.4 × 5.5 × 5.0 nm. It should be noted that the exact dimensions depend on the state of oxygenation of the molecule; the alteration in the size of haemoglobin with oxygenation/deoxygenation has been referred to by Perutz as "paradoxical breathing", since the molecule contracts on oxygenation and expands on deoxygenation. It is not sufficient merely to have a tetramer in order to have a functional haemoglobin. Haemoglobin H, for example, is tetrameric ($\beta_4$) and so is Hb Bart's ($\gamma_4$), but they have the same conformation in the oxygenated and deoxygenated state, suggesting that there is no such rearrangement of subunits such as occurs in normal adult haemoglobin ($\alpha_2\beta_2$) or foetal haemoglobin ($\alpha_2\gamma_2$) (see Fig. 3). Thus it appears that it is the combination of $\alpha$ and non-$\alpha$ chains to form the $\alpha_2\beta_2$ (or $\alpha_2\gamma_2$) tetramer which results in a molecule which exhibits a sigmoid oxygen dissociation curve, haem—haem interaction (co-operativity) and Bohr effect. To understand why this should be so, we must examine the quarternary structure in more detail.

From inspection of Fig. 7 it is clear that each $\alpha$-chain contacts the two $\beta$-chains along two different surfaces. If the subunits are designated $\alpha_1$, $\alpha_2$, $\beta_1$ and $\beta_2$, two interfaces between unlike subunits can be defined as $\alpha_1\beta_1$ and $\alpha_1\beta_2$ for reasons of symmetry, $\alpha_1\beta_1$ and $\alpha_2\beta_2$ contacts are identical, as are $\alpha_1\beta_2$ and $\alpha_2\beta_1$. During oxygenation the $\alpha_1\beta_1$ interface remains relatively fixed, whilst

Fig. 7. Model of the quaternary structure of haemoglobin showing (a) the $\alpha_1\beta_1$ contact and (b) the $\alpha_1\beta_2$ contact.

# STRUCTURE OF HAEMOGLOBIN

## TABLE I

(a) The α1β1 contact; the α contacts with the β-chain

| α1 Residue No. | Helical No. | Residue | β1 Residue No. | Helical No. | Residue |
|---|---|---|---|---|---|
| 30 | B11 | Glu | 124 | H2 | Pro |
| 31 | B12 | Arg | 127 | H5 | Gln |
|    |     |     | 124 | H2 | Pro |
|    |     |     | 123 | H1 | Thr |
|    |     |     | 122 | GH5 | Phe |
| 34 | B15 | Leu | 128 | H6 | Ala |
|    |     |     | 125 | H3 | Pro |
|    |     |     | 124 | H2 | Pro |
| 35 | B16 | Ser | 131 | H9 | Gln |
|    |     |     | 128 | H6 | Ala |
| 36 | C4 | Phe | 131 | H9 | Gln |
| 103 | G10 | His | 131 | H9 | Gln |
|    |     |     | 112 | G14 | Cys |
|    |     |     | 108 | G10 | Asn |
| 104 | G11 | Cys | 127 | H5 | Gln |
| 106 | G13 | Leu | 112 | G14 | Cys |
| 107 | G14 | Val | 127 | H5 | Gln |
|    |     |     | 115 | G17 | Ala |
|    |     |     | 112 | G14 | Cys |
| 111 | G18 | Ala | 119 | GH2 | Gly |
|    |     |     | 116 | G18 | His |
|    |     |     | 115 | G17 | Ala |
| 114 | GH2 | Pro | 116 | G18 | His |
| 117 | GH5 | Phe | 116 | G18 | His |
|    |     |     | 112 | G14 | Cys |
|    |     |     | 30 | B12 | Arg |
| 119 | H2 | Pro | 55 | D6 | Met |
|    |     |     | 51 | D2 | Pro |
|    |     |     | 33 | B15 | Val |
| 122 | H5 | His | 112 | G14 | Cys |
|    |     |     | 30 | B12 | Arg |
| 123 | H6 | Ala | 34 | B16 | Val |
| 126 | H9 | Asp | 35 | C1 | Tyr |
|    |     |     | 34 | B16 | Val |

*References p. 411*

## TABLE I

(b) The α1β1 contact; the β contacts with the α-chain

| β1 Residue No. | Helical No. | Residue | α1 Residue No. | Helical No. | Residue |
|---|---|---|---|---|---|
| 30 | B12 | Arg | 117 | GH5 | Phe |
|    |     |     | 122 | H5  | His |
| 33 | B15 | Val | 119 | H2  | Pro |
| 34 | B16 | Val | 123 | H6  | Ala |
|    |     |     | 126 | H9  | Asp |
| 35 | C1  | Tyr | 126 | H9  | Asp |
| 51 | D2  | Pro | 119 | H2  | Pro |
| 55 | D6  | Met | 119 | H2  | Pro |
| 108 | G10 | Asn | 103 | G10 | His |
| 112 | H14 | Cys | 103 | G10 | His |
|    |     |     | 106 | G13 | Leu |
|    |     |     | 107 | G14 | Val |
|    |     |     | 117 | GH5 | Phe |
|    |     |     | 122 | H5  | His |
| 115 | G17 | Ala | 107 | G14 | Val |
|     |     |     | 111 | G18 | Ala |
| 116 | G18 | His | 111 | G18 | Ala |
|     |     |     | 114 | GH2 | Pro |
|     |     |     | 117 | GH5 | Phe |
| 119 | GH2 | Gly | 111 | G18 | Ala |
| 122 | GH5 | Phe | 31  | B12 | Arg |
| 123 | H1  | Thr | 31  | B12 | Arg |
| 124 | H2  | Pro | 30  | B11 | Glu |
|     |     |     | 31  | B12 | Arg |
|     |     |     | 34  | B15 | Leu |
| 125 | H3  | Pro | 34  | B15 | Leu |
| 127 | H5  | Gln | 31  | B12 | Arg |
|     |     |     | 104 | G11 | Cys |
|     |     |     | 107 | G14 | Val |
| 128 | H6  | Ala | 34  | B15 | Leu |
|     |     |     | 35  | B16 | Ser |
| 131 | H9  | Gln | 35  | B16 | Ser |
|     |     |     | 36  | C1  | Phe |
|     |     |     | 103 | G10 | His |

TABLE I

(c) The α1β2 contact; the α contacts with the β-chain in the oxy conformation

| α1 | | | β2 | | |
|---|---|---|---|---|---|
| Residue No. | Helical No. | Residue | Residue No. | Helical No. | Residue |
| 38 | C3 | Thr | 97 | FG4 | His |
|    |    |     | 98 | FG5 | Val |
| 41 | C6 | Thr | 40 | C6  | Arg |
|    |    |     | 97 | FG4 | His |
| 42 | C7 | Tyr | 40 | C6  | Arg |
| 91 | FG3 | Leu | 40 | C6 | Arg |
| 92 | FG4 | Arg | 37 | C3 | Trp |
|    |    |     | 39 | C5 | Gln |
|    |    |     | 40 | C6 | Arg |
| 93 | FG5 | Val | 37 | C3 | Trp |
| 94 | G1 | Asp | 37 | C3 | Trp |
|    |    |     | 102 | G4 | Asn |
| 95 | G2 | Pro | 37 | C3 | Trp |
| 96 | G3 | Val | 99 | G1 | Asp |
|    |    |     | 101 | G3 | Glu |
| 140 | HC2 | Tyr | 36 | C2 | Pro |
|    |    |     | 37 | C3 | Trp |

there is considerable movement at the $\alpha_1\beta_2$ interface; thus the interatomic contacts made at this interface differ in the oxy- and deoxy-states (Table I).

The subunits interact with each other (see Table I) through weak, non-covalent bonds (mostly Van der Waals forces and some hydrogen bonds), there being no inter- or intra-chain disulphide bridges. Deoxyhaemoglobin is stabilised, in addition, by intra- and inter-subunit salt bridges (Fig. 8). Because of the importance of the $\alpha_1\beta_2$ contact in allowing movement on oxygenation (which is a vital feature of haem—haem interaction), many of the amino acids involved will be found to be relatively invariant when haemoglobins from numerous species are compared.

In contrast to the numerous contacts between unlike subunits,

*References p. 411*

## TABLE I

(d) The α1β2 contact; the α contacts with the β-chain in the deoxy conformation

| α1 | | | β2 | | |
|---|---|---|---|---|---|
| Residue No. | Helical No. | Residue | Residue No. | Helical No. | Residue |
| 37 | C2 | Pro | 146 | HC3 | His |
| 38 | C3 | Thr | 145 | HC2 | Tyr |
|    |    |     | 100 | G2  | Pro |
|    |    |     | 99  | G1  | Asp |
| 40 | C5 | Lys | 146 | HC3 | His |
| 41 | C6 | Thr | 98  | FG5 | Val |
|    |    |     | 97  | FG4 | His |
| 42 | C7 | Tyr | 99  | G1  | Asp |
|    |    |     | 40  | C6  | Arg |
| 44 | CD2 | Pro | 97 | FG4 | His |
| 92 | FG4 | Arg | 40 | C6  | Arg |
|    |     |     | 37 | C3  | Trp |
| 94 | G1 | Asp | 101 | G3 | Glu |
|    |    |     | 37  | C3 | Trp |
| 95 | G2 | Pro | 37  | C3 | Trp |
| 96 | G3 | Val | 101 | G3 | Glu |
| 140 | HC2 | Tyr | 37 | C3 | Trp |

contacts between similar chains ($\alpha_1\alpha_2$ and $\beta_1\beta_2$) are minimal; like-chains are separated one from the other by an internal cavity lined by polar amino acids and filled with water. The contacts, however, although not numerous, are nonetheless important. The two β-chains are widely separated in the deoxy-conformation, sufficiently so as to accommodate a molecule of the allosteric effector 2,3-DPG whilst in the oxy-conformation the gap is closed by the β-chains moving together and 2,3-DPG cannot get in. The $\alpha_1\alpha_2$ contact occurs only in deoxyhaemoglobin and plays an important part in haemoglobin function.

TABLE I

(e) The α1β2 contact; the β contacts with the α-chain in the oxy conformation

| β2 | | | α1 | | |
|---|---|---|---|---|---|
| Residue No. | Helical No. | Residue | Residue No. | Helical No. | Residue |
| 36 | C2 | Pro | 140 | HC2 | Tyr |
| 37 | C3 | Trp | 92 | FG4 | Arg |
|   |   |   | 93 | FG5 | Val |
|   |   |   | 94 | G1 | Asp |
|   |   |   | 95 | G2 | Pro |
|   |   |   | 140 | HC2 | Tyr |
| 39 | C5 | Gln | 92 | FG4 | Arg |
| 40 | C6 | Arg | 41 | C6 | Thr |
|   |   |   | 42 | C7 | Tyr |
|   |   |   | 91 | FG3 | Leu |
|   |   |   | 92 | FG4 | Arg |
| 97 | FG4 | His | 38 | C3 | Thr |
|   |   |   | 41 | C6 | Thr |
| 98 | FG5 | Val | 38 | C3 | Thr |
| 99 | G1 | Asp | 96 | G3 | Val |
| 101 | G3 | Glu | 96 | G3 | Val |
| 102 | G4 | Asn | 94 | G1 | Asp |

## 2.3. Haemoglobin function in relation to structure

2.3.1. $O_2$ affinity. The deoxyhaemoglobin tetramer is maintained in its particular conformation (the energy-rich "tense" (T) state) by constraining salt bridges which are successively broken during the molecule's reaction with oxygen (or other ligands such as carbon monoxide). In particular, the α- and β-chain C-terminal residues, α141 Arg and β146 His, are involved in intersubunit salt bridges as shown in Fig. 8; oxygenation-linked changes in ionisation of β82 Lys probably also take place. As each salt-bridge ruptures, a constraint holding the molecule in the deoxy form is removed, thereby tipping the equilibrium between the two alternative quaternary structures some way in favour of the oxy-conforma-

## TABLE I

(f) The α1β2 contact; the β contacts with the α-chain in the deoxy conformation

| β2 | | | α1 | | |
|---|---|---|---|---|---|
| Residue No. | Helical No. | Residue | Residue No. | Helical No. | Residue |
| 37 | C3 | Trp | 92 | FG4 | Arg |
|  |  |  | 94 | G1 | Asp |
|  |  |  | 95 | G2 | Pro |
|  |  |  | 140 | HC2 | Tyr |
| 40 | C6 | Arg | 92 | FG4 | Arg |
|  |  |  | 42 | C7 | Tyr |
| 97 | FG4 | His | 41 | C6 | Thr |
|  |  |  | 44 | CD2 | Pro |
| 98 | FG5 | Val | 41 | C6 | Thr |
| 99 | G1 | Asp | 42 | C7 | Tyr |
|  |  |  | 38 | C3 | Thr |
| 100 | G2 | Pro | 38 | C3 | Thr |
| 101 | G3 | Glu | 94 | G1 | Asp |
|  |  |  | 96 | G3 | Val |
| 145 | HC2 | Tyr | 38 | C3 | Thr |
| 146 | HC3 | His | 37 | C2 | Pro |
|  |  |  | 40 | C5 | Lys |

tion (and thereby raising the oxygen affinity because the subunits are no longer constrained by the salt bridges to maintain the deoxy-structure).

The oxygen tension at which the haemoglobin is half-saturated, is known as the $P_{50}$ value, this being a measure of oxygen affinity; the higher the $P_{50}$ value, the lower is the oxygen affinity of the haemoglobin. As Perutz has pointed out, there exists in nature a 250 000-fold variation in oxygen affinity, from *Ascaris* haemoglobin ($P_{50} = 4 \cdot 10^{-3}$ mmHg) to that of certain fish haemoglobins at acid pH ($P_{50} = 10^3$ mmHg), clearly showing the strong influence of globin on the chemical affinity of haem iron for $O_2$.

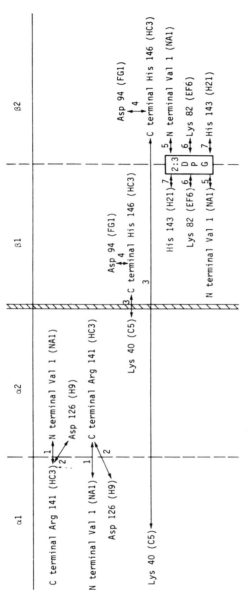

Fig. 8. The salt-bridges which maintain the haemoglobin molecule in the deoxy conformation.

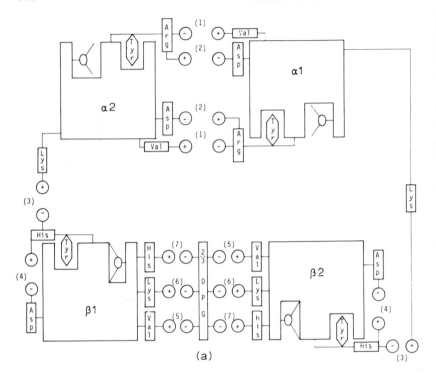

Fig. 9. Diagrammatic sketch showing a possible sequence of steps in the reaction of haemoglobin with oxygen. (a) The deoxygenated tetramer with all salt-bridges intact and one molecule of 2,3-DPG in place. (b) Oxygenation of the α haems. The penultimate tyrosines are expelled and the salt-bridges with the partner α-chains are broken. (c) Movement into the oxy-form. 2,3-DPG is expelled and the bridges between $\alpha_1$-$\beta_2$ and $\alpha_2$-$\beta_2$ are broken. (d) Oxygenation of the β-haems, accompanied by widening of the haem pockets, expulsion of the tyrosines and rupture of the internal salt-bridges between histidine β146 and aspartic acid β94.

*2.3.2. Co-operativity.* Fig. 9 shows in some detail a diagrammatic sketch of the possible sequence of steps in the reaction of haemoglobin with oxygen, as suggested by Perutz [12]. The oxygen reacts with the haem (not the globin) causing a transition in the iron atom, from the high spin to the low spin state, which results in a small change in its atomic radius. This is effectively amplified by the haem group into a large movement of the haem-linked histidine (F8). This is associated with small but significant changes in the

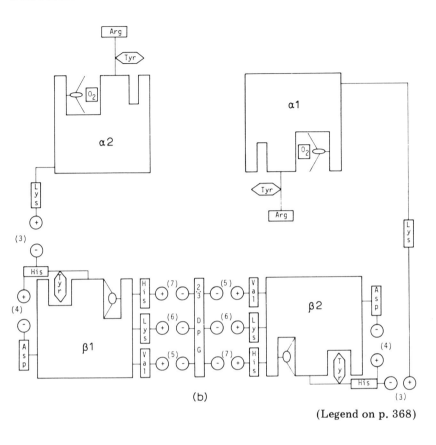

(Legend on p. 368)

tertiary structure of the reacting subunits, which include a movement of helix F towards the EF corner, narrowing the pocket between the helices. In deoxyhaemoglobin this pocket is occupied by the penultimate tyrosines of the chains. When the pocket is squeezed, the tyrosine is expelled and this in turn pulls the C-terminal residues with it, rupturing the salt bridges that had held the reacting subunit to its neighbours in the deoxy tetramer.

Thus the affinity of the tetramer for oxygen increases with the fractional saturation of the haemoglobin molecule, the arrival of each oxygen tipping the R-T (oxy-deoxy) equilibrium in favour of the oxy-conformation. This then means that if an oxygen molecule approaches two molecules of haemoglobin, one of which is

*References p. 411*

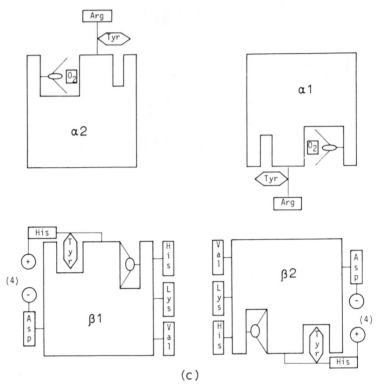

(Legend on p. 368)

partially oxygenated and the other not, there is a much higher chance that the former will be the species to which the incoming oxygen molecule is attached. Similarly, a fully oxygenated haemoglobin molecule is less likely to have an $O_2$ molecule withdrawn than one which is already partially deoxygenated. This, then, is the basis of haem—haem interaction, or co-operativity, and hence the sigmoid $O_2$ dissociation curve shown in Fig. 10. "Haem-haem interaction" is in some ways a misnomer, since the haems do not directly interact with each other; rather, the indirect reactions between the haems depend on rearrangements of the polypeptide chains during oxygenation.

The Hill equation [16] is an empirical expression for which there is no direct physical basis (because it was derived before it

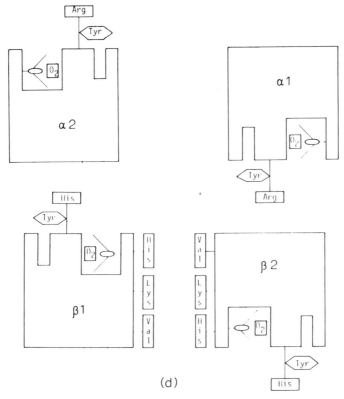

(d)

(Legend on p. 368)

was known that one molecule of haemoglobin binds four molecules of oxygen), of the reaction of haemoglobin with oxygen. It fits the middle range of the oxygen equilibrium data and provides a value, $n$, which is a useful overall measurement of co-operativity:

$$y = \frac{Kp^n}{1 + Kp^n}$$

($y$ is the fractional saturation with oxygen, $p$ is $O_2$ pressure and $K$ is an empirical constant).

The value of $n$ is an expression of haem—haem interaction and the $n$ values of most normal mammalian haemoglobins are of the order of 2.8 to 3.0. Non-cooperative globins or haemoglobins, such as myoglobin or haemoglobin H ($\beta_4$), have an $n$ value of unity.

*References p. 411*

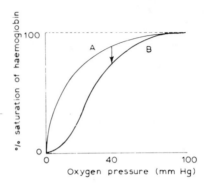

Fig. 10. Oxygen dissociation curves of myoglobin and haemoglobin. Note the increased efficiency of the sigmoid (haemoglobin) curve between 40 mm and 100 mm Hg $O_2$ pressure.

*2.3.3. The Bohr effect.* One aspect of the Bohr effect is the fact that oxygen affinity changes with pH (see Fig. 11). Obviously the important part of the curve in Fig. 11 is between pH 6.5 and 7.5 (the *alkaline* Bohr effect) and the physiological significance of this is fairly clear; in the tissues, where the hydrogen ion concentration is increased by lactic or carbonic acid formation, the oxygen affinity of haemoglobin will be decreased, thus unloading oxygen to the respiring tissues. Similarly, as $CO_2$ is expelled during circulation through the lungs there is a corresponding increase in pH

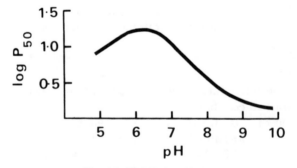

Fig. 11. The Bohr effect.

and a relative increase in haemoglobin oxygen affinity, which will favour the uptake of oxygen. The acid Bohr effect occurs below pH 6 and thus is of no physiological consequence.

In structural terms, most of the alkaline Bohr effect is caused by an interaction between positively charged weak bases (with p$K$ values around 7) and negatively charged groups in the deoxy-, but not the oxy-, structure. At pH 7 the weak bases will be charged and can interact with acid groups to form salt bridges which will stabilise the deoxy form, i.e., lower the oxygen affinity. At alkaline pH, however, these weak bases will be uncharged and no salt bridges will form; hence the stability of the deoxy-form is decreased and oxygen affinity is increased.

Some of the groups involved in the alkaline Bohr effect have been positively identified by X-ray crystallography [17] and by studies of mutant and chemically modified haemoglobins [18, 19]. The N-terminal valines of the $\alpha$-chains are responsible for 20—25% of the Bohr effect, in a somewhat indirect way. In the deoxy-structure of haemoglobin there is a chloride anion positioned between the $\alpha$-amino group of valine $\alpha$1 and the positively charged guanidinium of the C-terminal arginine $\alpha$141, whilst in the oxy-structure this salt bridge is broken and the $Cl^-$ is released, causing the p$K$ of $\alpha$1 valine to revert to normal. Thus it is clear also that inorganic anions can have a role in the Bohr effect, their removal decreasing the Bohr effect and increasing the oxygen affinity of haemoglobin. The C-terminal histidine residues of the $\beta$-chains ($\beta$146) contributed some 40% to the Bohr effect by forming a salt bridge between their imidazole groups and the carboxyl groups of $\beta$94 aspartate in the deoxy-, but not the oxy-, structure; measurements of the p$K$ of $\beta$146 by NMR has shown that it is 8.0 in deoxyhaemoglobin and 7.1 in carboxyhaemoglobin [20].

It seems that the groups responsible for the remaining 40% of the alkaline Bohr effect may be the $\beta$82 lysines. Haemoglobin variants which do not contain $\beta$82 Lys, but Met or Gln instead, have a diminished Bohr effect (Perutz et al. [166]). Hb Portland ($\zeta_2\gamma_2$) has a halved alkaline Bohr effect [21]; this may be due to a residue in the $\alpha$-chain which has been substituted for another, or one which is blocked in the $\alpha$-like $\zeta$-chain of Hb Portland. In the $\alpha$-chain, residue 1 participates in the Bohr effect, but in the $\zeta$-chain this residue is blocked.

*2.3.4. Carbon dioxide binding.* It has long been known that haemoglobin transports carbon dioxide directly and it is now recognised that it does so by a reversible combination of the N-terminal α-amino groups of the α- and β-chains with $CO_2$, to form carbamino compounds:

$$\alpha\text{-NH}_2 + CO_2 \rightleftharpoons \alpha\text{-NH}_2\ COO^{(-)} + H^{(+)}$$

In the absence of 2,3-diphosphoglycerate (see below), both the α- and β-chain α-amino groups bind $CO_2$ equally in liganded haemoglobin, but in deoxyhaemoglobin the β-chain has a three-fold higher affinity for $CO_2$ than the α-chain.

More $CO_2$ binds in total to the α-amino groups in the deoxy-structure than in the oxy-structure, which has the effect of lowering the oxygen affinity. (This lowering of the oxygen affinity is probably the main physiological role of $CO_2$ binding, the contribution of carbamino $CO_2$ to total $CO_2$ transport in the blood being only slight.) The structural explanation for this effect of $CO_2$ on oxygen affinity is probably a salt bridge between the carbamino group and a positively charged group in the deoxy structure.

*2.3.5. 2,3-Diphosphoglycerate binding.* Red cells contain 2,3-DPG which is known to bind to deoxyhaemoglobin stoichiometrically (1 mol of 2,3-DPG per haemoglobin tetramer) [22] and also to lower the oxygen affinity of a haemoglobin solution. The binding site for 2,3-DPG in deoxyhaemoglobin is shown in Fig. 12, from which it is clear that the ester can form extensive salt bridges with a number of groups from the two β-chains. On oxygenation the gap between the two β-chains closes and 2,3-DPG is expelled.

Haemoglobin F has serine at position γ143 instead of the charged histidine (see Fig. 3). This has no effect on the oxygen affinity of HbF (which has the same affinity as HbA) but it does decrease the binding constant of HbF for 2,3-DPG; thus, in foetal and adult blood where the 2,3-DPG concentration is the same, HbA will have a lower oxygen affinity than HbF, thereby facilitating oxygen transfer across the placenta.

Since both $CO_2$ and 2,3-DPG can bind to the β-chain α-amino group, their effect on oxygen affinity cannot be additive; similarly, the release of protons on $CO_2$ binding (which will decrease the

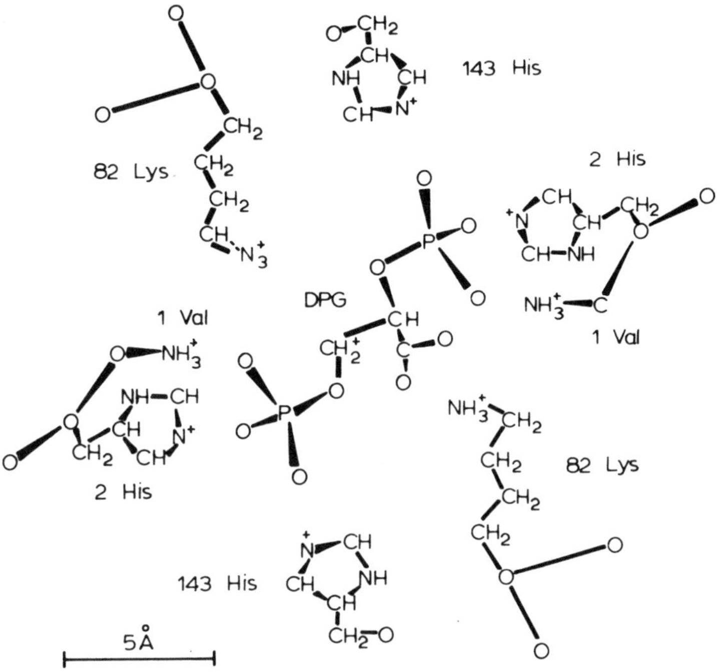

Fig. 12. The position of 2,3-DPG in the structure of human deoxyhaemoglobin.

alkaline Bohr effect) and the alteration of p$K$ values on 2,3-DPG binding (which at physiological pH increases the Bohr effect) will make different contributions to oxygen affinity, depending on the pH. A solution of purified haemoglobin in water has a higher oxygen affinity than a blood sample with the same haemoglobin concentration. It is the sum of the effects, in blood, of $CO_2$, 2,3-DPG and slightly alkaline pH which causes this difference.

*2.3.6. Allostery.* Several enzymes are classified as *allosteric* [23], showing cooperativity in the binding of substrate, the affinity for substrate at one binding site being enhanced by substrate binding at another site and leading to a sigmoid substrate binding curve. This cooperative binding of identical substrate molecules is called *homotropic* interaction, while the participation of allosteric modulators — either as activators or inhibitors — at sites other than the substrate binding site is known as *heterotropic* interaction. According to the elaborated theory of allostery proposed by Monod et al. [24] cooperative effects would result from an equilibrium between at least two different structures distinguished by the number and/or energy of the bonds between the subunits. They proposed that when the bonds are few and weak, the enzyme would be "relaxed" and fully active, but when they are strong it would be "tense" and less active. By direct analogy, haemoglobin may be regarded as an allosteric protein, made up of two different pairs of subunits, with oxygen as the substrate, haem as the coenzyme, and $H^+$ and 2,3-DPG as the allosteric modulators (effectors).

The oxy-structure, which has about the same oxygen affinity as the isolated subunits, is thus the relaxed form (R state) in the Monod sense, and the deoxy-structure, whose affinity for oxygen is several times lower, is the tense configuration (T state). The

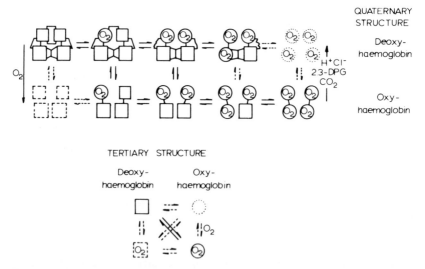

Fig. 13. Diagrammatic sketch of the allosteric mechanism of haemoglobin. The smaller subunits represent the α-, the larger ones the β-subunits. The clamps between them represent salt bridges; those between the β-subunits represent 2,3-DPG. Unstable forms have been drawn in broken lines. Loosening of salt bridges on ligand binding is indicated by wavy lines. The order of oxygenation and salt bridge rupture is arbitrary.

action of haemoglobin is to transduce stereochemical effects at the haem into electrostatic ones near the protein surface, which in turn regulate the allosteric equilibrium between the T and R states. It is fairly clear how this mechanism fits the two-state model of Monod et al. [24], and the model undoubtedly provides a useful framework in which to discuss interactions between haemoglobin and other ions and molecules. Account still needs to be taken, however, of the change in tertiary structure of subunits on ligation and the model needs to be extended to allow the subunits to influence each other within the same quaternary structure. Perutz et al. [25] have diagrammatically depicted the allosteric mechanism of haemoglobin as shown in Fig. 13, which embraces both the two-state Monod-Wyman-Changeux model and the "induced fit" theory of cooperativity of Koshland et al. [26].

*2.3.7. Methaemoglobin.* When the ferrous iron of the haem is oxidised to trivalent iron a chocolate brown pigment — methaemoglobin — results. Methaemoglobin is formed spontaneously *in vivo* at a rate of about 3% of the circulating haemoglobin per day. The red cells contain methaemoglobin reductase (or diaphorase) which is linked to NADH; it reduces cytochrome $b_5$ which latter then reacts directly with methaemoglobin. Thus diaphorase is strictly speaking a cytochrome, rather than a methaemoglobin reductase. A genetically determined deficiency of the enzyme causes methaemoglobinaemia, which can be treated successfully with ascorbic acid or with methylene blue, both of which act directly on the ferric haemoglobin. Methaemoglobinaemia can also be caused by structural abnormalities of the haem pocket resulting in the haemoglobins M (see below), these being permanent methaemoglobins which do not respond to the treatment with ascorbic acid or methylene blue.

Methaemoglobin is much more readily denatured than deoxy- or even oxyhaemoglobin, one reason for this being that methaemoglobin loses haem more easily and that free globin tends to precipitate.

*2.3.8. Haemichrome.* Methaemoglobin can also be denatured by forming a covalent link between the 6th ligand of the haem iron

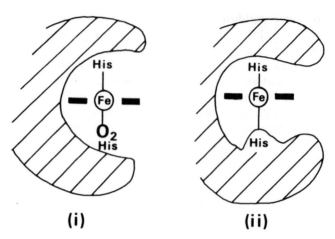

Fig. 14. Diagrammatic comparison of (i) oxyhaemoglobin and (ii) haemichrome I.

(which is normally free to combine with $O_2$) and the "distal" of the two haem-linked histidines. Normally in haemoglobin the only covalent link between iron and globin is that which involves the proximal of the two haem-linked histidines, $\alpha 87$ or $\beta 92$. In haemichrome, both the proximal and distal histidines form covalent bonds with the iron [27] (Fig. 14). As a consequence the globin is altered to haemichrome I, which can still be reconverted to haemoglobin. If, however, other globin residues displace the histidine the irreversible haemichrome II results, in which the molecular distortion causes destabilisation of the molecule followed by precipitation. The haemichromes are characterised by a specific absorption in the visible spectrum, and determination of electron paramagnetic (spin) resonance (EPR) has been a useful tool to follow their formation. In haemichrome I the bonding of the distal histidine to the oxidised haem causes the high-spin state of the iron to become low-spin. By reduction of the haem iron to the ferrous state, deoxyhaemoglobin can be restored. Further distortion of the molecule can be followed by observing the characteristic EPR spectra associated with such irreversible denaturation.

*2.3.9. Superoxide dismutase.* Oxygen is by itself a harmless gas but its reduced products, such as hydrogen peroxide and molecular oxygen in the excited state, are highly reactive and toxic to cells. The scavenging of active oxygen by reduced glutathione, ascorbic acid or α-tocopherol are means by which the cell protects itself. The first intermediate during the stepwise pathway of oxygen reduction is the superoxide radical ($O_2^-$). A family of enzymes, called the superoxide dismutases [28], minimise the potential toxicity of this radical. Copper is intrinsically involved in the oxidation of haemoglobin. The effect involves the β93 cysteine and thus removal of all traces of copper facilitates long-term storage of oxy- or deoxyhaemoglobin.

## 3. Variation in haemoglobin structure

### 3.1. Genetic variability

Haemoglobin is almost certainly the best-known example of genetic variation in protein structure, the first human variant (haemoglobin S) having been chemically identified some 20 years ago [30]. At present more than 300 variants of human haemoglobin alone are known. The list is growing almost weekly and the reader is referred to summaries published elsewhere [31, 32]. Although such genetic variation does not exist in man alone, variation in human haemoglobin structure has been studied most extensively. Those variants which involve no obvious change in the properties of the molecule (such as a Gly→Ala substitution in a non-critical part of the structure, for example), may remain undetected. Those, however, which result from a change in protein surface charge (as in Hb S, which has a Glu→Val replacement), will readily be detected by electrophoretic screening techniques; similarly, other variants may give rise to a change in haemoglobin stability or affinity for oxygen and consequently be detected from resulting clinical symptoms in the carrier, this being the concept of "molecular pathology" [33]. While the majority of variants are functionally of little consequence, they may often prove of value in anthropological studies [34].

*3.1.1. Single point mutants.* Most human haemoglobin variants represent single point nucleotide substitutions, compatible with the genetic code [35], within the appropriate globin gene. For instance, haemoglobins Riverdale-Bronx ($\beta$24 Gly→Arg) [36], Savannah ($\beta$24 Gly→Val) [37] and Moscva ($\beta$24 Gly→Asp) [38] can all be explained by single-base substitutions in the triplet GGU/C which codes for glycine $\beta$24. A few variants appear to possess amino acid substitutions at two different sites on the same subunit and curiously, all of the three known to be correctly identified so far have the Hb S ($\beta$6 Val→Glu) substitution as one of the two mutations [39—41].

Although nearly all of the observed point substitutions have been compatible with single nucleotide substitutions in a single corresponding codon of a *unique* mRNA sequence, there appears to be an exception to this scheme. The three haemoglobin variants Sydney ($\beta$67 Val→Ala) [42], M- Milwaukee ($\beta$67 Val→Glu) [43], and Bristol ($\beta$67 Val→Asp) [44] cannot arise from a single nucleotide base substitution in a *single* normal valine codon, as shown in Table II. On the basis of single point mutation, Hb Bristol must arise from a normal GUU/C codon while Hb M Milwaukee must come from GUA/G codon. Nucleotide sequence studies [45] have shown that, in the individuals studied so far, the codon for $\beta$67 is actually GUG. One therefore has to conjecture that either a double-base substitution (GUG→GAU or GAC) has occurred, or that a "silent polymorphism" exists at the codon for $\beta$67; in other words, some individuals exist with GUC or GUU instead of GUG as the codon for $\beta$67 (which still code for valine at this position). There is probably also a similar polymorphism at the codon for Thr $\beta$50 [45, 46].

TABLE II

Human haemoglobin variants at position 67

| Variant | Valine codons | Mutant codons |
|---|---|---|
| Hb Sydney $\beta$67 Val — Ala | GUU/C/A/G | GCU/C/A/G |
| Hb M-Milwaukee $\beta$67 Val — Glu | GUU/C/A/G | GAA/G |
| Hb Bristol $\beta$67 Val — Asp | GUU/C/A/G | GAU/C |

G, guanine; U, uridine; A, adenine; C, cytidine.

# GENETIC VARIABILITY

*3.1.2. Termination errors (elongated subunits).* A single point mutation in a termination codon will result in the insertion of an amino acid where the polypeptide chain should have finished, with subsequent "read-through" into, and synthesis from, the normally nontranslated 3' portion of the mRNA. This has been clearly illustrated by variants of the α-chain of human haemoglobin (which normally has a UAA termination codon [47]). Fig. 15 shows that there are six possible amino acids which can result from single nucleotide base substitutions in the codon UAA (α142) and of these six possible variants with α-chains elongated at their C-terminus, four (haemoglobins Constant Spring [48], Icaria [49], Seal Rock [50] and Koya Dora [51] have actually been indentified. The variants differ only in the nature of the amino acid inserted at position α142; they have α-chains which are elongated by the same 30 residues and which represent translation of the normally untranslated 3' α-mRNA sequence up to the next termination codon [52].

*3.1.3. Frameshift variants.* Since the genetic code is a triplet one, if one or two nucleotides are inserted into, or deleted from an mRNA nucleotide sequence — by, for instance, mispairing and unequal crossing over during meiosis — then the amino acid sequence which is translated from the mRNA will be, from the point of insertion or deletion, a nonsense sequence bearing only a chance relationship to the normal sequence. Such mutant proteins arising from such a change in the reading frame of the mRNA are known

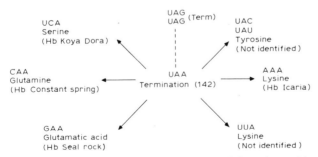

Fig. 15. Known termination codon variants and possible amino acids produced by point mutation of the termination codon of human haemoglobin α-chain (position α142).

*References p. 411*

as *frameshift* mutants. A frameshift close to the N-terminus of the globin chain gene will result in the synthesis of a polypeptide chain which cannot possibly make a functional haemoglobin molecule (although the sequence of the mRNA will hardly differ from the normal and might well behave as normal mRNA in hybridisation tests). If, however, the frameshift occurs at, or near, the C-terminus of the globin chain then the majority of the protein will possess the normal sequence and may be able to combine with the corresponding counterpart chains to form a haemoglobin tetramer of undefined stability and functional properties. Three such examples are known within human haemoglobins, one in the α-chain and two in the β-chain.

The α-chain of Hb Wayne [53] is normal up to position 138, but then continues with a sequence of eight new amino acids. Comparison of this sequence with the corresponding mRNA sequence for this region [47] (Fig. 16) suggests that Hb Wayne has arisen through the deletion of an adenine from the codon for α139 Lys, or one of the cytidines from the codon for α138 Ser, with subsequent change in reading frame and generation of the new sequence up until the next stop codon. Similarly, haemoglobins Tak [54] and Cranston [55] possess β-chains with elongations at their C-termini which have also arisen through a frameshift, this time through the deletion of one, or the insertion of two nucleotides.

*Hb A*

| Position | 137 | 138 | 139 | 140 | 141 | 142 | 143 |
|---|---|---|---|---|---|---|---|
| mRNA Amino acid | Thr ACC | Ser UCC | Lys AAA | Tyr UAC | Arg CGU | Term UAA | GCU |

*Hb Wayne*

| Position | 137 | 138 | 139 | 140 | 141 | 142 |
|---|---|---|---|---|---|---|
| mRNA Amino acid | Thr ACC | Ser UCC | Asn AAU | Thr ACC | Val GUU | Lys AAG |

Fig. 16. Generation of the amino acid sequence of Hb Wayne by frameshift mutation.

*3.1.4. Deletions and insertions in phase.* Just as one or two nucleotides (or four or five, etc.) can be deleted or inserted to produce a frameshift, so three (or a multiple of three) nucleotides can be deleted from, or inserted into, a globin mRNA sequence. The latter will, on translation, give rise to a globin chain which either lacks one or more amino acid residues (deletions), or possesses additional internal residues (insertions), but otherwise will be chemically normal. There are several known deletion variants in human $\beta$-globin chains, with from one to five amino acids missing. There is also one known $\alpha$-chain insertion variant, Hb Grady [56], in which the tripeptide sequence Glu-Pre-Thr ($\alpha$116—118) is repeated. It is likely that these types of variant arise by unequal crossing-over during meiosis, although it is possible that insertions or deletions could have arisen by some defect in DNA repair or replication or even in mRNA processing.

*3.1.5. Nonsense mutants.* In Hb McKees Rock [57] the two C-terminal $\beta$-chain residues ($\beta$145 and 146) are absent. Since residue $\beta$145 is tyrosine, it is possible that the codon for this residue (UAU) may have become a termination codon (UAA or UAG), a

*Hb A*

| Position | 144 | 145 | 146 | |
|---|---|---|---|---|
| mRNA | AAG | UAU | CAC | UAA |
| Amino acid | Lys | Tyr | His | Term |

  Substitution of A or G for U in UAU;
  *or*   deletion;
  *or*   frameshift.

*Hb McKees Rock*

| Position | 144 | |
|---|---|---|
| mRNA | AAG | $\overset{G}{U}$AA? |
| Amino acid | Lys | |

Fig. 17. Possible generation of the amino acid sequence of Hb McKees Rock.

*References p. 411*

type of mutation which is known as a *nonsense* mutation. (It may also be possible, however, that Hb McKees Rock is a frameshift mutant with an adenine inserted into the codon for tyrosine β145; Fig. 17.)

3.1.6. *Fusion variants.* Two types of fusion chain are known at present, and they have helped considerably in our understanding of the order of the non-globin genes on the appropriate human chromosome. The first type is that in which the non-α chain is a fused chain composed partly of a normal β-chain and partly of a normal δ-chain. If the mode of fusion is $NH_2$—δβ—COOH, the polypeptide is a Lepore globin chain, if $NH_2$—βδ—COOH, an anti-Lepore. Fig. 18 illustrates how such types can arise by nonhomol-

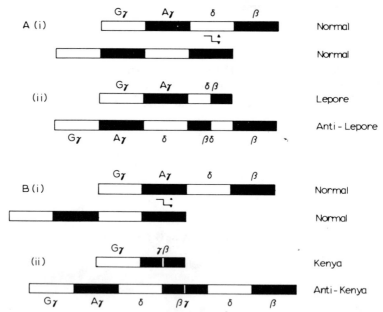

Fig. 18. Genetic mechanisms generating the fusion genes of Hb Lepore, Hb anti-Lepore, Hb Kenya and Hb anti-Kenya. A(i), B(i): normal arrangement of globin genes on chromosomes and misalignment of the chromatids at synapse. (Point of subsequent cross-over: ⇌). A(ii), B(ii): arrangement of globin genes on the chromosomes prior to meiosis, after breakage and crossover is completed. Compared with the normal, one chromosome has undergone a gene deletion, the other a gene amplification.

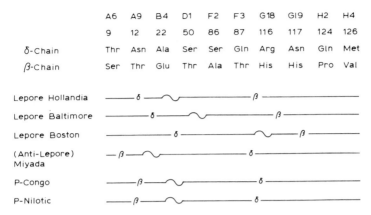

Fig. 19. The Lepore and anti-Lepore fusion chains. Since there are only ten points at which the β- and δ-chains differ, the exact points of fusion cannot be determined.

ogous crossing-over between the δ and β loci on complementary chromosomes. Such an event gives rise to two abnormal chromosomes; the Lepore chromosome will carry no normal β-chain gene but has instead only the δβ gene, while the anti-Lepore chromosome will have what is in effect a duplication, carrying both a normal β-chain gene, a fused βδ-chain gene, and a δ-chain gene.

The δ- and β-chains differ at only ten positions along their length, which sets an upper limit to the number of δβ or βδ fusion variants which can be detected by chemical means. In fact, three Lepore and three anti-Lepore types are known, varying at the point at which crossover has occurred (Fig. 19). To this list can probably be added Hb Coventry, which is chemically a β-chain with position 141 deleted but genetically is probably a βδ-chain [58] in which the crossover (at position β141) has occurred beyond the last point at which the δ- and β-chains differ (position 126).

Hb Kenya [59] appears to have arisen by nonhomologous crossing-over between the γ and β loci, giving rise to an $NH_2{-}\gamma\beta{-}COOH$ chain (Fig. 18). In heterozygotes for Hb Kenya, the HbF is composed entirely of $^G\gamma$-chains, indicating deletion of the $^A\gamma$- and δ-chain genes as a result of the cross-over.

Thus the study of fusion variants makes it highly likely that the arrangement of non-α globin genes on human chromosome is $^G\gamma^A\gamma\delta\beta$, and this has been borne out by gene mapping [89].

*References p. 411*

## 3.2. Non-genetic variation

An apparant variant, known as haemoglobin Koellicker [60], can arise artefactually as a result of haemolysis; the free haemoglobin found in the plasma is attacked by plasma carboxypeptidase, resulting in the removal of the carboxy-terminal arginine from normal $\alpha$-chains and the production of a "variant" which is not inherited.

## 3.3. Distribution of different haemoglobins in man

One can divide the human haemoblobin variants into those which occur in many tens of thousands of people, such as Hbs S, C, D and E, those which occur at some notable frequency in some limited areas such as Hbs G Korle Bu, G Philadelphia, O-Arab, Hasharon, and those which have been observed only a few times, or even only once.

HbS or sickle-cell haemoglobin ($\beta$6 Glu→Val) is mostly associated with tropical Africa where it may rise to an incidence of 40% in some East African tribes. This haemoglobin is also present at a lesser frequency in non-tropical Africa, in North but not South Africa, and in regions bordering that continent. There are foci of sickling in Italy, and in the Middle East. In India sickling is found in many pre-Dravidian tribes. Wherever African populations transferred to the New World they have also brought with them the sickling gene. Thus sickle-cell anaemia is an important public health problem in the West Indies, the U.S.A. and in tropical Latin America. The native American Indians do not possess a specific widespread haemoglobin variant.

Haemoglobin C ($\beta$6 Glu→Lys) is a West African genetic marker and reaches its highest incidence in Northern Ghana and Upper Volta. Haemoglobin D ($\beta$121 Glu→Gln) is present in 3% of Punjabis, and there is a lesser but regular incidence in neighbouring populations in Afghanistan and Iran. Haemoglobin E ($\beta$26 Glu→Lys) is found at high frequency in Burma, Malaysia, Thailand, Indonesia, Indochina, and to some extent in S.W. China but not in Melanesia.

These haemoglobin variants which occur in a large proportion of the population must confer some selective advantage, and this has been fully established for HbS where balanced polymorphism favours the sickle-cell heterozygote who does not die from homozygous sickle-cell anaemia and is better protected against malignant malaria than the normal homozygote.

Haemoglobins Korle-Bu ($\beta$73 Asp→Asn) and G Philadelphia ($\alpha$68 Asn→Lys) are distinct West African markers, and within West Africa the former is associated more with Ghana, and the latter with Nigeria. HbO-Arab ($\beta$121 Glu→Lys) first found in Arabs is in fact at its highest incidence in Bulgaria and neighbouring areas in Greece and Yugoslavia; it occurs occasionally in the Middle East, and in the Northern Sudan. Hb Hasharon ($\alpha$47 Asp→His) at first found occasionally and exclusively in Ashkenazi Jews is now seen also with some regularity in the Ferrara region of Italy and from time to time elsewhere in that country.

Even some of the variants seen only rarely may have anthropological indications. HbG Galveston ($\beta$43 Glu→Ala) was first found in North American Indians in the U.S.A. and Canada, and then seen in Taiwan and Japan. HbE Saskatoon ($\beta$22 Gly→Lys) was first seen in several Canadians with Scottish ancestry and later found in several families in the Orkneys, and in one in Edinburgh.

## 4. The effect of mutations on haemoglobin structure and function and the concept of "molecular pathology"

Virtually every possible class of protein variant is represented in the mutants of human haemoglobin. It is reasonable to expect that, in a molecule which occupies a physiologically critical position and which is extremely well adapted to a specialised function, a number of these mutations might well prove deleterious in some way to haemoglobin; this is indeed the case.

Broadly, structural mutants of haemoglobin can be classified as follows:

Those which have no obvious effect on the properties or the function of the molecule; examples would be neutral changes in non-critical regions of the molecule, such as an Asp→Glu, Gly→Ala, etc. Such variants will not be distinguishable from HbA (except by complete sequence determination of an apparently normal molecule!) and will escape detection.

Those which affect the stability of the tetramer, either directly (unstable globin chains) or indirectly (increased dissociation into dimers or monomers).

*References p. 411*

Those which affect the oxygen affinity of the molecule, be it by permanent formation of methaemoglobin, alteration in the $R \rightleftharpoons T$ equilibrium, changes in 2,3-DPG binding or Bohr effect, etc.

It should be noted that, apart from the last, there are no clear-cut categories into which a given haemoglobin variant will necessarily fall and very often a variant (particularly one in which the mutation is in the haem pocket) will have both altered stability and oxygen affinity.

From an extensive study of haemoglobin variants it has proved possible to draw some general conclusions concerning the effects of mutations at particular sites in the tetramer — that is, the *molecular pathology* [33] of the molecule. It is clear that the haemoglobin tetramer is relatively insensitive to replacements of most amino acids on its surface, but extremely sensitive to even quite small alterations of internal non-polar contacts, especially those near the haems, while replacements at the contacts between the $\alpha$- and $\beta$-subunits affect respiratory function. Listed below are some examples of each type of *structural* variant. It should be borne in mind, of course, that there also exist *non-structural* variants which can affect the rate of synthesis of globin chains and which will be discussed below in the section on thalassaemia.

### 4.1. Mutations which affect the stability of the tertiary or quaternary structure

There are a number of ways in which an unstable haemoglobin molecule may arise and these have been described in detail elsewhere [61—65]. Briefly they can be summarised as follows:

*4.1.1. Substitutions in the haem pocket.* A good example is Hb Hammersmith [66] in which the important non-polar phenylalanine CD1 is replaced by serine, which has a polar hydroxyl group side-chain; this probably allows water into the haem pocket, resulting in the loss of the haem group.

*4.1.2. Replacement of an important non-polar residue by a polar residue.* Unless the new polar side-chain can be pushed out to the (polar) surface of the molecule, or its charge can somehow be compensated internally by a proximal, oppositely charged group, gross distortion of the molecule will occur, resulting in instability. An example is Hb Volga ($\beta$27 [B9] Ala→Asp) [67].

*4.1.3. Replacements by proline.* The imino acid proline can be tolerated only in the first three residues of an $\alpha$-helix [68]; if it is introduced into an $\alpha$-helix beyond the third residue, distortion and consequent instability result from the disturbance of the normal secondary and tertiary structure. A good example is Hb Santa Ana ($\beta$88 [F4] Leu→Pro) [69].

*4.1.4. Replacements at tightly packed regions of the tetramer.* There is a region in the haemoglobin tetramer, at the point where the B and E helices come close to each other, which allows no room for a side chain and is normally occupied by glycine residues B6 and E8 (which have no side-chain). When one of these glycine residues is replaced by another residue, the spatial relationship of the E helix to the haem group (linked to E7) will be disturbed, resulting in instability. Hb Moscva ($\beta$24 [B6] Gly→Asp) [38] is such an example.

The physiological consequences of heterozygosity for an unstable haemoglobin can in some cases be quite striking and the severity of the clinical symptoms can often be related to the degree of instability; that is, clinical pathology can be related directly to molecular pathology. While the role of haemichrome and of "naked" globin (globin which has lost haem) in the precipitation of unstable haemoglobin has been studied in some detail [27], it is still not clear exactly how haemoglobin instability leads to the fairly well defined state of *unstable haemoglobin haemolytic anaemia* [65] and the underlying mechanism will probably differ from one variant to another. Essentially, however, the disease results from the production of a haemoglobin or a globin chain which is irreversibly precipitated within the red cell to form discrete bodies of insoluble protein (Heinz bodies). The presence of Heinz bodies causes red cell membrane damage and decreases the deformability of the red cell, leading to entrapment within the splenic sinusoids and an increased liability to haemolysis and subsequent anaemia.

A few variants are known in which substitutions at the tetrameric $\alpha_1\beta_1$ contact (see previously) result in an increased tendency to dissociate into dimers and, sometimes, to monomers; Hb Philly ($\beta$35 [Cl] Tyr→Phe) [70] is an example. Such variants tend to be only mildly unstable and are not associated with a severely anaemic state.

*4.2. Mutations which affect the oxygen affinity*

Since the oxygen-binding properties of haemoglobin depend on the interactions of many chemical groups in the haemoglobin molecule, it is not surprising that some mutations can upset the delicate balance and produce a molecule that has a higher, or lower, affinity for oxygen than normal. Clearly there will be a very large number of ways in which this can happen and although the atomic details of the explanation will vary from one case to another, such variants can be broadly classified as: Variants in which the haem iron is in a permanent ferric-state (haemoglobins M, permanent methaemoglobins); variants in which the oxygen affinity is lowered; variants in which the oxygen affinity is raised.

*4.2.1. Haemoglobins M.* Haemoglobins M, of which there are five known, are variants in which the amino acid substitution causes the haem groups to exist only in the ferric state. They are therefore valueless as respiratory pigment; carriers of HbM present with cyanosis due to this excess of methaemoglobin and occasionally also with mild haemolytic anaemia. In four of the five cases the proximal (F8) or distal (E7) His of the $\alpha$- or $\beta$-chain is replaced by tyrosine (see Table III). The phenolic side-chain of the tyrosine can react with the haem iron atom to form an iron-phenolate complex, the stability of which varies with the mutant; the net result, however, is the "freezing" of the iron in the $Fe^{3+}$ state (methaemoglobin). The iron-phenolate complex tends to be more stable in the $\alpha$-chain variants, the $\beta$-chain variants losing haem groups readily and becoming unstable. Both the $\alpha$-chain variants have very low oxygen affinities, negligible haem—haem interaction, no Bohr effect and are effectively stabilised in the quaternary deoxy structure. These extensive changes in quaternary structure

MUTATIONS AFFECTING OXYGEN AFFINITY 391

TABLE III

The haemoglobins M

| Haemoglobin | Amino acid substitution |
|---|---|
| Hb M Boston | α53 (E7) His → Tyr |
| Hb M Saskatoon | β63 (E7) His → Tyr |
| Hb M Iwate | α87 (F8) His → Tyr |
| Hb M Hyde Park | β92 (F8) His → Tyr |
| Hb M Milwaukee | β67 (E11) Val → Glu |

do not occur for the β-chain variants, which have α-chains with near-normal oxygen affinities and only slightly reduced allosteric interactions.

The fifth HbM, HbM Milwaukee, does not directly involve either proximal or distal histidines, but the group concerned (a glutamic acid at position β67 [E77]) points towards the haem group and its negatively charged carboxyl group interacts with the iron atom, displacing the distal histidine and stabilising the iron in the $Fe^{3+}$ state. The displacement of the distal histidine initiates changes in tertiary and quaternary structure which result in a reduced oxygen affinity and haem—haem interaction, but have no effect on the Bohr shift.

*4.2.2. Low-affinity variants.* Certain variants can give rise to structural alterations resulting in a haemoglobin which has a lower affinity for oxygen than normal without the formation of methaemoglobin. Depending on the nature of the variant, haem—haem interaction and the Bohr effect may or may not also be affected. The following afford good examples of the way in which apparently minor changes in structure can effect major changes in the properties of the haemoglobin molecule.

*4.2.2.1. Haemoglobins Kansas [71] and Titusville [72].* In the $\alpha_1\beta_2$ contact of normal oxyhaemoglobin there is a hydrogen bond, between aspartic acid α94 [G1] and asparagine β102 [G4], which lies in a non-polar environment from which water is largely excluded [11]; as a consequence this H-bond has a high bond energy because of the low local dielectric constant. Haemoglobins Kansas

*References p. 411*

($\beta$102 Asn→Thr) and Titusville ($\alpha$94 Asp→Asn) both dissociate much more readily into $\alpha\beta$ dimers in the *liganded* form than does HbA and this is almost certainly a consequence of their inability to form this particularly strong H-bond; in fact, the dissociation constant of Hb Kansas is two orders of magnitude higher than that of HbA.

The change in Hb Titusville, unlike that in Hb Kansas, is an almost perfect isomorphous replacement (Asp→Asn) which would not be expected to disturb the $\alpha$-chain tertiary structure. It is therefore likely that the properties of Hb Titusville result entirely from the *weakening* of this single hydrogen bond (since a poor H-bond can in fact form between $\beta$102 Asn and the new $\alpha$94 Asn); it is, then, striking that Hb Titusville shows a very low oxygen affinity, very little co-operativity and a very small Bohr effect which is further reduced in the presence of organic phosphates. Thus the weakening of this single hydrogen bond seems to result in the liganded form of Hb Titusville being heavily biased towards the T-state as well as a much increased dissociation into $\alpha\beta$ dimers.

*4.2.2.2. Haemoglobin Agenogi* [73, 74]. The substitution in Hb Agenogi, $\beta$90 [F6] Gln→Lys, occurs at the surface of the tetramer, where, generally speaking, changes tend to have little effect on haemoglobin function [33]. Hb Agenogi, however, shows a reduced oxygen affinity and it has been suggested that this arises from an effect of the new lysine residue on the salt bridges of deoxyhaemoglobin. It has already been mentioned that $\beta$FG1 aspartate forms a salt bridge with the imidazole ring of the C-terminal histidines of the same $\beta$-chain in the deoxy- but not the oxy-conformation. Position $\beta$F6, the site at which the substitution in Hb Agenogi occurs, is four residues away from $\beta$FG1 and, because of the nature of the $\alpha$-helix, the $\beta$F6 and $\beta$FG1 side-chains point in the same direction. Thus the new lysine side chain can strengthen the salt bridge between $\beta$FG1 Asp and $\beta$HC3 His, which will in turn bias the oxy-deoxy equilibrium towards the T state, leading to a low-affinity haemoglobin.

The clinical response to the presence of such low-affinity variants ranges from no obvious symptoms to, at worst, mild cyanosis due to arterial circulating deoxyhaemoglobin; there is no indication that life is shortened by heterozygosity for such variants. The haemoglobin level is lower than normal. There is less anoxia in the tissues because of the greater readiness of the pigment to give up oxygen, and therefore there is less stimulation of erythropoietin release. The result is a "relative" physiological anaemia.

*4.2.3. High-affinity variants.* Mutations which bias the R-T equilibrium towards the R state (by making the oxy-structure more stable or the deoxy-structure less stable than normal) will result in an increase in oxygen affinity and reduction of co-operativity. There are a number of ways in which this can arise, as shown by the following examples.

*4.2.3.1. $\alpha_1\beta_2$ interface mutations.* The importance of the $\alpha_1\beta_2$ interface in the deoxy-oxy transition has been stressed and, generally speaking, mutations at the $\alpha_1\beta_2$ contact tend to oppose the transition and "lock" the structure into the high-affinity oxy-form; the first such example to be described was Hb Chesapeake [75] ($\alpha$92 [FG4] Arg→Leu).

The penultimate tyrosine residues ($\alpha$140 and $\beta$145) play an important role in maintaining the correct orientation of the C-terminal arginine $\alpha$141 and histidine $\beta$146 residues to form the salt bridges found in deoxyhaemoglobin. Replacement of these tyrosines, as for example in Hb Bethesda [76] ($\beta$145 [HC2] Tyr→His), prevents the formation of the salt bridges, thus reducing the stability of the deoxy structure.

*4.2.3.2. Haem pocket mutations.* Most haem pocket variants are unstable and associated with haemolytic anaemia (see above). Haemoglobin Heathrow [77], however, ($\beta$103 [G5] Phe→Leu) is an exception and has a high oxygen affinity but is stable. It is not known why Hb Heathrow has a high oxygen affinity, but it is possible that the substitution of the smaller side chain of leucine causes a change in the inclination of the plane of the haem group more favourable to the oxy-structure.

*4.2.3.3. β-β Interface mutations.* In haemoglobin Hiroshima [78] ($\beta$146 [HC3] His→Asp) a pair of salt bridges is lost in the deoxy-structure, resulting in a low affinity for oxygen and, since histidine $\beta$146 contributes some 40% to the Bohr effect (see above); this results in a much reduced Bohr effect.

*4.2.3.4. 2,3-DPG-binding site mutations.* A number of abnormal haemoglobins are known in which replacements of positively charged residues by neutral ones result in weakened binding of 2,3-DPG and a consequent rise in oxygen affinity. An example is haemoglobin Rahere [79] (β82 [EF6] Lys→Thr).

The clinical symptoms resulting from the presence of a high-affinity haemoglobin are fairly clearly defined as so-called *polycythaemia*, that is, an excess of red cells in the circulation. Increased oxygen affinity of the haemoglobin leads to decreased release of oxygen to the tissues, resulting in tissue anoxia, which stimulates the production of the hormone *erythropoietin*, leading to polycythaemia. This is an interesting example of a molecular imperfection being expressed at the clinical level.

### 4.3. Sickle-cell haemoglobin (HbS) (β6 Glu→Val)

The nature of the sickling process has aroused particular interest because one can see here the possibility of a treatment at molecular level of sickle-cell anaemia [80]. It is known that sickling occurs on deoxygenation of sickle cells because of the low solubility of deoxyhaemoglobin S compared with that of normal deoxyhaemoglobin. The unidirectional haemoglobin crystals form spikes which deform the shape of cell (Fig. 20). The process can also be studied by following the gelling of haemoglobin solutions. A 26% solution of deoxyhaemoglobin S will gel at 37°C but will become a homogeneous solution at 4°C. Sickling is also pH-dependent and in vivo acidosis enhances, and alkalosis retards sickling [80]. Antisickling agents have been developed which are intended to modify the oxygen affinity of sickle-cell haemoglobin. Cyanate was thus introduced which by carbamylation of the N terminus of the $\beta^S$-chain interferes with the formation of deoxyhaemoglobin because the modified N-terminus cannot combine with 2,3-DPG. Unfortunately the cyanate caused undesirable side effects.

One can divide the formation of strands of sickle-cell haemoglobin molecules into nucleation, growth and alignment of the strands which form "ropes" which can be visualised by the electron microscope and X-ray crystallography [81, 82]. They are composed of aligned strands consisting of helically arranged haemoglobin molecules; each turn of the helix is thought to involve 6 or 7 molecules [81, 82]. The nucleation and growth of associating polymolecular complexes is probably not very different for haemoglobins A and S. The alignment of the strands which are so much more lasting with deoxyhaemoglobin S than with deoxyhaemo-

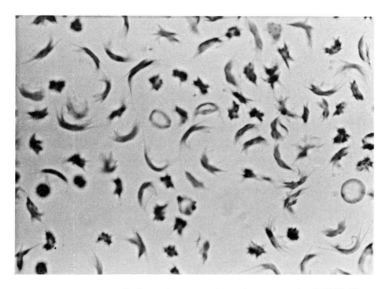

Fig. 20. Sickling observed through a scanning microscope (× 3000) (By courtesy of Dr. A. J. Salsbury, Brompton Hospital, London).

globin A is presumably due to the valine $\beta^S 6$ at the outside of the strand which must cause adhesion whereas failure to stick together must arise when that position is occupied in HbA by glutamic acid.

It is known that sickle-cell haemoglobin interacts well in the stacking of the bundles with some and not well with other haemoglobins [83]. Thus it seems that other residues besides $\beta 6$ must be involved in sickling. HbS sickles well when forming hybrids with Hbs, C, D and O-Arab. One of nature's anti-sickling variants is haemoglobin Korle-Bu ($\beta 73$ Asp→Asn) which either in a mixture of the two variants or in a combination of the HbS and the haemoglobin Korle-Bu mutation in the same $\beta$-chain (HbC Harlem) shows a remarkable protection against sickling. A most powerful antisickling haemoglobin is fetal haemoglobin which has not only a higher solubility than Hbs A and S, but also increases their solubility in a mixture [84].

## 5. Haemoglobin biosynthesis

Haemoglobin is composed of protein (globin chains) and prosthetic groups (haem) which are synthesised independently and combined in a coordinated fashion. The synthesis of globin chains has been most studied in man and rabbits and is considered here in relation to human globin genetics.

### 5.1. The genetics of normal human haemoglobin

The genetic control of normal human haemoglobin production is depicted diagrammatically in Fig. 21. Hb Gower 1 and Gower 2 are formed early in intrauterine life (see below); it is uncertain whether Hb Gower 1 is $\epsilon_4$ or $\zeta_2\epsilon_2$.

The $\beta$- and $\delta$-chains of HbA and HbA$_2$ are controlled by single structural genes, whilst there is direct evidence [85] (from restriction endonuclease mapping) for the original proposal [86] that the $\alpha$-chain locus is duplicated in most, if not all, human populations. There is clear genetic evidence that the $^A\gamma$- and $^G\gamma$-chains are the products of separate gene loci. A model has been proposed which incorporates four $\gamma$-chain genes per chromosome [87]. Direct measurements by cDNA/DNA hybridisation of the number of $\gamma$-chain genes per haploid genome, however, suggest that there are only two $\gamma$-chain loci [88], a conclusion which is confirmed by restriction endonuclease mapping [89]. The location of the $\zeta$- and $\epsilon$-genes is with the $\alpha$- and non-$\alpha$-genes, respectively. The human chromosomes carrying the $\alpha$-, $\beta$-, $\gamma$- and $\delta$-chain genes have been identified, by somatic cell fusion analysis, as 16 ($\alpha$-chain) and 11 (non-$\alpha$ chains) [90, 91]. The close linkage of the non-$\alpha$-chain genes has long been evident from the study of fusion variants (Hb Lepore, etc., see above) and it is clear now, from restriction mapping of human DNA, that the orientation of the non-$\alpha$ genes (with respect to transcription is $^G\gamma \rightarrow {}^A\gamma \rightarrow \delta \rightarrow \beta$ (Fig. 22). (see also Addendum)

### 5.2. Globin gene structure

The last few years have seen an explosion in our knowledge of the

GLOBIN GENE STRUCTURE 397

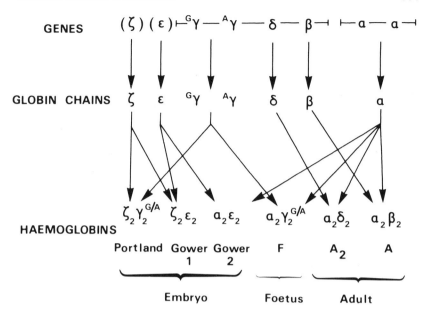

Fig. 21. The normal human haemoglobins and the genetic control of their production. Adapted from D. J. Weatherall and J. B. Clegg, Cell, 16 (1979) 467 [107].

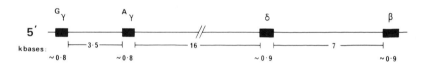

Fig. 22. Map of the human non α-globin loci. Adapted from D. J. Weatherall and J. B. Clegg, Cell, 16 (1979) 467 [107]. (see also Addendum)

globin genes, at the level of mRNA and DNA sequence, through both direct sequence studies [45, 47, 52, 92—96] and the use of recombinant DNA cloning and restriction endonuclease mapping [89, 97—99].

It has for some time been appreciated that the size of the mRNA for the α- and β-globin chains exceeds that required to

References p. 411

code for the amino acid sequence alone; that is, the mRNAs contain non-coding, or untranslated, 3'- and 5'-terminal sequences plus a 3' poly(A) "tail" (Fig. 23). What has only recently become clear, however, is that the DNA sequence from which the globin mRNA is produced is in turn longer than the mRNA, containing up to two intervening sequences (or *introns;* the coding, or expressed, sequences are called *exons*). The size and number of the introns varies according to the species under study. The larger is about 600 nucleotide base pairs (0.6 kb) long in the rabbit $\beta$-gene, 0.646 kb in the mouse $\beta$-gene and 0.8—1.0 kb in human $\beta$, $\delta$ and $\gamma$-genes; it is situated between the codons for amino acids $\beta 104$ and $\beta 105$ in mouse and human genes and in approximately the same region in the rabbit. The smaller intron, which has now also been observed in human globin genes, is 0.116 kb long in the mouse $\beta$-gene, probably the same in the rabbit $\beta$-gene and is situated in the mouse DNA between the codons for $\beta 30$ and $\beta 31$. There are introns in the mouse $\alpha$-chain gene, and there is also evidence for such regions in the human $\alpha$-genes.

The entire gene, including the coding, intervening and untranslated regions, is transcribed into precursor heterogeneous nuclear RNA (hnRNA) (of which only a small portion contains mRNA-specific sequences). This is followed by 5'-capping, addition of a 3'-sequence, excision of introns and ligation, association with proteins and transport from nucleus to cytoplasm (although not necessarily in that order) to produce the final 9—10S mRNA. It has been shown that a 27S RNA species and a 15S RNA (which latter contains a 5'-cap and a poly(A) sequence) are likely to be precursors to globin mRNA [100].

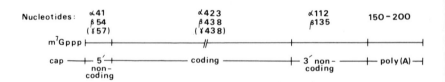

Fig. 23. Diagrammatic representation of the mRNA for human globin chains.

## 5.3. Ontogeny of haemoglobin synthesis

The changes in human globin synthesis during prenatal and neonatal development are shown in Fig. 24. ε-Chains can only be detected in the early foetus, disappearing by the 10th to 12th week of gestation. Synthesis of the ζ-chain, which is structurally similar to the α-chain, probably parallels that of ε-chains but must persist at a low level throughout foetal development, since trace amounts of Hb Portland ($\zeta_2\gamma_2$) are found in normal cord bloods. Synthesis of α-chain appears early and persists through foetal development while γ-chains appear in large quantity about the time that ε-chains disappear. Synthesis of δ-chain begins late in intra-uterine life and increases gradually during the first year of life to reach a normal adult $HbA_2$ level of about 3% of the total haemoglobin. β-Chains, however, are synthesised in small amounts throughout foetal development but appear in substantial amounts near term. After birth, β-chain synthesis normally increases rapidly as γ-chain synthesis decreases and virtually ceases. By 6 months of age β-chain synthesis has reached its maximal adult level and production of Hb F is reduced to about 1%, or less, of the total haemoglobin, this phenomenon being known as *the foetal switch*. Thus the expression of β-, γ- and δ-genes is not synchronous, despite their close linkage. While the timing and rate of changeover from γ- to β-chain synthesis has been fairly well characterised, the mechanism controlling this switch is unknown. Better knowledge of the mechanism(s) in-

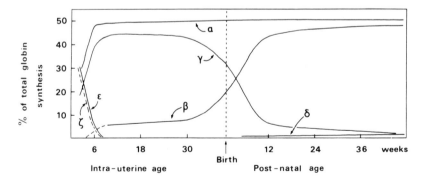

Fig. 24. Changes in globin-chain synthesis during human development. Adapted from W. G. Wood, Br. Med. Bull., 32 (1976) 282.

*References p. 411*

volved would be of some considerable practical interest in terms of manipulating the switch to increase the production of HbF in those severe haemoglobinopathies resulting from defective β-chain production (β-thalassaemia, see below) or from variant β-chains such as those of HbS.

### 5.4. The control of globin synthesis

The general pattern of protein synthesis in red cell precursors has been described fully elsewhere [102—104]. Most of the detailed studies of haemoglobin synthesis have been made on reticulocytes (which is hardly surprising, since over 90% of the protein made by reticulocytes is haemoglobin); the reticulocyte is the penultimate stage in the differentiation from a nucleated pluripotent stem cell to the mature red cell (which latter does not synthesise protein). Although such studies have provided much information on the control of haemoglobin synthesis in reticulocytes, little is known concerning the activation of haemoglobin genes during erythroid-precursor maturation, the repression/activation of globin genes in different types of cell line, the mechanisms for maintaining different levels of synthesis of different globin chains, the controls which balance α and non-α chain synthesis, or balance haem and globin syntheses; or even how haemoglobin synthesis is stopped when the intracellular content reaches a critical level.

There are several levels at which control of globin synthesis can occur: at the gene level (duplication, amplification); at the transcription level (rate of RNA production); at the level of RNA processing, transport, or stability; at the level of translation (peptide chain initiation, elongation and termination); and at the level of tetramer assembly (subunit interaction, and haem insertion) and stability.

*5.4.1. Control at the gene and transcriptional level.* Attempts to demonstrate specific amplification of globin genes (such as observed, for example, in amphibian rRNA genes) have not been successful and direct measurements of gene numbers [105] in man clearly indicate the presence of one or two copies of each globin gene per haploid genome. Studies on the specificity of chromatin transla-

CONTROL OF GLOBIN SYNTHESIS 401

tion have indicated that the nonhistone chromatin proteins may be responsible for the specificity of the sequences transcribed, since DNA and histones from liver or brain cell chromatin can be mixed with non-histone protein from erythroid cell chromatin to produce mRNA transcripts in a cell-free system [106]. The details of such control, however, are unknown. There also is evidence [99, 107] from studies of deletions in the non α-gene region that areas of the chromosomes carrying the β-, γ- and δ-chain genes are involved in the neonatal suppression, or activation, of the γ- or β-chain genes, respectively.

From the previous discussion of gene structure and globin mRNA production it is clear that there exist a very large number of potential points of control during the processing and transport of the initial transcript to give the final functional cytoplasmic mRNA species. Although very little is known about such control processes, studies on the synthesis of δ- and β-globin chains [108] have provided some useful clues. Despite the similarities in the structure of the δ- and β-chains, the production of δ-chains in the red cell is some 40-fold less than that of β-chains and it seems likely that this control is effected, at least in part, by the difference in stability between δ- and β-mRNAs; it also appears probable that additional control may also be effected by differential rates of transcription or hnRNA processing [109].

*5.4.2. Control at the translational level.* Haemoglobin is composed of equal numbers of α- and β-chains, the ratio being largely determined by the level of mRNA synthesis in early precursors [110]. There is no absolute requirement for the synthesis of one chain to make another [111], which suggests independent regulation of the synthesis of the two chains.

The relative number of ribosomes making α- and β-chains are approximately equal, but α-chains are made on smaller polysomes than β-chains [112, 113] this difference being accounted for by the slightly faster initiation rate of the β-chains relative to the α-chains; the rate of assembly of both chains is the same [114] and it seems that the αmRNA has a lower affinity for ribosomes than has βmRNA. Whether this represents a significant corrective control process is uncertain, but it does imply that there is more αmRNA than βmRNA in reticulocytes and it has been demonstrated that under

*References p. 411*

conditions where *elongation* of nascent chains is rate-limiting the cells do make 50% more α-chains than β-chains [115]. Under normal conditions there will be a slight excess of ribosome-free α-chain mRNA in reticulocytes, waiting to enter into polyribosomes [116] and a slight excess (of the order of 10—20%) of α-chains is produced in red cell precursors where it is, presumably, largely removed by proteolysis.

While there is little evidence for the co-ordinated synthesis of α- and β-chains, there is good evidence that haem and globin synthesis are mutually interdependent. If protein synthesis is inhibited, a decrease in haem synthesis ensues, probably through feedback inhibition of one of the enzymes of haem biosynthesis (δ-amino laevinulate synthase) by accumulated haemin [117]. The absence of haem inhibits globin chain synthesis and this phenomenon has proved difficult to explain completely. It is clear that an inhibitor forms in the absence of haemin and that this inhibitor can, in the presence of ATP, prevent the binding of the initiating tRNA (met-tRNAf) to the 40S ribosomal subunits, thus preventing globin chain initiation. It seems that the inhibitor is a kinase which is activated in the absence of haemin and which blocks the function of a specific initiation factor (eIF-2) by phosphorylating it [103, 150]. The details of the mechanism have still to be fully elucidated.

*5.4.3. Control at the level of tetramer assembly.* The exact timing of the combination of haem with globin chains is unknown, although it has been shown that globin does not combine with haem until after its release from polyribosomes [118]. The newly synthesised chains are rapidly and spontaneously incorporated into complete haemoglobin tetramers, probably through the αβ dimer as an intermediate. Complete haemoglobin molecules are much more stable than free globin chains or αβ dimers, whilst αα or ββ dimers are still less stable than αβ dimers.

Findings concerning the competition of normal β-chain and HbE or S β-chains for α-chains [119a, b] suggest that there may be some form of post-translation control over the assembly of HbA and HbE or S in some E or S trait erythroid precursor cells; it seems that the abnormal β-chains have a lower affinity than $\beta^A$-chains for α-chains, a suggestion which is substantiated by the finding of very low levels of HbE or S in cases of HbE or S/α-thalassaemia (where α-chains will be limiting). It is not known if this is a physiological control mechanism in normal cases.

### 5.5. Post-translational modification of haemoglobin

Haemoglobin undergoes non-enzymatic glycosylation *in vivo* [120]. The most prevalent glycosylated fraction HbA$_{Ic}$ contains glucose attached to the N-terminus of the β-chain by a keto-amine linkage. HbA$_{Ia}$ is likely to represent a mixture of adducts of haemoglobin with glucose-6-phosphate and fructose-1,6 diphosphate respectively, whilst the structure of the HbA$_{Ib}$ is still uncertain. The glycosylated haemoglobins are formed continuously throughout the life span of the erythrocytes, and the normal proportion of HbA$_{Ic}$ is about 10% of the total haemoglobin. In haemolytic anaemia, in which the life span of the red blood cells is reduced, the proportion of HbA$_{Ic}$ may be only half of that found normally. Of considerable practical interest is that HbA$_{Ic}$ is raised in diabetes mellitus. The rise is associated with the raised blood glucose in that disease and the level of HbA$_{Ic}$ can be used as an indicator of the likely blood sugar level over the last months. Whilst the blood sugar level of a diabetic patient will only indicate his present state, that of HbA$_{Ic}$ will yield valuable additional information on the course the disorder has steered over a longer period. In severe cases the level may rise to above 20% of the total haemoglobin.

## 6. The molecular basis of thalassaemia

There have been a number of authoritative reviews on thalassaemia [107, 121, 122] to which the reader is referred for detailed discussion of both its molecular basis and clinical consequences.

The thalassaemias are a *group* of disorders in which the rate of production of one or more of the globin chains of haemoglobin is reduced. There are several globin chains and numerous points at which their synthesis can go wrong; one would not then expect thalassaemia to be a single, well-defined disorder at the molecular level, and this is the case.

The word thalassaemia is derived from θαλασσα ("the sea"), reflecting the fact that early cases of the disorder were all of Mediterranean origin. Thalassaemias are classified broadly into α- and β-thalassaemia, depending on whether the rate of α- or β-chain synthesis is retarded. The molecular basis of the Asian form of α-thalassaemia is fairly well understood, and that of the β-thalassaemia is also being clarified.

## 6.1. α-Thalassaemia

At least three α-thalassaemia genes have been defined in Asians, these being α-thalassaemia 1 (or $\alpha^\circ$-thalassaemia), in which no α-chains are produced; α-thalassaemia 2, in which there is a reduced output of α-chains; and haemoglobin Constant Spring (see above), a structural variant (termination codon mutant) which is synthesised only in very small amounts. The latter two forms may be termed $\alpha^+$-thalassaemias. The interaction of these three genes produces the range of clinical forms of α-thalassaemia. Since the α-chains of both HbA and HbF are under the same genetic control, the α-thalassaemias cause defective production of HbA and HbF. Any excess of β- or γ-chains resulting from a reduced output of α-chains will give rise to tetrameric molecules with the formulae $\gamma_4$ (Hb Bart's) and $\beta_4$ (HbH).

In the most severe form of α-thalassaemia, the so-called *haemoglobin Bart's hydrops foetalis syndrome*, α-chain synthesis is totally absent and no fragments of α-chains are produced [123]. The molecular basis of $\alpha^\circ$-thalassaemia has been shown [124, 125] to be the absence of α-chain genes from the chromosomes of the affected infants and the hydrops foetalis syndrome therefore results from homozygosity for a deletion of the whole or a large part of the α-chain genes.

There is good evidence that the α-chain genes are duplicated in many populations [85, 86] and it had for some time been assumed that α-thalassaemia-2 could be the result of the deletion of one of a pair of linked α-chain genes; this has been confirmed by DNA hybridisation [126]. The interaction of α-thalassaemia 1 and 2 — that is, the absence of 3 of the 4 α-chain genes in a double heterozygote — gives rise to the clinical status of HbH disease.

In fact, about half of the instances of Asian HbH disease appear to be such deletions; the other half are double heterozygotes for α-thalassamia 1 and Hb Constant Spring which latter therefore behaves as an α-thalassaemia 2 gene. Hb Constant Spring is one of a class of α-chain variants in which a single base change in the termination codon has resulted in the insertion of an amino acid at position α142 and subsequent read-through into normally untranslated codons at the 3' end of the α-chain mRNA. Such chain-termination mutants are synthesised in only small amounts — hence the α-thalassaemia due to Hb Constant Spring — but it is not clear why this should be so. The α-Constant Spring-chains are synthesised actively only in nucleated red cells and there is no synthesis in reticulocytes, which suggests that the mRNA for α-Constant Spring-chains may be metabolically unstable [127].

Studies of HbH disease in some non-Asians [128] suggests that the molecular basis of α-thalassaemia here may be the presence of defective α loci, rather than the absence of α-chain genes. The usual α-thalassaemia in American blacks is either the heterozygous or the homozygous state for α-thalassaemia-2.

## 6.2. β-Thalassaemia

There is also considerable heterogeneity within the β-thalassaemias. Most of the work has been carried out on the fairly typical Mediterranean types of the disease, but milder forms also seem to exist in the Negro races. The β-thalassaemias can also be broadly grouped as $β^+$-thalassaemia, in which there is reduced β-chain synthesis and $β^o$-thalassaemia, where there is no β-chain synthesis at all. Excess α-chains cannot form tetramers, so there is no $α_4$ molecule produced in $β^o$-thalassaemia; the reduction in the synthesis of β-chains, however, results in an increase in the proportions of $HbA_2$ ($α_2δ_2$) and HbF ($α_2γ_2$).

### 6.2.1. $β^+$-Thalassaemia.
It has been shown that β-chain initiation, elongation and termination are normal in $β^+$-thalassaemia [129, 130] and present evidence indicates that the underlying molecular defect is a reduced production of β-chain mRNA [131–134] which may result from an abnormality of transcription or mRNA processing.

### 6.2.2. $\beta^0$-Thalassaemia.

This is a heterogeneous disorder at the molecular level. In some cases there appears to be no detectable $\beta$-chain mRNA present in the cytoplasm [135], in others inactive or abnormal $\beta$-chain mRNA has been demonstrated [136—138]. In the great majority of cases of $\beta^0$-thalassaemia investigated to date the $\beta$-chain genes seem to be intact, whilst in $\delta\beta^0$-thalassaemia there seems to have been a deletion of the $\beta$-chain and part of the $\delta$-chain genes [139, 140]. Thus $\delta\beta^0$-thalassaemia/$\beta^0$-thalassaemia falls into a number of groups, including: deletion of the $\beta$- and part of the $\delta$-chain genes; occasional deletion of part of the $\beta$-globin gene [141, 151]; presence of the genes but absence of mRNA production; production of non-functional mRNA; and production of functional mRNA which is not utilised because of a translational defect.

## 6.3. Haemoglobin Lepore ($\alpha_2\delta\beta_2$)

Haemoglobin synthesis studies have indicated that $\delta\beta$-chain synthesis takes place in bone marrow cells, but not in reticulocytes [142, 143]. Thus the kinetics of $\delta\beta$-chain synthesis parallel those of $\delta$-chains [108]. Since the formation of the $\delta\beta$ fusion gene may take place by chromosomal misalignment at meiosis (Fig. 18), Hb Lepore heterozygotes effectively have a $\beta$-chain gene replaced by a $\delta\beta$ one which produces only small amounts of $\delta\beta$-chains and thus they have mild thalassaemia.

## 6.4. Hereditary persistence of foetal haemoglobin (HPFH)

HPFH is a term which covers a variety of conditions, all associated with a persistent HbF production beyond the neonatal period [144]. Present evidence suggests that these disorders are mild forms of thalassaemia and there is direct evidence that in the Negro form of HPFH the $\delta$-chain and $\beta$-chain loci have been deleted [99, 145, 146]; thus there must be areas of DNA involved in the suppression of $\gamma$-chain production which have been deleted in this form of HPFH.

## 6.5. β-Thalassaemia due to structural variants

The way in which the α-chain variant Hb Constant Spring gives rise to α-thalassaemia has been described above. There is no such clear-cut case of β-thalassaemia arising from the presence of a β-chain variant, but it would seem that the β-chain of HbK Woolwich (β132 [H10] Lys→Gln) is inefficiently synthesised [147]; similarly, the β-chains of the frameshift variant Hb Tak has led to speculation [54, 148] on another potential cause of $\beta^0$-thalassaemia, since it is the product of a frame-shift mutation close to the C-terminal end of the β-chain. If such a frameshift took place near the 3' end of the structural gene instead, deleting just a single nucleotide, the mRNA produced from such a gene may or may not be metabolically unstable; it would however, be chemically almost normal but obviously incapable of being translated into a normal globin chain. Instead, despite the presence of an almost normal gene, a nonsense protein would be produced which would probably be rapidly degraded within the cell. Similarly, a point mutation to produce a termination codon early in the globin sequence would have much the same effect.

## 7. The evolution of haemoglobin

The structural genes for myoglobin and the haemoglobin α- and β-like chains probably arose by duplication of a single common ancestral gene. It is generally assumed that the initial duplication led to the evolution of individual functions — those of oxygen transport and storage — in the form of primitive haemoglobin and myoglobin, respectively. Further duplication and mutation of the primitive haemoglobin gene could lead to the formation of α- and non α-chains and, hence, the ability to form a tetramer. Further duplication of these genes appears to have happened in, for example, the case of the human α- and γ-chain genes; in man these duplications were then followed by the production of the two closely similar δ- and β-chain loci.

Studies of haemoglobin chains from a range of animals of varying phylogenetic position have shown that the number of amino acid substitutions by which homologous globin chains differ in different species is roughly proportional to the phylogenetic separation of the two species. For instance, the polypeptides of the haemoglobin of man differ from those of the gorilla and the rabbit by 2 and 22 amino acid residues, respectively. Clearly this implies that man and gorilla have diverged at a later date than man and rabbit, a supposition entirely consistent with phylogeny! Knowing the approximate time at which two species diverged and the number of globin amino acid differences, it is possible to measure the "average" rate of amino acid substitution; for haemoglobin, the figure seems to be about one amino acid substitution every seven million years. It should, however, be pointed out that apparent rates of evolutionary change will be affected by differences in generation time, that the rate of protein evolution is not constant and that it varies from protein to protein. Further, such calculations cannot take into account back-mutations, or "silent" mutations (those nucleotide substitutions which, through the degeneracy of the genetic code, do not result in an amino acid substitution), but do, nonetheless, give useful working values; more accurate estimates of mutation rates in globin genes will probably be obtained in the future from the DNA sequences of the genes themselves.

One of the striking features of haemoglobin evolution is that, despite the large differences in primary structure (up to 80% difference) a common tertiary structure has been maintained; thus, human myoglobin, human $\alpha$- and $\beta$-chains, and *Glycera dibranchiata* (annelid worm) haemoglobin all have the same overall three dimensional shape. Comparison of the known globin sequences shows that there are only a very few positions which never change — the so-called invariant residues — and these residues are at critical places in terms of three dimensional structure. Clearly, then, such evolutionary studies cannot be divorced from function. There is a limit to the number of amino acid differences that a given molecule can accept and it is likely that the variable sites are interdependent, because a new side chain will affect its neighbours in the tertiary structure and so put some constraint on the residues involved in the folding of the molecule. A good example of compensatory evolution at one position to allow for such a disturbance at another has been described in myoglobin [149].

## Addendum (December 1980 and October 1981)

This chapter was up to date in September 1979 when it was submitted for publication. There can, however, be few fields in which knowledge has increased at such a rate as that of globin gene structure and this addendum summarises the advances made over the past 15 months. The non-α chain genes have been completely sequenced, including the $\epsilon$ [152], $\gamma$ [153], $\delta$ [154] and $\beta$ [155] genes. Pseudogenes ($\psi$-genes) have been discovered [156, 158] — they are DNA sequences which are homologous to neighbouring globin genes but cannot be expressed as globin because of an accumulation of diverging mutations (usually frame-shifts generating new stop codons). Such $\psi$ genes are generally considered to be evolutionary failures; although it is also possible that they have a role in controlling neighbouring globin genes [158]. The arrangement of the non-α globin genes on chromosome 11 [157—160] is now known to be $\psi\beta_2-\epsilon-{}^G\gamma-{}^A\gamma-\psi\beta_1-\delta-\beta$ (cf. Fig. 22). Between these genes there are repetitive sequences which may be of regulatory importance.

The α-chain genes have been found to be at the 3' end of the $\zeta$-chain genes, again with a $\psi\alpha$-chain gene being involved [161, 162] the gene order being on chromosome 16 $\zeta-\zeta-\psi\alpha-\alpha-\alpha$. It is likely that the 3' $\zeta$-chain gene is in fact a $\psi\zeta$.

With minor exceptions, such as the discovery of a β-thalassaemia due to a mutation of a lysine codon to a termination codon at position β17 [163], the basic principles of the molecular basis of β-thalassaemia outlined in [107] remain unaltered. Evidence has come forward for the presence of the $(-/\alpha)$ mutation in Blacks which explains why they can have α-thalassaemia but do not show HbH disease $(--/-\alpha)$ or Bart's hydrops foetalis $(--/--)$. In addition, as predicted by analogy with the Lepore and anti-Lepore mutation, single (α) and triple $(\alpha-\alpha-\alpha)$ genes have been found.

An enormous step forward in the prenatal diagnosis of HbS disease as well as a new insight into the differential evolution of the sickle-cell gene in Africa was made with the discovery, by restriction enzyme analysis, of a Hb S-linked DNA polymorphism [164].

Finally, the globin region of the genome of *Xenopus laevis* has been shown to differ markedly from that of mammals in the close linkage of α and β loci and the apparent duplication of an α-β globin gene pair ($\alpha_1$-$\beta_1$ and $\alpha_2$-$\beta_2$) [165]; this is taken as evidence that vertebrate $\alpha$ and $\beta$ globin genes evolved by a tandem duplication of a single primordial globin gene.

A paper which conclusively summarises the identification of residues contributing to the acid and alkaline Bohr effects of human haemoglobin has been published by Perutz et al. [166].

Two valuable surveys of haemoglobin synthesis are those of Shaeffer et al. [167] and Proudfoot et al. [168]. The first deals with the assembly of the haemoglobin subunits and the rules which govern the proportion of normal and abnormal haemoglobins in the erythrocyte. The second describes the structure of the globin genes and their transcription. There are "consensus" sequences in the DNA of many proteins including those responsible for the globins. These sequences are found in the non-transcribed part of the DNA at a distance of 30 and of 70 nucleotides "upstream" to the mRNA capping site. They are called the ATA and CCAAT "boxes" respectively. Certainly the removal of the first of these deprives the DNA of its capacity to promote the transcription of the globin gene.

The intervening sequences or introns have been found to be the area where mutations can cause thalassaemia. For example, in one instance the sequencing of the complete $\beta$-globin gene of a patient with $\beta^+$ thalassaemia discovered only a single base substitution within the smaller intron. This altered the sequence to make it resemble the exon-intron border and could therefore generate an additional target for the enzyme removing the intron before the mRNA leaves the nucleus. This faulty mRNA would compete with the normal mRNA and as the result there would be the under production of normal globin [169]. An $\alpha$-thalassaemia was found to be associated with a pentanucleotide deletion within the 5' splice junction of the first intervening sequence [170].

## REFERENCES

1. L. J. Henderson, quoted in J. Barcroft, The Respiratory Function of the Blood, Vol. II, Cambridge University Press, Cambridge, 1928.
2. G. Wald, in E. S. Guzmann Barron (Ed.), Modern Trends in Physiology and Biochemistry, Academic Press, New York, 1952.
3. D. Keilin, Acta Biochem. Polon., 3 (1956) 439.
4. J. Barcroft, Physiol. Rev., 5 (1925) 596.
5. A. C. Redfield, Quart. Rev. Biol., 8 (1933) 31.
6. J. T. Ruud, Nature, 173 (1954) 848.
7. C. A. Appleby, in A. Quispel (Ed.), The Biology of Nitrogen Fixation, North-Holland, Amsterdam, 1974, p. 521.
8. R. Sidloi-Lumbroso, L. Kleiman and H. M. Schulman, Nature, 273 (1978) 558.
9. K. D. Nadler and Y. J. Avissar, Plant Physiol., 60 (1977) 433.
10. J. Barcroft, The Respiratory Function of the Blood, Cambridge University Press, 1928.
11. M. F. Perutz, H. Muirhead, J. M. Cox, L. C. G. Goaman, Nature, 219 (1968) 131.
12. M. F. Perutz, Nature, 228 (1970) 726.
13. L. Pauling, R. B. Corey and H. R. Branson, Proc. Natl. Acad. Sci. USA, 37 (1951) 205.
14. E. A. Padlan and W. E. Love, Nature, 220 (1968) 376.
15. M. A. Dayhoff. Atlas of Protein Sequence and Structure, Vol. 5, National Biomedical Research, Washington DC, 1972.
16. A. V. Hill, J. Physiol., 40 (1910) iv.
17. M. F. Perutz, H. Muirhead, L. Mazzarella, R. A. Crowther and J. V. Kilmartin, Nature, 222 (1969) 1240.
18. J. V. Kilmartin and L. Rossi-Bernardi, Nature, 222 (1969) 1243.
19. J. V. Kilmartin, Br. Med. Bull., 32 (1976) 209.
20. J. V. Kilmartin, J. Fodd, M. Luzzana and L. Rossi-Bernardi, J. Biol. Chem., 248 (1973) 7039.
21. S. Tuchinda, K. Nagai and H. Lehmann, FEBS Lett., 61 (1975) 148.
22. R. E. Benesch and R. Benesch, Adv. Protein Chem., 28 (1974) 211.
23. J. Monod. J. -P. Changeus and F. Jacob, J. Mol. Biol., 6 (1963) 306.
24. J. Monod, J. Wyman and J. -P. Changeux, J. Mol. Biol., 12 (1965) 88.
25. M. F. Perutz, Br. Med. Bull., 32 (1976) 195.
26. D. E. Koshland, Jr., G. Nemethy and D. Filmer, Biochemistry, 5 (1966) 365.
27. R. W. Carrell, C. C. Winterbourn and E. A. Rachmilewitz, Br. J. Haematol., 30 (1975) 259.
28. I. Fridovich, in O. Hayaishi and K. Asada (Eds.), Biochemical and Medical Aspects of Active Oxygen, University Park Press, London, 1977, p. 3.
29. C. C. Winterbourn and R. W. Carrell, Biochem. J., 165 (1977) 141.
30. V. M. Ingram, Biochim. Biophys. Acta, 28 (1958) 539.

31   H. F. Bunn, B. G. Forget and H. M. Ranney, Human Hemoglobins, Saunders, Philadelphia, 1977.
32   H. Lehmann and R. G. Huntsman, Man's Haemoglobins Including the Haemoglobinopathies and Their Investigation, North-Holland, Amsterdam, 1974.
33   M. F. Perutz and H. Lehmann, Nature, 219 (1968) 902.
34   A. E. Mourant, A. C. Kopec and K. Domaniewska-Sobczak, Distribution of the Human Blood Groups, Oxford Univ. Press, Oxford, 1976.
35   F. H. C. Crick, Cold Spring Harbor Symp. Quant. Biol., 31 (1966) 3.
36   H. M. Ranney, A. S. Jacobs, L. Udem and R. Zalusky, Biochem. Biophys. Res. Commun., 33 (1968) 1004.
37   T. H. J. Huisman, A. K. Brown, G. D. Efremov, J. B. Wilson, C. A. Reynolds, R. Uy and L. L. Smith, J. Clin. Invest., 50 (1971) 650.
38   L. I. Idelson, N. A. Didkovskii, R. Casey, P. A. Lorkin and H. Lehmann, Nature, 249 (1974) 768.
39   R. M. Bookchin, R. L. Nagel and H. M. Ranney, J. Biol. Chem., 242 (1967) 248.
40   J. G. Adams and P. Heller, Blood, 42 (1973) 990.
41   M. Goosens, M. C. Garel, J. Auvinet, P. Baset, P. F. Gomes and J. Rosa, FEBS Lett., 58 (1975) 149.
42   R. W. Carrell, H. Lehmann, P. A. Lorkin, E. Raik and E. Hunter, Nature, 215 (1967) 626.
43   P. S. Gerald and M. L. Efron, Proc. Natl. Acad. Sci. USA, 47 (1961) 1758.
44   J. H. Steadman, A. Yates and E. R. Huehns, Br. J. Haematol., 18 (1970) 435.
45   C. A. Marotta, J. T. Wilson, B. G. Forget and S. M. Weissman, J. Biol. Chem., 252 (1977) 5040.
46   P. A. M. Kynoch, FIST Thesis, London, 1979.
47   M. J. Proudfoot and J. I. Longley, Cell, 9 (1976) 733.
48   J. B. Clegg, D. J. Weatherall and P. F. Milner, Nature, 234 (1971) 337.
49   J. B. Clegg, D. J. Weatherall, I. Contopoulou-Griva, K. Caroutsos, P. Poungouras and T. Tsevrenis, Nature, 251 (1974) 245.
50   T. B. Bradley, R. C. Wohl and G. J. Smith, Clin. Res., 23 (1975) 131A.
51   W. W. W. De Jong, P. Meera Khan and L. F. Bernini, Am. J. Hum. Genet., 27 (1975) 81.
52   J. T. Wilson, J. K. De Riel, B. G. Forget, C. A. Marotta and S. M. Weissman, Nucl. Acids Res., 4 (1977) 2353.
53   M. Seid-Akhavan, W. P. Winter, R. K. Abramson and D. L. Rucknagel, Proc. Natl. Acad. Sci. USA, 73 (1976) 882.
54   H. Lehmann, R. Casey, A. Lang, R. Stathopoulou, K. Imai, S. Tuchinda, P. Vinai and G. Flatz, Br. J. Haematol., 31 (Suppl.) (1975) 119.
55   H. F. Bunn, G. J. Schmidt, D. N. Haney and R. G. Dluhy, Proc. Natl. Acad. Sci. USA, 72 (1975) 3609.
56   T. H. J. Huisman, J. B. Wilson, M. Gravely and M. Hubbard, Proc. Natl. Acad. Sci. USA, 71 (1974) 3270.
57   R. M. Winslow, M. -L. Swenberg, E. Gross and P. A. Chervenick, J. Clin.

Invest., 57 (1976) 772.
58  R. Casey, P. A. M. Kynoch, A. Lang, H. Lehmann, G. Nozari and N. K. Shinton, Br. J. Haematol., 38 (1978) 195.
59  T. H. J. Huisman, R. N. Wrightstone, J. B. Wilson, W. A. Schroeder and A. G. Kendall, Arch. Biochem. Biophys., 153 (1972) 850.
60  H. R. Marti, D. Beale and H. Lehmann, Acta Haematol., 37 (1967) 174.
61  H. Lehmann, R. G. Huntsman, R. Casey, A. Lang, P. A. Lorkin, in W. J. Williams, E. Beutler, A. J. Erslev and R. W. Rundles (Eds.), Hematology, 2nd ed., Chapter 57, McGraw-Hill, New York, 1977.
62  J. M. White, Clin. Haematol., 3 (1974) 333.
63  E. R. Huehns, in R. M. Hardisty and D. J. Weatherall (Eds.), Blood and its Disorders, Chapter 12, Blackwell, London, 1974.
64  J. M. White and J. V. Davie, in E. M. Brown and C. V. Moore (Eds.), Progress in Haematology, Chapter 3, Grune and Stratton, New York, 1971.
65  R. W. Carrell and H. Lehmann, Seminars in Hematology, 6 (1969) 116.
66  J. V. Dacie, N. K. Shinton, P. L. J. Gaffney Jr., R. W. Carrell and H. Lehmann, Nature, 216 (1967) 663.
67  L. I. Idelson, N. A. Didkovskii, A. V. Filippova, R. Casey, P. A. M. Kynoch and H. Lehmann, FEBS Lett., 58 (1975) 122.
68  M. F. Perutz, J. C. Kendrew and H. C. Watson, J. Mol. Biol., 13 (1965) 669.
69  R. W. Opfell, P. A. Lorkin and H. Lehmann, J. Med. Genet., 5 (1968) 292.
70  R. F. Rieder, F. A. Oski and J. B. Clegg, J. Clin. Invest., 48 (1969) 1627.
71  J. Bonaventura and A. Riggs, J. Biol. Chem., 243 (1968) 980.
72  R. G. Schneider, R. J. Atkins. T. S. Hosty, G. Tomlin, R. Casey, H. Lehmann, P. A. Lorkin and K. Nagai, Biochim. Biophys. Acta, 400 (1975) 365.
73  T. Miyaji, H. Suzuki, Y. Ohba and S. Shibata, Clin. Chim. Acta, 14 (1966) 624.
74  K. Imai, H. Morimoto, M. Kotani, S. Shibata, T. Miyaji and K. Matsumoto, Biochim. Biophys. Acta, 200 (1970) 197.
75  S. Charache, D. J. Weatherall and J. B. Clegg, J. Clin. Invest., 45 (1966) 813.
76  A. Hayashi. G. Stamatoyannopoulos, A. Yoshida and J. Adamson, Nature New Biol., 230 (1971) 264.
77  J. M. White, L. Szur, I. D. S. Gillies, P. A. Lorkin and H. Lehmann, Br. Med. J., 3 (1973) 665.
78  M. F. Perutz, P. Del Pulsinelli, L. Ten Eyck, J. V. Kilmartin, S. Shibata, I. Iuchi, T. Miyaji and H. B. Hamilton, Nature New Biol., 232 (1971) 147.
79  P. A. Lorkin, A. D. Stephens, M. E. J. Beard, P. F. M. Wrigley. L. Adams and H. Lehmann, Br. Med. J., 4 (1975) 200.
80  J. Dean and A. N. Schechter, New Eng. J. Med., 299 (1978) 752.
81  J. T. Finch, M. F. Perutz, J. F. Bertles and J. Döbler, Proc. Natl. Acad.

Sci. USA, 70 (1973) 718.
82  R. Josephs, H. S. Jarosch and S. J. Edelstein, J. Mol. Biol., 102 (1976) 409.
83  R. Josephs, H. S. Huehns, Br. Med. Bull., 32 (1976) 223.
84  J. A. Frier and M. F. Perutz, J. Mol. Biol., 112 (1977) 97.
85  S. H. Orkin, Proc. Natl. Acad. Sci. USA, 75 (1978) 5950.
86  H. Lehmann and R. W. Carrell, Br. Med. J., 4 (1968) 748.
87  T. H. J. Huisman, W. A. Schroeder, W. H. Bannister and J. L. Grech, Biochem. Genet., 7 (1972) 131.
88  J. Old, J. B. Clegg, D. J. Weatherall, S. Ottolenghi, P. Comi, B. Giglioni, J. Mitchell, P. Tolstoshev and R. Williamson, Cell 8 (1976) 13.
89  P. F. R. Little, R. A. Flavell, J. M. Kooter, G. Annison and R. Williamson, Nature, 278 (1979) 227.
90  A. Deisseroth, A. Nienhuis, P. Turner, R. Velez, W. F. Anderson, F. H. Ruddle, J. Lawrence, R. Creagan and R. Kucherlapati, Cell, 12 (1977) 205.
91  A. Deisseroth, A. Nienhuis, J. Lawrence, R. Giles, P. Turner and F. H. Ruddle, Proc. Natl. Acad. Sci. USA, 75 (1978) 1456.
92  F. E. Baralle, Cell, 12 (1977) 1085.
93  J. C. Chang, G. F. Temple, R. Poon, K. H. Neumann and Y. W. Kan, Proc. Natl. Acad. Sci. USA, 74 (1977) 5145.
94  J. C. Chang, R. Poon, K. H. Neumann and Y. W. Kan, Nucl. Acids Res., 5 (1978) 3515.
95  J. Van Den Berg, A. Van Ooyen, N. Mantei, A. Schamböck, G. Grosveld, R. A. Flavell and C. Weissman, Nature, 276 (1978) 37.
96  D. A. Konkel, S. M. Tilghman and P. Leder, Cell, 15 (1978) 1125.
97  D. C. Tiemeier, S. M. Tilghman, F. I. Polsky, J. G. Seidman, A. Leder, M. H. Edgell and P. Leder, Cell, 14 (1978) 237.
98  R. A. Flavell, J. M. Kooter, E. De Boer, P. F. R. Little and R. Williamson, Cell, 15 (1978) 25.
99  D. Tuan, P. A. Biro, J. K. De Riel, H. Lazarus and B. Forget, Nucl. Acids Res., 6 (1979) 2519.
100 R. N. Bastos and H. Aviv, Cell, 11 (1977) 641.
101 H. Kamuzora and H. Lehmann, Nature, 256 (1975) 511.
102 H. Borsook, Ann. N.Y.Acad. Sci., 119 (1964) 523.
103 T. Hunt, Br. Med. Bull., 32 (1976) 257.
104 T. Hunt and I. M. London, in W. J. Williams, E. Beutler, A. J. Erslev and R. W. Rundles (Eds.), Hematology, 2nd ed., Chapter 17, McGraw-Hill, New York, 1977.
105 R. Williamson, Br. Med. Bull., 32 (1976) 246.
106 J. Paul, R. S. Gilmour, N. Affara, G. Birnie, P. Harrison, A. Hell, S. Humphries, J. Windass and B. Young, Cold Spring Harbor Symp. Quant. Biol., 38 (1973) 885.
107 D. J. Weatherall and J. B. Clegg, Cell, 16 (1979) 467.
108 R. F. Rieder and D. J. Weatherall, J. Clin. Invest., 44 (1965) 42.
109 W. G. Wood, J. M. Old, A. V. S. Roberts, J. B. Clegg, D. J. Weatherall and N. Quattrin, Cell, 15 (1978) 437.

# REFERENCES

110 A. R. Hunter and R. J. Jackson, in H. Marti and L. Nowicki (Eds.), Synthesis, Structure and Function of Hemoglobin, Lehmann, Munich, 1972, p. 95.
111 G. R. Honig, B. Q. Rowan and R. G. Mason, J. Biol. Chem., 244 (1969) 2027.
112 R. T. Hunt, A. R. Hunter and A. J. Munro, Nature, 220 (1968) 481.
113 J. B. Clegg, D. J. Weatherall and C. E. Eunson, Biochim. Biophys. Acta, 247 (1971) 109.
114 H. F. Lodish and M. Jacobsen, J. Biol. Chem., 247 (1972) 3622.
115 H. F. Lodish, J. Biol. Chem., 246 (1971) 7131.
116 H. F. Lodish, Nature, 251 (1974) 385.
117 A. I. Grayzel, J. E. Fuhr and I. M. London, Biochem. Biophys. Res. Commun., 28 (1967) 705.
118 L. Felicetti, B. Colombo and C. Baglioni, Biochem. Biophys. Acta, 129 (1966) 380.
119a S. Tuchinda, D. Beale and H. Lehmann, Humangenetik, 3 (1961) 1.
119b J. R. Schaeffer, R. E. Kingston, M. J. McDonald and H. F. Bunn, Nature, 276 (1978) 631.
120 H. F. Bunn and M. J. McDonald, in W. S. Caughey (Ed.) Biochemical and Clinical Aspects of Hemoglobin Abnormalities, Academic Press, New York, 1978, p. 215.
121 D. J. Weatherall and J. B. Clegg, The Thalassaemia Syndromes, 3rd ed., Blackwell, Oxford, 1979.
122 D. J. Weatherall, in W. J. Williams, E. Beutler, A. J. Erslev and R. W. Rundles (Eds.), Hematology, 2nd ed., Chapter 39, McGraw-Hill, New York, 1977.
123 D. J. Weatherall, J. B. Clegg and H. B. Wong, Br. J. Haematol., 18 (1970) 357.
124 S. Ottolenghi, W. G. Lanyon, J. Paul, R. Williamson, D. J. Weatherall, J. B. Clegg, J. Pritchard, S. Pootrakul and H. B. Wong, Nature, 251 (1974) 389.
125 J. M. Taylor, A. Dozy, Y. W. Kan, H. E. Varmus, L. E. Lie-Injo, J. Ganeson and D. Todd, Nature, 251 (1974) 392.
126 Y. W. Kan, A. M. Dozy, H. E. Varmus, J. M. Taylor, J. Holland, L. E. Lie-Injo, J. Ganeson and D. Todd, Nature, 255 (1975) 255.
127 D. M. Hunt, quoted in D. J. Weatherall and J. B. Clegg, Cell, 16 (1979) 467.
128 S. H. Orkin, J. Old, H. Lazarus, C. Altay, A. Gurgey, D. J. Weatherall and D. G. Nathan, Cell, 17 (1979) 33.
129 J. B. Clegg, D. J. Weatherall, S. Na-Nakorn and P. Wasi, Nature, 220 (1968) 664.
130 D. G. Nathan, H. F. Lodish, Y. W. Kan and D. Housman, Proc. Natl. Acad. Sci. USA, 68 (1971) 2514.
131 E. J. Benz and B. G. Forget, J. Clin. Invest., 50 (1971) 2755.
132 A. W. Nienhuis and W. F. Anderson, J. Clin. Invest., 50 (1971) 2458.
133 D. Housman, B. G. Forget, A. Skoultschi and E. J. Benz, Proc. Natl. Acad. Sci. USA, 70 (1973) 1809.

134 D. L. Kacian, R. Gambino, L. W. Dow, E. Grossbard, C. Natta, F. Ramirez, S. Spiegelman, P. A. Marks and A. Bank, Proc. Natl. Acad. Sci. USA, 70 (1973) 1886.
135 P. Tolstoshev, J. Mitchell, G. Lanyon, R. Williamson, S. Ottolenghi, P. Comi, B. Giglioni, G. Masera, B. Modell, D. J. Weatherall and J. B. Clegg, Nature, 260 (1976) 95.
136 E. J. Benz, Jr., B. G. Forget, D. G. Hillman, M. Cohen-Solal, J. Pritchard, C. Cavallesco, W. Prensky and D. Housman, Cell, 14 (1978) 299.
137 J. M. Old, N. J. Proudfoot, W. G. Wood, J. I. Longley, J. B. Clegg and D. J. Weatherall, Cell, 14 (1978) 289.
138 Y. W. Kan, J. P. Holland, A. M. Dozy and H. E. Varmus, Proc. Natl. Acad. Sci. USA, 72 (1975) 5140.
139 S. Ottolenghi, P. Comi, B. Giglioni, P. Tolstoshev, W. G. Lanyon, G. J. Mitchell, R. Williamson, G. Russo, S. Musumeci, G. Schiliro, G. A. Tsistrakis, S. Charache, W. G. Wood, J. B. Clegg and D. J. Weatherall, Cell, 9 (1976) 71.
140 F. Ramirez, J. V. O'Donnell, P. A. Marks, A. Bank, S. Musumeci, G. Schiliro, G. Pizzarelli, G. Russo, B. Luppis and R. Gambino, Nature, 263 (1976) 471.
141 R. A. Flavell, R. Bernards, J. M. Kooter, E. De Boer, P. F. R. Little, G. Annison and R. Williamson, Nucl. Acids Res., 6 (1979) 2749.
142 A. V. Roberts, D. J. Weatherall and J. B. Clegg, Biochem. Biophys. Res. Commun., 47 (1972) 81.
143 J. W. White, A. Lang, P. A. Lorkin, H. Lehmann and J. Reeve, Nature New Biol., 235 (1972) 208.
144 G. M. Edington and H. Lehmann, Br. Med. J., 1 (1955) 1308.
145 Y. W. Kan, J. P. Holland, A. M. Dozy, S. Charache and H. H. Kazazian, Nature, 258 (1975) 162.
146 B. G. Forget, D. G. Hillman, H. Lazarus, E. F. Barell, E. J. Benz, Jr., C. T. Caskey, T. H. J. Huisman, W. A. Schroeder and D. Housman, Cell, 7 (1976) 323.
147 A. Lang, H. Lehmann and P. A. King-Lewis, Nature, 249 (1974) 467.
148 A. Lang and P. A. Lorkin, Br. Med. Bull., 32 (1976) 239.
149 A. E. Romero-Herrera, H. Lehmann, K. A. Joysey and A. E. Friday, Phil. Trans. Roy. Soc. B, 283 (1978) 61.
150 P. J. Farrell, T. Hunt and R. J. Jackson, Eur. J. Biochem., 89 (1978) 517.
151 S. H. Orkin, J. M. Old, D. J. Weatherall and D. G. Nathan, Proc. Natl. Acad. Sci. USA, 76 (1979) 2400.
152 F. E. Baralle, C. C. Shoulders and N. J. Proudfoot, Cell, 21 (1980) 521.
153 J. L. Slightom, A. E. Blechl and O. Smithies, Cell, 21 (1980) 627.
154 R. A. Spritz, J. K. deRiel, B. G. Forget and S. M. Weissman, Cell, 21 (1980) 639.
155 R. M. Lawn, A. Efstratiadis, C. O'Connell and T. Maniatis, Cell, 21 (1980) 647.
156 N. Proudfoot, Nature, (1980) 840.
157 Y. Nishioka, A. Leder and P. Leder, Proc. Natl. Acad. Sci. USA, 77

# REFERENCES

(1980) 2806.
158  E. F. Vanin, G. I. Goldberg, P. W. Tucker and O. Smithies, Nature, 286 (1980) 222.
159  E. F. Fritsch, R. M. Lawn and T. Maniatis, Cell, 19 (1980) 959.
160  A. Efstratiadis, J. W. Posakony, T. Maniatis, R. M. Lawn, C. O'Connell, R. A. Spritz, J. K. deRiel, B. G. Forget, S. M. Weissman, J. L. Slightom, A. E. Blechl, O. Smithies, F. E. Baralle, C. C. Shoulders and N. J. Proudfoot, Cell, 21 (1980) 653.
161  J. Lauer, C-K. J. Shen and T. Maniatis, Cell, 20, (1980) 119.
162  N. J. Proudfoot and T. Maniatis, Cell, 21, (1980) 537.
163  J. C. Chang and Y. W. Kan, Proc. Natl. Acad. Sci. USA 76 (1979) 2886.
164  Y. W. Kan and A. M. Dozy, Science, 209 (1980) 388.
165  A. J. Jeffreys, V. Wilson, D. Wood, J. P. Simons, R. M. Kay and J. G. Williams, Cell, 21 (1980) 555.
166  M. F. Perutz, J. V. Kilmartin, K. Nishikura, J. H. Fogg, P. J. G. Butler and H. S. Rollema, J. Mol. Biol., 138 (1980) 649—670.
167  J. R. Shaeffer, M. J. McDonald and H. F. Bunn, Trends Biochem. Sci., 6 (1981) 158.
168  N. J. Proudfoot, M. H. M. Shander, J. L. Manley, M. L. Gefter and T. Maniatis, Science, 209 (1980) 1329.
169  R. A. Spritz, P. Jagadeeswaran, P. V. Choudhari, P. A. Biro, P. A. Elder, J. T. de Riel, J. K. Manley, M. L. Gefter, B. G. Forget and S. M. Weissmann, Proc. Natl. Acad. Sci. USA, 18 (1981) 2455.
170  S. H. Orkin, S. C. Goff and R. L. Hechtman, Proc. Natl. Acad. Sci. USA, 78 (1981) 5041.

*Chapter 5*

# The Plasma Lipoproteins
# Their Formation and Metabolism

### J. PAUL MILLER and ANTONIO M. GOTTO JR.

*University of Manchester and
University Hospital of South Manchester, Manchester M20 8LR
(U.K.) and Department of Medicine, The Methodist Hospital and
Baylor College of Medicine, Houston, TX 77030 (U.S.A.)*

## 1. Introduction

Essentially all the plasma lipids, with the exception of free fatty acids which bind predominantly to albumin, travel in multimolecular complexes known as lipoproteins. The major lipoprotein classes (Table I) each contain cholesterol, cholesteryl esters, triglycerides, phospholipids and protein. They are distinguished from each other by the proportions of these constituents and the nature of the protein moiety. The physical differences which result allow the separation of lipoprotein classes by a variety of physical techniques including ultracentrifugation [1—4], electrophoresis [1, 5], gel filtration [6 –8] and differential precipitation [9, 10].

---

*Non-standard abbreviations used:* VLDL, very low density lipoproteins; IDL, intermediate density lipoproteins; LDL, low density lipoproteins; HDL, high density lipoproteins; Apo, apolipoprotein; VLDL-B, IDL-B, LDL-B, ApoB in VLDL, IDL, LDL; LPL, lipoprotein lipase; HTL, hepatic triglyceride lipase; LCAT, lecithin: cholesterol acyltransferase; ACAT, fatty acyl coenzyme A acyltransferase; HMG CoA reductase, 3-hydroxy-3-methyl glutaryl coenzyme A reductase; mRNA, messenger ribonucleic acid; FH cells, cells derived from patients homozygous for Familial Hypercholesterolemia; 4-APP, 4-aminopyrazolopyrimidine; TMU, Tetramethylurea; RER, rough endoplasmic reticulum; SER, smooth endoplasmic reticulum.

A major stimulus for the study of plasma lipoproteins has been the epidemiological association between elevated plasma lipid concentrations and coronary artery and peripheral vascular diseases [11]. The apolipoprotein B (apoB) containing lipoproteins (very low density and low density lipoproteins; VLDL and LDL) are believed to underly this relationship. The apoB-containing chylomicrons have in general not been implicated in the genesis of atherosclerosis but an interesting recent hypothesis suggests such a role for their partially metabolised remnants [12]. Recently much research has been directed towards the possibility that high density lipoporteins (HDL) may actually be protective against atherosclerosis [13].

Notwithstanding the putative role of lipoproteins in atherogenesis, their lipid components have important biological functions, as membrane constituents (cholesterol, phospholipids), as precursor for bile acids and steroid hormones (cholesterol) and as an energy source (triglycerides). The plasma lipoproteins also transport other lipid-soluble substances including vitamins [14], drugs [15] and toxins [16].

The study of the plasma lipoproteins is considerably complicated by the constant changes occurring within individual particles. In certain situations, such as the uptake of LDL by non-hepatic cells, the lipoprotein is metabolised as a whole, but frequently different lipoprotein constituents have different metabolic fates. Individual lipoprotein components undergo exchange between lipoprotein classes and in the case of unesterified cholesterol and phospholipids with cell membranes also [16A]. Specific exchange proteins are probably involved in cholesteryl ester, triglyceride and phospholipid transfer. Superimposed on these exchanges are the metabolic transformations mediated by the enzymes lipoproteinlipase (LPL) and lecithin: cholesteryl acyltransferase (LCAT).

Several recent reviews concerned with plasma lipoproteins have appeared [17—22]. This review deals predominantly with metabolism rather than structure and emphasizes the metabolism of the apolipoproteins.

## 2. Lipoprotein composition

Approximate compositions for the major plasma lipoprotein

Approximate compositions of major plasma lipoprotein classes

| | Chylo-microns | VLDL | IDL | LDL | HDL$_2$ | HDL$_3$ |
|---|---|---|---|---|---|---|
| Density (g/ml) | <0.95 | 0.95–1.006 | 1.006–1.019 | 1.019–1.063 | 1.063–1.125 | 1.125–1.210 |
| Flotation rate[a] | | | | | | |
| $S_f$ | >400 | 20–400 | 12–20 | 0–12 | — | — |
| $F_{1.21}$ | — | — | — | — | 4.8–6.1 | 1.7–4.1 |
| $F_{1.20}$ | — | — | — | — | 3.5–9.0 | 0–3.5 |
| Diameter (Å) | 800–5000 | 300–800 | 210–350 | 200–250 | 95–105 | 65–95 |
| Electrophoretic mobility | Origin | Pre-$\beta$ | Pre$\beta$-$\beta$ | $\beta$ | $\alpha$ | $\alpha$ |
| Percentage Composition (w/w)[b] | | | | | | |
| Triglyceride | 85–90 | 50–60 | 18–25† | 7–16 | 3–9 | 2–5 |
| Cholesterol | 2–6 | 14–22 | 30–38 | 40–50 | 19–30 | 13–22 |
| Phospholipid | 6–9 | 12–20 | 20–28 | 24–32 | 28–40 | 22–29 |
| Protein | 1–2 | 5–15 | 18–24 | 17–25 | 31–43 | 49–58 |
| Apoproteins | | | | | | |
| Major | apoB, apoC (apoA-I)[c] | apoB, apoC, apoE | apoB | apoB | apoA-I | apoA-I, apoA-II |
| Minor | apoA-II, apoA-IV (apoE)[d] | apoA-I, apoA-II | apoC, apoE | | apoA-II, apoC, apoE | apoC, apoD |

[a] Flotation rates for lipoproteins of d<1.063 are in Svedbergs ($S_f$, $10^{-13}$ cm sec$^{-1}$ dyn$^{-1}$ g$^{-1}$) and are measured at d 1.063. For lipoproteins of d>1.063 flotation rates have been variously measured at d 1.20 ($F_{1.20}$) and d 1.21 ($F_{1.21}$) [1, 40]. [b] Values are not normal ranges, but are intended to give an approximation to the range of mean normal values reported in the literature. [c] Apoprotein composition of lymph chylomicrons. In plasma, ApoA-I is a minor component and apoE a more major component. [d] Based on data for IDL (LpIII) of Type III patients [34].

classes are given in Table I. For a detailed review of their lipid composition consult ref. 23. There is a trend to higher density (i.e. lower lipid: protein ratio) as particle size decreases. By virtue of their composition, in normal post-absorptive plasma VLDL are the major determinants of plasma triglyceride concentration, while LDL are the major determinants of plasma cholesterol concentration. During the absorption of dietary triglyceride and in certain disease states chylomicrons also contribute significantly to plasma triglycerides, but have little influence on cholesterol concentrations, unless present in enormous quantities.

## 2.1. Chylomicrons

Chylomicrons are predominantly triglyceride. Their composition varies with their size and whether they are isolated from lymph or plasma. Smaller particles contain a greater proportion of relatively polar lipoprotein surface constituents (phospholipid, protein) and less triglyceride [24, 25]. The ratio of apoB to non-apoB apolipoproteins in chylomicrons from chylous effusions is greater in larger particles [25]. This contrasts with the situation in VLDL. The pattern of non-apoB proteins seems to be relatively independent of size [25]. VLDL isolated from chylous effusions contain relatively more apoB than small chylomicrons, an argument for regarding them as a distinct entity rather than part of a continuous spectrum with chylomicrons [25]. This is supported by the bimodal size distribution of particles in intestinal Golgi saccules [26]. Changes in chylomicron composition as they transfer from lymph to plasma are discussed in section 6.1.

The fatty acids of chylomicron triglycerides are very dependent on the composition of the diet. The fatty acid pattern in chylomicron phospholipids is much less sensitive to short-term dietary effects [27]. There is some disagreement as to the effect of dietary fat saturation on chylomicron size. It may be that chylomicron size is a function of the amount of apoprotein and other surface constituents available per cell for a given triglyceride flux. At high absorption rates more triglyceride could be transported by a given amount of apoprotein in larger particles simply from surface-volume considerations [28]. In accordance with this, inhibition of protein synthesis leads to an increase in chylomicron size [29].

The observation that palmitic acid requires a greater length of intestine for absorption than linoleic might then explain the increase in chylomicron size which some workers have reported with unsaturated dietary fatty acids [30].

### 2.2. Very low density lipoproteins

Very low density lipoproteins are heterogeneous and, like chylomicrons, their composition varies progressively through their size range of 300—750 Å. Smaller particles contain less triglyceride and relatively more phospholipid, protein and cholesterol [2, 6]. The protein component comprises an increasing proportion of apoB and a decreasing proportion of tetramethylurea-soluble apoproteins (mainly apoC proteins) as particle size decreases [2, 31].

### 2.3. Intermediate density lipoproteins

Intermediate density lipoproteins (IDL) of d 1.006—1.019 and $S_f$ 12—20 are referred to as $LDL_1$ in some of the older literature, and as $LDL_2$ by one group of workers [32]. They are present in low concentrations or are absent from the fasting plasma of normal individuals [1, 33, 34], but their concentration increases during alimentary lipemia [35], or after the intravenous injection of heparin [36]. A lipoprotein in this class (termed LpIII) accumulates in the plasma of type III hyperlipoproteinemia from which it may be isolated by rate zonal ultracentrifugation [34, 37].

The physical and chemical properties of IDL are generally intermediate between those of VLDL and LDL [34, 37] and IDL are believed to represent a stage in the metabolic conversion of VLDL to LDL [38, 39].

### 2.4. Low density lipoproteins

Low density lipoproteins are much more homogeneous with respect to size and composition than chylomicrons or VLDL, though some heterogeneity exists. ApoB makes up almost all the protein component of LDL, although other lipoprotein species, such as

*References p. 489*

Lp(a) (see below), which contain other apoproteins, may be isolated at densities overlapping the LDL range in certain individuals.

## 2.5. High-density lipoproteins

The major subfractions of HDL in human plasma are designated $HDL_2$ and $HDL_3$ (Table I). Of these, $HDL_3$ at least, are also heterogeneous [40]. Not only is the lipid: protein ratio of $HDL_2$ greater than that of $HDL_3$, befitting their differences in density, but there are also differences in apoprotein composition. Although not universally agreed, most authors have found the ratio of apoA-I: apoA-II to be higher in $HDL_2$ than $HDL_3$ [41]. ApoE can be traced to a specific electrophoretically separable minor subfraction of $HDL_2$ [42].

$HDL_1$ are a minor subfraction of $S_f$ 0—3 which may contaminate LDL floated at d 1.063. They have $\alpha_2$-mobility on electrophoresis [23, 33]. The ultracentrifugal fraction of d>1.210 contains the plasma proteins and a small amount of lipid in addition to free fatty acids. Lipoprotein complexes (termed very high density lipoproteins or VHDL) have been isolated from this fraction, though it is possible that their composition is, at least partially, an artefact of ultracentrifugation (see ref. 23 for discussion). Apolipoproteins, notably apoA-I and apoE [21], and LCAT are also demonstrable in the d 1.210 infranatant. Again it is uncertain to what extent their presence here is a consequence of ultracentrifugation, but it is notable that their concentrations may be increased in the bottom fraction by repeated centrifugations.

## 2.6. Lp(a) lipoproteins

Lp(a) constitute a separate class of lipoprotein particles isolated in the density range 1.05—1.08, thus overlapping LDL and $HDL_2$. The subject has recently been reviewed [43, 44]. Lp(a) consist of spherical particles with similar lipid composition to LDL. Lp(a) particles, however, are larger than LDL and have a lower lipid: protein ratio [45]. Thus they represent an exception to the usual tendency of lipoprotein particles of higher density (and lower lipid: protein ratio) to have smaller size. The protein component of Lp(a) comprises approximately 65% apoB, 20% the specific Lp(a) apoprotein and 15% albumin [46]. There is some doubt as to whether the albumin is an integral part of the lipoprotein [43, 47]. Lp(a) are particularly rich in carbohydrate and this is attributed to the specific Lp(a) protein [46], the amino acid composition of which has been reported [47]. Its $M_r$ determined by gel filtration is 35 000—40 000 [47].

Lp(a) have prebeta mobility on electrophoresis in 0.5% agarose and are believed to be identical to "sinking" prebeta lipoprotein [48] and prebeta-1 lipoproteins [49]. In general the presence of "sinking" prebeta lipoproteins (lipoproteins with prebeta mobility and d>1.006) or of prebeta-1 lipoproteins (an extra prebeta band of slower mobility than VLDL) are usually, but not always* accompanied by detectable Lp(a) as shown immunochemically [48—50]. Other individuals, however, may have sufficient Lp(a) to be detectable immunologically but not electrophoretically [48—50].

A tendency of prebeta-1 lipoproteins to be associated with clinical ischemic heart disease was first reported in 1971 [52] and this has been amply confirmed [43, 53, 54]. Possibly relevant to this association are the observations that Lp(a) can be demonstrated in the arterial wall [55] and that they can be precipitated by calcium ions in the presence and absence of glycosaminoglycans [56, 57]. Subjects with high Lp(a) concentrations also exhibit other metabolic differences from those with lower or zero concentrations such as a tendency to lower serum insulin concentrations during glucose tolerance tests [58].

Little is known of the metabolism of Lp(a). Their concentration is relatively constant in a given individual and they are probably metabolized independently of LDL, since several factors which altered plasma apoB concentration (cholesterol feeding, oral contraceptives, clofibrate) were without effect on Lp(a) [50, 54].

Family studies show little evidence for a major environmental role in the determination of Lp(a) concentrations [59] and show them to have a high degree of heritability [59, 60]. There is, however, some controversy over the distribution of Lp(a) concen-

---

*A second prebeta band which floats at d 1.006 may also occur "double floating prebeta" [51] which is distinct from the prebeta-1 or "sinking prebeta" due to Lp(a) lipoprotein.

tration in the population and the nature of its inheritance. This subject is discussed in ref. 43. In general, workers who have used double immunodiffusion have found about 30% of the population studied (mainly in Western Europe) to possess Lp(a). Others, using quantitative immunochemical methods, such as radial immunodiffusion [50, 59, 61], electroimmunoassay [55] and radioimmunoassay [54] have found a much higher proportion of the population to possess Lp(a) and have implied that with sufficiently sensitive methods virtually all normal individuals would be found to possess some Lp(a).

Those studies which have used electrophoretic or double diffusion techniques for detection of Lp(a) (and therefore classify individuals as either Lp(a+) or Lp(a—)) have concluded that the possession of Lp(a) is governed by a single autosomal dominant [43]. Among those who have quantitated Lp(a), some have observed a bimodal distribution and favour control of Lp(a) levels by a major genetic locus [62, 63], while others have found a continuously distributed variable with a positive skew [50, 54, 55] and favour polygenic inheritance, although it was pointed out that major gene effects could not be excluded [59].

### 2.7. Lipoprotein-X

It has been known for more than 100 years that patients with cholestatic (obstructive) liver disease have an elevation of serum cholesterol, and for more than 50 years that this cholesterol is predominantly unesterified. More recently, these changes have been shown to be largely due to an abnormal lipoprotein with the flotation characteristics of an LDL, but which does not react with antisera against apoLDL (apoB) [64, 65].

This abnormal lipoprotein has been termed lipoprotein-X [66] and unlike all other lipoproteins it migrates cathodically on agar gel electrophoresis. On agarose it has $\beta$-migration. The particle contains unesterified cholesterol and phospholipid, predominately phosphatidylcholine, in a 1:1 molar ratio. It possesses little of the neutral lipid (triglyceride and cholesteryl esters) normally found in the core of lipoprotein particles. Typical figures for the composition by weight of LP-X in cholestasis are phospholipid 65%, unesterified cholesterol 25%, cholesteryl esters 1%, triglycerides 3%

and protein 6% [67]. Although lacking apoB the protein component of LP-X contains apoC proteins, albumin, apoA-I and apoE [68]. The fact that the intact particle does not usually react with antisera against albumin while the delipidated protein component does, is probably explicable by the structure of LP-X. It appears to be made up of cholesterol-containing phospholipid bilayers in the form of vesicles 400—600 Å in diameter and about 100 Å thick [69]. It is presumed that albumin is trapped in an aqueous environment within the centre of the vesicles. While LP-X particles viewed by electron microscopy frequently appear as discs in rouleau formation, their behaviour on centrifugation or gel filtration suggests that they do not normally aggregate this way in solution [68]. Recently three different species of LP-X have been identified, but their metabolic relationships remain to be defined [68].

Although LP-X usually appears to be fairly specific for cholestatic liver disease [70], it is also present in patients with familial LCAT deficiency (who do not have liver disease) and it can be induced by cholesterol feeding in experimental animals such as the guinea-pig [67]. The formation of LP-X may represent an imbalance between the input of "surface lipid" (unesterified cholesterol and phospholipid; see next section) into plasma and the ability of LCAT to metabolise it [67]. In familial LCAT deficiency some of this "surface lipid" will be associated with dietary triglyceride in chylomicrons and dietary fat restriction has been found to reduce the concentration of LP-X [71]. Similarly in the cholesterol fed animal, LP-X may result from an excessive load of chylomicron "surface lipid" together with more or less normal LCAT activity, though other mechanisms are possible [67]. In the patients with cholestasis the excess lipid may derive from increased hepatic synthesis and regurgitation into plasma of biliary lipid [72]. In addition some patients will have reduced plasma LCAT activity, especially when obstruction is prolonged [73].

It appears therefore that LP-X may accumulate as a result of relative or absolute LCAT deficiency. Being rich in unesterified cholesterol and lecithin, and possessing the LCAT activators apoA-I (this has not been found by some authors) and apoC-I, one would predict that LP-X would be an LCAT substrate. Certainly LP-X containing plasma is a good substrate for LCAT and esterification is associated with the disappearance of LP-X [74]. Under these circumstances, however, the LP-X could be acting as a source

*References p. 489*

of substrate lipids for transfer to other lipoproteins. It is nonetheless capable of acting as an LCAT substrate when incubated with the partially purified enzyme in the absence of other lipoproteins [75].

## 3. The apolipoproteins

The functions of the individual apolipoproteins are not completely understood although much progress has been made in recent years. They appear to confer many of the specific properties possessed by the individual lipoprotein classes in which they occur. For example, secific apoproteins regulate the activity of the major enzymes involved in lipoprotein metabolism in plasma, and are necessary for the secretion of triglyceride-rich lipoproteins by both liver and intestine. The uptake and subsequent degradation of lipoproteins by cells in culture (and presumably *in vivo*) is also dictated by their apoprotein content. In addition, the apoproteins play an important structural role in lipoproteins. They are believed to occupy the outer part of a 22 Å shell on the surface of the particles where they are adjacent to the polar head groups of phospholipids. Free cholesterol is also believed to occupy this outer shell while non-polar cholesteryl esters and triglycerides are found within the hydrophobic interior of the particles (Fig. 1). Thus, the particles have structural analogies with the micelles of the intestinal lumen and both serve to transport water-insoluble lipid in an aqueous environment. In this context, the apolipoproteins have been termed the "bile salts of the blood" [76]. This property is conferred by the primary structure of the apoproteins which causes parts of the molecules to adopt an alpha-helical conformation with amphipathic properties [77, 78]. With this arrangement, the non-polar residues occupy one face of the helix and point towards the hydrophobic interior of the lipoprotein, probably interacting with the first few carbons of the fatty acyl chains of the phospholipids. The acidic residues (glutamic and aspartic acids) occupy the opposite face and are oriented towards the aqueous environment. The basic amino acids (lysine, arginine) occupy a position in the helix at the borders of the polar and non-polar faces. Experimental evidence for an important structural role for ionic interactions between the polar amino acids and the polar head groups of the phospholipids has not been produced.

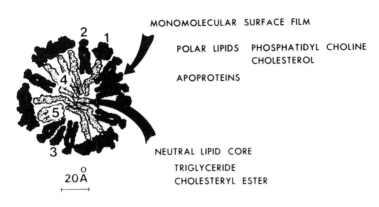

Fig. 1. Common features of lipoprotein structure. The monomolecular surface film is believed to consist of those constituents which have amphipathic properties, namely the apoproteins (1), phospholipids (2) and unesterified cholesterol (3). The core of the particles is made up of cholesteryl esters (4) and triglycerides (5).

Detailed discussion of lipoprotein and apolipoprotein structure can be found in refs. 18—20 and 79—82. The measurement of individual apoproteins within lipoproteins or plasma has also been recently reviewed [83, 84].

### 3.1. Apolipoprotein A-I

Apo A-I is secreted both by the intestine and the liver. Although appreciable quantities are secreted in association with triglyceride-rich lipoproteins (chylomicrons, VLDL), almost all the apoA-I in normal human plasma is associated with HDL, of which it is the major protein constituent (60—70% of apoHDL). ApoA-I activates LCAT (section 4.2.). It is a single polypeptide chain of $M_r \simeq 28\,000$. Two versions of its primary structure have been reported [85—88] and the differences between them remain to be resolved. ApoA-I contains no cysteine, cystine, isoleucine, or carbohydrate. It has carboxy-terminal glutamine and was formerly termed apoLp-Gln-I or R-Gln-I.

Only a small proportion (<20%) of the apoA-I present in HDL (or plasma) is normally detectable in immunoassays [89—92]. It is believed that this is the result of the masking of some of its immunoreactive sites by lipid or protein in intact HDL [93, 94]. Detection of the full apoA-I complement in HDL has generally required disruption of the particle with a variety of physical or chemical treatments [89—91], though a recent alternative approach has been the isolation by affinity chromatography of those antibody species which react only with the immunoreactive determinants of apoA-I normally exposed in HDL [92]. These "surface-specific" antibodies then react identically with intact HDL or apo-HDL.

### 3.2. Apolipoprotein A-II

Apo A-II (apoLp-Gln-II, R-Gln-II) is also secreted by liver and intestine and is found predominantly in plasma HDL of which it is the second most abundant protein constituent (approximately 20 –30% of apoHDL). In man it consists of two identical polypeptide chains, each of 77 amino acids, linked at cystine-6. The amino acid sequence has been determined and the calculated $M_r$ = 17 380 [95]. In several animal species, including the rat and the monkey, apoA-II exists as a single polypeptide chain with half the molecular weight of human apoA-II. Human apoA-II contains no histidine, arginine, tryptophan or carbohydrate [95]. Work with native and synthetic peptides has identified residues 12—31 and 47—77 as areas capable of binding phospholipids [96, 97]. Unlike apoA-I, all the apoA-II in HDL is detectable in immunoassays [93, 98].

### 3.3. Apolipoprotein B

ApoB is secreted by the liver and intestine and is a constituent of chylomicrons, VLDL and LDL. Indeed it appears to be necessary for the secretion of these particles since patients with abetalipoproteinemia lack these lipoproteins in plasma and accumulate lipid in the intestinal mucosa and liver [99]. Moreover, apoB specifically determines the binding and uptake of LDL by cells in tissue culture (Section 6.9.1.) and is likely therefore to be important in the metabolism of LDL in vivo. ApoB is also found in Lp(a) (Section 2.6.). It has recently become apparent that apoB exists in multiple forms in both man [99A] and the rat [99B].

Despite its relative abundance in plasma, the insolubility of the delipidated apoprotein has greatly hindered progress in the elucidation of its structure, which is one of the major current problems in the study of the plasma lipoproteins. It is a glycoprotein but there is little agreement about its molecular weight and its primary structure is unknown.

## 3.4. Apolipoprotein C

The apoC-proteins (apoC-I, apoC-II and the polymorphic apoC-III) are secreted predominantly by the liver although there is probably some intestinal contribution particularly for apoC-II (Section 5.1.3.). They are constituents of chylomicrons, VLDL, HDL and LP-X. A role for apoC-proteins in regulating the clearance of remnants of triglyceride-rich lipoproteins has been reported (Section 6.1.2.).

ApoC-I (apoLp-Ser, R-Ser), has a known primary structure and is a single polypeptide chain with 57 amino acids and a calculated $M_r$ of 6530 [100, 101]. It lacks histidine, cysteine, cystine and tyrosine. As a result of the absence of tyrosine, apoC-I is not labeled with radioiodine by many of the methods in common use for metabolic studies. Both cyanogen bromide fragments (residues 1—38 and 39—57) are capable of binding phospholipids [102]. ApoC-I, in addition to apoA-I, activates plasma LCAT (Section 4.2.). The complete apoC-I sequence has been synthesized by solid phase methods [103, 104]. The synthetic peptide binds phospholipid and activates LCAT similarly to the native peptide [103].

ApoC-II (apoLp-Glu, R-Glu) is a single polypeptide chain with 78 residues and a calculated $M_r = 8\ 837$. It lacks cystine, cysteine and histidine [105]. The ability of apoC-II to activate LPL (Section 4.1.1.) resides in the C-terminal 24 residues [106], while residues 43—55 are able to bind phospholipid [107].

ApoC-III (apoLp-Ala, R-Ala) is a single polypeptide chain of 79 residues and calculated $M_r = 8\ 751$. It possesses a carbohydrate side chain at threonine-74 consisting of galactose and galactosamine residues and 0—2 moles of sialic acid [115]. The variability in sialic acid content is responsible for the polymorphism of apoC-III on polyacrylamide gels (apoC-III$_0$, apoC-III$_1$, apoC-III$_2$). Lipid binding by apoC-III has been extensively studied [80]. When the molecule is cleaved by thrombin only the carboxy-terminal portion (residues 41—79) is capable of binding phospholipid [108].

### 3.5. Thin-line polypeptide

In double diffusion experiments between HDL and anti-HDL, a thin precipitin line can be identified close to the antigen well. Two reports of the isolation and partial characterisation of the thin-line polypeptide have appeared terming it variously apoD [109] and apoA-III [110]. The different nomenclature stems from uncertainty as to whether the thin-line polypeptide should be regarded as a constituent of a new lipoprotein family or not. The reported amino acid compositions of apoD and apoA-III differ and the two proteins may be distinct. The $M_r$ of apoA-III is approximately 20 000. Thin-line peptide has been reported to activate LCAT [111], but while other workers could not exclude this, they were unable to confirm it [112]. It has been suggested that apoD and the cholesteryl ester exchange protein are one and the same [113].

### 3.6. Apolipoprotein E

ApoE (arginine-rich protein) has about 9% of its amino acids as arginine. There is some disagreement as to the presence of half-cystine [32, 114, 116]. Reported $M_r$ values vary between 33 000 and 44 000 daltons [114]. Although first isolated from VLDL [32, 117], apoE has now been identified in all major density classes including the plasma protein fraction of d>1.21 [114, 118], though the latter may at least partly represent an artefact of ultracentrifugation. Others have detected little or no apoE in normal LDL or HDL [117, 119]. ApoE can be separated into several distinct isoforms by isoelectric focusing [120]. The structural difference between the different apoE isoforms is unknown. Although postulated that it might be due to differences in sialic acid content [121], this now seems unlikely since desialylation with neuraminidase does not influence the isoelectric focusing pattern of the various isoforms [122]. ApoE probably plays an important part in the removal of remnants of triglyceride-rich lipoproteins by the liver (Section 6.1.2.), and it enables lipoproteins to bind to the "LDL receptor" of fibroblasts and other cell types (Section 6.9.).

### 3.7. Proline-rich protein

A proline-rich protein has recently been identified in chylomicrons and in the ultracentrifugal fraction of d>1.21 [123].

## 4. Enzymes and exchange proteins

### 4.1. Plasma triglyceride hydrolases

Several reviews have been published [124–129, 129A]. Under normal circumstances human plasma contains little or no triglyceride hydrolase activity. There is a rapid (maximal within a few minutes) increase in such activity after the intravenous injection of heparin, which includes not only triglyceride, but also diglyceride and monoglyceride hydrolase and phospholipase activities. At least two enzymes or groups of enzymes are involved. These are lipoprotein lipase (LPL) released from a number of extrahepatic tissues (including heart, skeletal muscle, adipose tissue, lung and mammary gland), and hepatic triglyceride lipase (HTL) derived from the liver.

*4.1.1. Lipoprotein lipase.* LPL (or clearing factor lipase, EC 3.1.1.3) has been purified from a variety of tissues and species with apparent $M_r$ in the range 34 000—73 000 [20]. To what extent these differences are artefacts of isolation methods remains to be determined. There seems little doubt that different forms of lipoprotein lipase exist. Both high ($M_r$ = 69 250) and low ($M_r$ = 37 000) molecular weight forms can be isolated from rat post-heparin plasma [130, 131], and it seems that the former may correspond to the low affinity enzyme from adipose tissue, ($K_m'$ = 0.70 mM triglyceride) and the latter to the cardiac high affinity enzyme ($K_m'$ = 0.07 mM triglyceride) [132]. At normal plasma triglyceride concentrations

(<1.5 –2.0 mM), these kinetic differences imply that triglyceride clearance by the heart will be independent of substrate concentration whereas the clearance in adipose tissue will depend on availability of substrate. There are, however, immunological similarities between lipoprotein lipases of different tissues [128] and species [20]. LPL has a broad specificity, being active against triglycerides, diglycerides and 1(3) monoglycerides over a wide range of fatty acid chain lengths. The enzyme also has phopholipase activity. LPL and HTL have been reported to have identical peptide maps and amino acid compositions [133, 134], but this is now believed to be incorrect [135, 136]. The fact that they are immunologically distinct has frequently been used to separate their activities in post-heparin plasma.

LPL is readily inhibited by 1 M NaCl and protamine sulphate and requires apoC-II for maximal activity, whereas HTL exhibits none of these properties [137—139]. In vitro apoC-II forms a complex with the enzyme, but this is prevented by high salt concentrations [140]. As might be expected from this, NaCl quantitatively reverses the apoC-II-dependent activation of LPL, but it does not inhibit any basal apoC-II-independent activity, nor does it dissociate the enzyme-substrate complex [141]. The effect of salt is dependent on the anion and independent of the cation [141]. ApoC-II is a normal constituent of the triglyceride-rich chylomicrons and VLDL, the substrate lipoproteins for LPL. Reaction rates are greater with chylomicron triglyceride than with VLDL [126]. HDL are seen as a reservoir for apoC-II (and other apoC proteins) which transfer to newly secreted chylomicrons and VLDL, but which are returned to HDL during the course of lipolysis (Section 6.1.). ApoC-II maximally activates LPL in a 1:1 molar ratio [126, 142]. Studies with cyanogen bromide fragments and synthetic peptides suggest that residues 55—78 contain the minimal sequence necessary for LPL activation [106]. This accords with the observation that the C-terminal tryptic peptide (residues 50—78), but not the N-terminal, possesses activating activity [143]. An apoC-I stimulated LPL has been reported [144], though generally apoC-I and apoC-III inhibit purified LPL preparations [145]. The physiological roles of apoC-I and apoC-III in relation to LPL are unknown.

Recently patients have been described with apoC-II deficiency [146, 147] who have phenotypic type I hyperlipoproteinemia

[148]. Inheritance is believed to be autosomal recessive [147, 149]. The lack of apoC-II causes a functional defect in LPL activity and an accumulation of unhydrolysed chylomicrons. Transfusion of one of the patients with normal blood or plasma (which would not contain appreciable LPL activity, but would contain apoC-II) led to a rapid reduction in serum triglyceride levels even though plasma apoC-II concentrations were still very low [146]. This observation and the normal triglyceride concentrations found in heterozygotes despite low apoC-II levels [147] suggest that apoC-II concentrations are not limiting in triglyceride-clearance in normal individuals. Synthetic fragments of apoC-II (residues 50—78 and 43—78) can promote the hydrolysis of triglyceride in apoC-II-deficient VLDL by purified LPL in vitro [150]. It is possible that such fragments may in future be used as a rational therapy for apoC-II deficiency. These patients, for reasons which are unclear, also seem to have reduced HTL in post-heparin plasma [146, 147].

Other patients with familial type I hyperlipoproteinemia do not lack apoC-II, but are deficient in LPL or have an ineffective form of the enzyme [148, 151]. They do, however, possess appreciable, frequently normal, HTL activity in post-heparin plasma [152, 153].

A superficial site for LPL activity in relationship to the vascular endothelium is favoured by the rapidity with which enzyme is released into the circulation after the injection of heparin and the virtually immediate rise in plasma triglycerides provoked by the intravenous injection of an antibody to adipose-tissue LPL [154]. The kinetic properties of tissue-bond LPL in rat heart and the soluble enzyme from this source are similar also suggesting a superficial location [155]. Electron microscopic studies [125] show chylomicrons partially enmeshed in projections from the luminal surface of the capillary endothelium, but have not shown the intact particles crossing the endothelium. Cytochemical studies [125] suggest that lipolytic activity is present in a system of vesicles and vacuoles which permeate the endothelial cell and establish contact with the enmeshed chylomicrons.

Although LPL is apparently active in the vascular endothelium, there is evidence to suggest that it is secreted, at least in adipose tissue, in the adipocytes and that it migrates subsequently to the vasculature. A variety of models has been proposed for this process involving active and inactive forms of the enzyme [128]. It is

*References p. 489*

apparent, therefore, that assays of enzyme in whole tissue extracts may determine activity stored in the parenchyma, inaccessible to lipoproteins, as well as that located in the vascular endothelium. The latter is believed to be the source of heparin-releasable activity in vivo or in perfused organs. It appears for example that in the rat only a small part of the heart LPL, compared to that in adipose tissue, is directly active in triglyceride hydrolysis [132].

Rapid changes in LPL activity can occur in different tissues according to the metabolic and nutritional state of the animal. These changes may provide a mechanism for regulating the distribution of triglyceride fatty acids to different tissues or organs according to the metabolic needs of the whole organism [124]. For example, there are reciprocal changes in the activities of LPL in muscle and adipose tissue with the former being increased during fasting and the latter in the fed animal. Activity in the mammary gland is increased during lactation [124, 128]. Changes in activity are sufficiently rapid for a circadian rhythm to emerge in rat cardiac and skeletal muscle in parallel with the animal's nocturnal feeding habits. How these changes are mediated is unclear and it is notable that the rhythm persists for at least 48 h after the feeding stimulus is withdrawn [156]. The hormonal regulation of muscle LPL is poorly understood [128]. In adipose tissue, however, insulin in physiological concentrations stimulates LPL activity and significant correlations have been reported between adipose tissue LPL activity and plasma insulin concentrations, both in man and animals [128].

*4.1.2. Hepatic triglyceride lipase.* The HTL activity found in human plasma does not require an apoprotein cofactor for activity and it is resistant to the inhibitory effects of salt and protamine. These differences and the use of specific antibodies [152, 153, 157] or of sodium dodecyl sulphate to inhibit HTL [158] can be used to resolve plasma post-heparin lipolytic activity into its component LPL and HTL activities. It is important to do this since their activities frequently do not change in parallel in different metabolic and pathological states [152, 153, 159—163].

Little is known of the physiological role of HTL, though it has been suggested that it may be involved in the clearance of remnant lipoproteins or LDL by the liver. HTL of post-heparin plasma is presumed to be hepatic in origin, since lipolytic activity with similar properties can be released by heparin into the perfusate of the isolated liver. Although it can hydrolyse the triglycerides of chylomicrons and VLDL in vitro, there is little evidence to indicate that HTL is an important factor in the initial metabolism of triglyceride-rich lipoproteins, since intact chylomicrons are not effectively cleared by the isolated perfused liver (Section 6.1.2.). Moreover, patients with type I hyperlipoproteinemia frequently have normal post-heparin HTL activity [152, 153] and HTL activity does not show appropriate correlations with plasma triglyceride concentrations or clearance rates [164].

Recent immunofluorescent studies in the rat with antibodies to HTL have located it on the hepatic endothelial cells, from which it was released by heparin. No enzyme could be identified on the hepatic parenchymal or Kupffer cells [165]. When the anti-HTL antibody was injected intravenously, there was an accumulation of cholesterol and phospholipid in plasma LDL and HDL implying a role for HTL in the removal of these lipids. No differences were observed in any of the components of VLDL or in LDL protein [166].

## 4.2. Lecithin: cholesterol acyltransferase

Several reviews and symposia on lecithin: cholesterol acyltransferase (LCAT; EC 2.3.1.43) and the effects of its hereditary deficiency have appeared [67, 167—172]. The best authenticated source of the enzyme is the liver, the evidence being derived from studies of hepatectomy, the isolated perfused liver and patients with liver disease. Recently the enzyme has been shown to be secreted by rat hepatocytes in culture. The microtubular system is probably involved, since secretion can be inhibited with colchicine [173]. Some enzyme may also be secreted by the intestine [174]. Although secreted by the liver, little activity has been detected in homogenates of liver or of other tissues [173, 175] and the enzyme is believed to exert its effects mainly within the circulating plasma. Esterifying activity probably due to LCAT, albeit at low levels relative to plasma, is also detectable in lymph from a variety of sources [174, 176, 177].

*References p. 489*

The chemical substrates for LCAT in plasma are unesterified cholesterol and phosphatidylcholines (lecithin) and the products of the reaction are cholesteryl esters and lysolecithin. The fatty acid in the 2-position of lecithin is preferentially transferred. The human enzyme transesterifies in order of preference linoleic, oleic, arachidonic and saturated fatty acids. Different specificities were found for the enzyme from the rat and the rabbit [178]. Some fatty acid is also transferred from the 1-position of lecithin, the proportion depending on its degree of unsaturation. The fatty acid in the 1-position also influences the rate at which fatty acid is transferred from the 2-position [179]. The enzyme is able to esterify a variety of sterol substrates including cholestanol, desmosterol and β-sitosterol [180, 181]. Although unesterified cholesterol and lecithin substrates are present in all the major lipoprotein density classes, only those in HDL appear to be effective substrates when individual lipoprotein classes are incubated with partially purified enzyme [182]. $HDL_3$ are effective substrates and $HDL_2$ relatively ineffective [183]. The principal lipoprotein substrate however may be the discoid nascent HDL particles secreted by the liver and the intestine (Sections 5.1.2. and 5.2.2.).

An absolute requirement for cofactor(s) for LCAT has been demonstrated which can be supplied by $HDL_3$-proteins [183]. Subsequent work showed that both apoA-I [184] and apoC-I [112] are able to activate the enzyme. The degree of activation by apoA-I, but not by apoC-I, depends on the fatty acid composition of the phospholipid substrate [112]. Synthetic apoC-I activates LCAT as effectively as the native protein [103]. Synthetic fragments of apoC-I have also been tested and residues 17—57 are almost as effective as native apoC-I in activating LCAT. Residues 32—57 also contain a major phospholipid binding area [185]. ApoA-II impairs the ability of apoA-I to activate the enzyme [112, 184]. Apo A-III has been reported to activate LCAT, but in these studies activation due to apoA-I was not demonstrated [111]. The question has recently been re-examined using highly purified, apolipoprotein-free LCAT [186]. The activating effects of apoA-I and apoC-I were confirmed. ApoD could not be shown to activate the enzyme. There is, however, some doubt whether the proteins called apoD and apoA-III are truly identical (Section 3.5.).

LCAT is believed to catalyse the formation of most of the circulating cholesteryl esters [167], since the rate of formation and the fatty acid pattern of esters formed by LCAT in vitro are similar to calculated in vivo esterification rates and the fatty acid patterns of freshly isolated plasma cholesteryl esters. Most convincing is the very low proportion of esterified cholesterol in patients with familial LCAT deficiency. Those esters which are present probably originate in the intestine [170]. The patients provide further evidence of the importance of LCAT from the extensive clinical and lipoprotein abnormalities which they exhibit [170, 187]. Clinical manifestations include diffuse corneal opacities, mild anemia with the presence of target cells, proteinuria and the onset of renal failure in middle life. The lipoprotein abnormalities have been described in detail [67, 170, 187, 188] and will be discussed here (see below) only in so far as they have provided information about the role of LCAT in normal lipoprotein metabolism.

The role of LCAT in overall lipid transport is uncertain. One hypothesis proposes [167] that it is involved, together with HDL, in the net transport of cholesterol from peripheral tissues to the liver "reverse cholesterol transport"), which is necessitated by the ability of most tissues to synthesise cholesterol, but not to degrade it. The postulated mechanism is that unesterified cholesterol in the surface of HDL is in equilibrium with that of cell membranes and other lipoproteins. LCAT by removing some of this surface cholesterol and converting it to cholesteryl esters in the core of HDL, and secondarily in other lipoprotein fractions also, allows HDL to take up further cellular or lipoprotein cholesterol. The cholesteryl esters formed by LCAT were presumed to be removed by the liver. In vitro incubations with red cells show that LCAT-containing plasma can be used to deplete cellular cholesterol [189] and the presence of LCAT and lipoprotein in lymph [174, 176, 177, 190] make a similar process theoretically possible for most cell types. Moreover the red cells and tissues of LCAT-deficient patients contain excess unesterified cholesterol [172, 187]. However, it has not been possible to show that LCAT facilitates HDL-dependent removal of cholesterol from fibroblasts or smooth muscle cells in tissue culture [191]. The difficulties of examining the "reverse cholesterol transport" hypothesis using cells in culture have recently been summarised [172]. "Reverse cholesterol transport" based on the hypothetical mechanism outlined here might theoretically underly the negative relationship

*References p. 489*

between HDL concentrations and both tissue cholesterol pools and the incidence of coronary artery disease [192, 193].

Another hypothesis [194] of the physiological role of LCAT sees it as serving to metabolise and remove surface components (unesterified cholesterol and lecithin) from triglyceride-rich lipoproteins pari-passu with the removal of the triglyceride-core by LPL. In this way normal surface-volume relationships might be preserved. Since VLDL lipids are not a substrate for LCAT, the mechanism would probably have to involve HDL as the substrate lipoprotein and transfer of unesterified cholesterol and lecithin from VLDL to HDL and cholesterol esters from HDL to VLDL. This hypothesis does not exclude the hypothesis implicating LCAT in "reverse cholesterol transport", indeed there are some parallels between the two hypotheses if the surface lipid of triglyceride-rich lipoproteins is regarded as being analogous to cell membrane lipid. Support for an important role of LCAT in the metabolism of triglyceride-rich lipoproteins has been obtained from dietary experiments in patients with familial LCAT deficiency [71]. Restricting dietary fat caused major reductions in at least three of the abnormal circulating lipoprotein species. These were the large-sized VLDL and the large and intermediate-sized LDL. It is presumed, therefore, that these lipoproteins represent the abnormal accumulation of intermediates of chylomicron metabolism due to LCAT deficiency. The large VLDL with their irregular surfaces [188] had many of the ultrastructural features of chylomicron remnants generated experimentally [125]. Exactly how the large and intermediate-sized LDL might be formed from chylomicrons is unclear. The intermediate LDL resembles the abnormal LDL of cholestatic liver disease (LP-X, Section 2.7.). Attempts to identify circulating remnants of VLDL metabolism were made by feeding the patients a high carbohydrate diet to stimulate endogenous triglyceride production and VLDL secretion by the liver [71]. The results of these experiments were inconclusive.

The production of chylomicron surface lipid may exceed the capacity of LCAT to metabolise it, particularly, presumably, during peaks of triglyceride absorption and remnant formation [172]. Thus lipolysis and cholesterol esterification may not be tightly linked. LCAT probably works continuously to keep a rather variable pool of unesterified cholesterol, which is distributed between lipoproteins and cell membranes, within bounds [172].

Only if this pool becomes excessively large, will the lipoprotein and red cell membrane changes found in familial LCAT deficiency and liver disease become apparent.

Several studies have found correlations between triglyceride concentrations or turnover and plasma cholesterol esterification implying the existence of regulatory mechanisms related to the hypothesis outlined above [195—198]. No obligatory link could be found, however, between the hepatic secretion of LCAT and of triglyceride-rich lipoproteins [199, 200].

It is believed that HDL are secreted as disc-shaped structures by both the intestine [201] and the liver [202] (Sections 5.1.2. and 5.2.2.). In normal subjects such structures are not apparent in circulating HDL, having been replaced by spherical particles. The HDL of patients with familial LCAT deficiency, however, contain similar discoid structures [188] which are presumably nascent HDL secreted by the liver and intestine and which have accumulated because of the lack of LCAT to catalyse the formation of a cholesteryl ester core, thus normally transforming the particles into spheres. When these discoid HDL are incubated in vitro with partially purified LCAT, they are indeed replaced by spherical particles [188]. Moreover, LCAT may play an important role in the distribution of apolipoproteins between lipoproteins. Incubation of LCAT-deficient plasma with partially purified LCAT leads to the transfer of apoE from HDL to VLDL and the movement of apoC proteins in the opposite direction [203]. How this is interrelated with the similar transfer of apoC proteins associated with lipolysis is unclear. Presumably the LCAT-dependent change in HDL structure may alter the partitioning of apoC proteins between lipoproteins.

A major problem which is hindering the progress of work on the regulation of LCAT activity in plasma and related physiological studies is the lack of a suitable enzyme assay. Interpretation of the results with all the assays in current use is complicated by the presence of lipoproteins from the plasma of the subject under investigation, introducing variability in the composition of the substrates. Although the enzyme has been obtained in a highly pure state [186] no suitable immunoassay has been published. Even when one is available a sound chemical method of determining enzyme activity will still be required.

*References p. 489*

## 4.3. Cholesteryl ester exchange protein

Unesterified cholesterol exchanges rapidly between lipoproteins and membranes [204] probably by the transfer of individual molecules through the aqueous plasma rather than by the formation of collision complexes between lipoproteins [205]. Until recently, however, cholesteryl esters were widely believed not to undergo exchange [204] probably because exchange was not demonstrable between isolated lipoprotein fractions in vitro, although some reciprocal transfer of cholesteryl esters and triglycerides between lipoproteins had been reported when whole plasma was incubated, even with LCAT activity inhibited [206]. More recently [207], HDL and the d>1.21 plasma fraction in hypercholesterolemic rabbits were shown to promote the exchange of labeled cholesteryl esters between LDL and VLDL. A specific protein was believed to be responsible. No net mass transfer of cholesteryl esters was apparent in these experiments or in other in vitro incubations in which cholesteryl ester exchange between human LDL and HDL was studied [208]. The rate at which labeled cholesteryl esters exchange in vitro [208] and in vivo [209, 210] is as great or greater than the rate of formation of cholesteryl esters on HDL under the influence of LCAT. The in vivo studies, however, also did not establish mass transfer of cholesteryl esters though the potential seems to exist for the transport of esters to lipoproteins of lower density after their LCAT-mediated formation on HDL. Net transfer of cholesteryl esters from HDL to VLDL and LDL in exchange for triglycerides has been demonstrated in vitro and it has been tentatively suggested that apoD may be the exchange protein [113]. It has not been established that the exchange of one molecule of cholesteryl ester for another is mediated by the same mechanism as the exchange of cholesteryl ester for triglyceride.

Some mechanism for cholesteryl ester transfer has been assumed to exist in view of the evidence that LCAT catalyses the formation of most of the plasma cholesteryl esters in man even though VLDL and LDL are not substrates for the enzyme (Section 4.2.). The situations in the rat and man are probably very different with respect to the origin of plasma cholesteryl esters. The fatty acid pattern of VLDL esters differs from that in LDL and HDL in the rat and the VLDL esters are believed to be synthesised in the liver [211]. The relative deficiency of an exchange protein in the rat

[212] explains why the esters in all fractions do not come into equilibrium. In man, the VLDL esters more closely resemble those in LDL and HDL with respect to fatty acid pattern [213] and they have been reported to be identical, although others have noted small differences. These differences are probably genuine since kinetic studies suggest that not all VLDL cholesteryl esters in man originate in HDL [209].

A hypothesis summarising the overall economy of cholesteryl ester transport can be found in ref. 214.

### 4.4. Triglyceride exchange protein

Triglyceride enters plasma in chylomicrons carrying dietary fatty acids or in VLDL, carrying endogenously synthesised triglyceride, secreted by both the intestine and the liver (Section 5). Much of this triglyceride is rapidly removed from the circulation under the influence of lipoprotein lipase, but some appears in lipoproteins of d>1.006 [215]. That in LDL may be, at least partially, attributable to the metabolic conversion of VLDL to LDL (Section 6.2.). In addition, transfer of triglyceride from VLDL to HDL in exchange for cholesteryl esters has been reported (Section 4.3.). It is now apparent from in vitro studies with rabbit lipoproteins that labeled triglyceride can exchange without net transfer between LDL and HDL and that a thermolabile, ammonium sulphate precipitable agent from the d>1.21. fraction of plasma is necessary for this [216].

### 4.5. Phospholipid exchange proteins

Exchange of phospholipids has been demonstrated between individual lipoprotein fractions [204, 217, 218] in vivo and in vitro and between lipoproteins, platelets [219], red cell membranes [220] and a variety of other cell types and tissues in vitro [204]. The exchange is fairly rapid. For example, labeled phospholipid introduced into plasma in LDL or HDL equilibrated with respect to specific activity between both lipoprotein classes in about 4 h [217]. This was true for both lecithin and sphingomyelin. Sulphhydryl inhibitors did not prevent exchange but reduced the ex-

*References p. 489*

changeable pool of sphingomyelin, though not of lecithin. This was attributed to inhibition of LCAT causing reduced availability of sphingomyelin as a result of "masking" by lecithin. In a different system [218] [$^{32}$P]-phosphatidylcholine equilibrated completely between human VLDL and HDL in about 1 h, provided the plasma protein fraction of d>1.21 was also present. Very little exchange was observed in its absence. In both these systems, phospholipid exchange was highly temperature dependent [217, 218]. In the squirrel monkey the plasma fraction of d>1.21 promoted transfer of labeled lecithin between lipoproteins but this appeared to be due to net transfer rather than stimulation of the exchange process [221]. A supernatant fraction from rat liver cell homogenates stimulates phospholipid exchange between squirrel monkey lipoproteins but this is inhibited by sulphhydryl blocking agents and does not appear to be present in plasma [221]. Recently, a factor was identified in the d>1.21 fraction of human plasma which promotes phospholipid transfer in in vitro systems [222]. It is thermolabile and has an apparent $M_r > 100000$.

## 5. Lipoprotein formation and secretion

Lipoproteins are produced and secreted by two cell types derived from the embryological foregut, the hepatic parenchymal cell (hepatocyte) and the mucosal cell of the small intestine (enterocyte). Present evidence indicates that while triglyceride-rich lipoproteins of d<1.006 (chylomicrons and VLDL), and HDL are produced by these cell types, LDL are usually not secreted as such in normal man, but are the metabolic product of plasma VLDL metabolism (Section 6.2.). Both the intestine and the liver esterify fatty acids, of dietary or endogenous origin respectively, to produce triglycerides which they secrete in the triglyceride-rich lipoproteins. There are several morphological similarities in the ways in which they do this [223].

### 5.1. The intestine

Many studies on secretion of lipoproteins by the intestine involve examination of the lipoproteins recovered from mesenteric lymph.

In interpreting these studies it has to be remembered that the presence of a substance within mesenteric lymph, whether lipid or protein, does not necessarily imply its origin within the enterocyte [224]. Mesenteric lymph may contain lymph that has bathed tissues other than the intestine and may acquire plasma constituents by filtration. Even a substance identified within the enterocyte may have been synthesised elsewhere.

When radioactive amino acids are given intraluminally, a proportion will enter the portal circulation and be incorporated into lipoproteins in the liver. After secretion into the plasma some of these labeled lipoproteins may be filtered into intestinal lymph making it unsafe to infer that all labeled lymph apoproteins in this situation are necessarily synthesised within the intestine.

Several approaches have been used in an attempt to overcome these difficulties. These include the use of isolated enterocytes [225—227], mucosal biopsies [228, 229], the hepatectomised animal [230], the isolated perfused intestine [231], and a preparation in which labeled amino acids are placed intraluminally in a localised segment of intestine from which the entire venous drainage is collected [232] thus preventing the labeled amino acids from gaining access to the liver. Alternatively, agents which block hepatic lipoprotein secretion may be used, such as orotic acid [233] or 4-aminopyrazolopyrimidine (4APP) [234]. Moreover, plasma lipoproteins may be radiolabeled to gain an estimate of their filtration into lymph [235].

Even when lipoproteins within intestinal lymph have originated within the intestine they may well have been considerably modified since secretion from the enterocyte. Exchange of lipoprotein constituents between lipoprotein classes is rapid and lymph contains LCAT, one of the enzymes believed to be important in the metabolism of lipoproteins within plasma.

*5.1.1. Chylomicrons and VLDL.* The digestion of lipid, its absorption into the enterocyte and secretion in chylomicrons and VLDL are reviewed in ref. 223, 224, 236—239.

The major products of the hydrolysis of dietary triglyceride are fatty acids and 2-monoglycerides which pass into the enterocyte by an energy-independent process. Triglyceride is resynthesised within the smooth endoplasmic reticulum (SER) in the apex of the cell and from this point secretion of triglyceride-rich lipopro-

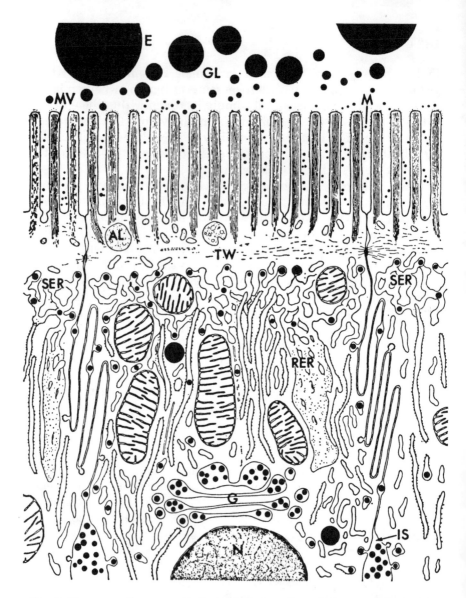

Fig. 2. Diagrammatic representation of apical portion of an enterocyte during triglyceride absorption. Based on electron micrographs. E, triglyceride emulsion particle; M, micelle; GL, gut lumen; MV, microvillus; AL, apical lysosome; TW, terminal web; SER, smooth endoplasmic reticulum; RER, rough endoplasmic reticulum; G, Golgi complex; N, nucleus; IS, intercellular space. The small black circles within the SER, Golgi and intercellular spaces represent osmiophilic particles believed to be nascent triglyceride-rich lipoproteins in the course of secretion. (From [237]. Reproduced with permission.)

teins (seen as osmiophilic particles) can be followed with the electron microscope (Fig. 2). Such particles are visible even within the enterocytes of fasting rats [26, 240, 241], and man [241, 242] and probably derive their lipid from bile and mucosal cells shed into the intestinal lumen, since they disappear with biliary diversion or treatment with resins which bind bile acids [241, 243]. These particles* are similar in size to plasma VLDL to which they probably contribute since the hepatectomised dog is still able to generate some plasma VLDL [230].

Within minutes of fat feeding there is a rapid increase in the number and size of osmiophilic particles within the enterocyte in the apical SER. These particles are believed to be the precursors of chylomicrons (so named because they are derived from chyle and are visible by dark-field microscopy) in the intestinal lymph and ultimately the systemic blood. Such particles are not usually seen within the terminal web beneath the microvillous membrane of the enterocyte and are rarely visible in the rough endoplasmic reticulum (RER), which may transform into SER during fat absorption [244]. Electron microscopic studies [26, 240—246] suggest that from the SER the particles pass to the Golgi apparatus in the supranuclear portion of the cell. The Golgi apparatus is probably involved with the attachment of carbohydrate to apoproteins synthesised within the intestine [247]. From the Golgi the osmiophilic particles pass in Golgi-derived vesicles to the plasmalemma in the lateral and basal aspects of the cell where they are expelled into the intercellular space by reverse pinocytosis (exocytosis). Specialised areas, known as coated pits, in the membranes of the Golgi vesicles have been identified which seem to become incorporated into the plasma membrane at the site of fusion [246]. Their function is uncertain.

Lipoprotein secretion from the enterocyte is believed to involve the cytoskeleton since colchicine, which binds to tubulin and inhibits its polymerisation to form microtubules, interferes with transport of triglyceride out of the mucosal cell [249, 250].

From the intercellular spaces, triglyceride-rich lipoproteins pass through gaps in the basement membrane into the lamina propria

---

*Particles in the VLDL size range which are secreted by the intestine are variously known as intestinal VLDL or small chylomicrons [248].

*References p. 489*

of the intestinal villus and thence between interdigitations of adjacent endothelial cells into intestinal lymphatics. They are not secreted into portal venous blood [231, 232].

*5.1.2. LDL and HDL.* Kinetic data indicate that under normal circumstances virtually all the apoB in LDL (LDL-B) is derived from that in VLDL (VLDL-B) in both rats and man (Section 6.4.). Small amounts of radioactivity are incorporated into lymph lipoproteins of LDL density by hepatectomised dogs [230] and the isolated perfused rat intestine [231]. In compositional terms, however, these particles may not represent true LDL since they contain HDL peptides [231]. It may be that little or no LDL is secreted by the intestine.

Radioactive amino acids are incorporated, however, into lipoproteins of HDL density by hepatectomised dogs and the isolated perfused intestine [230—232]. This material is secreted both into lymph and portal blood. It is likely that a significant proportion of this originates in the intestine and does not simply derive from filtration of plasma HDL into lymph, since intestinal lymph contains discoid HDL [201] similar to the nascent HDL produced by the liver [202]. Only spherical HDL were isolated from the plasma.

*5.1.3. Apolipoprotein secretion.* Although the secretion of individual apolipoproteins is discussed here this is purely for convenience of presentation, and is not intended to imply that they are secreted in lipoprotein-free form.

*5.1.3.1. Apolipoprotein A-I.* ApoA-I has been detected in the enterocytes of animals [227, 228, 251] and man [252, 253] and in human intestinal lymph recovered from the thoracic duct and chylous urine or effusions [25, 254—256]. While this could be derived, at least in part, from exchange with plasma lipoproteins, there is evidence that this is not so in the rat, since apA-I is still readily detectable in mesenteric lymph even in animals treated with 4-APP which markedly reduces plasma lipid levels by blocking hepatic lipoprotein secretion [234]. When radioiodinated HDL are infused intravenously into rats, the specific activity of apoA-I in lymph chylomicrons is only about 5% of that in plasma HDL, implying that only a small proportion of lymph chylomicron apoA-I is derived from exchange with plasma lipoproteins [235].

Fat feeding is followed by increases in plasma apoA-I in man [252, 257] and by a rapid increase in immunofluorescence due to apoA-I in enterocytes [227, 228, 234, 252] which is well maintained unless protein synthesis is simultaneously inhibited. As with apoB even enterocytes from bile-diverted animals, which will not be producing triglyceride-rich lipoproteins [241], still contain some apoA-I [234].

Incorporation of radioactive amino acids into apoA-I of mesenteric lymph lipoproteins has been demonstrated [234, 235, 251] and persists even when hepatic lipoprotein secretion is inhibited with 4-APP [234]. The isolated perfused rat intestine [231, 232], rat small bowel mucosal scrapings [258] and human small intestinal biopsies [229] also incorporate radioactive amino acids into apoA-I.

Although apoA-I is secreted by both the liver and intestine little is known of their relative contributions and the factors which regulate them. In the rat, bile-diversion for 3 days, which depletes intestinal lymph of d<1.006 lipoproteins, reduced plasma ApoA-I by 17%. Such animals, however, may well still be secreting apoA-I of intestinal origin into plasma in HDL [234]. Calculations based on lymphatic transport and plasma turnover of apoA-I in fasted rats suggest that the former may contribute about 23% of the latter [235]. Differential labeling experiments using [$^3$H]leucine and [$^{14}$C]leucine suggest that between 25 and 62% of apoA-I production is due to the intestine [259].

*5.1.3.2. Apolipoprotein A-II.* There are few data on the synthesis of apoA-II by the intestine. In the rat, apoA-II consists of a single polypeptide chain ($M_r \simeq 8500$) which migrates with the apoC proteins on SDS polyacrylamide gels [260]. It is a minor component of rat intestinal lymph and does not contain radioactivity after [$^3$H]leucine has been administered intra-luminally [232]. Human enterocytes, however, do appear to contain apoA-II [261] and human small bowel biopsies have been shown to incorporate radioactive leucine into apoA-II recovered in VLDL and HDL [229].

*5.1.3.3. Apolipoprotein A-IV.* ApoA-IV, a major apoprotein in the rat [262] also now identified in man [256], is secreted by rat intestine and is found in intestinal lymph chylomicrons, VLDL and d>1.006 lipoproteins [232, 235, 259, 263].

*5.1.3.4. Apolipoprotein B.* ApoB has been identified in chylous effusions and thoracic duct lymph in man [25, 254, 255] and by the use of RIA or fluorescent antibodies, in the enterocytes of both the rat [226—228] and man [225]. It is present in the apical portion of the enterocytes even in bile-diverted animals [226] in which it is unlikely to be associated with lipoprotein particles [241]. Its location would, however, favour an early association with lipid as the developing particle traverses the SER and Golgi. This is supported by the finding that in abetalipoproteinemia lipid accumulates within the endoplasmic reticulum at a stage before it would normally reach the Golgi apparatus [264]. Also apoB-like material occurs in lipoprotein particles isolated from rat intestinal Golgi [26].

Fat feeding was associated with decreased apoB in human enterocytes [225], but led to a marked increase in immunofluorescence in the rat [226, 227]. The fluorescence appears to surround (presumably lipid) cytoplasmic droplets [228] and becomes depleted after the animal is treated with inhibitors of protein synthesis [227]. That the apoB is synthesised within the intestine rather than taken up from the circulation is supported by the ability of the isolated perfused intestine [231], microsomes derived from rat enterocytes [265] and human small bowel biopsies [229] to incorporate labeled amino acids into apoB. Also, orotic acid, which inhibits secretion of apoB-containing lipoproteins by the liver and greatly diminishes plasma apoB concentrations in the rat, does not markedly affect the appearance of $\beta$-lipoproteins detected by immunoelectrophoresis in intestinal lymph [233].

Experiments in which [$^3$H]leucine was given intraduodenally and [$^{14}$C]leucine intravenously to rats suggest that the intestine secretes 5—21% of apoB output and the liver the remainder [259].

5.1.3.5. *C-apolipoproteins*. Little or no radioactivity is incorporated into the small molecular weight apoproteins of lymph lipoproteins when labeled amino acids are administered intraluminally or into the perfusate of the isolated perfused intestine [231, 235, 251]. Particles of d<1.006 isolated from rat small intestinal Golgi do not appear to contain the small molecular weight apoproteins [26]. Data of this type have led to the conclusion that the apoC proteins are not synthesised to any major extent by the intestine. However, the double isotope technique already described [259] and another recent study [232] suggest that some apoC-II and apoC-III$_0$ are synthesised in the intestine.

*5.1.3.6. Apolipoprotein E.* There is very little incorporation of radioactive amino acids into apoE of lipoproteins in intestinal lymph [232, 235, 259, 263], leading to the conclusion that little or no apoE is synthesised by the enterocyte.

A recent study has identified immunofluorescence due to apoE in rat enterocytes, but its distribution was different from that due to apoA-I and apoB, being most marked in the lower third of the villus in the fasting state [228]. After fat feeding there was an increase in fluorescence due to apoE which was now more extensive. Unlike apoA-I and apoB, however, it did not surround droplets (presumed to be nascent lipoproteins) within the mucosal cells. Preliminary evidence of intestinal polyribosomes holding nascent peptide chains of apoE has been reported [22].

## 5.2. The liver

*5.2.1. Very low density lipoproteins.* An increase in the free fatty acid concentration in the perfusate of the isolated perfused liver is accompanied by an increase in the number and size of osmiophilic particles in the hepatocytes. These appear initially (after 2 min perfusion) in the SER and Golgi and are only occasionally seen within the RER. The reason for the apparent lack of osmiophilic particles in the RER is unknown. One suggestion is that particles are present but have small size and are relatively poor in triglyceride making them difficult to observe [266].

Labelled glycerol is much more rapidly incorporated into lipoprotein triglyceride and phospholipid isolated from the RER and SER than from the Golgi [266]. Since the Golgi lacks the enzymes necessary for both the de novo synthesis of lecithin and the synthesis of triglycerides from 1,2-diglyceride and acyl-CoA esters [267], it seems that triglyceride and phospholipid synthesis are achieved in the SER and RER. The apoproteins, in common with other secreted proteins, are believed to be synthesised on the RER [268, 269]. The electron microscopical and biochemical studies described above have been interpreted as implying that synthesis of apoproteins occurs on the RER and lipid in both the RER and SER, with nascent lipoproteins becoming richer in triglyceride and relatively poorer in protein as they pass from RER to SER to Golgi [266].

Recently, experiments [269] using the antigen-binding fragments (Fab) from anti-rat VLDL and anti-rat LDL conjugated with horseradish peroxidase located apoB (and possibly other apoproteins contained in VLDL) in the RER and Golgi of rat liver. Although apoB was also found in the smooth-surfaced ends of the RER it was not observed in the SER. These authors critically reviewed the previous literature and on the basis of their findings have proposed that apoproteins are synthesised on the RER and lipid in the SER, coming together only in the smooth ends of the RER. From here the nascent lipoproteins would pass through special tubules to the Golgi. It is uncertain if it is obligatory for all particles to pass through the Golgi. Subsequently, the particles are found in vesicles approaching the cell membrane and they appear to be discharged by reverse pinocytosis into the space of Disse [270—272].

Particles isolated at d<1.006 from a cell-free, Golgi-rich fraction of rat liver have similar composition to plasma VLDL and react identically with anti-VLDL [273]. The pattern of apoproteins in these Golgi-derived VLDL is also very similar to that in plasma VLDL [274] making the evidence that these particles are indeed nascent plasma VLDL very strong.

*5.2.2. LDL and HDL.* Although some material may be secreted in the LDL density range by the isolated liver [231, 275—277], it is generally small in amount and may at least partially represent small VLDL, as judged by apoprotein composition, rather than true LDL [277]. This is compatible with the notion that most plasma LDL are not normally secreted as such, but are derived from the catabolism of plasma VLDL. There are, however, probably some situations in which appreciable LDL secretion occurs (Section 6.2.).

Isolated liver preparations also secrete material in the HDL density range [231, 275—277], but these particles appear to undergo considerable modification before becoming typical plasma HDL. When HDL are isolated from the perfusate of rat livers perfused with a plasma-free medium containing an LCAT inhibitor, the particles appear as discs (190 Å × 46 Å) whereas plasma HDL are spherical (diameter 113 Å) [202]. These nascent discoid HDL are deficient in non-polar lipids (cholesteryl esters, triglycerides) and rich in polar lipids (unesterified cholesterol, phospholipids) relative to plasma HDL. They resemble HDL from patients with hereditary LCAT deficiency or cholestasis [67]. Although orotic acid inhibits secretion of VLDL by the liver, material continues to be secreted in the HDL density range [278].

### 5.2.3. Apolipoprotein secretion

*5.2.3.1. Apolipoprotein A-I.* ApoA-I has been identified in rat hepatocytes by immunofluorescence, though the fluorescence was much less intense than that for apoE or apoB [228]. ApoA-I is secreted into serum-free perfusates by the isolated rat liver [202, 276, 277], which also incorporates radioactive amino acids into the apoprotein [277]. The liver is estimated to contribute 38—75% of total apoA-I production [259]. ApoA-I production is 12—30% of apoE production in the isolated rat liver [276].

Most of the apoA-I recovered from perfusates is found in the HDL density range, but there may be changes in distribution after secretion, mediated by LCAT. In the presence of an LCAT inhibitor more apoA-I is found in the $d>1.21$ fraction [276]. However, this may simply mean that apoA-I is more readily displaced from discoid HDL which have not been acted upon by LCAT [203]. Studies of the plasma of patients with alcoholic hepatitis suggest that nascent HDL are secreted devoid of apoA-I and acquire it subsequently [279]. $HDL_2$ and $HDL_3$ from isolated liver perfusates have a low ratio of apoA-I/apoE compared to plasma HDL, a difference which is accentuated by LCAT inhibitors [276].

*5.2.3.2. Apolipoprotein A-IV.* The isolated perfused rat liver secretes and incorporates radioactive amino acids into apoA-IV [276, 277], and the pattern of its secretion seems to parallel that for apoA-I [276]. The liver has been estimated to account for 17—50% of apoA-IV production [259].

*5.2.3.3. Apolipoprotein B.* ApoB has been recognised within hepatocytes by immunofluorescence [228] and using Fab fragments of anti-LDL conjugated with horseradish peroxidase [269]. It appears to be present in particles isolated at $d<1.006$ from the hepatocyte Golgi apparatus [274] and is secreted by rat hepatocytes in culture [280]. The isolated perfused liver secretes apoB into the perfusate [276, 277] and incorporates radioactive amino acids into perfusate apoB [231, 275, 277]. Under specific experimental conditions the liver has been estimated to contribute 79—95% of total apoB production [259]. ApoB is secreted almost entirely in VLDL where it constitutes 40—60% of the total VLDL apoprotein, which is appreciably more than the proportion in plasma VLDL [276, 277]. Feeding rats 1—2% orotic acid specifically blocks the secretion of apoB-containing lipoproteins by the liver [281], but not the intestine [233]. It seems, however, that the apoprotein can still be identified within the hepatocyte [278].

*5.2.3.4. C-apolipoproteins.* Until recently it was assumed that the liver is responsible for all apoC protein synthesis, though there is now evidence that at least some apoC-II and apoC-III$_0$ are produced in the intestine [232, 259]. Nevertheless it is likely that the vast majority of synthesis of apoC proteins is accomplished by the liver [259].

These small molecular weight proteins can be identified on SDS-polyacrylamide gels used to separate apoproteins from the perfusates of the isolated liver, which incorporates radioactive amino acids into these bands. The apoC proteins are recovered in both VLDL and HDL [202, 231, 276, 277]. It is uncertain if they are actually secreted in both these density classes, but the presence of apoC proteins in the plasma of patients with abetalipoproteinemia suggests that at least some are secreted with HDL [99]. They also appear to be present in particles of d<1.006 isolated from rat liver Golgi [274].

*5.2.3.5. Apolipoprotein E.* Fluorescence due to apoE has been identified in rat hepatocytes [228] and apoE is secreted into the perfusion medium by the isolated rat liver at a rate appreciably in excess of that for apoA-I [202, 276, 277]. The isolated liver also incorporates radioactive amino acids into apoE [277] and it is estimated that essentially all synthesis occurs in the liver [259]. ApoE is recovered from both the VLDL and HDL fractions of hepatic perfusates [276, 277]. The presence of an LCAT inhibitor reduces the amount in VLDL and increases that in HDL$_3$. Under these circumstances approximately 27% of plasma apoE is recovered in VLDL, 40% in HDL$_2$, 12% in HDL$_3$ and 16% in the d>1.21 fraction [276]. Plasma HDL from patients with familial LCAT deficiency, which resemble nascent HDL, are also relatively rich in apoE [119].

*5.2.4. Molecular biology of apolipoprotein synthesis.* Work is now beginning to appear on the cellular events leading to apolipoprotein synthesis. In particular this involves isolation of messenger ribonucleic acid (mRNA) specific for individual apolipoproteins and its translation in in vitro systems to apolipoprotein precursors. Estrogen injection in cockerels causes the accumulation of mRNA for a major VLDL apoprotein (apoVLDL-II) but not for apoA-I or albumin. The translation products of the mRNAs for apoVLDL-II and apoA-I are larger than the corresponding plasma proteins by about 20—30 amino acid residues, presumably due to signal sequences common to most secretory proteins [282, 283]. The complete amino acid sequence of the signal peptide for apoVLDL-II has been determined [284]. The mRNA for the apolipoprotein has been purified, and its corresponding double-stranded complementary DNA has been synthesised in vitro. The DNA has been successfully inserted into a bacterial plasmid pBR322, and amplified in an $E.$ $coli$ X1776 host. Nucleotide sequence determination of the cloned DNA insert demonstrates that a faithful copy of the structural gene for apoVLDL-II has been amplified [284].

Protein synthesised from rat liver mRNA in an in vitro translation system and precipitated by purified antibodies to apoE comigrated with apoE on polyacrylamide gels suggesting that they were of similar size. Studies of the N-terminal amino acid sequence of the translation product, however, suggest that it does possess a signal sequence of about 18 residues [285]. Partial purification of hepatic mRNA by precipitation with anti-apoE antibodies suggests that the antibodies bind to polysomes containing 9—12 ribosomes, compatible with a translation product of $M_r$ = 30 000—40 000 [285, 286]. Similar experiments for apoB suggest a monomeric $M_r$ in the range 12 000—30 000 [286]. Cholesterol-feeding appears to increase plasma apoE concentrations at least partially by increases in synthesis, since it has been shown to result in the accumulation of mRNA specific for apoE [287].

## 6. Lipoprotein metabolism

### 6.1. Triglyceride-rich lipoproteins

Little is known of changes in chylomicron and intestinal VLDL

*References p. 489*

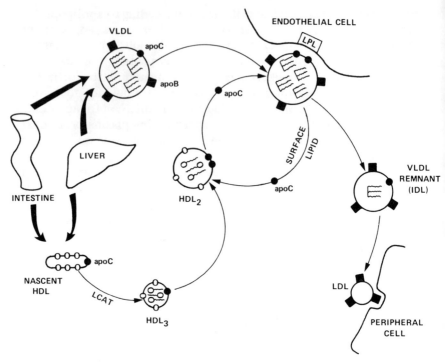

Fig. 3. Partial scheme for metabolism of plasma lipoproteins. VLDL and discoid nascent HDL are secreted by both liver and intestine. ApoC proteins (●) are secreted predominantly by the liver but also to a small extent by the intestine. They are shown here as being secreted on both VLDL and HDL, although it is uncertain if this is the case. The action of LCAT on nascent HDL removes unesterified cholesterol (○) and lecithin from the surface of the particle and by producing cholesteryl esters (o⟶) expands the particle core to produce spherical plasma HDL (shown here as $HDL_3$). HDL act as a reservoir for apoC proteins which are transferred to newly secreted VLDL. $HDL_2$ are shown here as the principal donors of apoC. VLDL are depleted of most of their triglyceride by the action of lipoprotein lipase (LPL) on the vascular endothelium of peripheral tissues. At the same time VLDL lose surface material (apoC and phospholipids) which is transferred to HDL and is instrumental in the conversion of $HDL_3$ to larger, less dense $HDL_2$. The fate of the VLDL remnants (which include IDL) differs in different species and various pathological states. They are shown here as being quantitatively converted to LDL by the loss of further triglyceride and surface material. This situation probably approximates to that in normal man. It should be noted that the triglyceride-rich lipoprotein retains its complement of apoB (■) throughout its metabolism. In hypertriglyceridemic man and in the rat significant quantities of VLDL remnants are cleared from plasma without conversion to LDL. LDL are shown here binding to the high affinity receptor of a peripheral cell, be-

*References p. 489*

composition which occur after they leave the enterocyte and before they are isolated from mesenteric lymph or chylous effusions. They may gain some protein (e.g. apoC proteins and apoE) from small amounts of plasma lipoproteins (especially HDL) filtered into lymph.

There are striking differences between the compositions of chylomicrons isolated from mesenteric lymph (or chylous urine and effusions in man) and plasma. The former contain apoC proteins, apoB, apoA-I and apoA-IV as major components, while apoA-II and apoE are relatively minor components (<5% of total apoprotein) [248, 254—256]. Plasma chylomicrons, however, contain little or no apoA-I and have apoB, apoE and apoC proteins as their major apoproteins in man [252, 256].

Lymph chylomicrons incubated with plasma lose phospholipid [248, 288], but gain unesterified cholesterol [288, 289] and total protein [248, 256, 289, 290]. Part of the gain in protein is due to apoC proteins including the LPL-activator protein, apoC-II [25, 248, 256, 290]. Similarly, apoC proteins are transferred from plasma to phospholipid-stabilised triglyceride emulsions "intralipid" in vitro [291, 292]. During alimentary lipemia in man, the fat-rich meal is associated with little change in the content of LPL activator in whole serum, but it is markedly increased in d<1.006 lipoproteins predominantly at the expense of its content in $HDL_2$ [291].

Lymph chylomicrons, intestinal VLDL or triglyceride emulsions incubated with plasma gain not only apoC proteins (Fig. 3) but also apoE [248, 256, 292]. It seems that enzymatic activity is not essential for transfer of apoE to chylomicrons since it still occurs at 4°C [248, 256] and in the presence of LCAT inhibitors [289]. It is uncertain how this non-enzymatic transfer of apoE is related to the LCAT-dependent transfer of apoE from HDL to VLDL (Section 4.2.).

Compositional studies already mentioned leave no doubt that the gain in protein due to apoE and apoC proteins when lymph chylomicrons are incubated with plasma is partially offset by loss

---

fore their uptake and degradation by the LDL pathway (Fig. 4). Some LDL is probably also degraded by the liver. In many respects the above scheme applies to the metabolism of chylomicrons carrying dietary triglyceride from the intestine. It is likely however that most chylomicron remnants are cleared from the circulation without making a major contribution to LDL.

of apoA-I. There is, however, some disagreement about mechanisms. Chylomicrons derived from human urine [256] and rat intestinal lymph [289] incubated with plasma lost most of their apoA-I. Similar changes were seen at 4°C [256] but did not occur in control incubations with saline [256, 289]. The presence of an LCAT inhibitor had no appreciable influence on the observed changes in apoA-I [289]. Other workers using small chylomicrons (intestinal VLDL) from rat intestinal lymph [248] found they lost apoA-I to a similar extent whether incubated with saline or the d>1.006 fraction of plasma. This loss was attributed to the second ultracentrifugation needed to reisolate chylomicrons after incubation. Thus it was concluded that loss of apoA-I occurring in vivo must occur in connection with the metabolism of chylomicrons.

The subsequent metabolism (Fig. 3) of chylomicrons and intestinal or plasma VLDL is probably rather similar at least in the rat. Even though the clearance rates from plasma of the individual lipid components of these particles may be similar [293, 294], it is known that their fates are very different. The initial process is the removal of most (>70%) of the triglyceride from the particles by lipoprotein lipase (Section 4.1.1.) situated in the vascular endothelium of extrahepatic tissues. Associated with this is removal of surface components (phospholipid, protein and possibly unesterified cholesterol) from the particles which enables the remnant lipoproteins to retain a spherical shape during depletion of their triglyceride core [295, 296].

A variety of metabolic fates may await the triglyceride-depleted remnants of chylomicrons and VLDL. In the rat, chylomicron remnants are probably cleared to a major extent by the liver (Section 6.1.2.) and the same may be true in man, in whom chylomicron constituents do not appear to make a major contribution to circulating LDL (Section 6.4.). In normal human subjects VLDL remnants (or at least their apoB moiety) are more or less quantitatively converted to LDL (Fig. 3, Section 6.4.). In hypertriglyceridemic man and in the rat, however, a substantial part of the circulating VLDL or their remnants are cleared from plasma without conversion to LDL (Section 6.4.). Not only the liver but also extrahepatic tissues may be implicated in this (Section 6.9.4.).

*6.1.1. Remnant formation.* Several lines of evidence suggest that the liver is not involved to a major extent in the initial metabolism of triglyceride-rich lipoproteins. These include tracer studies in intact animals [297, 298], the rapid clearance of triglycerides observed in hepatectomised animals [294, 295, 299] and the poor clearance of intact chylomicrons by the isolated perfused liver [300—303]. This contrasts with the rapidity with which other isolated organs and tissues, such as the heart [296, 304, 305] and adipose tissue [305] hydrolyse triglycerides in chylomicrons and VLDL.

The precise characteristics of remnant particles depend on the system used to generate them, whether hepatectomised animal, isolated perfused organ or soluble lipoprotein lipase in vitro [126]. Remnants prepared in functionally hepatectomised rats from chylomicrons, intestinal VLDL or plasma VLDL have similar dimensions and chemical compositions [295]. The majority are in the VLDL size range*, though larger and smaller particles are encountered [295, 306—309]. Despite removal of most of the triglyceride from the original particle, this remains the major constituent of remnants ($\sim$ 60—70%) so that the majority still float at d 1.006 although some may be recovered at higher densities. As a result, remnants are difficult to separate from unmetabolised VLDL. Contamination of remnants by unmetabolised VLDL can be minimised in rats by reducing the secretion of the latter by pretreatment with 4-APP [295]. As a result of triglyceride depletion, remnant particles contain increased proportions of protein, cholesterol, cholesteryl esters, and in some systems phospholipid [295, 306]. In terms of mass, however, remnants are formed by removal of triglyceride, phospholipid and tetramethylurea (TMU)-soluble proteins from the precursor lipoproteins, the original complement of the TMU-insoluble apoB being retained [295, 306, 307] (Fig. 3). Various reports on the mass changes of unesterified cholesterol and cholesteryl esters during remnant formation have been made [295, 306—309].

HDL apparently act as acceptors for surface components removed from triglyceride-rich lipoproteins during lipolysis (Fig. 3). This is true for chylomicrons and VLDL of both human and rat origin [36, 291, 306—308, 310]. The presence of HDL is not essential,

---

*Since it seems that the particles retain a spherical shape a large reduction in volume can be achieved by a relatively small reduction in diameter.

*References p. 489*

however, for the removal of these components from VLDL [296, 309]. They may be relased in the absence of VLDL as discoid particles comprising unesterified cholesterol, phospholipid and protein [296, 309, 310]. Loss of phospholipid during remnant formation involves both transfer of intact molecules and hydrolysis of phosphatidylcholine to lysophosphatidylcholine, probably by lipoprotein lipase which also has phospholipase activity [296]. The lysophosphatidylcholine becomes associated with serum albumin. ApoC proteins appear to be conserved in plasma by a process of recycling. Newly secreted triglyceride-rich lipoproteins acquire apoC proteins from HDL [291], but these are rapidly returned to HDL as remnants are formed by lipolysis (Fig. 3).

During their formation, remnants are probably repeatedly bound and released from the sites of lipolysis on the vascular endothelium [126] so that a spectrum of particles with varying degrees of triglyceride-depletion circulates at a given time [36, 311]. Particles which retain more than about 30% of their initial triglyceride component remain good substrates for lipoprotein lipase [311] and intact chylomicrons, VLDL and their partially lipolysed remnants are thought to compete for lipolytic sites. Why remnants become less good substrates for lipoprotein lipase when more than about 75% of their triglyceride has been removed is uncertain [126]. When rat chylomicrons enter plasma they bind a considerable excess of lipoprotein lipase activator (apoC-II) [290]. During lipolysis this is recycled back to HDL and by the time 75—80% of triglyceride has been hydrolysed, the apoC-II content of triglyceride-rich lipoproteins may be rate-limiting [126].

*6.1.2. Remnant removal by the liver.* In the rat at least remnant particles are rapidly cleared by the liver. When chylomicrons or VLDL labeled with [$^{14}$C]-cholesterol are injected intravenously into rats the label is rapidly taken up by the liver [293, 312], while in functionally hepatectomised animals cholesteryl ester-rich, triglyceride-depleted particles accumulate [295, 299, 306]. These remnants, however, are rapidly taken up if they are injected into the circulation of intact animals [299] or into the perfusion medium of an isolated perfused liver [300, 301, 303, 313]. Unesterified cholesterol, cholesteryl esters and triglycerides of chylomicron remnants appear to be taken up by the liver in proportion to their concentrations in the perfusate leading to the notion that remnants may be cleared as intact particles [293, 301, 303]. Remnant removal is a saturable process [303, 313]. Inhibitors of microtubular (colchicine) and microfilament function (vinblastin) do not impair remnant uptake though they do reduce the subsequent slow hydrolysis of cholesteryl esters [314]. Although in the rabbit cholesterol feeding inhibits remnant removal from plasma [315], in the rat remnant uptake is uninfluenced by factors which alter hepatic cholesterogenesis. Remnant uptake, however, reduces hepatic cholesterol synthesis [301, 316]. This contrasts with the uptake of LDL by fibroblasts where not only is cholesterol synthesis inhibited, but also further LDL uptake, by reduced receptor synthesis (Section 6.9.1.).

Until recently the factors which cause chylomicron remnants, but not their intact precursors, to be rapidly taken up by the liver were unknown [317]. Size or gross lipid composition do not seem to be critical, since remnants formed from large chylomicrons can be similar in these respects to intact small chylomicrons, and yet have very different rates of uptake by the liver. Remnant lipoproteins have been found to have lipoprotein lipase associated with them unlike intact chylomicrons and it has been suggested that this is the hepatic recognition factor [300]. It might be possible to explore this possibility by investigating the effect on remnant uptake of reacting the particles with an antiserum to lipoprotein lipase.

Recent work from several centres [317A–D] suggests that it is apoprotein composition which regulates hepatic clearance of circulating remnants. ApoE (particularly the E-3 and E-4 isoforms) appears to promote remnant clearance, while the apoC proteins inhibit clearance even in the presence of adequate amounts of apoE. It is not clear from existing data to what extent this is a general property of apoC proteins or whether it is confined to a specific apoC protein. The picture which is emerging suggests that nascent triglyceride-rich lipoproteins gain both apoE and apoC proteins on entering plasma. ApoC-II promotes triglyceride removal at extrahepatic sites while at the same time some or all of the apoC proteins prevent premature hepatic clearance. There comes a stage during lipolysis at which loss of apoC proteins limits further lipolysis but allows remnant removal by the liver, facilitated by a receptor which recognises apoE.

*References p. 489*

An alternative fate, of unknown quantitative importance, for the partially metabolised remnants of triglyceride-rich lipoproteins is uptake and degradation by extrahepatic tissues (Section 6.9.4.). Studies of the metabolism of cholesterol-rich chylomicrons by the isolated rat heart show it to be capable of removing cholesteryl esters independently of protein from chylomicron remnants [304]. Such a pathway might be involved in the proposed atherogenic role of chylomicron remnants [12].

## 6.2. Formation of low density lipoproteins

Discussion of the metabolism of lipoproteins in terms of the intact particles is frequently difficult because of the rapidity with which their individual constituents are exchanged or undergo metabolism. In the case of LDL the situation is somewhat simplified by the finding that apoB, which comprises almost all the protein in human LDL, is not readily exchanged and acts as a marker for the particles during their metabolism [21].

We have already seen that the isolated liver and intestine in the rat secrete little material into the LDL density range, and that some of this may not represent true plasma LDL. Cell-free preparations of rat small bowel mucosa incorporate radioactive amino acids into lipoproteins in the LDL density range [265], but one cannot necessarily extrapolate from this to the intact animal. Nevertheless LDL secretion has been demonstrated by the isolated pig liver [318], some of which represents re-release of preformed LDL and some of which is LDL synthesised de novo. In the squirrel monkey, a small fraction of the circulating LDL may also be directly secreted [319].

Kinetic studies of apoB metabolism in normal man (Section 6.4.) suggest that LDL-B is primarily derived from VLDL-B rather than secreted de novo. It is uncertain if the whole of this transformation, involving progressive delipidation and loss of soluble apoproteins from VLDL, is achieved within the circulation. Studies in man [36] have drawn attention to the rapidity with which VLDL-B is transformed to IDL-B after the intravenous injection of heparin and the relative delay in the transformation of IDL-B to LDL-B, leading to the idea that the initial process was mediated by lipolysis within the plasma compartment and that a different mechanism was involved in the second stage. Work in the rat in which remnants of triglyceride-rich lipoproteins formed by lipolysis are rapidly taken up by the liver (Section 6.1.2.) raise the possibility that IDL are removed by the liver, modified and re-released as LDL. However there is no direct evidence for this in man and the kinetics of apoB metabolism are very different in the rat and normal man (Section 6.4.).

Recently an attempt was made to see if LDL could be generated completely within the circulation by incubating human VLDL in vitro with purified bovine lipoprotein lipase and albumin [320]. The particles generated had many similarities to native LDL in that they were enriched with cholesteryl esters and apoB, but important differences remained. The in vitro lipoproteins were larger than native LDL and there were some differences in composition. While some of the latter might be due to experimental techniques and the presence of a small number of liposome-like and discoid particles of d 1.019—1.063, the in vitro LDL contained an increased mass of cholesteryl esters which could not readily be accounted for. To what extent the addition of other plasma constituents (e.g. HDL, LCAT, other plasma proteins) to the incubation system would modify the structure of the in vitro LDL is still uncertain.

### 6.3. The fate of plasma low density lipoproteins

The metabolic fate of plasma LDL is uncertain. The controversy surrounds the relative amounts of LDL which are degraded by the liver and by extrahepatic cells. Many of the studies which will be described have used LDL labeled with radioactive iodine and therefore are primarily concerned with the fate of the protein component of LDL (apoB).

Because the liver is the major site of cholesterol catabolism and because in man and some animals, though not in the rat, LDL carry most of the plasma cholesterol, it has been tempting to assume that the liver would prove to be the major site of LDL catabolism. Indeed comparison of clearance rates of [$^{125}$I]-LDL from plasma by intact rats and from the perfusate of the isolated liver suggested that the liver was a major source of LDL catabolism [321]. However, the LDL used in these studies probably included IDL (d 1.006—1.019) which are presumed to be remnants of triglyceride-rich lipoproteins and are rapidly taken up by the liver (Section 6.1.2.). When [$^{131}$I]-LDL of narrow density range are used, their fractional catabolic rate in the isolated perfused liver is only about 7% of that in the intact rat [322]. Measurements of tissue radioactivity in the pig [323] and in the rat [324], after the intravenous injection of [$^{125}$I]-LDL, identified the liver as the major site where label is detected although this was only a small proportion ($\sim$ 3%) of the injected isotope.

There are several difficulties in interpreting data obtained by measurements of tissue radioactivity in this way. The labeled lipoprotein can be taken up into a tissue, degraded in the lysosomes and the labeled amino acids are re-excreted (Section 6.9.1.) so that the radioactivity detected in the tissue represents an underestimate of the amount of lipoprotein degraded. Experimental approaches to circumvent this problem include the use of chloroquine or $^{14}$C-labeled sucrose covalently linked to LDL (see below). However, even when chloroquine is used to inhibit lysosomal degradation of LDL, only a small proportion of the injected [$^{125}$I]-LDL, which have been cleared from the plasma, accumulate in rat liver [325]. Another difficulty is that an appreciable part of the LDL pool after the injection of labeled LDL is extravascular [326]. It is likely, in the swine at least, that a major part of the extravascular pool is in the extracellular space of the liver [327]. Indeed hepatectomy in the swine converts a biexponential decay of intravenously administered [$^{125}$I]-LDL into a mono-exponential decay [328]. Thus, much of the [$^{125}$I]-LDL associated with the liver in tissue distribution studies may simply be sequestered in this extravascular pool and not necessarily undergo degradation. Consonant with this is the finding that much of the $^{125}$I associated with the liver is still precipitable with trichloroacetic acid and may be part of intact lipoproteins [323, 324, 327].

The use of hepatectomy to study the role of the liver in LDL catabolism has yielded variable results. The fractional catabolic rate of $^{125}$I-labeled LDL was increased by hepatectomy in the swine [328], but decreased by partial hepatectomy in the rat [329].

Neither does the use of cultured cells yield a clear answer. Internalisation and degradation of human LDL by cultured rat hepatocytes was observed at rates in excess of those explicable by fluid endocytosis alone, suggesting the involvement of a receptor mechanism [330]. Moreover, a protein in liver plasma membranes has been identified which binds LDL [331]. Calculations based on the rates of LDL degradation by extrahepatic cells in culture, however, suggest that they could account for the total LDL catabolism by the whole organism [332]. It is unknown if such extrapolations from tissue culture to the intact animal are quantitatively justifiable.

When lipoprotein concentration differences were measured across the splanchnic bed in man (between arterial and hepatic venous blood at cardiac catheterisation) it appeared that LDL-cholesterol was being removed, but no significant concentration difference for LDL-protein was found [333]. When an antibody to HTL was given intravenously in rats, LDL-cholesterol but not LDL-protein accumulated in plasma [166].

Perhaps the data reviewed thus far can best be summarised by saying that there is little evidence that the liver plays a major role in the catabolism of LDL-protein, though it may modify LDL by selective removal of lipid components. A new approach to the problem has recently been described, however, which suggests that both liver and extrahepatic tissues play an important role in LDL catabolism [334]. This involves covalently binding [$^{14}$C]-sucrose to LDL. The labeled complex is taken up by pinocytosis and the sucrose is only released in the secondary lysosome where the LDL are degraded, the degradation products being excreted from the cell (Section 6.9.1.). The sucrose, however, is largely retained within the lysosome, the accumulation of $^{14}$C being an index of the amount of LDL degraded within a given tissue. The advantage of this approach is that it should enable lipoprotein degradation to be studied in the whole animal under a variety of physiological conditions. So far the technique has only been exploited in the study of LDL catabolism in the swine [335]. About 75—80% of the injected $^{14}$C was recovered from the tissues of pigs examined at 24—48 h. All tissues and organs examined contained $^{14}$C and therefore, presumably contributed to LDL degradation. These included liver, skin, adrenal, spleen, lymph nodes and kidney. Quantitatively the most important in terms of overall LDL degradation were the liver (which accounted for about half of the $^{14}$C recovered), adipose tissue, muscle and small intestine. The most active tissues in LDL degradation on a unit weight basis were the adrenal, liver and spleen in that order.

*References p. 489*

### 6.4. Kinetics of apolipoprotein B metabolism

Mean plasma concentrations of apoB are about 80—90 mg/dl in normal subjects with standard deviations of 15—30% of the mean [84]. There is some disagreement as to whether the concentrations are the same in men and women. As would be expected, hypertriglyceridemic subjects with increased concentrations of apoB-containing triglyceride-rich lipoproteins had increased total plasma apoB concentrations, but the highest values (approximately 4 times normal) were found in subjects who are homozygous for familial hypercholesterolemia [336].

Although it has been known for some time that there is some sort of precursor-product relationship between the protein moieties of VLDL and LDL [36, 337], it is only recently that the quantitative aspects have started to become clear. In normal man it appears that all LDL-B is derived from VLDL-B since the specific activity-time curve for LDL-B is intersected at its maximum by the decaying curve for VLDL-B after the injection of $^{131}$I-labeled VLDL [338]. Moreover a large part and perhaps all of VLDL-B appears to be converted to LDL-B in normal subjects [338]. Thus, although many other kinetic situations may exist in pathological states and in animals, the statement that all VLDL-B is converted to LDL-B and that all LDL-B is derived from VLDL-B in normal man probably represents a reasonable approximation to the truth.

In human subjects with hypertriglyceridemia the synthetic rate for VLDL-B is frequently abnormally high [339] and exceeds that for LDL-B in similar or the same subjects [340, 341] leading to the conclusion that a significant proportion of VLDL-B is cleared from the circulation before it can appear as LDL-B as a result of the progressive delipidation of VLDL. In some cases of hypertriglyceridemia the flux of apoB was similar in $S_f$ 60—400 and $S_f$ 12—60 lipoproteins suggesting that it may be from the latter range (probably $S_f$ 20—60) that apoB, not destined for LDL ($S_f$ 0—12), leaves the plasma [341]. The studies already described have assumed either a mono- [338, 342] or biexponential [341] decay of specific-radioactivity in VLDL-B in the analysis of their data. Since the major portion of the radioactivity in VLDL-B is removed during the course of the initial exponential(s) the errors from doing this are small. Recently, a more complex multicompartmental computer model of VLDL-B metabolism has been described

*References p. 489*

[21, 343]. In normal subjects the data are described by a model in which VLDL-B passes through a series of pools ($\alpha_2$-VLDL pathway) as a result of progressive delipidation, via IDL, to LDL. In hypertriglyceridemic patients (other than type III) some apoB is lost from plasma without conversion to LDL [343], in broad agreement with the studies already described. In Type III patients a major fraction (10—29%) of VLDL-B passes through a separate VLDL compartment ($\beta$-VLDL) which is cleared from plasma without conversion to IDL or LDL. Similarly 28—62% of IDL-B appeared to be directly degraded. In addition there seemed to be considerable direct IDL-B or LDL-B synthesis in type III hyperlipoproteinemia. The fate of chylomicron apoB has been little studied, but it is probable that a major fraction of it in man is not converted to LDL-B [255] and that chylomicron remnants are cleared from plasma in the VLDL density range [344].

Study of VLDL-B kinetics in a heterogeneous group of normal and hypertriglyceridemic subjects revealed expansion of VLDL-B pool size associated with both increased input and decreased fractional removal. The possibility that the latter was a consequence of increased pool size (by saturation of removal processes) could not be excluded. Body weight appeared to be an important determinant of removal rate [345]. There is probably considerable variability in the relationship between VLDL-B and VLDL-triglyceride kinetics due to newly secreted VLDL carrying variable amounts of triglyceride for a given apoB content. Sucrose feeding for example, which has been shown to increase triglyceride production, increased VLDL-B pool size, but this appeared to be more closely associated with changes in removal rate rather than changes in flux [346].

In the rat only a small proportion of VLDL-B ($\sim$ 5%) appears in LDL-B, most of the remainder being recovered in the liver [347]. Specific activity-time curves for VLDL-B and LDL-B suggest that the latter is derived from the former [347], but recent work [348] suggests a more complex situation in which different LDL subfractions have different origins. The predominant subfraction (d 1.04—1.063, $S_f$ 0—5) appeared to have a major independent source of apoB, while apoB in $S_f$ 5—12 and $S_f$ 12—20 could have been derived from VLDL-B [348].

The most striking kinetic abnormality of apoB metabolism in subjects with familial hypercholesterolemia is a decrease in its fractional catabolic rate [326, 349] which is independent of LDL pool size [350]. This is readily explicable by the impaired binding and the resultant impairment of uptake and degradation of LDL in the non-hepatic cells of such subjects (Section 6.9.1.). ApoB synthesis appears to be increased in homozygous subjects [326, 350], but it is less clear if this is so in heterozygotes [349, 351]. The mechanism for such an increase in apoB synthesis is unknown. In homozygotes, the synthetic rate for LDL-B considerably exceeds that for VLDL-B in the same individual making it likely that some apoB is secreted directly into the circulation in IDL or LDL [342]. This abnormal pathway is promoted when LDL levels are acutely lowered by plasmapheresis [352]. In heterozygotes the apoB fluxes are generally similar in VLDL and LDL [342].

Cholestyramine and nicotinic acid which have been used to lower LDL concentrations in these subjects, probably do so by different mechanisms, the former increasing clearance of [$^{125}$I]-LDL and the latter decreasing apoB synthesis [353].

## 6.5. Metabolism of high density lipoproteins

The recent re-evalution [192] of the long-known [354], but largely forgotten, observation that elevated high-density lipoprotein concentrations are associated with reduced risk of ischemic heart disease, has led to an explosion of interest in HDL metabolism [355]. Modern epidemiological studies have amply confirmed the strength of the relationship [356—360], and several recent reviews and symposia are devoted to HDL, their metabolism and clinical significance [13, 76, 361—364, 364A].

The secretion by both liver and intestine [201, 202] of nascent HDL as flattened discoid particles, comprising mainly phospholipid, unesterified cholesterol and protein, but devoid of cholesteryl esters, has already been described. In addition the surface components of triglyceride-rich lipoproteins, which are removed during the course of lipolysis, also contribute to the mass of circulating HDL (Section 6.1.1.), although the precise mechanism involved is uncertain [76]. This phenomenon provides a basis for the negative correlation between plasma triglycerides (or VLDL-lipids) and HDL-cholesterol concentrations [192]. Factors which reduce plasma triglycerides and increase HDL-cholesterol include weight loss [365], exercise [366], clofibrate [365] or nicotinic acid [367] administration, intravenous heparin [368] and insulin-treatment of diabetic ketoacidosis [368]. Carbohydrate feeding has opposite effects [365]. Several of these perturbations have been shown to be associated with altered plasma or adipose tissue LPL activity including heparin, clofibrate [162], insulin [368] and exercise [368].

The conversion of the discoid nascent HDL particles to native spherical plasma HDL is believed to be the result of the action of plasma LCAT activity (Section 4.2.). The substrate lipids of LCAT (unesterified cholesterol and phosphatidylcholine) are plentiful in the bilayered nascent HDL and in the surface of spherical HDL. As the reaction proceeds, unesterified cholesterol is converted to non-polar cholesteryl esters which seek the hydrophobic interior of the particle thus generating a core of neutral lipid at the expense of the depletion of surface material. Lysolecithin the other product of the reaction leaves the lipoprotein and binds to albumin.

HDL do not represent a single population but consist of two major subfractions (Table I). Of these, $HDL_3$, at least, are also heterogeneous [40]. The higher $HDL_2$ levels in women and the observation that much of the variability in HDL concentrations is attributable to $HDL_2$ has led to the proposal that $HDL_2$ may be more closely related to protection against ischemic heart disease than total HDL [369]. Recent work *in vitro* shows that it is possible to transform $HDL_3$ into $HDL_2$-like particles by incubating them with VLDL and purified bovine milk LPL [308]. Intravascular lipolysis in vivo promoted by heparin has also been shown to be accompanied by an increase in the peak flotation rate of HDL with more material appearing in the $HDL_2$ range [36, 370]. This was accompanied by a reduction in $HDL_3$ concentration [370]. These studies have led to the concept of a precursor-product relationship between $HDL_3$ and $HDL_2$ (Fig. 3) mediated by the transfer of surface material from triglyceride-rich lipoproteins to $HDL_3$ during lipolysis [308]. Details of the mechanism remain to be established, but the concept is strengthened by observations that situations in which lipolytic activity is enhanced tend to be accompanied by higher $HDL_2$ concentrations and vice-versa. Examples include clofibrate administration [162, 371], physical exercise [366, 368] and type I hyperlipoproteinemia [372].

*References p. 489*

The metabolic fate of HDL, like that of LDL, is controversial and again the difficulty relates to the relative contributions made by the liver and extrahepatic tissues to HDL degradation. It seems that both rat liver parenchymal and non-parenchymal cells are capable of taking up and metabolising HDL [373—375]. When tissue associated radioactivity is measured shortly after the intravenous injection of [$^{125}$I]-HDL, the liver is associated with by far the most radioactivity on an organ basis [375], though not all of this need necessarily be undergoing hepatocellular uptake or degradation (Section 6.3.). The adrenal is associated with most radioactivity on a unit weight basis. Liver associated [$^{125}$I]-HDL uptake greatly exceeded that for [$^{125}$I]-polyvinylpyrrolidine (an index of fluid endocytosis), thus implying a specific mechanism for HDL accumulation by the liver [375].

No arteriovenous difference for HDL-cholesterol was detected across the splanchnic bed in man [333], though at high flow rates a very small difference could represent appreciable uptake. In the rat intravenous injection of antiserum to HTL led to the accumulation of HDL-cholesterol and HDL-phospholipid in plasma. HDL-protein was not studied [166]. Thus in the rat HTL may play a role in the removal of HDL lipid by the liver and perhaps in the conversion of $HDL_2$ to $HDL_3$ [375A].

Experiments in which circulating HDL concentrations were varied in vivo by intravenous infusion showed that HDL could increase hepatic cholesterol content and inhibit cholesterol synthesis, but that chylomicrons were very much more effective in this respect [376]. Comparison of the fractional catabolic rates of [$^{131}$I]-HDL in the isolated perfused liver and in the intact rat suggest only a minor role for the liver in the degradation of HDL protein [377], and this is supported by failure of removal of two-thirds of the liver to influence [$^{131}$I]-HDL catabolism in the rat [378].

It seems likely that the liver contributes to the degradation of HDL proteins in vivo but this may be a relatively minor contribution overall. It has been calculated that extrahepatic cells could account for all the HDL catabolism found in the intact organism [379], but this does not necessarily mean that they do so. The liver may modify HDL composition by selective removal of lipid even when the whole lipoprotein is not taken up.

## 6.6. Kinetics of apolipoprotein A-I and A-II metabolism

Mean normal plasma apoA-I concentrations are in the range 100—150 mg/dl with standard deviations of 15—50% of the mean [84]. Published reports are divided between those which find higher plasma apoA-I concentrations in women than in men [90, 91, 380] and those which find no significant difference [89, 381, 382]. Mean apoA-I concentrations were found to be normal in hyperlipidemia except in men and women with Type V and women with Type IV hyperlipidemia, who had low values [383]. Other situations associated with reduced plasma apoA-I concentrations include ischemic heart disease [380], diets enriched with polyunsaturated fat [384] and treatment with the anabolic steroid oxandrolone, which has androgenic properties [385]. Nicotinic acid [367], cholestyramine [386], estrogen [380] treatments and pregnancy [89] have been found to be associated with increased apoA-I concentrations.

Mean normal plasma apoA-II concentrations are usually found to be about 25—40 mg/dl [367, 381, 383, 387, 388], although others report values of about 80 mg/dl [389]. In general apoA-II concentrations are normal in patients with hyperlipidemia, though low values were found in men with the type V abnormality [383]. Reduced plasma apoA-II concentrations have been reported in association with ischemic heart disease [388] and nicotinic acid treatment [367], while cholestyramine [386] was without effect. Estrogen treatment was associated with increased apoA-II concentrations [387]. A tendency to higher apoA-I but not apoA-II concentrations in women compared to men [383] may be due to increased concentrations in women of $HDL_2$, which probably has a higher apoA-I: apoA-II ratio than $HDL_3$. The kinetic changes underlying some of these alterations in plasma apoA-I and apoA-II concentrations are discussed below.

Three methods have been used for the radiolabeling of apoA-I and apoA-II in HDL for the study of their metabolism. The whole HDL particle may be labeled with radioactive iodine, in which case all the apoprotein constituents of HDL (apoA-I, apoA-II and apoC proteins) become labeled necessitating the separation of these apoproteins from each timed HDL sample if their individual kinetics are to be studied [41, 390]. Alternatively, the individual apoproteins may be isolated, labeled with radioiodine and reincorporated into HDL by incubation [391, 392]. With this technique the labeled apoproteins exchange rapidly between $HDL_2$ and $HDL_3$ [393]. ApoA-II seems to be metabolised identically whichever method is used, but the half-life of apoA-I in plasma is about 20% shorter when the labeled apoprotein is subsequently incorporated into HDL rather than being labeled in the intact lipoprotein [392]. The labeled free apoproteins have also been injected intravenously, when a major proportion rapidly becomes associated with HDL [394, 395]. This last method may perhaps be regarded as the in vivo counterpart of the in vitro method already described [391], although it has been suggested that a part of the injected apoprotein may be cleared abnormally rapidly from the circulation [382].

Although it was originally reported that apoA-I and apoA-II were cleared synchronously from plasma [390], it may be that apoA-I is in fact cleared rather more rapidly than apoA-II whichever labeling technique is used [397]. In hypertriglyceridemic subjects, however, apoA-II has a higher fractional catabolic rate than apoA-I [397A].

Several factors have been identified which influence the metabolism of HDL protein. Catabolism of HDL protein is increased in the nephrotic syndrome [337] and in hypertriglyceridemia [396]. In these early studies apoHDL was studied as an entity since its heterogeneity was not yet appreciated. Catabolism of apoA-I and A-II is increased by carbohydrate feeding [390] and decreased by nicotinic acid treatment [367, 390]. When the in vitro technique of exchanging [$^{121}$I]-apoA-I into HDL is used, nicotinic acid reduces the fractional catabolic rate of unlabeled apoA-I, but not of the labeled apoA-I [367]. Other kinetic studies support the existence of more than one pool of apoA-I within the plasma compartment [398]. Nicotinic acid also reduced the rate of apoA-II synthesis [367]. Other therapeutic measures used in the treatment of hyperlipoproteinemias which influence apoA-I metabolism include feeding a diet rich in polyunsaturated fat, which reduces plasma apoA-I concentration [384], and the bile acid sequestrant, cholestyramine, which increases it [386]. Both these changes appear to be the result of altered synthesis. No significant differences in apoA-I metabolism between the sexes were found [381] in association with the higher ratio of $HDL_2$ : $HDL_3$ frequently observed in women.

Patients who are homozygous for the rare hereditary disorder known as Tangier Disease [99] have very low circulating levels of HDL and A-apoproteins. Decay of radioactivity from plasma in these patients after injection of homologous [$^{125}$I]-HDL is monoexponential [253] rather than biexponential as in normal subjects. Both apoA-I and apoA-II radioactivities decay rapidly with half-lives of hours compared to normal values of about 4—6 days. Moreover apoA-I decays more rapidly than apoA-II. The patients show normal fluorescence patterns for apoA-I and apoA-II in their small bowel mucosa [252, 253] which is compatible with the finding of synthetic rates for the two apoproteins which approach normality [399]. The major defect appears to be one of increased apoA-I catabolism for reasons which are so far unknown. Newly synthesised apoA-I appears unable to associate with infused normal HDL in these patients [253, 400].

## 6.7. Metabolism of the C-apolipoproteins

Several aspects of apoC protein metabolism have already been summarised. Briefly they are synthesised predominantly in the liver and are recovered in both VLDL and HDL in isolated liver perfusates (Section 5.2.3.), though this need not represent their density distribution when secreted from the hepatocyte. Intestinal triglyceride-rich lipoproteins are secreted which are poor in apoC proteins although they probably contain some apoC-II and apoC-III$_0$ (Section 5.1.3.). They acquire apoC proteins, predominantly from HDL, during their transport through intestinal lymphatics and after entry into plasma (Section 6.1.). During lipolysis of triglyceride-rich particles the apoC proteins are transferred back to HDL (Section 6.1.1.) and are thus conserved in plasma and can presumably be recycled back to chylomicrons and VLDL again.

The concentrations of apoC-II and apoC-III (sum of all isoforms) in normal plasma are approximately 4 and 15 mg/dl respectively with about 45% of apoC-II and 35% of apoC-III being present in the d<1.006 fraction. Most of the remainder resides in HDL. Hypertriglyceridemia is associated with an increase in the plasma concentrations of both apoproteins and with the proportions carried in lipoproteins of d<1.006 [401—403]. The ratio of apoC-II/apoC-III is variable implying some differences in metabolism. For example the ratio is higher in d<1.006 than in d>1.006 lipoproteins [403] and is increased in VLDL by carbohydrate feeding [404] and decreased in pregnancy [405].

Iodine-labeled apoC proteins introduced into plasma in VLDL rapidly equilibrate with apoC proteins in HDL and vice-versa [21, 406]. It has been proposed that these apoproteins distribute between lipoproteins through a water-soluble form in the aqueous phase of plasma, with relatively high association constants for triglyceride-rich lipoproteins and HDL and a low constant for LDL [21]. The multi-compartmental computer model used for the study of apoB metabolism was also employed for the study of labeled apoC proteins introduced into plasma in [$^{125}$I]-VLDL [21, 343]. It was not possible to determine whether newly synthesised apoC proteins entered plasma in VLDL or HDL. ApoC protein synthesis in normal individuals was in the range of 175—660 mg/day depending on the particle with which apoC was assumed to enter plasma. These values may be compared with production rates of about 77 mg/day for apoC-II and 175 mg/day for apoC-III$_1$ (both calculated for a 70 kg individual) determined by the use of isoelectric focussing to separate individual apoC proteins [406]. Overall apoC and apoC-II production rates were relatively similar in a variety of hyperlipidemic states [343, 406]. Transfer of total apoC from HDL to VLDL was computed to be about 400—900 mg/day in normal subjects with a fractional catabolic rate in plasma of 1.5—2.5 pools/day [21, 343].

### 6.8. Metabolism of apolipoprotein E

Much remains to be discovered about the metabolism of apoE. It is known to be secreted almost exclusively by the liver in rats, and perfusion experiments using LCAT inhibitors suggest that nascent HDL particles are the major secretory vehicle (Section 5.2.3.). It seems that LCAT plays a role in the transfer of apoE from HDL to triglyceride-rich lipoproteins. Patients with hereditary [119] or acquired [119, 279] LCAT deficiency were found to have disc-shaped HDL with apoE as the major apoprotein, whereas in normal subjects apoE is at best a minor component of total HDL. When the whole plasma of a patient with hereditary LCAT deficiency was incubated with partially purified LCAT there was a marked increase in apoE in VLDL and a decrease in HDL [203]. Enzyme-independent transfer of apoE from HDL to triglyceride-rich lipoproteins of intestinal origin has, however, also been well documented (Section 6.1.).

The kinetics of apoE metabolism in VLDL have been studied [407] and the decay of VLDL-E shows irregularities similar to those seen for the apoC proteins. Possibly a recycling mechanism between VLDL and HDL exists resembling that of the apoC proteins. If so, it appears that apoE leaves the parent triglyceride-rich lipoprotein less readily than the apoC proteins during remnant formation since the remnants contain an increased ratio of apoE: apoC proteins [295]. A similar situation is found in man where those subfractions within the VLDL density range which are likely to contain circulating remnants have an increased ratio of apoE: apoC proteins. This includes the smaller subfractions of normal VLDL isolated by gel filtration [31], the $\beta$-VLDL of type III hyperlipoproteinemia [116] and the slower migrating subfraction of subjects with "double pre-beta lipoproteinemia" [51]. The situation is complex, however, and the relative enrichment of $\beta$-VLDL with apoE cannot simply be due to the loss of apoC proteins, if that is the mechanism for the increases in the proportion of apoB, since the ratio of apoE: apoB is higher in $\beta$-VLDL than it is in prebeta VLDL [116].

The mean normal plasma apoE concentration has been reported as 10 mg/dl [114] and 24 mg/dl [118]. The total plasma concentration is increased in type III hyperlipoproteinemia [114, 118] and possibly in type V hyperlipoproteinemia [114], although this was not confirmed in a small number of subjects [118]. On isoelectric focussing the most cationic isoforms of apoE are markedly reduced or absent in type III hyperlipoproteinemia [51, 408, 409], but this deficiency is not necessarily always accompanied by the type III lipoprotein abnormalities [409].

Other situations associated with abnormalities of apoE metabolism include hypothyroidism [410], where an increased proportion of apoE is found in apoVLDL, and cholesterol feeding [411]. In the swine and *Patas* monkey, both of which are susceptible to spontaneous atherosclerosis, cholesterol feeding leads to the appearance of $\beta$-VLDL together with increased concentrations of IDL and LDL. In addition a fraction known as $HDL_c$ appears, so called because it usually contains apoA-I and lacks apoB, even though its high ratio of cholesteryl ester to protein causes it to float into the LDL density range. These lipoproteins are enriched with apoE [411].

Current work suggests that apoE promotes the uptake of remnant lipoproteins (Section 6.1.2.) and apoE-containing subfractions of HDL [317B, 411A] by the liver. It also has a high affinity for the LDL receptor and can deliver cholesterol to non-hepatic cells (Section 6.9.1.). ApoE has been reported to inhibit LPL [412], but this could not be confirmed [467].

### 6.9. Metabolism of lipoproteins by cultured cells

Reviews of lipoprotein metabolism by cells in tissue culture can be found in references 332, 413—417.

*6.9.1. Low density lipoproteins.* Elucidation of the pathway for LDL metabolism by non-hepatic cells has been one of the major growth areas in the study of lipoprotein physiology in recent years. The human skin fibroblast has been the cell most investigated [414] and these studies will form the basis of the present summary, although similar pathways exist in a variety of extrahepatic cells, including arterial endothelium, intimal smooth muscle cells and blood lymphocytes [417, 418]. Studies of LDL metabolism in these in vitro systems have uncovered the biochemical basis for familial hypercholesterolemia.

In summary (Fig. 4), it can be said that LDL bind with high affinity to specialised areas of fibroblast plasma membrane. These are internalised by endocytosis with the attached LDL and fuse with lysosomes to form secondary lysosomes. Within these apoLDL is degraded to amino acids, which are excreted from the cell, while cholesteryl esters, the major lipid fraction of LDL, are hydrolysed. The liberated unesterified cholesterol serves as a source of cellular membrane cholesterol and exerts three regulatory actions on the LDL pathway: reduction of endogenous cholesterol synthesis by lowering the activity of 3-hydroxy-3-methylglutaryl CoA reductase (HMG CoA reductase: EC 1.1.1.34), stimulation of cholesterol esterification by fatty-acyl-CoA acyltransferase (ACAT, EC 2.3.1.26) and inhibition of synthesis of the plasma membrane LDL receptor.

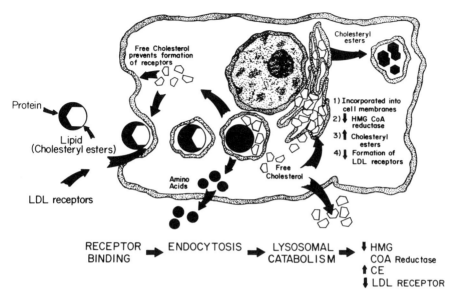

Fig. 4. Summary of LDL metabolism by extrahepatic cells as typified by the human skin fibroblast. LDL bind to specialised receptors in specific areas (coated pits) of the cell membrane whence they are internalised by endocytosis. The endocytotic vesicles fuse with primary lysosomes to form secondary lysosomes within which the LDL are degraded. The protein moiety of LDL is hydrolysed to its constituent amino acids or small peptides, which are excreted from the cell. Lipoprotein cholesteryl esters are hydrolysed to release unesterified cholesterol which exerts a variety of metabolic effects on cellular cholesterol metabolism. These include inhibition of endogenous cholesterol synthesis by reducing HMGCoA reductase activity, stimulation of the re-esterification of cholesterol by ACAT and limitation of further LDL uptake by decreasing the number of membrane LDL receptors.

*6.9.1.1. LDL binding.* The high affinity LDL receptor is probably protein since its activity is destroyed by proteolytic enzymes such as pronase and its regeneration can be prevented by cycloheximide [419, 420]. Binding of LDL to the receptor, which is largely independent of temperature between 4 and 37°C [379, 421] is maximal at about 25—50 µg LDL protein/ml [421].

The ability of LDL to bind to the receptor is attributed to their apoB content, since unlabeled VLDL and LDL, but not serum from a patient with abetalipoproteinemia will compete effectively with [$^{125}$I]-LDL for binding [422]. Moreover, a variety of modifications of apoB interfere with binding (see below). The arginine-rich protein (apoE) also enables lipoproteins to bind to the LDL receptor and to initiate the series of events which regulate cellular cholesterol metabolism (see below). This became apparent from studies with $HDL_c$, an apoE-containing lipoprotein isolated from cholesterol-fed swine [423]. Recently, a preparation of canine $HDL_c$ containing apoE as its only protein constituent was found to be effective in binding to human fibroblasts [424]. In fact apoE-containing $HDL_c$ bind much more effectively than apoB-containing LDL as judged by competitive binding assays [425]. The delipidated apoprotein is ineffective in binding, but its activity can be restored by complexing it with dimyristoylphosphatidylcholine [426]. This may be the result of the alteration in the conformation of apoE produced by complexing it with phospholipid [427]. A similar situation probably exists with LDL, in that LDL extracted with heptane, which have lost most of their cholesterol and triglyceride but which retain their apoprotein and phospholipid, maintain their ability to bind normally [428]. Incorporating apoA-I, apoC-II or apoC-III into phospholipid complexes did not enable them to interact with the LDL receptor [426].

A minor component of human $HDL_2$ has recently been identified which contains both apoE and a disulphide-linked complex of apoE and apoA-II [42]. This HDL subfraction binds to the fibroblasts' LDL receptor mainly by virtue of tis "free" apoE. The (E-A-II) complex is apparently incapable of binding unless split into its constituent apoproteins [429].

Charge appears to be an important factor in the binding of lipoproteins to the high affinity LDL receptor. Polyanionic compounds, such as heparin and dextran sulphate, can both prevent LDL binding and detach pre-bound LDL from the receptor [421]. It has been postulated that binding to the receptor depends on positive charges in the apoproteins concerned and that heparin may compete with the receptor for interaction with these groups. Some support for this is gained by the observation that a variety of positively charged proteins (including platelet factor 4) can inhibit [$^{125}$I]-LDL binding, though the mechanisms involved are uncertain [430]. Moreover, modification of the positively charged amino acids, arginine and lysine, in apoB and apoE interferes with the binding of LDL and HDL and has led to the conclusion that both lysine and arginine are required for binding [424, 431].

The situation is complex, however, since reductive methylation of lysyl residues, which preserves their positive charge also prevents binding [431]. These studies in which apoprotein composition is modified provide further evidence of the importance of the apoprotein component in determining binding of lipoproteins to cultured cells.

By the use of LDL labeled with ferritin cores, the LDL receptor has been localised in electron microscopic studies to specialised thickened areas of cell membrane known as coated regions or coated pits [432] which, although they occupy only 1.4% of the visualised linear membrane surface, contained more than 70% of the bound LDL in normal fibroblasts. The cell membranes of fibroblasts from a patient with homozygous familial hypercholesterolemia, however, did not bind LDL-ferritin in the region of the coated pits [432], in accordance with the absence of high-affinity [$^{125}$I]-LDL binding also observed in such subjects. Heterozygotes exhibit binding intermediate between that of normal subjects and homozygotes [433]. It is now apparent that at least three abnormalities can give rise to phenotypic familial hypocholesterolemia [414]. In addition to the receptor-negative type just described, patients who are receptor-defective (cells possess 5—20% of normal number of receptors) have been identified. A third type of patient has also been described whose cells bind LDL normally but fail to internalise the bound lipoprotein.

A number of hormones, known to influence lipoprotein metabolism in vivo, also modify the binding and degradation of [$^{125}$I]-LDL when applied to fibroblast cultures in physiological concentrations. Insulin [434, 435] and triiodothyronine [436] increase binding and degradation, while hydrocortisone has the opposite effect [437]. In each case the change in binding seems to be mediated by an alteration in receptor number. Regulation of LDL binding by cellular cholesterol content is further discussed below.

*6.9.1.2. LDL uptake and degradation.* Although LDL binds effectively to fibroblasts at 4°C, their metabolism proceeds no further. When cells, to which LDL have bound at 4°C, are warmed to 37°C, the bound lipoprotein is internalised as evidenced by acculumation of lipoprotein which is not releasable by heparin [415, 421]. Over a similar time-course, LDL-ferritin complexes bound to the putative high-affinity receptor in the coated regions of the cell membrane can be seen by electron microscopy to be

*References p. 489*

internalised by invagination of the membrane and to be incorporated into secondary lysosomes [415]. Within the lysosome LDL are degraded, their protein component being excreted into the incubation medium as amino acids and small peptides [419], while their cholesteryl esters are hydrolysed [438]. There seems little doubt that these processes occur in the lysosome since they are dependent on an acid pH and are inhibited by chloroquine, which inhibits lysosomal degradative processes but has no effect on LDL binding by the cell [420, 439]. Moreover, in patients with Wolman syndrome or cholesteryl ester storage disease, who lack a lysosomal acid lipase, the cultured cells accumulate cholesteryl esters of a fatty acid composition similar to those in plasma lipoproteins [440, 441].

*6.9.1.3. Regulation of cellular cholesterol metabolism.* The effect of LDL uptake by cultured cells on endogenous cholesterol synthesis appears to be through changes in the synthesis of the rate-limiting enzyme, HMGCoA reductase, since LDL produces decay of enzyme activity at the same rate as cycloheximide [442]. Moreover, the increase in enzyme activity which follows removal of lipoproteins from the culture medium can be blocked by cycloheximide [443]. In addition, LDL do not inhibit the enzyme in cell-free systems, militating against a direct inhibitory effect [442]. It is likely that cholesterol delivered to the cell by LDL is, in some unknown way, the mediator of suppression of enzyme activity since a similar effect can be produced by lipoprotein-free cholesterol dissolved in ethanol both in normal cells and in cells from patients with Familial Hypercholesterolemia (FH cells) [442]. Moreover, LDL, from which all the cholesterol and triglyceride have been removed by heptane extraction, are normally bound, internalised and degraded by fibroblasts, but are without effect on endogenous cholesterol synthesis [428]. A variety of cholesterol analogues, such as 7-keto- and 25-hydroxy-cholesterol, in lipoprotein-free form can also inhibit sterol synthesis, several of them more effectively than cholesterol [444]. For LDL to exert their effects on cellular cholesterol synthesis they have not only to be taken up by the cell, they have to be degraded within the lysosomes, which involves hydrolysis of their cholesteryl esters [438]. When lysosomal degradation of LDL is blocked with chloroquine, no effect on cholesterol synthesis is observed [439].

The inhibition of HMG-CoA reductase which follows uptake and degradation of LDL by fibroblasts is accompanied by a reciprocal increase in cholesterol esterifying activity within the cell due to a membrane bound fatty acyl coenzyme A: cholesterol acyltransferease (ACAT) [445, 446]. As with the regulation of HMG-CoA reductase activity the increase in ACAT activity depends on LDL uptake and degradation [439] and is not, therefore, produced by LDL in FH cells [445]. Once again, however, the effect can be produced by lipoprotein-free sterols in both normal and FH cells [446]. Regulation of ACAT activity is said to occur even when protein synthesis is inhibited by cycloheximide [414]. The fatty acid pattern of the cholesteryl esters produced differs from that in LDL with a predominance of mono- rather than polyunsaturated fatty acids [440].

The third regulatory event which follows LDL uptake is a decrease in subsequent LDL binding by the cell. This is thought to be due to an alteration in the number of available receptors since the affinity of binding is unaltered and the time-course ($t_{\frac{1}{2}} = 25$ h) of the change in binding is similar whether produced by LDL or cycloheximide treatment [420]. Similar effects on LDL binding to those produced by preincubation with unlabeled LDL can be produced by lipoprotein-free cholesterol and 25-hydroxycholesterol. The lipoprotein-mediated effect on binding is also blocked by chloroquine [420]. Recently, a more rapid (maximal at 4 h) transient change in LDL binding, also preventable by cycloheximide, has been observed when cells are transferred from fetal calf serum to lipoprotein-deficient serum. This seems to be due to an acute efflux of cholesterol from the cell, probably mediated by lipid-apoprotein complexes of d>1.25 present in "lipoprotein-deficient serum." $HDL_3$ (d 1.12—1.21) was able to promote a similar effect [447].

Thus, it appears that the high affinity receptor mechanism functions as a way of delivering cholesterol to cells. This can be by-passed, however, by using lipoprotein-free sterols in vitro. Although FH cells lack the mechanism for specific uptake of LDL, the regulatory mechanisms involving cellular cholesterol metabolism normally invoked by LDL uptake are intact and can be elicited in culture by lipoprotein-free sterols.

*References p. 489*

*6.9.1.4. The LDL pathway in vivo.* The observations that freshly isolated blood lymphocytes transferred into lipoprotein-deficient medium progressively increase their capacity to bind, take up and degrade LDL and the finding that whole body cholesterol synthesis is increased in abetalipoproteinemia have been cited as evidence for the LDL pathway to function in vivo [414]. The latter point, however, is far from established [448]. Recent studies in which the arginyl residues of apoLDL were modified with cyclohexanedione suggest that about 33—50% of overall LDL catabolism occurs via the specific LDL receptor in normal man [449] and animals [449A].

Patients who lack the LDL receptor, and as a result suffer from homozygous familial hypercholesterolemia, have a defect in LDL and apoB catabolism [21, 326], and suffer severe premature atherosclerosis [148]. It is easy to see that the former abnormality could be, and tempting to assume that the latter abnormality might be, related to the absence of a functional LDL pathway. Such subjects might also be expected to show an increased whole body cholesterol synthesis secondary to the inability of LDL to suppress cholesterol synthesis in peripheral cells. Such an abnormality is not a sine qua non of these patients and it may be that it is only present in younger patients, some other adaptive response coming into play with the passage of time [450].

It has been suggested [414] that the function of the LDL pathway is to supply cholesterol to peripheral cells at lipoprotein concentrations which are non-atherogenic. The pathway would seem to be well capable of this. Most peripheral cells, with the exception of vascular endothelium, are not exposed to full plasma LDL concentrations. There is a paucity of data about the lipoprotein concentrations in lymph bathing various peripheral tissues. Such data as are available [176] suggest that cholesterol and apoB concentrations are about 10% of those in plasma. Thus, for an individual who has an LDL-cholesterol of 100 mg/dl*, the LDL cholesterol in lymph is likely to be of the order of 10 mg/dl. This corresponds approximately to an LDL-protein of about 50 µg/ml

---

*This is the sort of LDL-cholesterol which may be seen in an individual with a total plasma cholesterol of 160 mg/dl; a low concentration by Western standards, and one rarely associated with atherosclerotic disease [11].

which is sufficient to saturate the LDL receptor. Much of the regulatory function of the receptor may therefore be lost in vivo, at least in man.

Even though the function of the LDL pathway may be to supply cholesterol to non-hepatic cells it has to be remembered that these cells generally possess the capacity for de novo cholesterol synthesis themselves [451, 452] — a capacity which is only fully realized when they are grown in a lipoprotein-free environment. Since with few exceptions, these cells lack the ability to further metabolise cholesterol, in the steady state the net flux of cholesterol will be, if anything, away from peripheral cells and towards the liver, where it can be excreted into the bile, either unchanged or as bile acids.

Certain cell types however, in particular those synthesising steroid hormones, are able to further metabolise cholesterol and excrete it from the cell. In order to perform this function maximally, at least in vitro, they require the presence of lipoproteins in the medium [453], their endogenous sterol synthesis being insufficient. In these in vitro systems LDL, but not HDL, can support sterol synthesis [453, 454]. This contrasts with findings in incubated tissues freshly removed from rats subjected to a variety of metabolic interventions [452, 454], which suggest that HDL-cholesterol, rather than LDL-cholesterol, is the more important regulator of endocrine-cell cholesterol metabolism and steroid hormone production. The apparent discrepancy may be explained on the basis of two independent receptor mechanisms, the one for HDL not being expressed in cultured adrenal cells [456]. The reason for this failure to detect the HDL pathway in vitro is unknown but a variety of hypotheses is available [456]. These experiments emphasize some of the difficulties which may be encountered when trying to extrapolate findings from in vitro culture systems to the intact organism. Some of the inconsistencies may be due to species differences.

*6.9.1.5. Non-specific uptake of LDL.* Not all LDL enters cultured cells by means of the high-affinity LDL receptor. In practice, receptor-independent uptake of LDL (non-specific uptake) could result from more than one process. This uptake was originally assumed to be due to bulk-phase pinocytosis of LDL molecules in solution in the medium [419], but it probably also includes adsorptive endocytosis of molecules which may be bound diffusely to the cell surface and which are not localised to the high-affinity LDL receptors [332, 457]. Non-specific uptake of LDL occurs in both normal and FH cells [419, 457]. It does not exhibit saturation kinetics and therefore can become quantitatively important at high LDL concentrations [419, 422]. As might be expected, destruction of the LDL receptor in normal cells with pronase does not prevent non-specific LDL uptake [419]. LDL taken up by the non-specific process appear to be degraded by the cells, and the TCA-soluble products of apoLDL digestion within the lysosomes are excreted into the medium, as with high affinity uptake [419, 457]. In the case of non-specific uptake, however, cellular cholesterol content and metabolism are not influenced [419, 445, 458]. The reasons for this are unknown.

*6.9.2. High density lipoproteins.* On a particle basis human HDL (d 1.09—1.21) bind almost as well to cultured fibroblasts as LDL [379]. For a given degree of binding however, they are much more slowly internalised and degraded and they do not increase cellular cholesterol or influence its metabolism [379, 442, 458]. It is likely that the bulk of these HDL preparations do not bind to the LDL receptor since binding is not markedly diminished in cells from patients with familial hypercholesterolemia. Indeed it can be increased [457], and it is uninfluenced by pronase treatment or preincubation of cells with lipoprotein-free sterols [379, 459]. It is calculated that HDL uptake can be accounted for by bulk-phase pinocytosis and non-specific adsorptive endocytosis, without the need to postulate specific localized receptors as for LDL [379].

Some of the results described above may represent the net effect of HDL preparations containing subfractions with different individual effects. Recent work with HDL subfractions separated by rate zonal ultracentrifugation shows that $HDL_2$ suppress fibroblast HMG CoA reductase activity while $HDL_3$ stimulate it [460]. The suppressive action of $HDL_2$ is attributable to their protein moiety (probably through an apoE-containing subfraction) and is likely to be mediated through the LDL receptor since it is absent in FH cells. The stimulatory effect of $HDL_3$ could be reproduced by extracted $HDL_3$ lipids and $HDL_2$ lipids were also stimulatory, possibly because of an ability to withdraw cholesterol from cultured cells.

*References p. 489*

*6.9.3. Interaction between high and low density lipoproteins.* HDL preparations have been found to be capable of reducing [$^{125}$I]LDL binding to cultured cells [422, 461]. This effect seems to be mediated through both the high-affinity and non-specific processes and HDL in sufficient concentrations can reduce cellular accumulation of cholesterol due to LDL [461]. A phenomenon of this sort may underly observations that HDL-cholesterol concentrations are negatively correlated with tissue cholesterol pools and the incidence of ischemic heart disease [192, 193, 359]. The mechanism for HDL-mediated inhibition of LDL binding to fibroblasts and other cell types is unclear. Although most of the HDL molecules in these preparations do not appear to bind to the LDL receptor it is possible that the effect is at least partly due to an HDL subfraction containing apoE. Such subfractions do bind to the high affinity receptor as judged by their ability to influence fibroblast cholesterol metabolism in normal subjects, but not in subjects with familial hypercholesterolemia [423, 429].

In longer term incubations ($\geq$ 48 h) the effect of HDL on LDL binding is reversed [462] and [$^{125}$I]-LDL are bound to a greater extent when HDL are present. This seems to be due to the ability of HDL to promote cholesterol efflux from the cells [447, 462]. This too has been proposed as a mechanism for the observed negative relationship between HDL-cholesterol concentration and tissue cholesterol pools [193]. In vivo where cells are presumably exposed to relatively constant concentrations of lipoproteins it may be that the longer term effect of HDL on cellular cholesterol content will outweigh the initial effect of HDL in reducing LDL-binding when lipoprotein-deficient serum is replaced by medium containing lipoproteins.

The interaction of individual HDL subfractions with LDL metabolism in cultured cells has not been studied but it seems possible that the initial inhibition of LDL binding by mixed HDL populations is due to apoE-containing HDL particles competing for the LDL receptor while the secondary increase in LDL binding results from the removal of cellular cholesterol by HDL lipids.

*6.9.4. Very low density lipoproteins.* The metabolism of triglyceride-rich lipoproteins by fibroblasts in culture has been relatively little studied. VLDL have been found to compete with [$^{125}$I]-

LDL for binding and are presumably taken up and degraded since they suppress HMG CoA reductase activity [442]. However, the VLDL used in these studies were obtained from hypertriglyceridemic patients. Moreover, VLDL isolated by preparative ultracentrifugation even from normal subjects can be contaminated with IDL [463]. When VLDL, free of IDL, are isolated from normal plasma by rate zonal ultracentrifugation they no longer suppress HMG CoA reductase activity, though those similarly prepared from patients with types III, IV and V hyperlipoproteinemia are as suppressive as LDL [463].

Detailed studies using individual VLDL subfractions ($S_f$ 100—400, 60—100, 20—60) from the plasma of hypertriglyceridemic subjects show that they are all effective in enzyme suppression in normal cells but not in FH cells implying interaction with the LDL receptor. The protein moiety of VLDL seems to be involved in the effect since modification of arginyl residues with 1,2-cyclohexanedione renders the lipoprotein non-suppressive [464, 465]. It appears, therefore, that in hypertriglyceridemic subjects the ability of VLDL to suppress HMG CoA reductase is not confined to the flotation range traditionally associated with remnant particles ($S_f$ 12—60). In normal subjects VLDL of $S_f > 60$ are non-suppressive, but become so if incubated with bovine milk lipoprotein lipase in vitro [466] leading to the conclusion that all the necessary determinants for lipoprotein uptake and enzyme suppression are present in the orginal particle and that structural changes induced by lipolysis allow them to be expressed. Why hypertriglyceridemic VLDL of $S_f$ 60—400 are suppressive is uncertain, but it may be that these particles include chylomicron remnants which have been partially degraded.

Although the quantitative aspects are unclear these in vitro studies suggest a possible metabolic alternative for triglyceride-rich lipoproteins to the removal of remnants by the liver and may account for at least a part of the VLDL-B which leaves plasma in hypertriglyceridemic subjects without appearing as LDL-B (Section 6.4.).

## 7. Conclusion

Major advances have been made in our understanding of lipopro-

## CONCLUSION

tein metabolism during the last decade. These have stemmed in large part from the identification and isolation of the main plasma apolipoproteins. The primary structure of several of these is now known (though that of apoB and apoE remain major unsolved problems) allowing their laboratory synthesis in whole or in part. This has permitted a better understanding of their roles in lipid binding and the modulation of the activities of enzymes important in lipoprotein metabolism. Simultaneously, immunochemical assays for most of the apolipoproteins have been developed and these are now being widely used in studies of lipoprotein metabolism in health and disease.

Partially purified preparations of hepatic and lipoprotein lipases have been obtained and a variety of chemical and immunological methods have been developed to separate their activities. Nevertheless, the metabolic role of hepatic triglyceride lipase, especially, remains uncertain and a proper understanding of the mechanism of action of this enzyme and lipoprotein lipase will probably only follow the determination of their primary structures. Although lecithin:cholesterol acyltransferease has been obtained in pure form, we still await determination of its primary structure and the development of better chemical and immunoassays to improve our understanding of the regulation of LCAT activity in different metabolic situations.

Recently, a variety of factors in the plasma protein fraction which facilitate the exchange of non-polar lipids and phospholipids between lipoproteins has been identified. The cholesteryl ester exchange protein may be apoD but the other factors remain to be characterised. Doubtless considerable effort will be devoted to this in the near future as well as to developing a better understanding of the metabolic importance of these exchange proteins.

The complex metabolic interrelationships between different lipoprotein classes are only now becoming clear. The progressive delipidation of VLDL to yield LDL (possibly entirely within the circulation) also leads to increases in HDL mass by virtue of surface components simultaneously lost from VLDL. It is likely that this leads to the intraplasmatic conversion of $HDL_3$ to $HDL_2$. Understanding of these metabolic pathways and of ways of manipulating them may be of great clinical importance in view of the relationships identified epidemiologically between these lipoprotein fractions and vascular disease.

The study of lipoprotein metabolism in cultured cells has identified the metabolic defect in familial hypercholesterolemia and also a major potential site for the catabolism of lipoproteins in vivo especially for LDL but possibly also for remnants of triglyceride-rich lipoproteins and HDL. Much more information is required on the relative roles of the liver and extrahepatic tissues in lipoprotein catabolism and on factors which permit the identification and uptake of triglyceride-rich lipoproteins and their remnants at these sites.

Fruitful as the last decade may have been, there remains no shortage of unsolved problems for the future.

## 8. Acknowledgements

This review was completed during November, 1979. A few modifications and a small number of additional references were incorporated in December 1980.

We are grateful to Ms. Sharon Bonnot, Kay Halfant, Barbara Allen, Susan Davies and Debbie Mason for preparing the manuscript and to Doctors William A. Bradley, Iain F. Craig, Sandra H. Gianturco, Joan B. Karlin, Joel D. Morrisett, Josef R. Patsch, Henry J. Pownall, Louis C. Smith and James T. Sparrow for helpful discussion. We also thank those workers who sent us manuscripts accepted for publication before their appearance in print.

During the preparation of this review J. P. M. was on leave of absence from the University Hospital of South Manchester, Manchester, U.K. and he is grateful to the Medical Research Council of England for a Travelling Fellowship.

Work from the Department of Medicine, Baylor College of Medicine and The Methodist Hospital, Houston, Texas, described in this review was supported in part by a grant from the National Heart, Lung and Blood Institute for the National Heart and Blood Vessel Research and Demonstration Center (HL 17269) and a grant from the Fritz Thyssen Stiftung.

## REFERENCES

1 F. T. Lindgren, L. C. Jensen and F. T. Hatch, in G. J. Nelson (Ed.), Blood Lipids and Lipoproteins: Quantitation, Composition and Metabolism, Wiley-Interscience, New York, 1972, pp. 181—274.
2 W. Patsch, J. R. Patsch, G. M. Kostner, S. Sailer and H. Braunsteiner, J. Biol. Chem., 253 (1978) 4911.
3 J. R. Patsch, S. Sailer, G. Kostner, F. Sandhofer, A. Holasek and H. Braunsteiner, J. Lipid Res., 15 (1974) 356.
4 F. T. Lindgren, M. Shen, R. M. Krauss and T. M. Sargent, in K. Lippel (Ed.), HDL. Report of the High Density Lipoprotein Methodology Workshop, U.S.D.H.E.W., N.I.H. Publication 79—1661, 1979, pp. 45—51.
5 R. M. Krauss, F. T. Lindgren, A. Wong, B. Anderson, S. S. Pan and S. B. Lewis, in K. Lippel (Ed.), HDL. Report of the High Density Lipoprotein Methodology Workshop, U.S.D.H.E.W., N.I.H. Publication 79—1661, 1979, pp. 114—123.
6 T. Sata, R. J. Havel and A. L. Jones, J. Lipid Res., 13 (1972) 757.
7 T. Sata, D. L. Estrich, P. D. S. Wood and L. W. Kinsell, J. Lipid Res., 11 (1970) 331.
8 L. L. Rudel, J. A. Lee, M. D. Morris and J. M. Felts, Biochem. J., 139 (1974) 89.
9 M. Burstein and H. R. Scholnick, Adv. Lipid Res., 11 (1973) 68.
10 G. R. Warnick, M. C. Cheung and J. J. Albers, Clin. Chem., 25 (1979) 596.
11 A. M. Gotto, Atherosclerosis Reviews, 4 (1979) 17.
12 D. B. Zilversmit, Circulation, 60 (1979) 473.
13 F. T. Lindgren, A. V. Nichols and R. M. Krauss (Eds.), Symposium: High Density Lipoproteins (HDL), American Oil Chemists Society, Champaign, IL, 1979, Individual articles published in Lipids, Vols. 13—14.
14 E. C. McCormick, D. G. Cornwell and J. B. Brown, J. Lipid Res., 1 (1960) 221.
15 Z. Chen and A. Danon, Biochem. Pharm., 28 (1979) 267.
16 T. C. Chen, W. A. Bradley, A. M. Gotto and J. D. Morrisett, FEBS Lett., 104 (1979) 236.
16A F. P. Bell, Prog. Lipid Res., 17 (1978) 207.
17 S. Eisenberg and R. I. Levy, Adv. Lipid Res., 13 (1975) 1.
18 R. L. Jackson, J. D. Morrisett and A. M. Gotto, Physiol. Rev., 56 (1976) 259.
19 J. C. Osborne and H. B. Brewer, Adv. Prot. Chem., 31 (1977) 253.
20 L. C. Smith, H. J. Pownall and A. M. Gotto, Ann. Rev. Biochem., 47 (1978) 751.
21 E. J. Schaefer, S. Eisenberg and R. I. Levy, J. Lipid Res., 19 (1978) 667.
22 G. S. Getz and R. V. Hay, in A. M. Scanu, R. W. Wissler, and G. S. Getz (Eds.), The Biochemistry of Atherosclerosis, Dekker, New York, 1979, pp. 151—188.
23 V. P. Skipski in G. J. Nelson (Ed.), Blood Lipids and Lipoproteins:

Quantitation, Composition and Metabolism, Wiley-Interscience, New York, 1972, pp. 471—583.
24   A. Yokoyama and D. B. Zilversmit, J. Lipid Res., 6 (1965) 241.
25   R. M. Glickman and K. Kirsch, Biochim. Biophys. Acta., 371 (1974) 255.
26   R. W. Mahley, B. D. Bennett, J. Morré, M. E. Gray, W. Thistlethwaite and V. S. Lequire, Lab. Invest., 25 (1971) 435.
27   D. B. Zilversmit, J. Clin. Invest., 44 (1965) 1610.
28   D. B. Zilversmit in J. M. Dietschy, A. M. Gotto and J. A. Ontko (Eds.), Disturbances in Lipid and Lipoprotein Metabolism, American Physiological Society, Bethesda, MD, 1978, pp. 69—81.
29   R. M. Glickman, K. Kirsch and K. J. Isselbacher, J. clin. Invest., 51 (1972) 356.
30   R. K. Ockner, J. P. Pittman and J. L. Yager, Gastroenterology, 62 (1972) 981.
31   J. P. Kane, T. Sata, R. L. Hamilton and R. J. Havel, J. clin. Invest., 56 (1975) 1622.
32   V. G. Shore and B. Shore, Biochemistry, 12 (1973) 502.
33   M. Barclay in G. J. Nelson (Ed.), Blood Lipids and Lipoproteins: Quantitation, Composition and Metabolism, Wiley-Interscience, New York, 1972, pp. 585—704.
34   J. R. Patsch, S. Sailer and H. Braunsteiner, Europ. J. clin. Invest., 5 (1975) 45.
35   R. Fellin, B. Agostini, W. Rost and D. Seidel, Clin. Chim. Acta., 54 (1974) 325.
36   S. Eisenberg, D. W. Bilheimer, R. I. Levy and F. T. Lindgren, Biochim. Biophys. Acta, 326 (1973) 361.
37   J. R. Patsch, S. Sailer, H. Braunsteiner and T. Forte, Europ. J. clin. Invest., 6 (1976) 307.
38   R. I. Levy, D. W. Bilheimer and S. Eisenberg in R. M. S. Smellie (Ed.), Plasma Lipoproteins, Biochemical Society Symposia No. 33, Academic Press, London, 1971.
39   S. Eisenberg in C. E. Day and R. S. Levy (Eds.), Low Density Lipoproteins, Plenum Press, New York, 1976, pp. 73—92.
40   W. Patsch, G. Schonfeld, A. M. Gotto and J. R. Patsch, J. Biol. Chem., 255 (1980) 3178.
41   E. J. Schaefer, D. Foster, L. Jenkins, F. Lindgren, M. Berman, R. Levy and H. Brewer, Lipids, 14 (1979) 511.
42   K. H. Weisgraber and R. W. Mahley, J. Biol. Chem., 253 (1978) 6281.
43   K. Berg in A. M. Scanu, R. W. Wissler and G. S. Getz (Eds.), The Biochemistry of Atherosclerosis, Dekker, New York, 1979, pp. 419—490.
44   G. M. Kostner in C. E. Day and R. S. Levy (Eds.), Low Density Lipoproteins, Plenum Press, New York, 1976, pp. 229—269.
45   K. Simons, C. Enholm, O. Renkonen and B. Bloth, Acta Pathol. Microbiol. Scand., B78 (1970) 459.
46   C. Enholm, H. Garoff, O Renkonen and K. Simons, Biochemistry, 11 (1972) 3229.
47   G. Jürgens and G. M. Kostner, Immunogenetics, 1 (1975) 560.

| | |
|---|---|
| 48 | A. K. Rider, R. I. Levy and D. S. Frederickson, Circulation, 42 (1970) III—10. |
| 49 | K. Berg, G. Dahlen and M. H. Frick, Clin. Genet., 6 (1974) 230. |
| 50 | J. J. Albers, V. G. Cabana, G. R. Warnick and W. R. Hazzard, Metabolism, 24 (1975) 1047. |
| 51 | A. Pagnan, R. J. Havel, J. P. Kane and L. Kotite, J. Lipid Res., 18 (1977) 613. |
| 52 | G. Dahlen, C. Ericson, C. Furberg, L. Lundkvist and K. Svärdsudd, Acta Med. Scand. Suppl., 531 (1972) 1. |
| 53 | M. H. Frick, G. Dahlen, K. Berg, M. Valle and P. Hekali, Chest, 73 (1978) 62. |
| 54 | J. J. Albers, J. L. Adolphson and W. R. Hazzard, J. Lipid Res., 18 (1977) 331. |
| 55 | K. W. Walton, J. Hitchens, H. N. Magnani and M. Khan, Atherosclerosis, 20 (1974) 323. |
| 56 | C. Ericson, G. Dahlen and K. Berg, Clin. Genet., 11 (1977) 433. |
| 57 | G. Dahlen, C. Ericson and K. Berg, Clin. Genet., 14 (1978) 115. |
| 58 | G. Dahlen, and K. Berg, Clin. Genet., 16 (1979) 418. |
| 59 | J. J. Albers, P. Wahl and W. R. Hazzard, Biochem. Genet., 11 (1974) 475. |
| 60 | D. Hewitt, J. Milner, C. Breckenridge and G. Maguire, Clin. Genet., 11 (1977) 224. |
| 61 | J. J. Albers and W. R. Hazzard, Lipids, 9 (1974) 15. |
| 62 | J. S. Schultz, D. C. Shreffler, C. F. Sing and N. R. Harvie, Ann. Hum. Genet., 38 (1974) 39. |
| 63 | C. F. Sing. J. S. Schultz and D. C. Shreffler, Ann. Hum. Genet., 38 (1974) 47. |
| 64 | H. A. Eder, E. M. Russ, R. A. R. Pritchett, M. M. Wilber and D. P. Barr, J. clin. Invest., 34 (1955) 1147. |
| 65 | S. Switzer, J. clin. Invest., 46 (1967) 1855. |
| 66 | D. Seidel, P. Alaupovic and R. H. Furman, J. clin. Invest., 48 (1969) 1211. |
| 67 | J. A. Glomset and K. R. Norum, Adv. Lipid. Res., 11 (1973) 1. |
| 68 | J. R. Patsch, K. C. Aune, A. M. Gotto and J. D. Morrisett, J. biol. Chem., 252 (1977) 2113. |
| 69 | R. L. Hamilton, R. J. Havel, J. P. Kane, A. E. Blaurock and T. Sata, Science, 172 (1971) 475. |
| 70 | D. Seidel, H. Gretz and C. Ruppert, Clin. Chem., 19 (1973) 86. |
| 71 | J. A. Glomset, K. R. Norum, A. V. Nichols, W. C. King, C. D. Mitchell, K. R. Applegate, E. L. Gong and E. Gjone, Scand. J. clin. Lab. Invest., 35 (1975) Suppl. 142, 3—30. |
| 72 | E. Manzato, R. Fellin, G. Baggio, S. Walch, W. Neubeck and D. Seidel, J. clin. Invest., 57 (1976) 1248. |
| 73 | D. S. Harry, R. C. Day, J. S. Owen, J. Agorastas, A. Y. Foo and N. McIntyre, Scand. J. clin. Lab. Invest., 38 (1978) Supp. 150, 223—227. |
| 74 | W. Patsch. J. R. Patsch, F. Kunz, S. Sailer and H. Braunsteiner, Europ. J. clin. Invest., 7 (1977) 523. |

75 J. R. Patsch, A. K. Soutar, J. D. Morrisett, A. M. Gotto and L. C. Smith, Europ. J. clin. Invest., 7 (1977) 213.
76 A. R. Tall and D. M. Small, New Engl. J. Med., 299 (1978) 1232.
77 J. P. Segrest, R. L. Jackson, J. D. Morrisett and A. M. Gotto, FEBS Lett., 38 (1974) 247.
78 R. L. Jackson, J. D. Morrisett, A. M. Gotto and J. P. Segrest, Mol. Cell. Biochem., 6 (1975) 43.
79 T. Forte and A. V. Nichols, Adv. Lipid Res., 10 (1972) 1.
80 J. D. Morrisett. R. L. Jackson and A. M. Gotto, Biochim. Biophys. Acta, 472 (1977) 93.
81 W. A. Bradley and A. M. Gotto, in J. M. Dietschy, A. M. Gotto and J. A. Ontko (Eds.), Disturbances in Lipid and Lipoprotein Metabolism, American Physiological Society, Bethesda, MD, 1978, pp. 111—137.
82 C. Edelstein, F. J. Kezdy, A. M. Scanu and B. W. Shen, J. Lipid Res., 20 (1979) 143.
83 J. B. Karlin, D. J. Juhn, R. Goldberg and A. H. Rubenstein, Ann. Clin. Lab. Sci., 8 (1978) 142.
84 J. B. Karlin and A. H. Rubenstein, in A. M. Scanu, R. W. Wissler and G. S. Getz (Eds.), The Biochemistry of Atherosclerosis, Dekker, New York, 1979, pp. 189—227.
85 H. N. Baker, T. Delahunty, A. M. Gotto and R. L. Jackson, Proc. nat. Acad. Sci., 71 (1974) 3631.
86 T. Delahunty, H. N. Baker, A. M. Gotto and R. L. Jackson, J. biol. Chem., 250 (1975) 2718.
87 H. N. Baker, A. M. Gotto and R. L. Jackson, J. biol. Chem., 250 (1975) 2725.
88 H. B. Brewer, T. Fairwell, A. LaRue, R. Ronan, A. Houser and T. J. Bronzert, Biochem. Biophys. Res. Commun., 80 (1978) 623.
89 G. Schonfeld and B. Pfleger, J. clin. Invest., 54 (1974) 236.
90 J. B. Karlin, D. J. Juhn, J. I. Starr, A. M. Scanu and A. H. Rubenstein, J. Lipid Res., 17 (1976) 30.
91 J. P. Miller, S. J. T. Mao, J. R. Patsch and A. M. Gotto, J. Lipid Res., 21 (1980) 775.
92 S. J. T. Mao, J. P. Miller, A. M. Gotto and J. T. Sparrow, J. biol. Chem., 255 (1980) 3448.
93 G. Schonfeld, J. -S. Chen and R. G. Roy, J. biol. Chem., 252 (1977) 6651.
94 G. Schonfeld, J. -S. Chen and R. G. Roy, J. biol. Chem., 252 (1977) 6655.
95 H. B. Brewer, S. E. Lux, R. Ronan and K. M. John, Proc. nat. Acad. Sci., 69 (1972) 1304.
96 S. J. T. Mao, J. T. Sparrow, E. B. Gilliam, A. M. Gotto and R. L. Jackson, Biochemistry, 16 (1977) 4150.
97 T. C. Chen, J. T. Sparrow, A. M. Gotto and J. D. Morrisett, Biochemistry, 18 (1979) 1617.
98 S. J. T. Mao, A. M. Gotto and R. L. Jackson, Biochemistry, 14 (1975) 4127.
99 P. N. Herbert, A. M. Gotto and D. S. Fredrickson, in J. B. Stanbury, J.

B. Wyngaarden and D. S. Fredrickson (Eds.), The Metabolic Basis of Inherited Disease, McGraw-Hill, New York, 1978, pp. 544—588.
99A J. P. Kane, D. A. Hardman and H. E. Paulus, Proc. nat. Acad. Sci., 77 (1980) 2465.
99B K. V. Krishnaiah, L. F. Walker, J. Borensztajn, G. Schonfeld and G. S. Getz, Proc. nat. Acad. Sci., 77 (1980) 3806.
100 R. L. Jackson, J. T. Sparrow, H. N. Baker, J. D. Morrisett, O. D. Taunton and A. M. Gotto, J. Biol. Chem., 249 (1974) 5308.
101 R. S. Schulman, P. N. Herbert, K. Wehrly and D. S. Fredrickson, J. Biol. Chem., 250 (1975) 182.
102 R. L. Jackson, J. D. Morrisett, J. T. Sparrow, J. P. Segrest, H. J. Pownall, L. C. Smith, H. F. Hoff and A. M. Gotto, J. biol. Chem., 249 (1974) 5314.
103 G. F. Sigler, A. K. Soutar, L. C. Smith, A. M. Gotto and J. T. Sparrow, Proc. Natl. Acad. Sci., 73 (1976) 1422.
104 D. R. K. Harding, J. E. Battersby, D. R. Husbands and W. S. Hancock, J. Amer. Chem. Soc., 98 (1976) 2664.
105 R. L. Jackson, H. N. Baker, E. B. Gilliam and A. M. Gotto, Proc. nat. Acad. Sci., 74 (1977) 1942.
106 P. K. J. Kinnunen, R. L. Jackson, L. C. Smith, A. M. Gotto and J. T. Sparrow, Proc. nat. Acad. Sci., 74 (1977) 4848.
107 J. T. Sparrow and A. M. Gotto, Circulation, 58 (1978) II—77.
108 J. T. Sparrow, H. J. Pownall, F. -J. Hsu, L. D. Blumenthal, A. R. Culwell and A. M. Gotto, Biochemistry, 16 (1977) 5427.
109 W. J. McConathy and P. Alaupovic, FEBS Lett., 37 (1973) 178.
110 G. M. Kostner, Biochim. Biophys. Acta, 336 (1974) 383.
111 G. M. Kostner, Scand. J. clin. Lab. Invest., 33 (1974) Suppl. 137, 19—21.
112 A. K. Soutar, C. W. Garner, H. N. Baker, J. T. Sparrow, R. L. Jackson, A. M. Gotto and L. C. Smith, Biochemistry, 14 (1975) 3057.
113 T. Chajek and C. J. Fielding, Proc. nat. Acad. Sci., 75 (1978) 3445.
114 M. D. Curry, W. J. McConathy, P. Alaupovic, J. H. Ledford and M. Popovic, Biochim. Biophys. Acta, 439 (1976) 413.
115 H. B. Brewer, R. Shulman, P. Herbert, R. Ronan, K. Wehrly, J. biol. Chem., 249 (1974) 4975.
116 R. J. Havel and J. P. Kane, Proc. nat. Acad. Sci., 70 (1973) 2015.
117 F. A. Shelburne and S. H. Quarfordt, J. biol. Chem., 249 (1974) 1428.
118 R. S. Kushwaha, W. R. Hazzard, P. W. Wahl and J. J. Hoover, Ann. Intern. Med., 87 (1977) 509.
119 G. Utermann, H. J. Menzel and K. H. Langer, FEBS Lett., 45 (1974) 29.
120 G. Utermann, Hoppe-Seyler's Z. Physiol. Chem., 356 (1975) 1113.
121 K. H. Weisgraber, R. W. Mahley and G. Assmann, Atherosclerosis, 28 (1977) 121.
122 G. Utermann, W. Weber and U. Beisiegel, FEBS Lett., 101 (1979) 21.
123 T. Sata, R. J. Havel, L. Kotite and J. P. Kane, Proc. nat. Acad. Sci., 73 (1976) 1063.
124 D. S. Robinson, in M. Florkin and E. H. Stotz (Eds.), Comprehensive Biochemistry, Vol. 18, Elsevier, Amsterdam, 1970, pp. 51—116.
125 R. W. Scow, E. J. Blanchette-Mackie and L. C. Smith, Circ. Res., 39 (1976) 149.

126 C. J. Fielding, in J. M. Dietschy, A. M. Gotto and J. A. Ontko (Eds.), Disturbances in Lipid and Lipoprotein Metabolism, American Physiological Society, Bethesda, MD, 1978, pp. 83—98.
127 M. H. Tan, Can. Med. Assoc. J., 118 (1978) 675.
128 J. Borensztajn, in A. M. Scanu, R. W. Wissler and G. S. Getz (Eds.), The Biochemistry of Atherosclerosis, Dekker, New York, 1979, pp. 231—245.
129 J. Augustin and H. Greten, Atherosclerosis Reviews, 5 (1979) 91.
129A P. Nilsson-Ehle, A. S. Garfinkel and M. C. Schotz, Ann. Rev. Biochem., 49 (1980) 667.
130 P. E. Fielding, V. G. Shore and C. J. Fielding, Biochemistry, 13 (1974) 4318.
131 P. W. Fielding, V. G. Shore and C. J. Fielding, Biochemistry, 16 (1977) 1896.
132 C. J. Fielding, Biochemistry, 15 (1976) 879.
133 J. Augustin, H. Freeze, J. Bogerg and W. V. Brown, in H. Greten (Ed.), Lipoprotein Metabolism, Springer-Verlag; Berlin, Heidelberg, New York, 1976, pp. 7—12.
134 J. Augustin, H. Freeze, P. Tejada and W. V. Brown, J. biol. Chem., 253 (1978) 2912.
135 A. -M. Östlund-Lindquist and J. Boberg, FEBS Lett., 83 (1977) 231.
136 A. -M. Östlund-Lindquist, Biochem. J., 179 (1979) 555.
137 R. J. Havel, V. G. Shore, B. Shore and D. M. Bier, Circ. Res., 27 (1970) 595.
138 J. C. LaRosa, R. I. Levy, P. Herbert, S. E. Lux and D. S. Fredrickson, Biochem. Biophys. Res. Commun., 41 (1970) 57.
139 R. M. Krauss, H. G. Windmueller, R. I. Levy and D. S. Fredrickson, J. Lipid Res., 14 (1973) 286.
140 A. L. Miller and L. C. Smith, J. biol. Chem., 248 (1973) 3359.
141 C. J. Fielding and P. E. Fielding, J. Lipid Res., 17 (1976) 248.
142 J. Chung and A. M. Scanu, J. biol. Chem., 252 (1977) 4202.
143 T. A. Musliner, E. C. Church, P. N. Herbert, M. J. Kingston and R. S. Schulman, Proc. nat. Acad. Sci., 74 (1977) 5358.
144 D. Ganesan and H. B. Bass, FEBS Lett., 53 (1975) 1.
145 A. Bensadoun, C. Enholm, D. Steinberg and W. V. Brown, J. biol. Chem., 249 (1974) 2220.
146 W. C. Breckenridge, J. A. Little, G. Steiner, A. Chow and M. Poapst, New Engl. J. Med., 298 (1978) 1265.
147 T. Yamamura, H. Sudo, K. Ishikawa and A. Yamamoto, Atherosclerosis, 34 (1979) 53.
148 D. S. Fredrickson, J. L. Goldstein and M. S. Brown, in J. B. Stanbury, J. B. Wyngaarden and D. S. Fredrickson (Eds.), The Metabolic Basis of Inherited Disease, McGraw Hill, New York, 1978, pp. 604—655.
149 D. W. Cox, W. C. Breckenridge and J. A. Little, New Engl. J. Med., 299 (1978) 1421.
150 A. M. Gotto, P. K. J. Kinnunen, J. T. Sparrow, A. L. Catapano, R. L. Jackson, L. C. Smith, W. C. Breckenridge and J. A. Little, Clin. Res., 26 (1978) 483A.

# REFERENCES

151 R. J. Havel and R. S. Gordon, J. Clin. Invest., 39 (1960) 1777.
152 R. M. Krauss, R. I. Levy and D. S. Fredrickson, J. clin. Invest., 54 (1974) 1107.
153 H. Greten, R. DeGrella, G. Klose, W. Rascher, J. L. De Gennes and E. Gjone, J. Lipid Res., 17 (1976) 203.
154 I. P. Kompiang, A. Bensadoun and M. -W. Wang Yang, J. Lipid Res., 17 (1976) 498.
155 C. J. Fielding and J. M. Higgins, Biochemistry, 13 (1974) 4324.
156 T. J. Kotlar and J. Borensztajn, Amer. J. Physiol., 233 (1977) E316.
157 J. K. Huttunen, C. Enholm, P. K. J. Kinnunen and E. A. Nikkilä, Clin. Chim. Acta, 63 (1975) 335.
158 M. L. Baginsky and W. V. Brown, J. Lipid Res., 20 (1979) 548.
159 C. Enholm. J. K. Huttunen, P. J. Kinnunen, T. A. Miettinen and E. A. Nikkilä, New Engl. J. Med., 292 (1975) 1314.
160 G. Klose, J. Windelband, A. Weizel and H. Greten, Eur. J. clin. Invest., 7 (1977) 557.
161 E. A. Nikkilä, J. K. Huttunen and C. Enholm, Diabetes, 26 (1977) 1.
162 H. Greten, V. Laible, G. Zipperle and J. Augustin, Atherosclerosis, 26 (1977) 563.
163 R. Mordasini, F. Frey, W. Flury, G. Klose and H. Greten, New Engl. J. Med., 297 (1977) 1362.
164 J. K. Huttunen, C. Enholm, E. A. Nikkilä and M. Ohta, Europ. J. clin. Invest., 5 (1975) 435.
165 T. Kuusi, E. A. Nikkilä, I. Virtanen and P. K. J. Kinnunen, Biochem. J., 181 (1979) 245.
166 T. Kuusi, P. K. J. Kinnunen and E. A. Nikkilä, FEBS Lett., 104 (1979) 384.
167 J. A. Glomset, J. Lipid Res., 9 (1968) 155.
168 J. A. Glomset, in G. J. Nelson (Ed.), Blood Lipids and Lipoproteins: Quantitation, Composition and Metabolism, Wiley-Interscience, New York, 1972, pp. 745—787.
169 E. Gjone and K. R. Norum (Eds.), Scand. J. Clin. Lab. Invest., 33 (1974) Suppl. 137.
170 E. Gjone, K. R. Norum and J. A. Glomset, in J. B. Stanbury, J. B. Wyngaarden and D. S. Fredrickson (Eds.), The Metabolic Basis of Inherited Disease, McGraw-Hill, New York, 1978, pp. 589—603.
171 E. Gjone (Ed.), Scand. J. clin. Lab. Invest., 38 (1978) Suppl. 150.
172 J. A. Glomset, in A. M. Scanu, R. W. Wissler and G. S. Getz (Eds.), The Biochemistry of Atherosclerosis, Dekker, New York, 1979, pp. 247—273.
173 G. Nordby, T. Berg. M. Nilsson and K. R. Norum, Biochim. Biophys. Acta, 450 (1976) 69.
174 S. B. Clarke and K. R. Norum, J. Lipid Res., 18 (1977) 293.
175 J. A. Glomset and D. M. Kaplan, Biochim. Biophys. Acta, 98 (1965) 41.
176 D. Reichl, L. A. Simons, N. B. Myant, J. J. Pflug and G. L. Mills, Clin. Sci. Mol. Med., 45 (1973) 313.
177 K. T. Stokke, N. B. Fjeld, T. H. Kluge and S. Skrede, Scand. J. clin. Lab. Invest., 33 (1974) 199.

178 D. S. Sgoutas, Biochemistry, 11 (1972) 293.
179 G. Assmann, G. Schmitz, N. Donath and D. Lekim, Scand. J. clin. Lab. Invest., 38 (1978) Suppl. 150, 16—20.
180 S. Skrede and K. T. Stokke, Scand. J. clin. Lab. Invest., 33 (1974) 97.
181 G. Nordby and K. R. Norum, Scand. J. clin. Lab. Invest., 35 (1975) 677.
182 Y. Akanuma and J. Glomset, J. Lipid Res., 9 (1968) 620.
183 C. J. Fielding and P. E. Fielding, FEBS Lett., 15 (1971) 355.
184 C. J. Fielding, V. G. Shore and P. E. Fielding, Biochem. Biophys. Res. Commun., 46 (1972) 1493.
185 A. K. Soutar, G. F. Sigler, L. C. Smith, A. M. Gotto and J. T. Sparrow, Scand. J. clin. Lab. Invest., 38 (1978) Suppl. 150, 53—58.
186 J. J. Albers, J. Lin and G. P. Roberts, Artery, 5 (1979) 61.
187 E. Gjone, Scand. J. clin. Lab. Invest., 33 (1974) Suppl. 137, 73—82.
188 T. Forte, A. Nichols, J. Glomset and K. Norum, Scand. J. clin. Lab. Invest., 33 (1974) Suppl. 137, 121—132.
189 J. R. Murphy, J. Lab. clin. Med., 60 (1962) 86.
190 D. Reichl, A. Postiglione, N. B. Myant, J. J. Pflug and M. Press, Clin. Sci. mol. Med., 49 (1975) 419.
191 O. Stein, R. Goren and Y. Stein, Biochim. Biophys. Acta, 529 (1978) 309.
192 G. J. Miller and N. E. Miller, Lancet, 1 (1975) 16.
193 N. E. Miller, P. J. Nestel and P. Clifton-Bligh, Atherosclerosis, 23 (1976) 535.
194 V. N. Schumaker and G. H. Adams, J. Theoret. Biol., 26 (1970) 89.
195 P. J. Nestel, Clin. Sci., 38 (1970) 593.
196 H. G. Rose and J. Juliano, J. Lab. clin. Med., 88 (1976) 29.
197 J. P. Miller, A. Chait and B. Lewis, Clin. Sci. mol. Med., 49 (1975) 617.
198 L. Wallentin, B. Angelin, K. Einarsson and B. Leijd, Scand. J. clin. Lab. Invest., 38 (1978) Suppl. 150, 103—110.
199 J. P. Miller, Scand. J. clin. Lab. Invest., 38 (1978) Suppl. 150, 138—141.
200 G. Nordby and K. R. Norum, Scand. J. clin. Lab. Invest., 38 (1978) Suppl. 150, 111—114.
201 P. H. R. Green, A. R. Tall and R. M. Glickman, J. clin. Invest. 61 (1978) 528.
202 R. L. Hamilton, M. C. Williams, C. J. Fielding and R. J. Havel, J. clin. Invest., 58 (1976) 667.
203 K. R. Norum, J. A. Glomset, A. V. Nichols, T. Forte, J. J. Albers, W. C. King, C. D. Mitchell, K. R. Applegate, E. L. Gong, V. Cabana and E. Gjone, Scand. J. clin. Lab. Invest., 35 (1975) Suppl. 142, 31—55.
204 F. P. Bell, in C. E. Day and R. S. Levy (Eds.), Low Density Lipoproteins, Plenum Press, New York, 1976, pp. 111—133.
205 Y. J. Kao, S. C. Charlton and L. C. Smith, Fed. Proc., 36 (1977) 936.
206 A. V. Nichols and L. Smith, J. Lipid Res., 6 (1965) 206.
207 D. B. Zilversmit, L. B. Hughes and J. Balmer, Biochim. Biophys, Acta, 409 (1975) 393.
208 P. J. Barter and M. E. Jones, Atherosclerosis, 34 (1979) 67.
209 P. J. Nestel, M. Reardon and T. Billington, Biochim. Biophys. Acta, 573 (1979) 403.

# REFERENCES

210  J. I. Lally and P. J. Barter, J. Lab. clin. Med., 93 (1979) 570.
211  L. I. Gidez, P. S. Roheim and H. A. Eder, J. Lipid Res., 6 (1965) 377.
212  P. J. Barter and J. I. Lally, Biochim. Biophys. Acta, 531 (1978) 233.
213  D. S. Goodman, Physiol. Rev., 45 (1965) 747.
214  C. J. Fielding and P. E. Fielding, in A. M. Gotto, L. C. Smith and B. Allen (Eds.), Atherosclerosis V, Springer-Verlag, New York, 1980, pp. 379—382.
215  R. J. Havel and J. P. Kane, Fed. Proc., 34 (1975) 2250.
216  P. J. Barter, J. I. Lally and D. Wattchow, Metabolism, 28 (1979) 614.
217  D. R. Illingworth and O. W. Portman, J. Lipid Res., 13 (1972) 220.
218  S. Eisenberg, J. Lipid Res., 19 (1978) 229.
219  G. Béréziat, J. Chambaz, G. Trugnan, D. Pépin and J. Polonovski, J. Lipid Res., 19 (1978) 495.
220  C. F. Reed, M. Murphy and G. Roberts, J. clin. Invest., 47 (1968) 749.
221  D. R. Illingworth and O. W. Portman, Biochim. Biophys. Acta, 280 (1972) 281.
222  M. E. Brewster, J. Ihm, J. R. Brainard and J. A. K. Harmony, Biochim. Biophys. Acta, 529 (1978) 147.
223  R. L. Hamilton, Adv. Exp. Med. Biol., 26 (1972) 7.
224  W. J. Simmonds, in G. J. Nelson (Ed.) Blood Lipids and Lipoproteins: Quantitation, Composition and Metabolism, Wiley-Interscience, New York, 1972, pp. 705—743.
225  D. Rachmilewitz, J. J. Albers and D. R. Saunders, J. clin. Invest., 57 (1976) 530.
226  R. M. Glickman, J. Khorana and A. Kilgore, Science, 193 (1976) 1254.
227  R. M. Glickman, A. Kilgore and J. Khorana, J. Lipid Res., 19 (1978) 260.
228  G. Schonfeld, E. Bell and D. H. Alpers, J. clin. Invest., 61 (1978) 1539.
229  D. Rachmilewitz, J. J. Albers, D. R. Saunders and M. Fainaru, Gastroenterology, 75 (1978) 677.
230  P. S. Roheim, L. I. Gidez and H. A. Eder, J. clin. Invest., 45 (1966) 297.
231  H. G. Windmueller, P. N. Herbert and R. I. Levy, J. Lipid Res., 14 (1973) 215.
232  A. -L. Wu, H. G. Windmueller, J. biol. Chem., 253 (1978) 2525.
233  H. G. Windmueller and R. I. Levy, J. biol. Chem., 243 (1968) 4878.
234  R. M. Glickman and P. H. R. Green, Proc. Nat. Acad. Sci., 74 (1977) 2569.
235  K. Imaizumi. R. J. Havel, M. Fainaru and J. -L. Vigne, J. Lipid Res., 19 (1978) 1038.
236  J. M. Johnston, Handbook of Physiology, Section 6, Alimentary Canal, Volume III, Intestinal Absorption, Am. Physiol. Soc., Washington, D.C., 1968, pp. 1353—1375.
237  J. M. Johnston, in M. Florkin and E. H. Stotz (Eds.), Comprehensive Biochemistry, Vol. 18, Elsevier, Amsterdam, 1970, pp. 1—18.
238  E. W. Strauss, Handbook of Physiology, Section 6, Alimentary Canal, Volume III, Intestinal Absorption, Amer. Physiol. Soc., Washington, D. C., 1968, pp. 1377—1406.

239 A. Gangl and R. K. Ockner, Gastroenterology, 68 (1975) 167.
240 S. L. Palay and L. J. Karlin, J. biophys. Biochem. Cytol., 5 (1959) 373.
241 A. L. Jones and R. K. Ockner, J. Lipid Res., 12 (1971) 580.
242 G. N. Tytgat, C. E. Rubin and D. R. Saunders, J. clin. Invest., 50 (1971) 2065.
243 H. P. Porter, D. R. Saunders, G. Tytgat, O. Brunser and C. E. Rubin, Gastroenterology, 60 (1971) 1008.
244 R. R. Cardell, S. Badenhausen and K. R. Porter, J. Cell Biol., 34 (1967) 123.
245 R. A. Jersild, Amer. J. Anat., 118 (1966) 135.
246 S. M. Sabesin and S. Frase, J. Lipid Res., 18 (1977) 496.
247 J. I. Kessler, P. Narcessian and D. P. Mauldin, Gastroenterology, 68 (1975) 1058.
248 K. Imaizumi, M. Fainaru and R. J. Havel, J. Lipid Res., 19 (1978) 712.
249 C. A. Arreaza-Plaza, V. Bosch and M. A. Otayek, Biochim. Biophys. Acta, 431 (1976) 297.
250 R. M. Glickman, J. L. Perrotto and K. Kirsch, Gastroenterology, 70 (1976) 347.
251 R. M. Glickman and K. Kirsch, J. clin. Invest., 52 (1973) 2910.
252 R. M. Glickman, P. H. R. Green, R. S. Lees and A. Tall, New Engl. J. Med., 299 (1978) 1424.
253 G. Assmann, A. Capurso, E. Smooz and U. Wellner, Atherosclerosis, 30 (1978) 321.
254 G. Kostner and A. Holasek, Biochemistry, 11 (1972) 1217.
255. E. J. Schaefer, L. L. Jenkins and H. B. Brewer, Biochem. Biophys. Res. Commun., 80 (1978) 405.
256 P. H. R. Green, R. M. Glickman, C. D. Saudek, C. R. Blum and A. R. Tall, J. clin. Invest., 64 (1979) 233.
257 J. M. Falko and G. Schonfeld, Clin. Res., 26 (1978) 318A.
258 J. A. Rooke and E. R. Skinner, Biochem. Soc. Trans., 4 (1976) 1144.
259 A. -L. Wu and H. G. Windmueller, J. biol. Chem., 254 (1979) 7316.
260 P. N. Herbert, H. G. Windmueller, T. P. Bersot and R. S. Shulman, J. biol. Chem., 249 (1974) 5718.
261 D. E. Schwartz, L. Liotta, E. Schaefer and H. B. Brewer, Circulation, 58 (1978) Suppl. II—90.
262 J. B. Swaney, F. Braithwaite and H. A. Eder, Biochemistry, 16 (1977) 271.
263 P. R. Holt, A. -L. Wu and S. B. Clark, J. Lipid Res., 20 (1979) 494.
264 W. O. Dobbins, Gastroenterology, 50 (1966) 195.
265 J. I. Kessler, J. Stein, D. Dannacker and P. Narcessian, J. biol. Chem., 245 (1970) 5281.
266 H. Glaumann, A. Bergstrand and J. L. E. Ericsson, J. Cell Biol., 64 (1975) 356.
267 L. M. G. Van Golde, B. Fleischer and S. Fleischer, Biochim. Biophys. Acta, 249 (1971) 318.
268 J. J. B. De Jong and J. B. Marsh, J. biol. Chem., 243 (1968) 192.
269 C. A. Alexander, R. L. Hamilton and R. J. Havel, J. Cell Biol., 69 (1976) 241.

# REFERENCES

270 R. L. Hamilton, D. M. Regen, M. E. Gray and V. S. LeQuire, Lab. Invest., 16 (1967) 305.
271 A. L. Jones, N. B. Ruderman and M. G. Herrera, J. Lipid Res., 8 (1967) 429.
272 A. Claude, J. Cell Biol., 47 (1970) 745.
273 R. W. Mahley, R. L. Hamilton and V. S. LeQuire, J. Lipid Res., 10 (1969) 433.
274 R. W. Mahley, T. P. Bersot, V. S. LeQuire, R. I. Levy, H. G. Windmueller and W. V. Brown, Science, 168 (1970) 380.
275 S. -P. Noel and D. Rubinstein, J. Lipid Res., 15 (1974) 301.
276 T. E. Felker, M. Fainaru, R. L. Hamilton and R. J. Havel, J. Lipid Res., 18 (1977) 465.
277 J. B. Marsh, J. Lipid Res., 17 (1976) 85.
278 L. A. Pottenger and G. S. Getz, J. Lipid Res., 12 (1971) 450.
279 J. B. Ragland, P. D. Bertram and S. M. Sabesin, Biochim. Biophys. Res. Commun., 80 (1978) 81.
280 R. A. Davis, S. C. Engelhorn, S. H. Pangburn, D. B. Weinstein and D. Steinberg, J. biol. Chem., 254 (1979) 2010.
281 H. G. Windmueller and R. I. Levy, J. biol. Chem., 242 (1967) 2246.
282 L. Chan, R. L. Jackson and A. R. Means, Circulat. Res., 43 (1978) 209.
283 L. Chan, L. D. Snow, R. L. Jackson and A. R. Means, in T. H. Hamilton, J. H. Clark and W. A. Saddler (Eds.), Ontogeny of Receptors and Reproductive Hormone Action, Raven Press, New York, 1979, pp. 331—351.
284 L. Chan, W. A. Bradley, A. Dugaiczyk and A. R. Means, Ann. N.Y. Acad. Sci., 348 (1980) 427.
285 G. Getz, R. V. Hay and C. Reardon, in A. M. Gotto, L. C. Smith and B. Allen (Eds.), Atherosclerosis V, Springer-Verlag, New York, 1980, pp. 156—159.
286 R. Hay and G. S. Getz, J. Lipid Res., 20 (1979) 334.
287 Y. Lin and L. Chan, in Proceedings of the 6th International Congress of Endocrinology, Melbourne, Australia, February, 1980.
288 O. Minari and D. B. Zilversmit, J. Lipid Res., 4 (1963) 424.
289 S. F. Robinson and S. H. Quarfordt, Biochim. Biophys. Acta, 541 (1978) 492.
290 C. J. Fielding and P. E. Fielding, J. Lipid Res., 17 (1976) 419.
291 R. J. Havel, J. P. Kane and M. L. Kashyap, J. clin. Invest., 52 (1973) 32.
292 S. F. Robinson and S. H. Quarfordt, Lipids, 14 (1979) 343.
293 D. S. Goodman, J. clin. Invest., 41 (1962) 1886.
294 P. J. Nestel, R. J. Havel and A. Bezman, J. clin. Invest., 42 (1963) 1313.
295 O. D. Miøs, O. Faergeman, R. L. Hamilton and R. J. Havel, J. clin. Invest., 56 (1975) 603.
296 T. Chajek and S. Eisenberg, J. clin. Invest., 61 (1978) 1654.
297 E. N. Bergman, R. J. Havel, B. M. Wolfe and T. Bohmer, J. clin. Invest., 50 (1971) 1831.
298 P. J. Nestel, R. J. Havel and A. Bezman, J. clin. Invest., 41 (1962) 1915.
299 T. G. Redgrave, J. clin. Invest., 49 (1970) 465.

300   J. M. Felts, H. Itakura and R. T. Crane, Biochem. Biophys. Res. Commun., 66 (1975) 1467.
301   A. D. Cooper, Biochim. Biophys. Acta, 488 (1977) 464.
302   S. -P. Noel, P. J. Dolphin and D. Rubinstein, Biochem. Biophys. Res. Commun., 63 (1975) 764.
303   B. C. Sherrill and J. M. Dietschy, J. biol. Chem., 253 (1978) 1859.
304   C. J. Fielding, J. clin. Invest., 62 (1978) 141.
305   C. J. Fielding, J. P. Renston and P. E. Fielding, J. Lipid Res., 19 (1978) 705.
306   T. G. Redgrave and D. M. Small, J. clin. Invest., 64 (1979) 162.
307   S. Eisenberg and D. Rachmilewitz, J. Lipid Res., 16 (1975) 341.
308   J. R. Patsch, A. M. Gotto, T. Olivecrona and S. Eisenberg, Proc. Nat. Acad. Sci., 75 (1978) 4519.
309   S. Eisenberg and T. Olivecrona, J. Lipid Res., 20 (1979) 614.
310   A. R. Tall, P. H. R. Green, R. M. Glickman and J. W. Riley, J. clin. Invest., 64 (1979) 977.
311   J. M. Higgins and C. J. Fielding, Biochemistry, 14 (1975) 2288.
312   O. Faergeman and R. J. Havel, J. clin. Invest., 55 (1975) 1210.
313   A. D. Cooper and P. Y. S. Yu, J. Lipid Res., 19 (1978) 635.
314   A. Nilsson, Biochem. Biophys. Res. Commun., 66 (1975) 60.
315   A. C. Ross and D. B. Zilversmit, J. Lipid Res., 18 (1977) 169.
316   J. M. Andersen, F. O. Nervi and J. M. Dietschy, Biochim. Biophys. Acta, 486 (1977) 298.
317   B. C. Sherrill, in J. M. Dietschy, A. M. Gotto and J. A. Ontko (Eds.), Disturbances in Lipid and Lipoprotein Metabolism, American Physiological Society, Bethesda, MD, 1978, pp. 99—109.
317A  F. Shelburne, J. Hanks, W. Meyers and S. Quarfordt, J. clin. Invest., 65 (1980) 652.
317B  B. C. Sherrill, T. L. Innerarity and R. W. Mahley, J. biol. Chem., 255 (1980) 1804.
317C  E. Windler, Y. Chao and R. J. Havel, J. biol. Chem., 255 (1980) 8303.
317D  R. J. Havel, Y. Chao, E. E. Windler, L.Kotite and L. S. S. Guo, Proc. nat. Acad. Sci., 77 (1980) 4349.
318   N. Nakaya, B. H. Chung, J. R. Patsch and O. D. Tauton, J. biol. Chem., 252 (1977) 7530.
319   D. R. Illingworth, Biochim. Biophys. Acta, 388 (1975) 38.
320   R. J. Deckerbaum, S. Eisenberg, M. Fainaru, Y. Barenholz and T. Olivecrona, J. biol. Chem., 254 (1979) 6079.
321   R. V. Hay, L. A. Pottenger, A. L. Reingold, G. S. Getz and R. W. Wissler, Biochem. Biophys. Res. Commun., 44 (1971) 1471.
322   G. Sigurdsson, S. -P. Noel and R. J. Havel, J. Lipid Res., 19 (1978) 628.
323   G. D. Calvert, P. J. Scott and D. N. Sharpe, Atherosclerosis, 22 (1975) 601.
324   S. Eisenberg, H. G. Windmueller and R. I. Levy, J. Lipid Res., 14 (1973) 446.
325   Y. Stein, V. Ebin, H. Bar-on and O. Stein, Biochim. Biophys. Acta, 486 (1977) 286.
326   L. A. Simons, D. Reichl, N. B. Myant and M. Mancini, Atherosclerosis, 21 (1975) 283.

# REFERENCES

327 A. D. Sniderman, T. E. Carew and D. Steinberg, J. Lipid Res., 16 (1975) 293.
328 A. D. Sniderman, T. E. Carew, J. G. Chandler and D. Steinberg, Science, 183 (1974) 526.
329 A. Van Tol, F. M. Van't Hooft and T. Van Gent, Atherosclerosis, 29 (1978) 449.
330 S. H. Pangburn and D. B. Weinstein, Fed. Proc., 37 (1978) 1482.
331 P. S. Bachorik, P. O. Kwiterovich and J. C. Cooke, Biochemistry, 17 (1978) 5287.
332 D. Steinberg, Adv. Exp. Med. Biol., 109 (1978) 3.
333 A. Sniderman, D. Thomas, D. Marpole and B. Teng, J. clin. Invest., 61 (1978) 867.
334 R. C. Pittman and D. Steinberg, Biochem. Biophys. Res. Commun., 81 (1978) 1254.
335 R. C. Pittman, A. D. Attie, T. E. Carew and D. Steinberg, Proc. nat. Acad. Sci., 76 (1979) 5345.
336 G. J. Bautovich, L. A. Simons, P. F. Williams and J. R. Turtle, Atherosclerosis, 21 (1975) 217.
337 D. Gitlin, D. G. Cornwell, D. Nakasato, J. L. Oncley, W. L. Hughes and C. A. Janeway, J. clin. Invest., 37 (1958) 172.
338 G. Sigurdsson, A. Nicoll and B. Lewis, J. clin. Invest., 56 (1975) 1481.
339 G. Sigurdsson, A. Nicoll and B. Lewis, Europ. J. clin. Invest., 6 (1976) 167.
340 G. Sigurdsson, A. Nicoll and B. Lewis, Europ. J. clin. Invest., 6 (1976) 151.
341 M. F. Reardon, N. H. Fidge and P. J. Nestel, J. clin. Invest., 61 (1978) 850.
342 A. K. Soutar, N. B. Myant and G. R. Thompson, Atherosclerosis, 28 (1977) 247.
343 M. Berman, M. Hall, R. I. Levy, S. Eisenber, D. W. Bilheimer, R. D. Phair and R. H. Goebel, J. Lipid Res., 19 (1978) 38.
344 T. G. Redgrave and L. A. Carlson, J. Lipid Res., 20 (1979) 217.
345 P. J. Nestel, M. F. Reardon and N. H. Fidge, Circulat. Res., 45 (1979) 35.
346 P. J. Nestel, M. Reardon and N. H. Fidge, Metabolism, 28 (1979) 531.
347 O. Faergeman, T. Sata, J. P. Kane and R. J. Havel, J. clin. Invest., 56 (1975) 1396.
348 N. H. Fidge and P. Poulis, J. Lipid Res., 19 (1978) 342.
349 T. Langer, W. Strober and R. I. Levy, J. clin. Invest., 51 (1972) 1528.
350 G. R. Thompson, T. Spinks, A. Ranicar and N. B. Myant, Clin. Sci. mol. Med., 52 (1977) 361.
351 C. J. Packard, J. L. H. C. Third, J. Shepherd, A. R. Lorimer, H. G. Morgan and T. D. V. Lawrie, Metabolism, 25 (1976) 995.
352 A. Soutar, N. Myant and G. Thompson, Circulation, 58 (1978) II-39.
353 R. I. Levy and T. Langer, Adv. Exp. med. Biol., 26 (1972) 155.
354 D. P. Barr, E. M. Russ and H. A. Eder, Amer. J. Med., 11 (1951) 480.
355 D. Steinberg, Europ. J. clin. Invest., 8 (1978) 107.
356 G. C. Rhoads, C. L. Gulbrandsen and A. Kagan, New Engl. J. Med., 294 (1976) 293.

357 N. E. Miller, D. S. Thelle, O. H. Førde and O. D. Miøs, Lancet, 1 (1977) 965.
358 W. P. Castelli, J. T. Doyle, T. Gordon, C. G. Hames, M. C. Hjortland, S. B. Hulley, A. Kagan and W. J. Zukel, Circulation, 55 (1977) 767.
359 T. Gordon, W. P. Castelli, M. C. Hjortland, W. B. Kannel and T. R. Dawber, Amer. J. Med., 62 (1977) 707.
360 T. Gordon, W. P. Castelli, M. C. Hjortland, W. B. Kannel and T. R. Dawber, Ann. Int. Med., 87 (1977) 393.
361 A. M. Gotto, N. E. Miller and M. F. Oliver (Eds.), High Density Lipoproteins and Atherosclerosis, Elsevier/North Holland, Amsterdam, 1978.
362 E. A. Nikkilä, Europ. J. clin. Invest., 8 (1978) 111.
363 K. Lippel (Ed.), Report of the High Density Lipoprotein Methodology Workshop, U.S.D.H.E.W., N.I.H. Publication 79—1661, 1979.
364 A. Nicoll, N. E. Miller and B. Lewis, Adv. Lipid Res., 17 (1980) 53.
364A A. R. Tall and D. M. Small, Adv. Lipid Res., 17 (1980) 1.
365 D. E. Wilson and R. S. Lees, J. clin. Invest., 51 (1972) 1051.
366 P. D. Wood and W. L. Haskell, Lipids, 14 (1979) 417.
367 J. Shepherd, C. J. Packard, J. R. Patsch, A. M. Gotto and O. D. Taunton, J. clin, Invest., 63 (1979) 858.
368 E. A. Nikkilä, in A. M. Gotto, N. E. Miller and M. F. Oliver (Eds.), High Density Lipoproteins and Atherosclerosis, Elsevier/North Holland, 1978, pp. 177—192.
369 D. W. Anderson, Lancet, 1 (1978) 819.
370 T. M. Forte, R. M. Krauss, F. T. Lindgren and A. V. Nichols, Proc. nat. Acad. Sci., 76 (1979) 5934.
371 J. R. Patsch, D. Yeshurun, R. L. Jackson and A. M. Gotto, Amer. J. Med., 63 (1977) 1001.
372 D. S. Fredrickson, R. I. Levy and F. T. Lindgren, J. clin. Invest., 47 (1968) 2446.
373 T. Nakai, P. S. Otto, D. L. Kennedy and T. F. Whayne, J. biol. Chem., 251 (1976) 4914.
374 T. J. C. Van Berkel and A. Van Tol, Biochim. Biophys. Acta, 530 (1978) 299.
375 L. Ose, T. Ose, K. R. Norum and T. Berg, Biochim. Biophys. Acta, 574 (1979) 521.
375A A. Van Tol, T. Van Gent and H. Jansen, Biochem. Biophys. Res. Commun., 94 (1980) 101.
376 J. M. Andersen, S. D. Turley and J. M. Dietschy, Proc. nat. Acad. Sci., 76 (1979) 165.
377 G. Sigurdsson, S. -P. Noel and R. J. Havel, Circulation, 56 (1977) III-4.
378 A. Van Tol, T. Van Gent, F. M. Van't Hooft and F. Vlaspolder, Atherosclerosis, 29 (1978) 439.
379 N. E. Miller, D. B. Weinstein and D. Steinberg, J. Lipid Res., 18 (1977) 438.
380 J. J. Albers, P. W. Wahl, V. G. Cabana, W. R. Hazzard and J. J. Hoover, Metabolism, 25 (1976) 633.
381 J. Shepherd, C. J. Packard, J. R. Patsch, A. M. Gotto and O. D. Taunton, Europ. J. clin. Invest., 8 (1978) 115.

## REFERENCES

382 M. J. Caslake, E. Farish and J. Shepherd, Metabolism, 27 (1978) 437.
383 G. Schonfeld, A. Bailey and R. Steelman, Lipids, 13 (1978) 951.
384 J. Shepherd, C. J. Packard, J. R. Patsch, A. M. Gotto and O. D. Taunton, J. clin. Invest., 61 (1978) 1582.
385 T. Hara, J. P. Miller, A. M. Gotto and J. R. Patsch, Lipids, in press.
386 J. Shepherd, C. J. Packard, H. G. Morgan, J. L. H. C. Third, J. M. Stewart and T. D. V. Lawrie, Atherosclerosis, 33 (1979) 433.
387 M. C. Cheung and J. J. Albers, J. clin. Invest., 60 (1977) 43.
388 J. J. Albers, M. C. Cheung and W. R. Hazzard, Metabolism, 27 (1978) 479.
389 M. D. Curry, P. Alaupovic and C. A. Suenram, Clin. Chem., 22 (1976) 315.
390 C. R. Blum, R. I. Levy, S. Eisenberg, M. Hall, R. H. Goebel and M. Berman, J. clin. Invest., 60 (1977) 795.
391 J. Shepherd, A. M. Gotto, O. D. Taunton, M. J. Caslake and E. Farish, Biochim. Biophys. Acta, 489 (1977) 486.
392 J. Sheperd, C. J. Packard, A. M. Gotto and O. D. Taunton, J. Lipid Res., 19 (1978) 656.
393 J. Shepherd, J. R. Patsch, C. J. Packard, A. M. Gotto and O. D. Taunton, J. Lipid Res., 19 (1978) 383.
394 T. Nakai and T. F. Whayne, J. Lab. clin, Med., 88 (1976) 63.
395 L. A. Zech, E. J. Schaefer and H. B. Brewer, Circulation, 58 (1978) II-40.
396 R. H. Furman, S. S. Sanbar, P. Alaupovic, R. H. Bradford and R. P. Howard, J. Lab. clin. Med., 63 (1964) 193.
397 N. Fidge, P. Nestel, T. Ishikawa, M. Reardon and T. Billington, Metabolism, 29 (1980) 643.
397A S. N. Rao, P. J. Magill, N. E. Miller and B. Lewis, Clin. Sci., 59 (1980) 359.
398 N. Fidge, P. J. Nestel, M. Reardon and T. Ishikawa, Circulation, 58 (1978) II-40.
399 E. J. Schaefer, C. B. Blum, R. I. Levy, L. L. Jenkins, P. Alaupovic, D. M. Foster and H. B. Brewer, New Engl. J. Med., 299 (1978) 905.
400 E. J. Schaefer and H. B. Brewer, Clin. Res., 26 (1978) 532A.
401 M. D. Curry, P. Alaupovic and W. J. McConathy, Circulation, 54 (1976) II-134.
402 M. L. Kashyap, L. S. Srivastava, C. Y. Chen, G. Perisutti, M. Campbell, R. F. Lutmer and C. J. Glueck, J. clin. Invest., 60 (1977) 171.
403 G. Schonfeld, P. K. George, J. Miller, P. Reilly and J. Witztum, Metabolism, 28 (1979) 1001.
404 G. Schonfeld, S. W. Weidman, J. L. Witztum and R. M. Bowen, Metabolism, 25 (1976) 261.
405 A. Montes, R. H. Knopp and J. Humphrey, J. clin. Endocrinol. Metab., 45 (1977) 1060.
406 N. Fidge, P. J. Nestel and H. W. Huff, in A. M. Gotto, L. C. Smith and B. Allen (Eds.), Atherosclerosis V, Springer-Verlag, New York, 1980, pp. 820—823.

407 S. H. Quarfordt, Biochim. Biophys. Acta., 489 (1977) 477.
408 G. Utermann, M. Jaeschke and J. Menzel, FEBS Lett., 56 (1975) 352.
409 G. R. Warnick, C. Mayfield, J. J. Albers and W. R. Hazzard, Clin. Chem., 25 (1979) 279.
410 B. Shore, V. Shore, A. Salel, D. Mason and R. Zelis, Biochem. Biophys. Res. Commun., 58 (1974) 1.
411 R. W. Mahley, in J. M. Dietschy, A. M. Gotto and J. A. Ontko (Eds.), Disturbances in Lipid and Lipoprotein Metabolism, American Physiological Society, Bethesda, MD, 1978, pp. 181—197.
411A S. Quarfordt, J. Hanks, R. Scott Jones and F. Shelburne, J. biol. Chem., 255 (1980) 2934.
412 D. Ganesan, H. B. Bass, W. J. McConathy and P. Alaupovic, Metabolism, 25 (1976) 1189.
413 J. L. Goldstein and M. S. Brown, Curr. Top. Cell. Regul., 11 (1976) 147.
414 J. L. Goldstein and M. S. Brown, Ann. Rev. Biochem., 46 (1977) 897.
415 J. L. Goldstein, M. S. Brown and R. G. W. Anderson, in B. R. Brinkley and K. R. Porter (Eds.), International Cell Biology 1976—1977, Rockefeller University Press, New York, 1977, pp. 639—648.
416 R. W. Mahley and T. L. Innerarity, Adv. Exp. Med. Biol., 109 (1978) 99.
417 Y. Stein and O. Stein, in A. M. Scanu, R. W. Wissler and G. S. Getz (Eds.), The Biochemistry of Atherosclerosis, Dekker, New York, 1979, pp. 313—344.
418 Y. K. Ho, M. S. Brown, D. W. Bilheimer and J. L. Goldstein, J. Clin. Invest., 58 (1976) 1465.
419 J. L. Goldstein and M. S. Brown, J. biol. Chem., 249 (1974) 5153.
420 M. S. Brown and J. L. Goldstein, Cell, 6 (1975) 307.
421 J. L. Goldstein, S. K. Basu, G. Y. Brunschede and M. S. Brown, Cell, 7 (1976) 85.
422 M. S. Brown and J. L. Goldstein, Proc. nat. Acad. Sci., 71 (1974) 788.
423 T. P. Bersot, R. W. Mahley, M. S. Brown and J. L. Goldstein, J. biol. Chem., 251 (1976) 2395.
424 R. W. Mahley, T. L. Innerarity, R. E. Pitas, K. H. Weisgraber, J. H. Brown and E. Gross, J. biol. Chem., 252 (1977) 7279.
425 T. L. Innerarity and R. W. Mahley, Biochemistry, 17 (1978) 1440.
426 T. L. Innerarity, R. E. Pitas and R. W. Mahley, J. biol. Chem., 254 (1979) 4186.
427 R. I. Roth, R. L. Jackson, H. J. Pownall and A. M. Gotto, Biochemistry, 16 (1977) 5030.
428 D. Steinberg, P. J. Nestel, D. B. Weinstein, M. Remaut-Desmeth and C. M. Chang, Biochim. Biophys. Acta, 528 (1978) 199.
429 T. L. Innerarity, R. W. Mahley, K. H. Weisgraber and T. P. Bersot, J. biol. Chem., 253 (1978) 6289.
430 M. S. Brown, T. F. Deuel, S. K. Basu and J. L. Goldstein, J. Supramol. Struct., 8 (1978) 223.
431 K. H. Weisgraber, T. L. Innerarity and R. W. Mahley, J. biol. Chem., 253 (1978) 9053.

# REFERENCES

432 R. G. W. Anderson, J. L. Goldstein and M. S. Brown, Proc. nat. Acad. Sci., 73 (1976) 2434.
433 M. S. Brown and J. L. Goldstein, Science, 185 (1974) 61.
434 A. Chait, E. L. Bierman and J. J. Albers, Biochim. Biophys. Acta, 529 (1978) 292.
435 A. Chait, E. L. Bierman and J. J. Albers, J. clin. Invest., 64 (1979) 1309.
436 A. Chait, E. L. Bierman and J. J. Albers, J. clin. Endocrinol. Metab., 48 (1979) 887.
437 K. Henze, A. Chait, J. J. Albers and E. L. Bierman, Proceedings of the Fifth International Symposium on Atherosclerosis, Houston, TX, 1979.
438 M. S. Brown, S. E. Dana and J. L. Goldstein, Proc. nat. Acad. Sci., 72 (1975) 2925.
439 J. L. Goldstein, G. Y. Brunschede and M. S. Brown, J. biol. Chem., 250 (1975) 7854.
440 J. L. Goldstein, S. E. Dana, J. R. Faust, A. L. Beaudet and M. S. Brown, J. biol. Chem., 250 (1975) 8487.
441 M. S. Brown, M. K. Sobhani, G. Y. Brunschede and J. L. Goldstein, J. biol. Chem., 251 (1976) 3277.
442 M. S. Brown, S. E. Dana and J. L. Goldstein, J. biol. Chem., 249 (1974) 789.
443 M. S. Brown, S. E. Dana and J. L. Goldstein, Proc. nat. Acad. Sci., 70 (1973) 2162.
444 M. S. Brown and J. L. Goldstein, J. biol. Chem., 249 (1974) 7306.
445 J. L. Goldstein, S. E. Dana and M. S. Brown, Proc. nat. Acad. Sci., 71 (1974) 4288.
446 M. S. Brown, S. E. Dana and J. L. Goldstein, J. biol. Chem., 250 (1975) 4025.
447 J. Oram, J. J. Albers and E. L. Bierman, J. biol. Chem., 255 (1980) 475.
448 N. B. Myant, D. Reichl and J. K. Lloyd, Atherosclerosis, 29 (1978) 509.
449 C. J. Packard, S. Bicker, H. G. Morgan, T. D. V. Lawrie and J. Shepherd, Proceedings of the Fifth International Symposium on Atherosclerosis, Houston, TX, 1979.
449A R. W. Mahley, K. H. Weisgraber, G. W. Melchior, T. L. Innerarity and K. S. Holcombe, Proc. nat. Acad. Sci., 77 (1980) 225.
450 K. B. Schwartz, J. Witztum, G. Schonfeld, S. M. Grundy and W. E. Connor, J. clin. Invest., 64 (1979) 756.
451 J. M. Dietschy and J. D. Wilson, New Engl. J. Med., 282 (1970) 1128, 1179, 1241.
452 J. Andersen and J. M. Dietschy, J. biol. Chem., 252 (1977) 3652.
453 P. T. Kovanen, J. R. Faust, M. S. Brown and J. L. Goldstein, Endocrinology, 104 (1979) 599.
454 P. T. Kovanen, S. K. Basu, J. L. Goldstein and M. S. Brown, Endocrinology, 104 (1979) 610.
455 J. M. Andersen and J. M. Dietschy, J. biol. Chem., 253 (1978) 9024.
456 P. T. Kovanen, W. J. Schneider, G. M. Hillman, J. L. Goldstein and M. S. Brown, J. biol. Chem., 254 (1979) 5498.
457 N. E. Miller, D. B. Weinstein and D. Steinberg, J. Lipid Res., 19 (1978) 644.

458 M. S. Brown, J. R. Faust and J. L. Goldstein, J. clin. Invest., 55 (1975) 783.
459 T. Koschinsky, T. E. Carew and D. Steinberg, J. Lipid Res., 18 (1977) 451.
460 W. H. Daerr, S. H. Gianturco, J. R. Patsch, L. C. Smith and A. M. Gotto, Biochim. Biophys. Acta, 619 (1980) 287.
461 N. E. Miller, D. B. Weinstein, T. E. Carew, T. Koschinsky and D. Steinberg, J. clin. Invest., 60 (1977) 78.
462 N. E. Miller, Biochim. Biophys. Acta, 529 (1978) 131.
463 S. H. Gianturco, A. M. Gotto, R. L. Jackson, J. R. Patsch, H. D. Sybers, O. D. Taunton, D. L. Yeshurun and L. C. Smith, J. clin. Invest., 61 (1978) 320.
464 S. H. Gianturco, C. J. Packard, J. Shepherd, L. C. Smith, A. L. Catapano, H. D. Sybers and A. M. Gotto, Lipids, 15 (1980) 456.
465 S. H. Gianturco, S. G. Eskin, L. T. Navarro, C. J. Lahart, L. C. Smith and A. M. Gotto, Biochim. Biophys. Acta, 618 (1980) 143.
466 A. Catapano, S. H. Gianturco, P. K. J. Kinnunen, S. Eisenberg, A. M. Gotto and L. C. Smith, J. biol. Chem., 254 (1979) 1007.
467 C. J. Fielding and P. E. Fielding, in J. Polonovski (Ed.), Cholesterol Metabolism and Lipolytic Enzymes, Masson, New York, 1977, pp. 165—172.

*Chapter 6*

# The Chromosomal Proteins*

JOHN M. WALKER, GRAHAM H. GOODWIN,
BRYAN J. SMITH and ERNEST W. JOHNS

*Institute of Cancer Research, Royal Cancer Hospital, Chester Beatty Research Institute, Fulham Road, London SW3 6JB (U.K.)*

## 1. Introduction

### 1.1. Historical

A historical introduction to a chapter entitled "Chromosomal Proteins" is likely to be anachronistic since it is only relatively recently that many of the well-characterised nuclear proteins have been shown to be associated with DNA. Indeed, the definition of a chromosomal protein presents some difficulties (see below). Therefore, with hindsight we shall select those areas of research which, in our opinion (admittedly not impartial), have led to a better understanding of the structure and function of the major well-characterised proteins associated with DNA and which are being investigated at the present time.

The initial discovery by Friedrich Miescher [131] that the acidic substance in salmon sperm (later called nucleic acid) was combined with an organic base, was possibly the beginning of this field of study. He named this organic base "protamine". Ten years later, Albrecht Kossel [107] carrying out a similar investigation,

---

*This chapter was written in early 1979.

but using goose erythrocyes, found that the basic substance associated with the nucleic acid was more complex than the simple protamine described by Miescher. Kossel isolated this basic substance by extracting the goose erythrocyte nuclei with dilute hydrochloric acid and wrote "Ich schlage für diese Substanz den Namen Histon vor". Since this first isolation and naming of the histones they have been shown to exist, associated with DNA, in all the animal and plant somatic cells so far examined, and it is now generally accepted that they are present in all the somatic cells of multicellular organisms and in many unicellular lower eukaryotes. However, at that time and for about 50 years subsequent to their discovery, the histones were thought of as a single protein. It was not until Edgar and Ellen Stedman working in Edinburgh began their classic investigations into nuclear proteins that the first evidence was produced showing that the histones were indeed a mixture of proteins with similar properties [185, 186].

Until 1942 it had also been assumed that nucleic acid and histone or protamine were the only constituents of the cell nucleus. However, Mayer and Gullick [128] in 1942 showed that 20% of the mass of the nucleus of calf thymocytes consisted of a protein with an isoelectric point of about 6. Stedman and Stedman [184] also described an acidic protein, "chromosomin" which they had isolated from cod sperm nuclei. They even questioned the importance of DNA at this time, thinking that the chromosomin was the major constituent of the chromosomes, but this idea was vigorously criticised by Callan [26] who asserted the importance of DNA. However, this early work clearly established that there were other proteins in the nucleus apart from the basic proteins, and all are probably associated with the DNA. For those interested in this early work on the chromosomal proteins and chromatin, there are a number of reviews to which reference can be made for further details [95, 115]. From the Stedman's pioneering work, it was apparent that the histones were complex, but it was to take approximately 15 years of work by a number of groups throughout the world to establish that there were in fact only 5 major histones in mammalian somatic cell nuclei, and one extra specific histone in nucleated erythrocytes. The mammalian histones are now named H1, H2A, H2B, H3 and H4, and the nucleated erythrocyte-specific histone is named H5 [15]. Opinion was divided for many years as to whether there were very many histones [161] or only a few [93, 97], and the various nomenclatures used by various laboratories created considerable confusion. Probably the two most important developments which helped clarify this confused situation were two methods for looking more directly at the complexity of mixtures of proteins. These were electrophoresis in stabilising media, first utilised for histones by Neelin and Connell [138] and N-terminal group analysis first applied to histones by Luck et al. [116] and Phillips [156]. These two techniques, modified in many cases by subsequent workers, proved invaluable for distinguishing and characterising the many fractions obtained. The following review articles cover this period of the histone chemistry [25, 82, 136].

Progress during this period (1950—1965) on the non-histone chromosomal proteins was extremely slow. A few groups were prominent in this field and probably the publications of the groups directed by Professors H. Busch, A. E. Mirsky and T. Y. Wang contributed most to the knowledge of the non-histone chromosomal proteins during the 15 year period. In retrospect it can be seen that the major problem was one of definition. Initially extracts of nuclei were made which contained all the soluble nuclear proteins as well as the chromatin-associated proteins. Even when the chromatin-associated proteins only were isolated it was not certain how many of the proteins had become adsorbed to the DNA during the isolation procedures, and how many, by virtue of their function, became associated with chromatin only during specific periods of the cell cycle [100]. Even now, chromatin is difficult to define, and the term "non-histone chromosomal protein" is almost meaningless, since it means everything associated with chromatin except the histones. For these reasons, current research in this field is largely restricted either to proteins with a definite function (polymerases, etc.) or to other well-defined groups which appear to have structural relationships similar to those of the histones with the DNA.

The possible importance of some of the non-histone proteins in specific gene derepression has also become a subject of considerable interest in recent years since the early work of Paul and Gilmour [51, 154] (for review see Stein et al. [197]).

However, within the past 6 years our understanding of the importance of the histones in the structure of the basic unit thread

*References p. 566*

of chromatin has been considerably clarified. This has come about because of three major discoveries which occurred at approximately the same time and which complemented each other to a considerable extent: These were:

(1) The discovery that certain endogenous nucleases in rat liver when allowed to auto-digest the chromatin, did not split the DNA at random but went more specifically for positions approximately 200 base pairs apart, leading the authors to suggest a simple basic repeating sub-structure [81].
(2) The discovery that two each of four of the five histones (H2A, H2B, H3 and H4) combined together to form a very specific octamer around which it was suggested that DNA diad helix wound itself to form a compact unit structure [105, 140, 194].
(3) The visualisation in the electron microscope of sperical particles approximately 100 Å in diameter spaced at regular intervals along the chromatin unit thread [145, 149, 221].

Together these discoveries laid the foundation for the basic structure of chromatin as we understand it at the present time. It is thought to be organised in repeating units called nucleosomes,

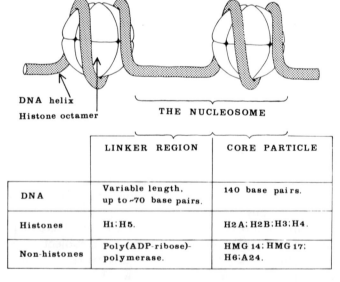

|  | LINKER REGION | CORE PARTICLE |
|---|---|---|
| DNA | Variable length, up to ~70 base pairs. | 140 base pairs. |
| Histones | H1; H5. | H2A; H2B; H3; H4. |
| Non-histones | Poly(ADP-ribose)-polymerase. | HMG 14; HMG 17; H6; A24. |

Fig. 1. The structure of the nucleosome. diagrammatic representation to show location of components (not drawn to scale, H1 molecule not shown).

each repeat containing about 200—250 base pairs of DNA, one molecule of histone H1 and 2 each of the other four histones. The repeat itself consists of about 140—145 base pairs of DNA, associated with 2 molecules each of histones H2A, H2B, H3 and H4 usually referred to as the core particle and some linker DNA thought to be associated with histone H1. The length of the linker DNA appears to vary from 0 to 60 base pairs depending on the tissue and species (for review see Kornberg [100]). This basic chromatin structure (see Fig. 1) will be referred to in subsequent sections.

*1.2. Definition*

Strictly, the term chromosomal protein means "the proteins of the chromosomes", but since the chromosomes only exist for a brief period within the cell cycle, it is normally extended to encompass those proteins associated with the DNA at any time during the cell cycle, i.e., "chromatin proteins". However, as mentioned above the term "chromatin" is not easily defined, and its protein complement almost certainly varies qualitatively and quantitatively during the cell cycle. For these reasons, work in this field has been confined mainly to those proteins which can be shown to co-isolate with DNA under conditions which should not disrupt protein-DNA linkages. The problem of artefacts arising by adsorption of protein from the cytoplasm and/or nuclear sap when the ionic environment is changed is, of course, an important one and many attempts have been made to determine how many of the DNA-associated proteins are in fact truly associated with the DNA in vivo. This question has not been answered satisfactorily, and when one sees the multiplicity of peaks from polyacrylamide gel electrophoresis scans of so called non-histone chromosomal proteins reproduced in the literature, one yearns for strict definition. However, such a definition will not come about until the functions of all the chromatin proteins are known. Therefore, for the purposes of this review, we shall confine ourselves to the proteins which have been well-characterised and which are generally accepted as being associated with the DNA in the nucleus. There are two major groups of proteins in this category, the basic proteins, the histones, and a group of non-histone chromosomal proteins which have been called the high mobility group proteins (see below). A few other proteins have also been characterised and these will be dealt with individually. We do not intend to discuss the protamines in this review since they have a very specialised function being found only in association with the condensed DNA of spermatozoa. They are relatively simple proteins and are replaced by the histones after fertilisation. Not all DNA in spermatozoa is found associated with protamines. In some cases, e.g., sea urchin sperm, histone-like proteins are found. This is, thus, a fairly complex subject and the reader is referred to a review by Meistrich et al. [130].

## 2. The histones

### 2.1. Occurrence and specificity

The nuclear DNA of all eukaryotic cells (except germ cells) is complexed with a group of very basic, low molecular weight proteins, the histones. The group has been divided into 5 main types: H1, H2A, H2B, H3, H4. They are remarkable for the conservation of their primary structures during evolution, as has been reviewed by Von Holt et al. [200]. H4 has generally been very strictly conserved so that there are only two amino acid differences between pea and calf, and these changes are conservative. The H4 from a species of *Tetrahymena*, however, seems to be exceptional in having even within its N-terminal 66 residues, 15 amino acid differences from the equivalent part of calf H4, and also an unblocked N-terminal residue [54]. The structure of H3 has also been evolutionally stable. Pea H3 and calf H3 differ by only four amino acids. Histones H2A and H2B show more species variation than do H3 and H4, so that these proteins in plants and animals have different solubility, chromatographic and electrophoretic properties [45, 182]. It is now recognised that all histones except H4 can, in some situations at least, be microheterogeneous, which is to say that a number of primary structural variants can exist within one organism at the same time (see Section 2.4.7.). The microheterogeneity of histones H4, H3, H2A and H2B mirrors their evolutionary stability in that while the first two show no or only minor differences between variants, the last two show quite extensive differences (reviewed by Von Holt et al. [200]. These histone variants seem to arise during development and differentiation according to a fixed programme, so they must have some functional significance, though its nature is as yet unknown. There is little information to date concerning the tissue specificity of variants of histones H2A, H2B and H3.

The fifth major histone type, H1, shows less evolutionary stability than even H2A or H2B. This is so marked that in Fungi, which are amongst the lowest of eukaryotes, while other histones can be recognised as such, the presumptive H1 species have analyses as much like calf thymus HMG proteins 1 and 2 as like calf thymus H1 [152, 181]. For example, see the amino acid analysis of a protein which could equally well be labelled "yeast H1", or as in Table VI, a "yeast HMG" protein. However, despite some considerable species variation of H1 in respect of amino acid content and sequences in some parts of the molecule, the hydrophobic central part of H1 has been more strictly conserved, between rabbit and trout at least, as Cole [30] has pointed out. This fact may have some functional significance. For some years it has been known that H1 consists of a family of structural variants or sub-fractions which co-exist within the same cell nucleus, and more is known about them than about variants of other histones. The amounts of H1 subfractions, relative to each other, are organ-specific [102], they change during differentiation [22, 164] and with metabolic activity [83], but at the moment their function(s) is uncertain (see Section 2.6.).

Thus, the above discussion may be summarised by saying that the tissue specificities of the major histones are only qualitative, due to differences in sub-fractions. There may be an exception to this general rule, however, in *Tetrahymena pyriformis*, which seems to lack H1 and H3 in its micronucleus, although both of these proteins are present in the macronucleus of the same cell [73, 74]. While it may be said that negative evidence is never very convincing, data such as these, and those concerning *Tetrahymena* H4 (see above), may be important for appreciating chromatin structure and function, and also the divergence of organisms during evolution.

Two other histone classes are known which are somewhat H1-like in their amino acid contents and chromatographic and solubility properties, and might even be considered as H1 sub-fractions. These are H1° and H5. H1°, or proteins like it, have been found in many animal organs, most particularly those undertaking little or

*References p. 566*

no DNA synthesis [122, 151]. Little else is known about H1° except that it exists in multiple forms which have slightly different chromatographic and electrophoretic mobilities [129, 179]. More is known about H5: it is found only in the transcriptionally-inactive nucleated erythrocytes of amphibians, birds and fish; it is located within the linker region of the nucleosome [9]; its molecular weight is similar to that of H1 (Section 2.4.6.); it shows species variation in amino acid content [82] and has sequence homologies with H1 from the same species [31]; variant forms of H5 are known [77]. Another protein, originally called "histone H6" [217] but now classed as an HMG protein, is dealt with in Sections 3.4 and 3.5.

A feature of the histones which is worthy of note is their stoichiometry. Firstly, they are the most abundant of nuclear proteins, and their combined weight is about equal to that of the DNA with which they are complexed. One exception to this is the unfertilised frog egg, which has an approximately 50 000-fold excess of histones stored ready to complex with the DNA which is rapidly synthesised after fertilisation has occurred [3, 222]. Secondly, as may be deduced from the nucleosome model (Section 1.1.), for every one molecule of H1 there are two each of H2A, H2B, H3 and H4. Where H5 occurs, it reaches levels equal to or double that of H1 [146, 212] in molecular terms. This stoichiometry is of interest in some cases, such as that of yeast H1: although its amino acid content is similar to both H1 and HMG proteins of higher eukaryotes (see above), the quantities of this protein relative to the other histones, suggest that it is more like H1 in function [181].

A point of interest is that while the nuclear DNA of eukaryote cells is complexed with histones, so too are the genomes of some viruses which infect these cells, such as the DNA-containing viruses polyoma and SV40. The intracellular SV40 genome bears all 5 main histones, arranged as they are on their host's DNA, and during maturation this "mini-chromosomes" becomes wrapped up in viral proteins to yield the infectious virion [137]. The synthesis of these histones is controlled by host cell-derived mRNA, and this is not co-ordinated with synthesis of DNA (of either host or virus) [101], as histone synthesis usually is (as reviewed by Elgin and Weintraub [43]).

## 2.2. Preparation

It is not intended to give any details of the many isolation procedures for the various histone fractions, as there are recent publications concerned only with this problem [98, 99].

In general, the isolation of the individual histones has been carried out in two different ways. The first procedure is to isolate the total histone from chromatin by extraction with dilute acids or strong (2 –3 M) salt solution, and then to separate the individual fractions from the mixture by ion-exchange or gel filtration columns. The second approach, which allows for larger quantities of material to be isolated, depends upon selective extraction of individual fractions, or groups of fractions, with acid or salt in the presence of organic solvents and their subsequent separation and purification by precipitation with organic solvents.

However, the need for preparing histones in large quantities by either of these two approaches has decreased considerably over the past few years since many micro-techniques can now be applied to the individual fractions as separated by electrophoresis in the various stabilising media. The two approaches mentioned above gave rise during the 1960s to two different nomenclatures which have now been combined to give the histone nomenclature generally accepted today [15]. Whilst this has been extremely useful in simplifying what was a confused situation, the nomenclature based on the methods of preparation gave some information about the fractions which is now generally overlooked. These are:-
  (1) That histones H2A and H4 are very similar proteins. They are selectively removed together from chromatin [96]; they have similar internal sequences and many other regions of the protein chain in common. This would suggest similar functions in the histone octamer.
  (2) That histone H2B was often selectively extracted with histone H1 indicating that it was much less hydrophobic than histones H3, H2A and H4, which were selectively removed together [93].

From the points above it would seem that the differences in the function of the two complexes (H2A, H2B) and (H3, H4) within the octamer lie in the differences between the histones H2B and H3. Interestingly these are the only two histone fractions that have free N-terminal amino acids.

*References p. 566*

## 2.3. Amino acid composition

Table I provides data on the amino acid compositions of the major histone classes and H1° and H5. Note that these analyses are of whole fractions, not of particular sub-fractions of, for example, H1, which differ slightly in their total amino acid contents (e.g. see ref. 177). Note also that these analyses are those which were

### TABLE I

Amino acid analyses of purified histone fractions

Analyses given as moles per 100 moles of all amino acids recovered. No corrections for hydrolytic losses. Lys includes all ϵ-substituted lysines. Sources of histones and refs.: H4, H3, H2B, H2A, H1 — calf thymus [61], H5 — chicken erythrocyte [61], H1° — calf lung [223] with added data * from calf liver [49] which otherwise has similar analysis.

| Amino Acid | H4 | H3 | H2B | H2A | H1 | H1° | H5 |
|---|---|---|---|---|---|---|---|
| Asp (A) | 5.2 | 4.2 | 5.0 | 6.6 | 2.5 | 3.3 | 1.7 |
| Thr | 6.3 | 6.8 | 6.4 | 3.9 | 5.6 | 7.7 | 3.2 |
| Ser | 2.2 | 3.6 | 10.4 | 3.4 | 5.6 | 8.5 | 11.9 |
| Glu (A) | 6.9 | 11.5 | 8.7 | 9.8 | 3.7 | 4.2 | 4.3 |
| Pro | 1.5 | 4.6 | 4.9 | 4.1 | 9.2 | 9.4 | 4.7 |
| Gly | 14.9 | 5.4 | 5.9 | 10.8 | 7.2 | 4.3 | 5.3 |
| Ala | 7.7 | 13.3 | 10.8 | 12.9 | 24.3 | 16.8 | 16.3 |
| Val | 8.2 | 4.4 | 7.5 | 6.3 | 5.4 | 5.2 | 4.2 |
| ½-Cys | 0.0 | 1.0 | 0.0 | 0.0 | 0.0 | 0.0 | 0.0 |
| Met | 1.0 | 1.1 | 1.5 | 0.0 | 0.0 | (∼0.5)* | 0.4 |
| Ile | 5.7 | 5.3 | 5.1 | 3.9 | 1.5 | 1.9 | 3.2 |
| Leu | 8.2 | 9.1 | 4.9 | 12.4 | 4.5 | 2.1 | 4.7 |
| Tyr | 3.8 | 2.2 | 4.0 | 2.2 | 0.9 | 1.1 | 1.2 |
| Phe | 2.1 | 3.1 | 1.6 | 0.9 | 0.9 | 0.9 | 0.6 |
| Lys ⎫ | 11.4 | 10.0 | 14.1 | 10.2 | 26.8 | 31.3 | 23.6 |
| His ⎬ (B) | 2.2 | 1.7 | 2.3 | 3.1 | 0.0 | 0.6 | 1.9 |
| Arg ⎭ | 12.8 | 13.0 | 6.9 | 9.4 | 1.8 | 2.6 | 12.4 |
| Acidics (A) | 12.1 | 15.7 | 13.7 | 16.4 | 6.2 | 7.7 | 6.0 |
| Basics (B) | 26.4 | 24.7 | 23.3 | 22.7 | 28.6 | 34.5 | 37.9 |
| B/A** | 2.2 | 1.6 | 1.7 | 1.4 | 4.6 | 4.6 | 6.3 |
| Lys/Arg | 0.9 | 0.8 | 2.0 | 1.1 | 15.0 | 12.0 | 0.8 |
| Gly × Arg | 190.0 | 70.0 | 41.0 | 102.0 | 13.0 | 11.2 | 65.7 |
| N-terminal group | Acetyl | Alanine | Proline | Acetyl | Acetyl | Blocked* | Acetyl |

** No allowance has been made for amides in these figures.

obtained experimentally, not those calculated from particular primary sequences, since those do not account for the presence of other variants in fractions obtained in practice, and can give slightly different results (e.g. see ref. 98).

Two salient features of histones are their high content of basic residues, particularly lysine and arginine, and their lack of tryptophan. The main histone types have been grouped according to the ratio of their lysine content to their arginine content. Thus, H3 and H4, having slightly more arginine than lysine, have been called arginine-rich; H2A and H2B, with slightly more lysine than arginine, have been called slightly lysine-rich; and H1, with a greater excess of lysine, has been called very lysine-rich. H1° is like H1 in its lys/arg ratio, but H5, which is like H1 in other respects, is richer in arginine than lysine. Such terminology is, therefore, of doubtful usefulness when discussing all of the histone classes. The individual histones have other characteristics which are helpful in identification, such as the relatively high serine content of H2B (although similar levels are attained in H1° and H5), the high glycine content of H4, the high alanine and proline contents of H1, the cysteine content of H3 (unique in this respect) and the blockage of the N-terminal residue, usually by an acetyl group (that for H1° is not known).

## 2.4. Primary structures

The structure of the nucleosome, where two each of the four histones H2A, H2B, H3 and H4 form an octamer around which the DNA helix is folded [15], is discussed in Sections 1.1. and 2.6. When discussing the primary structures of the histones, therefore, one must consider the relationship between nucleosome structure and histone structure.

One question that the nucleosome model poses is: How do the histones bind to one another and to the DNA chain? It has been shown that when chromatin is subjected to mild tryptic digestion, only 20—30 residues are cleaved from the N-terminal of each of the histones H2A, H2B, H3 and H4, the remainder of the chain being resistant [211]. This suggests that the N-terminal regions of these proteins are on the outside of the nucleosome bound to the phosphate groups of DNA. The remaining 70—100 residues from the central and C-terminal regions of the molecules are presumably interacting with each other within the core. As we will see from the sequence data, the N-terminal regions of these four histones are very basic (and therefore likely to bind to DNA), and contain a high proportion of residues such as proline, serine and glycine which favour extended chain structures [16, 18]. In contrast, the carboxyl halves of H3 and H4 and the central regions of H2A and H2B contain a high proportion of apolar and other residues which favour helix formation [16, 18]. It is these apolar regions which form well-defined structures (secondary and tertiary structures) which then act as sites of histone-histone interactions, while the N-terminal basic segments are thought to be the sites of interaction with the DNA phosphate groups. The fifth histone, H1, is found in the linker region between nucleosomes, and is believed to play a role in generating higher order structures of the chromatin. The primary structures of the individual histones will be discussed separately.

2.4.1. *Histone H4.* Histone H4 is the smallest histone, of $M_r \simeq$ 11 280 (102 amino acid residues). The amino acid sequence was originally determined for calf thymus H4 [34, 142], and is shown in Fig. 2. Subsequent sequence analysis has shown that the sequence of H4 from all those mammals investigated so far is identical with that of calf thymus. The sequence of pea-seedling H4 differs in only two positions from the calf thymus sequence. Valine at position 60 in the calf histone is replaced by isoleucine in the pea histone and lysine at position 77 is replaced by arginine [35]. Both these changes are highly conservative; valine and isoleucine are both hydrophobic amino acids, and lysine and arginine are both basic amino acids. These changes are, therefore, unlikely to have any structural effects on the molecule. This strict conservation of sequence makes histone H4 the most evolutionarily stable protein known, and suggests that virtually any mutation to this histone is lethal. The implication, therefore, is that the entire sequence of this histone, and not just an active centre or site, is important for its function. Such a requirement for sequence conservation could arise if H4 is required to exist in two or more configurations at different stages during the cell cycle.

Consideration of the distribution of amino acids in histone H4 shows that the N-terminal region is almost completely lacking in

```
                                                           10
Ac — Ser — Gly — Arg — Gly — Lys — Gly — Gly — Lys — Gly — Leu —
                                                           20
      Gly — Lys — Gly — Gly — Ala — Lys — Arg — His — Arg — Lys —
                                                           30
      Val — Leu — Arg — Asp — Asn — Ile — Gln — Gly — Ile — Thr —
                                                           40
      Lys — Pro — Ala — Ile — Arg — Arg — Leu — Ala — Arg — Arg —
                                                           50
      Gly — Gly — Val — Lys — Arg — Ile — Ser — Gly — Leu — Ile —
                                                           60
      Tyr — Glu — Glu — Thr — Arg — Gly — Val — Leu — Lys — Val —
                                                           70
      Phe — Leu — Glu — Asn — Val — Ile — Arg — Asp — Ala — Val —
                                                           80
      Thr — Tyr — Thr — Glu — His — Ala — Lys — Arg — Lys — Thr —
                                                           90
      Val — Thr — Ala — Met — Asp — Val — Val — Tyr — Ala — Leu —
                                                          100
      Lys — Arg — Gln — Gly — Arg — Thr — Leu — Tyr — Gly — Phe —

      Gly — Gly
```

Fig. 2. The amino acid sequence of calf thymus histone H4.

hydrophobic amino acids, but is high in basic residues and helix-destabilising residues. Residues 1—45 have an overall charge of +16. On the other hand, the C-terminal portion (46—102) only has a net charge of +3 and contains most of the hydrophobic residues. It is presumably this portion of the molecule that is involved in binding to other histones in the core of the nucleosome (see Section 1.1.).

2.4.2. *Histone H3*. Histone H3 has an $M_r \doteq 15\,300$ (135 amino acids). The amino acid sequence of the calf thymus protein [36, 37] is shown in Fig. 3. Histone H3 is the only histone to contain the amino acid cysteine. H3 isolated from most mammals has two cysteines, whereas H3 from plants, invertebrates, vertebrates, and some lower mammals (including mouse and rat), has only one cysteine residue, residue 96 being serine. However, after careful analysis, it has been shown that heterogeneity exists in these samples. It has now been shown that calf thymus H3 in fact con-

```
                                                   10
Ala — Arg — Thr — Lys — Gln — Thr — Ala — Arg — Lys — Ser —
                                                   20
Thr — Gly — Gly — Lys — Ala — Pro — Arg — Lys — Gln — Leu —
                                                   30
Ala — Thr — Lys — Ala — Ala — Arg — Lys — Ser — Ala — Pro —
                                                   40
Ala — Thr — Gly — Gly — Val — Lys — Lys — Pro — His — Arg —
                                                   50
Tyr — Arg — Pro — Gly — Thr — Val — Ala — Leu — Arg — Glu —
                                                   60
Ile  — Arg — Arg — Tyr — Gln — Lys — Ser — Thr — Glu — Leu —
                                                   70
Leu — Ile  — Arg — Lys — Leu — Pro — Phe — Gln — Arg — Leu —
                                                   80
Val — Arg — Glu — Ile  — Ala — Gln — Asp — Phe — Lys — Thr —
                                                   90
Asp — Leu — Arg — Phe — Gln — Ser — Ser — Ala — Val — Met —
                                                   100
Ala — Leu — Gln — Glu — Ala — Cys — Glu — Ala — Tyr — Leu —
                                                   110
Val — Gly — Leu — Phe — Glu — Asp — Thr — Asn — Leu — Cys —
                                                   120
Ala — Ile  — His — Ala — Lys — Arg — Val — Thr — Ile  — Met —
                                                   130
Pro — Lys — Asp — Ile  — Gln — Leu — Ala — Arg — Arg — Ile  —

Arg — Gly — Glu — Arg — Ala
```

Fig. 3. The amino acid sequence of calf thymus histone H3.

tains about 20% of molecules having only one cysteine residue. In lower animals the situation is reversed and about 20% of molecules have two cysteine residues. These cysteine residues are obviously capable of forming disulphide bridges within the nucleosome or possibly with -SH groups on neighbouring non-histone proteins. Like H4, the sequence of H3 has been highly conserved throughout evolution. The only difference between carp histone H3 and calf thymus H3 is that cysteine at position 96 of calf thymus is replaced by serine in the carp [84]. Pea histone H3 and calf thymus H3 differ at only 4 positions, 3 of these differences being conservative changes [153]. Again, like H4, the N-terminal sequence is highly basic and lacks hydrophobic residues. The amino terminal 53 residues have an overall charge of +18. The

# PRIMARY STRUCTURES

carboxyterminal region (residues 54–135) has an overall charge of only +4 (although the region 113–135 is also rather basic), and contains most of the hydrophobic residues.

*2.4.3. Histone H2A.* Histone H2A has an $M_r$ = 15 000 (129 residues). The complete sequence of calf thymus H2A [226] is shown in Fig. 4. The first 9 residues of this protein are identical with those of histone H4. H2A is characterised by two hydrophobic sequences which contain no basic residues. The first region consists of residues 43–70, the second of residues 100–117. Unlike H3 and H4, histone H2A has basic residues in both the N- and C-terminal regions. Comparison of the sequences of H2A from calf thymus and trout [8] shows there to be 4 substitutions and 2 deletions, most of which are conservative. These changes are

```
                                                         10
Ac  — Ser — Gly — Arg — Gly — Lys — Gln — Gly — Gly — Lys — Ala —
                                                         20
    Arg — Ala — Lys — Ala — Lys — Thr — Arg — Ser — Ser — Arg —
                                                         30
    Ala — Gly — Leu — Gln — Phe — Pro — Val — Gly — Arg — Val —
                                                         40
    His — Arg — Leu — Leu — Arg — Lys — Gly — Asn — Tyr — Ala —
                                                         50
    Glu — Arg — Val — Gly — Ala — Gly — Ala — Pro — Val — Tyr —
                                                         60
    Leu — Ala — Ala — Val — Leu — Glu — Tyr — Leu — Thr — Ala —
                                                         70
    Glu — Ile — Leu — Glu — Leu — Ala — Gly — Asn — Ala — Ala —
                                                         80
    Arg — Asp — Asn — Lys — Lys — Thr — Arg — Ile — Ile — Pro —
                                                         90
    Arg — His — Leu — Gln — Leu — Ala — Ile — Arg — Asn — Asp —
                                                         100
    Glu — Glu — Leu — Asn — Lys — Leu — Leu — Gly — Lys — Val —
                                                         110
    Thr — Ile — Ala — Gln — Gly — Gly — Val — Leu — Pro — Asn —
                                                         120
    Ile — Gln — Ala — Val — Leu — Leu — Pro — Lys — Lys — Thr —
    Glu — Ser — His — His — Lys — Ala — Lys — Gly — Lys
```

Fig. 4. The amino acid sequence of calf thymus histone H2A.

*References p. 566*

principally in the basic C- and N-terminal regions and none are involved in the hydrophobic regions. We see, therefore, that the basic regions can be slightly altered during evolution, but the central hydrophobic regions are unaltered, which again presumably reflects the importance of these central regions in binding to other histones at the centre of the nucleosome. Although the sequence of this histone is less conserved than that of H3 or H4, in comparison with other proteins, H2A still shows a very high conservation of sequence over a long evolutionary period. In addition to evolutionary heterogeneity, limited heterogeneity of H2A within a single cell type has been observed. Rat chloroleukaemia has been shown to contain two forms of H2A differing by two amino acid substitutions [108].

                                                                10
Pro — Gln — Pro — Ala — Lys — Ser — Ala — Pro — Ala — Pro —
                                                                20
Lys — Lys — Gly — Ser — Lys — Lys — Ala — Val — Thr — Lys —
                                                                30
Ala — Gln — Lys — Lys — Asp — Gly — Lys — Lys — Arg — Lys —
                                                                40
Arg — Ser — Arg — Lys — Glu — Ser — Tyr — Ser — Val — Tyr —
                                                                50
Val — Tyr — Lys — Val — Leu — Lys — Gln — Val — His — Pro —
                                                                60
Asp — Thr — Gly — Ile — Ser — Ser — Lys — Ala — Met — Gly —
                                                                70
Ile — Met — Asn — Ser — Phe — Val — Asn — Asp — Ile — Phe —
                                                                80
Glu — Arg — Ile — Ala — Gly — Glu — Ala — Ser — Arg — Leu —
                                                                90
Ala — His — Tyr — Asn — Lys — Arg — Ser — Thr — Ile — Thr —
                                                               100
Ser — Arg — Glu — Ile — Gln — Thr — Ala — Val — Arg — Leu —
                                                               110
Leu — Leu — Pro — Gly — Glu — Leu — Ala — Lys — His — Ala —
                                                               120
Val — Ser — Glu — Gly — Thr — Lys — Ala — Val — Thr — Lys —
Tyr — Thr — Ser — Ser — Lys

Fig. 5. The amino acid sequence of calf thymus histone H2B.

## PRIMARY STRUCTURES

*2.4.4. Histone H2B.* Histone H2B consists of a single polypeptide chain of 125 amino acids ($M_r \simeq 13\,800$). The complete sequence of the calf thymus protein [87] is shown in Fig. 5. As with the three histones previously mentioned, histone H2B also has a highly basic N-terminal region. There are 15 basic residues in the first 35 residues, but only one hydrophobic residue. The sequence of trout H2B differs from the sequence of calf thymus H2B by the deletion of two residues and substitution of 7 other residues [104]. All the observed changes, except one, occur in the N-terminal half of the molecule in the first 37 residues. Therefore, as with H2A, evolutionary changes have been permitted in the N-terminal region of the molecule, but the remainder of the molecule has remained essentially invariant suggesting a highly specific function for this part of the molecule, presumably in binding to the other histones. The mutational rate for H2B is identical with the mutational rate shown for H2A and offers further evidence for the pairwise behaviour of these two histones within the subunit structure of chromatin (see Introduction).

*2.4.5. Histone H1.* Calf thymus histone H1 is characterised by a high lysine (27%) and alanine (24%) content. It is much larger than the other four calf thymus histones, having an $M_r \simeq 21\,000$, and comprises a closely related group of proteins (see Section 2.1). For example, rabbit thymus H1 gives 5 overlapping peaks on ion-exchange chromatography with shallow gradients [30]. Small, yet distinct differences in primary structure, mainly in the N-terminal half of the molecule, exist between these various fractions. For example, comparison of the N-terminal 73 residues of rabbit thymus fractions 3 and 4 shows 10% of the residues to be different [30]. Because of this heterogeneity and the difficulty in separating the sub-fractions, the complete sequence of only one H1 sub-fraction from rabbit thymus has been determined [30], (see Fig. 6), although considerable sequence data for a large part of a number of other calf and rabbit thymus sub-fractions are also available [30]. Unlike the situation in the other four histones, it is the C-terminal region of H1 (residues 115—225) which is most basic (net charge +40). This region consists largely of lysyl, alanyl and prolyl residues, which together account for 85% of the residues in this region. Similarly, 75% of the first 40 amino acids are Lys,

*References p. 566*

| | | | | | | | | | | |
|---|---|---|---|---|---|---|---|---|---|---|
| RTL-3 | Ac | — | Ser — | Glu — | Ala — | Pro — | Ala — | Glu — | Thr — | Ala — Ala — |
| Trout | Ac | — | Ala — | Glu — | Ala — | Pro — | Ala — | Glu — | Val — | Ala — O — |
| RTL-3 | | | O — | Lys — | Lys — | Lys — | Ala — | Ala — | Lys — | Lys — Pro — |
| Trout | | | Pro — | Lys — | Lys — | Lys — | Ala — | Ala — | Ala — | Lys — Pro — |
| RTL-3 | | | Gly — | Pro — | Pro — | Val — | Ser — | Glu — | Leu — | Ile — Thr — |
| Trout | | | Gly — | Pro — | Ala — | Val — | Gly — | Glu — | Leu — | *Ile* — *Gly* — |
| RTL-3 | | | Leu — | Ser — | Leu — | Ala — | Ala — | Leu — | Lys — | Lys — Ala — |
| Trout | | | Val — | Ser — | Leu — | Ala — | Ala — | Leu — | Lys — | Lys — Ser — |
| RTL-3 | | | Asn — | Ser — | Arg — | Ile — | Lys — | Leu — | Gly — | Leu — Lys — |
| Trout | | | *Asn* — | Ser — | Arg — | Val — | Lys — | Ile — | Ala — | Val — Lys — |
| RTL-3 | | | Lys — | Gly — | Thr — | Gly — | Ala — | Ser — | Gly — | Ser — Phe — |
| Trout | | | Lys — | Gly — | Thr — | Gly — | Ala — | Ser — | Gly — | Ser — Phe — |
| RTL-3 | | | Lys — | Pro — | Lys — | Pro — | O — | Lys — | Lys — | Ala — Gly — |
| Trout | | | Lys — | O — | Lys — | Pro — | Ala — | Lys — | Lya — | Ala — Ala — |
| RTL-3 | | | O — | Ala — | Thr — | Pro — | Lys — | Lys — | Pro — | Lys — Lys — |
| Trout | | | Pro — | Ala — | Ala — | Ala — | Lys — | Lys — | Pro — | Lys — Lys — |
| RTL-3 | | | Lys — | Thr — | Pro — | Lys — | Lys — | Ala — | Pro — | Lys — Pro — |
| Trout | | | Lys — | Ser — | Pro — | Lys — | Lys — | Ala — | O — | Lys — O — |
| RTL-3 | | | Lys — | Ser — | Pro — | Ala — | Lys — | *Val* — | Ala — | Lys — Ser — |
| Trout | | | O — | Thr — | Pro — | Lys — | Lys — | Ala — | Ala — | Lys — Ser — |
| RTL-3 | | | Ala — | Ala — | Lys — | Pro — | Lys — | O — | Ala — | Pro — Lys — |
| Trout | | | Ala — | Ala — | Lys — | Pro — | Lys — | Lys — | Ala — | Ala — Lys — |
| RTL-3 | | | *Ala* — | Lys — | Lys — | Lys — | Lys — | OH | | |
| Trout | | | Ala — | Lys — | Lys — | OH | | | | |

# PRIMARY STRUCTURES

|  | 10 |  |  |  |  |  |  |  |  | 20 |
|---|---|---|---|---|---|---|---|---|---|---|
| Pro — | Ala — | Pro — | Ala — | Glu — | Lys — | Ser — | Pro — | Ala — | Lys — | O — |
| Pro — | Ala — | Pro — | Ala — | Ala — | Ala — | Pro — | Ala — | Ala — | Lys — | Ala — |
| 30 |  |  |  |  |  |  |  |  |  | 40 |
| Gly — | O — | Ala — | Gly — | Ala — | Ala — | Lys — | Arg — | Lys — | Ala — | Ala — |
| Lys — | Lys — | Ala — | Gly — | O — | O — | O — | O — | O — | O — | O — |
| 50 |  |  |  |  |  |  |  |  |  | 60 |
| Lys — | Ala — | Val — | Ala — | Ala — | Ser — | Lys — | Glu — | Arg — | Asn — | Gly — |
| Lys — | Ala — | Val — | Ala — | Ala — | Ser — | Lys — | Glu — | Arg — | Ser — | Gly — |
| 70 |  |  |  |  |  |  |  |  |  | 80 |
| Leu — | Ala — | Ala — | Gly — | Gly — | Tyr — | Asp — | Val — | Glu — | Lys — | Asn — |
| Leu — | Ala — | Ala — | Gly — | *Gly* — | *Tyr* — | Asp — | Val — | Glu — | Lys — | Asn — |
| 90 |  |  |  |  |  |  |  |  |  | 100 |
| Ser — | Leu — | Val — | Ser — | Lys — | Gly — | Thr — | Leu — | Val — | Glu — | Thr — |
| Ser — | Leu — | Val — | Thr — | Lys — | Gly — | Thr — | Leu — | Val — | Glu — | Thr — |
| 110 |  |  |  |  |  |  |  |  |  | 120 |
| Lys — | Leu — | Asp — | Lys — | Lys — | Ala — | Ala — | Ser — | Gly — | Glu — | Ala — |
| Lys — | Leu — | Asn — | Lys — | Lys — | Ala — | O — | O — | Val — | Glu — | Ala — |
| 130 |  |  |  |  |  |  |  |  |  | 140 |
| Ala — | Ala — | Lys — | Pro — | Lys — | Lys — | Pro — | Ala — | Gly — | O — | O — |
| Ala — | Pro — | Lys — | Ala — | Lys — | Lys — | Val — | Ala — | Ala — | Lys — | Lys — |
| 150 |  |  |  |  |  |  |  |  |  | 160 |
| Ala — | Ala — | Gly — | Ala — | Lys — | Lys — | Ala — | Val — | O — | O — | Lys — |
| Val — | Ala — | O — | Ala — | Lys — | Lys — | Ala — | Val — | Ala — | Ala — | Lys — |
| 170 |  |  |  |  |  |  |  |  |  | 180 |
| Lys — | Ala — | Ala — | *Ala* — | *Lys* — | Pro — | Lys — | *Val* — | Ala — | *Lys* — | Pro — |
| O — | O — | O — | O — | Lys — | Pro — | O — | O — | Ala — | O — | O — |
| 190 |  |  |  |  |  |  |  |  |  | 200 |
| Pro — | *Lys* — | Lys — | *Ala* — | O — | Lys — | Ala — | Val — | Lys — | Pro — | Lys — |
| Pro — | Lys — | Lys — | Ala — | Thr — | Lys — | Ala — | Ala — | Lys — | Pro — | Lys — |
| 210 |  |  |  |  |  |  |  |  |  | 220 |
| O — | Pro — | Lys — | Ala — | *Ala* — | Lys — | *Ala* — | Lys — | Lys — | *Thr* — | *Ala* — |
| Ser — | Pro — | Lys — | Lys — | Val — | Lys — | O — | Lys — | Pro — | Ala — | Ala — |

Fig. 6. The amino acid sequence of rabbit thymus histone H1 subfraction RTL-3, compared with that of trout histone H1. Positions where polymorphism has been detected are in italics. Deletions are indicated by — O —.

*References p. 566*

Ala and Pro. Both terminal regions may, therefore, be involved in binding to DNA. The central region has a low basic amino acid content, no proline residues, but is quite rich in hydrophobic amino acids. This region is, therefore, capable of forming a globular structure and is presumably available for interaction with other components of chromatin other than DNA. The picture we have for H1 is, therefore, of a globular hydrophobic core of about 70 residues sandwiched between two random-coiled basic regions, the N-terminal 40 residues and the C-terminal 110 residues or so.

In contrast to the small sequence differences that occur between the individual sub-fractions from a given species, the sequence of H1 is quite variable between species, especially in the C-terminal half of the molecule. Although a precise function for H1 is not known, the greater variation in sequence of this histone between different species suggests interactions with components of chromatin which vary from species to species (e.g., non-histone proteins). In contrast with rabbit and calf thymus H1, trout testis H1 appears to lack the microheterogeneity that these other H1 histones possess. Comparison of the trout testis H1 sequence [119] with the rabbit RTL3 sequence (Fig. 6) shows only about 70% sequence similarity between the two proteins, (although the overall architecture of the molecules is preserved). There is, therefore, far less conservation of sequence during the evolution of the H1 histones than there has been for the nucleosome "core" histones H4, H3, H2A, H2B.

*2.4.6. Histone H5.* Some species, such as fish, birds and reptiles, possess nucleated erythrocytes which are inactive in transcription and replication of DNA. In these inert erythrocyte nuclei, histone H5 largely, but not completely, replaces H1. The complete amino acid sequence of histone H5 has yet to be determined. The first 111 residues are known [5, 169] and are shown in Fig. 7 together with the composition of the remaining approximately 87 residues. H5 is closely related to H1. They are approximately of the same size (H5 has $\sim$ 197 residues, H1 has $\sim$ 225), and are both highly basic (although H5 contains more arginine $-$ 11% c.f. 2%; lysine is $\sim$ 25% in both). There are fairly hydrophobic regions in the central portions of both molecules, and both C-terminal regions are rich in basic and alanyl residues. However, they differ some-

Thr — Glu — Ser — Leu — Val — Leu — Ser — Pro — Ala — Pro $^{10}$ —
Ala — Lys — Pro — Lys — $\genfrac{}{}{0pt}{}{\text{Gln}}{\text{Arg}}$ — Val — Lys — Ala — Ser — Arg $^{20}$ —
Arg — Ser — Ala — Ser — His — Pro — Thr — Tyr — Ser — Glu $^{30}$ —
Met — Ile — Ala — Ala — Ala — Ile — Arg — Ala — Glu — Lys $^{40}$ —
Ser — Arg — Gly — Gly — Ser — Ser — Arg — Gln — Ser — Ile $^{50}$ —
Gln — Lys — Tyr — Ile — Lys — Ser — His — Tyr — Lys — Val $^{60}$ —
Gly — His — Asn — Ala — Asp — Leu — Gln — Ile — Lys — Leu $^{70}$ —
Ser — Ile — Arg — Arg — Leu — Leu — Ala — Ala — Gly — Val $^{80}$ —
Leu — Lys — Gln — Thr — Lys — Gly — Val — Gly — Ala — Gly $^{90}$ —
Ser — Ser — Phe — Arg — Leu — Ala — Lys — Ser — Asp — Lys $^{100}$ —
Ala — Lys — Arg — Ser — Pro — Gly — Lys — Lys — Lys — Ala $^{110}$ —
Lys —(Thr$_3$, Ser$_8$, Pro$_{10}$ Gly$_2$, Ala$_{16}$ Val$_3$, Lys$_{32}$ Arg$_{13}$)

Fig. 7. The partial amino acid sequence of chicken erythrocyte histone H5.

what in the N-terminal regions since the N-terminal of H5 is not particularly basic. It has been suggested that the globular region of H5 is located in the N-terminal region [169] which is a close parallel to the conformation of H1. An important difference between H5 and H1, however, is that in H5 the N-terminal tail is about 10 residues shorter. In H1, by contrast, the first 40 residues are highly basic and not globular. Similar to H1, and unlike the other four histones, H5 shows species variability, i.e., the sequence has changed considerably during evolution.

*2.4.7. Histone variants.* It has recently been shown that cells involved in differentiation processes (e.g., embryonic cells) contain a number of variants of the histones H2A and H2B, which appear in a programmed fashion during embryonic development [200]. Data on the primary structures of histone H2B variants from sea urchins have shown that the C-terminal hydrophobic part of the molecule, which is involved in the histone-histone interactions that are essential for the production of the nucleosome structure, is essentially constant, whereas a large variety of N-terminal sequences exist among different variants. These variable N-terminal sequences, which are involved in the histone-DNA interactions, are thought to provide a variety of chromatin structures suitable for specific regulatory processes. The subject of histone variants has recently been reviewed by Von Holt et al. [200].

### 2.5. Post-synthetic modifications

*2.5.1. Acetylation.* Histones H1, H2A and H4 all have N-terminal serine residues which are acetylated irreversibly during protein synthesis. In addition to this form of acetylation, all the histones, with the exception of H1, undergo extensive and reversible acetylation of certain lysine residues as a post-translational event. Depending on the histone source, $\epsilon$-N-acetyl-lysine has been found at residues 5 in H2A, 5, 10, 13 and 18 in H2B, 9, 14, 18 and 23 in H3, and 5, 8, 12 and 16 in H4 (for reviews see refs. 39, 165). It should, however, be pointed out that only partial acetylation is ever observed at a given site. All these acetylation sites occur in the basic N-terminal regions of the molecules (i.e., those regions of the molecule that are thought to be involved in DNA binding), and mask the positive charge on the lysine side-chain. It has, therefore, been suggested that the reversible modification of histones by acetylation is part of a complex set of mechanisms which control chromatin structure and influence DNA-template function [5]. Recently evidence has been produced which supports this hypothesis. Exposure of cells to sodium butyrate has been shown to lead to an accumulation of multi-acetylated forms of histone H3 and H4 in the cell. Brief digestion of chromatin isolated from these cells with DNAse I (which is known to degrade preferentially transcribed sequences) has revealed that chromatin containing hyperacetylated histones is preferentially digested [139, 199]. In other words, the DNA in those nucleosomes that are associated with acetylated histones must be held in an altered configuration which makes it more susceptible to DNAse I digestion.

Further information has been provided by a comparative analysis of histone H4 "transcriptionally active" and "transcriptionally inactive" chromatin prepared by the method of Gottesfeld [75]. Highly acetylated histone H4 was found to be predominantly associated with the DNA fraction which is enriched in transcribing gene sequences [33]. It was, therefore, concluded that histone H4 of active genes is present in a highly acetylated state. It has also been suggested that histone acetylation is associated with the deposition of histones on replicating DNA [39, 165].

*2.5.2. Methylation.* The methylations of arginine, histidine and lysine residues of histones are known to occur as post-translational events. However, information on the actual positioning of these modifications is limited. In Histone H4 of calf thymus, lysine at position 20 is found in both mono- and dimethyl forms [34, 35] and calf thymus H3 has two principal methylation sites at lysine residues at positions 9 and 27 [38]. Although, like the acetylation, methylation sites are restricted to the basic regions of the histones, it is unlikely that methylation has an analogous role to acetylation, since, unlike acetylation, methylation does not alter the overall charge on the lysine residue.

*2.5.3. Phosphorylation.* Specific sites of phosphorylation have been identified on all the histones, although histone H1 is more heavily phosphorylated than the rest. H1 is phophorylated at serine residues at positions 37 and 106, at a threonine residue at position 16, and at three other positions as yet unidentified (one threonine, two serines) in the C-terminal half of the molecule [110]. Phosphate groups are found on histone H2A at serine residues 1 and 19, on H2B at serines 6, 14 and 36, on histone H3 at serines 10 and 28, and on histone H4 at serine 1 [40, 218]. Histone H5 is phosphorylated at serines 3 and 7 [192], but other as yet unidentified phosphorylation sites are known to exist in the C-terminal region of the molecule [192]. It should be noted that nearly all these modifications occur in the basic regions of these molecules and, therefore, presumably play a part in modulating protein-DNA interactions.

The phosphorylation of histone H1 has received the most attention in recent years. Three major H1 kinases have been isolated

*References p. 566*

and characterised from mammalian cells; kinase-A, kinase-B and kinase-G [19, 80, 109, 110, 170, 176]. Kinases A and B are found in non-growing cells, kinase-G in very actively growing cells. Each kinase has a different specificity. Kinase A phosphorylates serine 37 in calf thymus H1, kinase-B phosphorylates serine 106 in calf thymus H1 and kinase-G phosphorylates the four other sites mentioned above. Cell-cycle studies have shown that histone H1 is phosphorylated from late $G_1$ phase through mitosis [78], but at differing places in the H1 amino acid sequence and to differing extents depending on the stage of the cell cycle. A comprehensive study of histone phosphorylation in Chinese hamster cells [79] has shown that during interphase, and the chromatin aggregation stages of mitosis (preprophase and telophase), histone H1 has 1—3 phosphates per molecule. During the second stage of mitosis (prophase, metaphase and anaphase) when chromosome structures are fully condensed, H1 exists as a super-phosphorylated molecule containing 3—6 phosphate groups. Exit of cells from anaphase has been correlated with the dephosphorylation of H1 to a molecule containing 0—3 phosphate groups. Further dephosphorylation giving unphosphorylated H1 occurred as the cells left telophase and entered $G_1$. In *Physarum polycephalum*, the time of mitosis can be advanced by the addition of a histone H1 kinase [19, 20]. This suggested that phosphorylation of histone H1 may condense chromatin during late $G_2$ phase acting as a "mitotic trigger". Phosphorylation of H1 could, therefore, play a role in the initiation or maintenance of condensation of chromosomes during mitosis.

*2.5.4. Poly ADP-ribosylation.* In recent years a new enzyme bound to chromatin, poly(ADP-ribose)-polymerase, has been isolated and characterised [121, 143]. The calf thymus enzyme has an $M_r \sim 120\ 000$ [121] and catalyses the formation of a homopolymer of ADP-ribose units linked by 1'-2' glycosidic bonds. The substrate for the reaction is NAD, and in the presence of DNA the enzyme successively adds ADP-ribose units onto an initial ADP-ribose residue which has been reported to be covalently attached to various nuclear proteins, particularly the histones. There is disagreement over the sub-chromatin localisation of this enzyme. One report has claimed the enzyme to be associated primarily with template active regions (euchromatin), and to be much reduced in transcriptionally inert chromatin fractions [134]. However, a second report claimed no preferential localisation in transcriptionally active chromatin [227]. But there does seem to be a consensus of opinion that the enzyme is located in the linker region of the nucleosome [53]. Since the ADP-ribose linkage is labile under alkaline conditions, all protein isolation techniques have to be carried out below a pH of about 6.0, if undamaged material is to be isolated. Two types of protein-ADP-ribose linkages have been demonstrated: one which is labile to hydroxylamine and one which is hydroxylamine insensitive [23]. Altough the linkage region of the ADP-ribose-protein conjugate is not known, the lability of one of the types of bonds to hydroxylamine suggests an ester glycoside type of linkage for this bond. Supporting evidence for an ester bond involving glutamic acid residues in histone H1 has been reported [41]. However, a different linkage of ADP-ribose to histone H1, involving a serine phosphate group, has also been reported [180].

Although there is considerable confusion in the literature at present concerning the length of the ADP-ribose units and the number and nature of the chromosomal proteins so modified, evidence for the ADP-ribosylation of some histones, in particular H1, is convincing.

ADP-ribosylation of intact nuclei shows histones H1 and H2B to be the major acceptors, H2A and H3 being modified to a lesser extent [52]. No ADP-ribosylation of H4 has been detected. Analysis of ADP-ribosylated histone H1 isolated from *intact* HeLa cells shows most H1 molecules to carry a single ADP-ribose unit linked by a hydroxylamine insensitive bond [2]. Less than a quarter of the total ADP-ribose residues were in the form of oligomeric or polymeric chains. By contrast, analysis of histone H1 from incubated *isolated* HeLa nuclei showed H1 to contain predominantly polymeric ADP-ribose residues linked by both hydroxylamine sensitive and insensitive bonds [2]. However, in both cases only 1—2% of the total protein-bound ADP-ribose residues present in the nucleus was bound to histone H1. A complex of two H1 molecules, held together by covalent and non-covalent association with a poly(ADP-ribose) chain of ~ 15 ADP-ribose units, has also been described in HeLa cell nuclei [191]. It has been suggested that this type of cross-linking reaction could be involved in chromosome condensation. None of the exact positions of these ADP-ribose linkage to histones has been determined.

*References p. 566*

## 2.6. Function

It is generally held that if the DNA of a mammal was in a linear form it would be of the order of a metre long. In order to fit inside the cell nucleus the DNA has to be folded in some way, and it is its association with proteins, histones in particular, which accomplishes this. Much work over the last few years has enabled the elaboration of a model (see Fig. 1) for the basic structure of chromatin in which DNA and the 5 main classes of histone are arranged in repeating units called nucleosomes, as mentioned in the Introduction (Section 1.1.), and as covered in various reviews [46, 50, 106, 197]. In this model the smaller histones are found within the core particle and H1, and/or H5, if present, is found in the linker region, possibly together with other protein(s) such as poly(ADP-ribose)-polymerase [53]. This, the basic nucleosomal thread, is folded into higher order structures (as reviewed by Georgiev et al. [50]), firstly becoming coiled into solenoids or superbeads, and then becoming further coiled or looped, so as to achieve the final degree of packing which is at its greatest in the metaphase chromosome. The nucleosomal thread is thought to exist throughout all or most of the chromatin, throughout the cell cycle, and there is evidence for the existence of some higher order structure during interphase [162, 224]. But probably the highest order(s) of structure is necessary for only maximal chromatin condensation, at cell division.

The problems of how DNA is replicated and transcribed when it is held within a nucleosomal thread have not yet been fully resolved. It is known that syntheses of DNA and histones are generally co-ordinated [43], that newly-made DNA rapidly becomes associated with histones, and that in the nucleosomes so formed the core particles and repeat lengths are different from those in mature chromatin [135, 173]. But data on exactly how pre-existing and newly synthesised histones segregate on daughter strands are conflicting (see review by Seale [172]). Actively transcribing genes, with the possible exception of ribosomal genes (as discussed by Lilley [114] and by Pedersen [155]), are associated with histones in nucleosome-like structures, so one might ask how the RNA polymerase molecule, which is about the same size as the core particle itself, gains proper access to the DNA. One answer could be that the histone octamer can slide along the DNA, ahead of

FUNCTION 533

the progressing polymerase molecule. While some data support the idea that nucleosomes can slide along DNA, at least when depleted in H1 (e.g., refs. 13, 188), it is not yet clear whether or not this phenomenon occurs in vivo (as discussed by Georgiev et al. [50]. An alternative explanation is that the nucleosome remains stationary on the DNA but opens out its structure in some way, to allow the polymerase access. Studies with deoxyribonucleases have shown that DNA is more accessible, to enzymatic degradation at least, when it is in a conformation which can allow transcription [48, 213, 225], but the exact mechanism of this process is obscure.

Thus, chromatin is composed of repeated basic units, the nucleosomes. A cell's population of nucleosomes is, however, not homogeneous or constant. Firstly, structural variants of histones occur (see Section 2.4.7). They have been particularly well studied in the developing sea urchin and it has been found that their occurrence is programmed during differentiation [22]. It has, therefore, been suggested that they play some role in this complex process [200]. Secondly, post-synthetic reversible histone modification can create heterogeneity in the nucleosome population, and alter interactions between proteins or interaction between proteins and DNA. It could be by such modification of core histones that the nucleosome structure is opened out to allow transcription to proceed through it. Acetylation of H3 and H4 in particular has been positively correlated with gene activation (see Section 2.5.1). Protein A24 might be regarded as an unusual variant or modified form of H2A (see Section 4.1), but the ubiquitin moiety apparently does not affect the contribution of H2A to the nucleosome's structure [124]. Functions tentatively suggested for A24 have been as a recognition signal for other proteins [124] and as an element involved in maintenance of higher order chromatin structures [57]. The distribution of these various protein types is currently unknown.

The heterogeneity of the nucleosome population is further complicated by the variability in linker DNA length. The significance of the length of the linker region is uncertain, but in Worcel's [223] model of chromatin structure it can govern higher order structures and packing ratios. That the linker length is related to some function is indicated by its alteration during erythropoiesis in chick [212], and in postnatal development in the rat [44, 228], and also by the observation that the distribution of the different

*References p. 566*

lengths of linker in the chromatin of one cell is not random - linkers of similar size are clustered [123]. On the other hand, published data concerning correlation between longer linkers and lesser transcriptional activity have been conflicting. For example, chicken liver has a shorter average linker length than does the transcriptionally inactive nucleated chicken erythrocyte [133], but the linker lengths in transcribed ribosomal RNA genes and in non-transcribed satellite DNA in the mouse are the same [76].

As mentioned above, the linker region is thought to bear various proteins, but in particular, H1. In Worcel's model of chromatin structure [223], the H1 has an important role in maintaining higher order structures by interacting with H1 molecules on neighbouring coils of the solenoid and so holding these coils together. Although in appropriate ionic conditions the DNA of the linker region may interact with the core histones [157] and so contribute to chromatin packing, it seems very likely that the H1 does play a major role in this process. This is because when chromatin is depleted of H1 it swells [17], and the post-synthetic phosphorylation and dephosphorylation of H1 correlates temporally with the condensation and decondensation (respectively) of chromatin during the cell cycle [79]. Studies of H1 phosphorylation in *Physarum* have led to the proposal that phosphorylation of H1 is a trigger for mitosis [19] (see Section 2.5.3.). Different data can be collected to support the idea that H1 can goven metabolic activity. For example, there are tissue-related differences in the relative abundance of H1 sub-fractions [102]; there are changes amongst H1 sub-fractions during development [22, 164]; there are changes in the H1 sub-fraction content of the same cell in different functional states [83]. In this context we may also consider the H1-like proteins H1° and H5. The occurrence of H1° correlates inversely with cell division and DNA synthesis in developing, regenerating and normal differentiated tissue [122, 151, 198], and the function suggested for this protein has, therefore, been suppression of DNA synthesis. The appearance of H5 in developing nucleated erythrocytes correlates with the suppression of genetic activity [43], (and aggregation of the chromatin and extension of the nucleosome repeat length), and its disappearance correlates with genetic reactivation in heterokaryons involving nucleated erythrocytes [7]. H5 is, therefore, generally considered to function as a repressor of transcription. It has also been suggested that the

H1 molecule dictates the length of the linker DNA on which it resides. Two observations indicate that this could be the case. Firstly, lower eukaryotes, which have H1 species which are less basic than are those of higher eukaryotes have shorter linker lengths [132, 141] while chromatin which bears the very basic H5 molecule has a longer linker [32]. Secondly, removal of H1 from chromatin can destroy the pattern of repeat lengths given by nuclease digestion of whole chromatin [103] although this may be reflecting nucleosome sliding, which might be an artefact.

The points outlined above are not incompatible if one holds that transcription can be regulated, at a coarse level, by chromatin structure, which can be regulated by the length of the linker region. The H1 molecule might, therefore, fulfil a regulatory role, with the different functional states of the chromatin brought about by different H1 sub-fractions, each holding the chromatin in a particular way which helps or hinders progress of the RNA polymerase molecule over the chromatin. In support of this hypothesis, it is known that different H1 sub-fractions react differently with DNA [216]. Alternatively, or additionally, the H1 might mediate the action of non-histone proteins which bind to it and so alter chromatin structure and function.

So, in summary, the histones appear to play a major role in packaging DNA into a conveniently-sized form, but while looking to non-histone proteins for regulation of specific genes [187]. Histones seem to have a part in this *general* process by holding the DNA in various configurations so that it can be reached and recognised (or not) by specific gene regulators and/or polymerases.

## 3. The HMG proteins

### 3.1. Introduction

Although the histones are the major chromosomal structural protein, folding the DNA into nucleosomes, super-coils or super-beads, etc. it is quite likely that there are other structural proteins which are required to modify these structures at different stages of the cell cycle and in different regions of the chromatin. The high mobility group (HMG) proteins are probably examples of such proteins. The HMG proteins are a set of rather intriguing proteins with some very unusual properties but at the same time they have a number of properties in common with the histones. They are present in the chromatin in smaller quantities than the histones which accounts for the fact they were not discovered until much later. Two of the HMG proteins, later numbered HMG1 and HMG2, were first detected by us as impurities in calf thymus histone fractions [93, 94]. These two proteins were then "rediscovered" during the fractionation of the non-histone proteins that are loosely bound to chromatin, proteins which are extracted from the chromatin with 0.35 M NaCl [61, 62].

Extraction of calf thymus chromatin with 0.35 M NaCl releases a heterogeneous mixture of non-histone proteins which can be fractionated into two groups by the addition of trichloroacetic acid (TCA) to the extract. A complex mixture of high molecular weight proteins precipitate with 2% TCA, leaving a relatively simple mixture of proteins in the supernatant which have higher electrophoretic mobilities than the proteins precipitated by 2% TCA. This high mobility group (HMG) of proteins was found to consist of 16 polypeptides (HMG 1—16) when analysed electrophoretically [61]. Later an additional protein, HMG17, was discovered [63, 167]. These proteins were then the subject of extensive charaterisation at the Chester Beatty Research Institute [70].

During the course of this work it became apparent that some of the HMG protein bands initially seen on polyacrylamide gels were proteolytic degradation products of other chromosomal proteins, e.g., HMG3 is a proteolytic degradation product of HMG1, and

TABLE II

The high mobility group (HMG) proteins

| | |
|---|---|
| 1 | Four main proteins: HMG 1, HMG 2, HMG 14 and HMG 17. |
| 2 | Constitute approx. 3% of the total chromosomal protein (thymus). |
| 3 | Loosely bound to chromatin; can be extracted from chromatin with 0.35 M NaCl. |
| 4 | Extracted with 5% perchloric acid. |
| 5 | Small (M = 9000—26 500 mol.wt.), highly-charged molecules. |
| 6 | Bind to DNA. |
| 7 | Ubiquitous; found in the nuclei of numerous tissues and species. |

# INTRODUCTION

HMG8 a degradation product of histone H1 [68]. In preparations where proteolysis is completely inhibited there are only four main HMG proteins in most mammalian tissues, HMG1, 2, 14 and 17. Some of the properties of these proteins are summarised in Table II.

The HMG proteins are found to be associated with a sub-population of nucleosomes and it was suggested that this HMG-containing

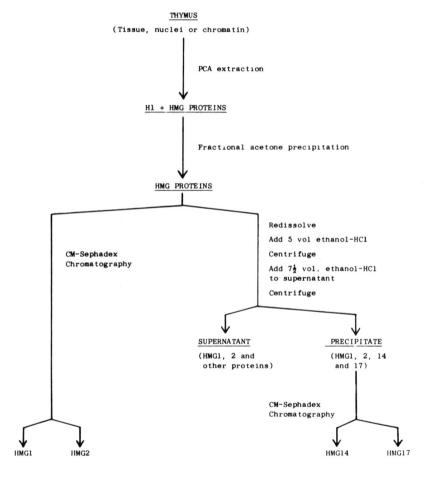

Fig. 8. Schematic diagram of the isolation of the four calf thymus HMG proteins.

*References p. 566*

set of nucleosomes are associated with transcribed sequences [67]. Considerable interest in the HMG proteins has now been kindled by the recent evidence pointing to such an association of HMG proteins with transcriptionally active genes (see below).

### 3.2. Isolation

There are basically two methods for the extraction of the 4 HMG proteins from the nucleus or chromatin. The first is the 0.35 M

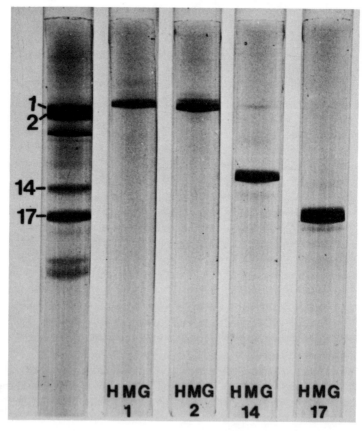

Fig. 9. Low pH polyacrylamide gel electrophoresis patterns of the perchloric acid-extracted total HMG Proteins (left gel), and the four purified proteins.

NaCl extraction method followed by 2% TCA precipitation [61, 63]. The other is based on the finding that 5% perchloric acid (PCA) selectively extracts the HMG proteins together with histone H1. The H1 can then subsequently be removed by fractional acetone precipitation [93, 166].

The four HMG proteins can be separated from one another and obtained in a pure form by the procedures outlined in Fig. 8 which involves (in the final stages), ion-exchange chromatography on CM-Sephadex or CM-cellulose columns at pH 8.8—9.0 [62, 63, 168]. Fig. 9 shows the electrophoretic analysis of the 4 purified HMG proteins from calf thymus. For further details the reader is referred to two reviews [66, 70].

## 3.3. Properties of the calf proteins

The four calf HMG proteins, HMG1, HMG2, HMG14 and HMG17, are best considered as two pairs of closely related proteins, HMG1 and 2 forming one pair, and HMG14 and 17 the other. Thus, HMG1 and HMG2 are about the same size ($M_r \sim 26\,000$—$26\,500$) [64, 175], have similar amino acid compositions, primary sequences, and secondary and tertiary structures. HMG14 and HMG17 are smaller proteins (HMG14 has $M_r \sim 11\,000$ and HMG17 $\sim 9\,000$) [203, 204] with similar compositions, sequences, and both are random coil proteins when free in solution.

*3.3.1. Amino acid analyses.* The HMG proteins have rather unusual amino acid analyses (Table III). Like the histones they have high contents of basic amino acids (19—24% lysine) but in addition, have high contents of the acidic amino acids, glutamic and aspartic acid (22 –29%) so that about half of the residues of the proteins are negatively or positively charged.

HMG1 and HMG2 are very similar in composition but are distinguished by the slightly differing serine, glutamic acid, and proline contents. In addition to the data given in Table III, where no allowances were made for hydrolytic losses, sequence data show that HMG1 and 2 have two cysteines, two tryptophans and four methionines. HMG14 and HMG17, in addition to having high basic and acidic amino acids, have high alanine contents;

## TABLE III
Amino acid analyses (moles %) of calf thymus HMG proteins

| Amino acid | HMG1 | HMG2 | HMG14 | HMG17 |
|---|---|---|---|---|
| Asp | 10.7 | 9.3 | 8.3 | 12.0 |
| Thr | 2.5 | 2.7 | 4.1 | 1.2 |
| Ser | 5.0 | 7.4 | 8.0 | 2.3 |
| Glu | 18.1 | 17.5 | 17.5 | 10.5 |
| Pro | 7.0 | 8.9 | 8.1 | 12.9 |
| Gly | 5.3 | 6.5 | 6.4 | 11.2 |
| Ala | 9.0 | 8.1 | 14.8 | 18.4 |
| Val | 1.9 | 2.3 | 4.0 | 2.0 |
| ½-Cys | Trace | Trace | — | — |
| Met | 1.5 | 0.4 | — | — |
| Ile | 1.8 | 1.3 | — | — |
| Leu | 2.2 | 2.0 | 2.0 | 1.0 |
| Tyr | 2.9 | 2.0 | — | — |
| Phe | 3.6 | 3.0 | — | — |
| Lys | 21.3 | 19.4 | 21.1 | 24.3 |
| His | 1.7 | 2.0 | — | — |
| Arg | 3.9 | 4.7 | 5.4 | 4.1 |

— indicates values close to zero.

HMG17 has a high glycine and proline content also. HMG14 and 17 are characterised by not having aromatic amino acids, cysteine, methionine and isoleucine. Also, they have very few hydrophobic amino acids. This paucity of hydrophobic amino acids and their high content of the helix destabilisers, glycine, serine and proline account for their lack of secondary and tertiary structures (see below).

*3.3.2. Microheterogeneity and modifications.* Just as heterogeneity has been observed in the histone fractions as a result of sequence differences and post-synthetic modificatons, so microheterogeneity has been observed in some of the HMG proteins, though at the moment it is not known whether the microheterogeneity seen is due to sequence or post-synthetic differences. The most striking example is exhibited in HMG2. The calf protein has four main sub-fractions which can be separated by ion-exchange chromatography (Fig. 10) and which differ in their isoelectric points (Fig.

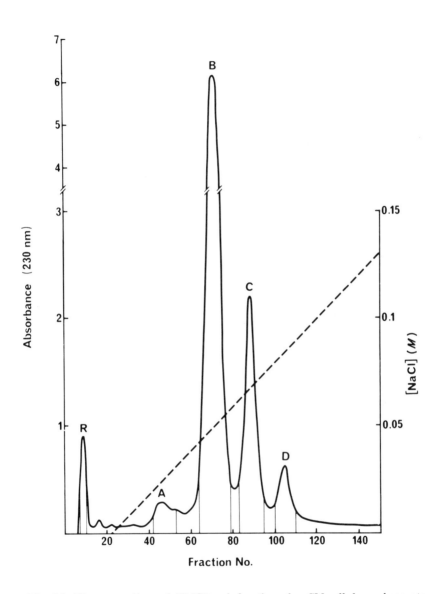

Fig. 10. The separation of HMG2 sub-fractions by CM-cellulose chromatography. Peaks A, B, C and D are the four HMG2 sub-fractions. Peak R is some HMG1 contamination.

11) [65]. The amino acid compositions of the four sub-fractions are very similar. The N-terminals are all the same (glycine), but the peptide maps show a few differences. HMG1 shows multiple bands on isoelectric focusing gels also, but this may be due to aggregation of the protein near its isoelectric point. HMG14 and 17 may also be microheterogeneous since they both give doublets when analysed by neutral pH polyacrylamide gel electrophoresis.

Post-synthetic modifications may account for the microheterogeneity seen in the HMG proteins. Thus HMG1 and 2 have been found to be phosphorylated [47], acetylated [190] and possibly modified by ADP-ribose [27]. HMG14 and HMG17 may be modified by poly(ADP-ribose) since this group has been found attached to the similar protein in trout testis, protein H6 [219] (see Section 3.5.3. below).

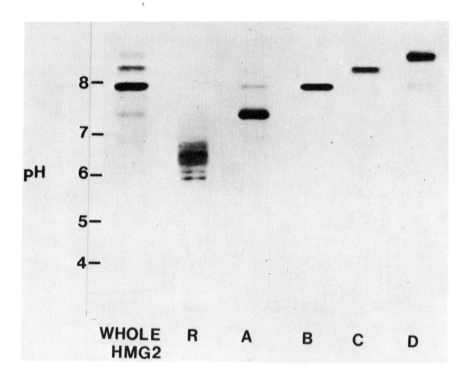

Fig. 11. Isoelectric focusing analysis of the HMG2 sub-fractions.

*3.3.3. Secondary and tertiary structures.* The two larger HMG proteins have highly ordered structures, behaving as typical globular proteins. Thus, they have 40—50% α-helix content and fold to form specific tertiary structures at neutral pH [11, 28, 89]. No β-sheet structure has been detected in HMG1 and HMG2.

In contrast to the highly ordered structures seen with HMG1 and 2, HMG14 and 17 appear to be devoid of any secondary and tertiary structure (as revealed by n.m.r., circular dichroisim, infra-red spectroscopy and neutron scattering) [1, 90, 158]. This result is perhaps not surprising in view of the unusual compositions of HMG14 and 17, i.e., having few hydrophobic amino acids and having high contents of α-helix destabilisers. It is of interest that there are portions of the sequence of histone H1 which are homologous with the HMG14 and HMG17 (see Section 3.5. below) and these lie within the basic random coils sections of H1 [70]. Herein is maybe the clue to the function(s) of HMG14 and HMG17.

*3.3.4. Interactions with DNA.* All four HMG proteins bind to DNA, the binding being primarily ionic in nature between the phosphate groups of the DNA and the basic amino acids of the protein. HMG1 and 2 bind to to both single and double stranded DNA [64, 175] and from n.m.r. studies of HMG1 - DNA complexes it would appear that the lysines and aromatic residues of the basic N-terminal half of the protein are involved in DNA-binding, leaving the acidic C-terminal free [28]. The binding of HMG1 and 2 to circular double-stranded DNA causes a change in the linking number by unwinding the double-helix [89, 91] and this is probably due to their preferential binding to single-stranded DNA [14]. Thus, HMG1 and 2 appear to be typical DNA-unwinding proteins. HMG14 and 17 on the other hand do not change the linking number on binding to DNA [92].

Nuclear magnetic resonance studies of HMG17 - DNA complexes show that the principal, DNA-binding region of the molecule is located in the region between about residues 15 and 40 [1]. This region encompasses the proline-lysine-rich region near the centre of the molecule and also the sequence Arg-Arg-Ser-Ala-Arg- (see Fig. 14). It would appear that arginines are important for DNA-binding since this sequence is also found in the DNA-binding region of HMG14 (residues 17—60) [158] and is also found in

*References p. 566*

the trout protein H6 (which is similar to the mammalian HMG14 and 17) (see Section 3.5.). The acidic C-terminals of HMG14 and 17 do not bind to DNA and are presumably free to interact with other chromosomal proteins.

### 3.4. Occurrence and specificity

Most of the work on HMG proteins to data has been carried out on calf thymus and trout testis tissues. It is apparent that the HMG proteins are not confined to these tissues but are widespread throughout, and are present in the tissues of vertebrates, invertebrates, plants and lower eukaryotes.

A comprehensive analysis of a number of different calf tissues (kidney, liver, spleen) reveal that the four HMG proteins, HMG 1, 2, 14 and 17 are all present [159]. Judging by the amino acid analyses, electrophoretic mobilities, chromatographic behaviour and isoelectric focusing patterns, the four HMG proteins are almost indistinguishable from the calf thymus proteins. Even the HMG2 sub-fractions in the four tissues are identical. The HMG proteins have been isolated from the tissues of a number of other mammals (pigs, rabbits, sheep) and are very similar to those four HMG proteins in calf, though the HMG2 sub-fractions differ somewhat.

The HMG proteins in avian tissues although similar to the calf proteins, do differ in a number of respects [127, 160, 189]. In avian erythrocytes there are again two pairs of proteins: HMG1 and 2, HMG14 and 17. HMG1 and 2 have analyses similar to those of the corresponding calf proteins (Table IV) but the differences between the 4 or 5 chicken HMG2 sub-fractions appear to be more pronounced than found between the calf HMG2 sub-fractions. Thus, at least one of the chicken sub-fractions has an alanine N-terminal, whilst the others have the usual glycine N-terminal. The amino acid analyses of the partially separated sub-fractions differ, as do their electrophoretic mobilities [127, 189]. An interesting finding is that there are quantitative differences in the HMG2 sub-fractions in dfferent avian tissues: the chicken erythrocyte has more of the acidic sub-fractions than chicken thymus [127].

The chicken erythrocyte HMG14 and 17 proteins have amino acid compositions characteristic of this pair of proteins (Table IV), having little or no aromatic amino acids and having high proline, alanine, lysine and high contents of the acidic amino acids.

The one example (so far) of a tissue-specific HMG protein has been furnished by the chicken oviduct which has, in addition to HMG1, 2, 14 and 17, a high molecular weight protein ($M_r$ = 95 000) which is extractable with 0.35 M NaCl and has high contents of basic and acidic amino acids (Table IV) [193].

Going further back in evolution to fish, there are basically two HMG-like proteins in trout testis, called HMG-T and H6 [86, 125, 208]. A third 0.3 M NaCl extactable protein identified as ubiquitin (see Sections 4.1. and 4.3.) is not classed with the HMG proteins [209]. From the amino acid compositions (Table V) and sequence data (see Section 3.5), it is apparent that HMG-T is very similar to HMG1 and 2 in mammals and birds, whilst H6 is similar to HMG14 and 17. Most of the characterisation of trout HMG proteins has been on those from testis but the little work that has been done

TABLE IV

Amino acid analyses (moles %) of chicken erythrocyte HMG1, 2, 14 and 17 and the oviduct specific 95K protein

| Amino acids | HMG1 | HMG2 | HMG14 | HMG17 | 95K |
|---|---|---|---|---|---|
| Asp | 11.5 | 12.9 | 9.3 | 9.1 | 10.00 |
| Thr | 2.7 | 2.5 | 4.6 | 3.0 | 7.20 |
| Ser | 5.8 | 5.7 | 5.2 | 4.3 | 5.35 |
| Glu | 18.1 | 15.3 | 15.6 | 11.7 | 9.85 |
| Pro | 5.1 | 6.1 | 10.5 | 12.1 | 12.85 |
| Gly | 5.9 | 7.2 | 5.6 | 10.0 | 11.60 |
| Ala | 8.7 | 10.0 | 18.0 | 17.2 | 3.55 |
| Val | 3.5 | 3.5 | 0.3 | 2.2 | 5.25 |
| ½-Cys | 0.9 | 0.4 | — | — | 4.65 |
| Met | 0.5 | 0.4 | — | — | 1.95 |
| Ile | 1.8 | 1.7 | — | — | 2.10 |
| Leu | 2.7 | 2.3 | 1.1 | 1.2 | 2.15 |
| Tyr | 2.2 | 2.5 | 0.2 | 0.2 | — |
| Phe | 4.3 | 4.9 | 0.1 | 0.1 | 0.25 |
| Lys | 18.2 | 17.3 | 24.0 | 23.6 | 14.35 |
| His | 1.7 | 0.9 | 1.1 | 0.2 | 1.85 |
| Arg | 4.0 | 4.2 | 4.1 | 4.6 | 6.35 |
| N-terminal amino acid | Gly | Gly | Pro | Pro | |

— indicates values close to zero.

*References p. 566*

## TABLE V
Amino acid analyses (moles %) of trout testis HMG proteins

| Amino acid | HMG-T [181] | H6 [179] |
|---|---|---|
| Asp | 11.3 | 6.7 |
| Thr | 3.0 | 1.6 |
| Ser | 4.5 | 5.6 |
| Glu | 9.1 | 6.1 |
| Pro | 7.6 | 12.3 |
| Gly | 17.4 | 7.4 |
| Ala | 8.3 | 25.4 |
| Val | 3.8 | 3.4 |
| ½-Cys | 0.8 | — |
| Met | 1.9 | — |
| Ile | 1.5 | — |
| Leu | 2.6 | 1.2 |
| Tyr | 2.3 | — |
| Phe | 3.4 | — |
| Lys | 15.5 | 23.1 |
| His | 0.4 | — |
| Arg | 5.3 | 7.2 |

— indicates values close to zero.

on other tissues suggests that HMG-T and H6 are present in trout liver and spleen [158]. The situation in trout may be more complex than this since, recently, two more proteins like HMG14 and HMG17 have been isolated from trout liver [158]. These have yet to be sequenced before they can be related to trout H6 and the vertebrate HMG14 and HMG17.

HMG-like proteins have been found in invertebrates, plants and yeast (Table VI), but these have not been sufficiently characterised to be able to relate them to the vertebrate HMG proteins. The yeast protein which was isolated by two groups of workers [181, 183] has an amino acid composition and electrophoretic mobility close to that of calf HMG1 and 2 and hence may be the analogous protein in this organism. However, it also has a close similarity to histone H1.

In addition to the four main HMG proteins, HMG 1, 2, 14 and 17, a number of smaller HMG-like proteins are also observed in perchloric acid extracts of thymus nuclei [72]. These proteins

# PRIMARY STRUCTURES

## TABLE VI

Amino acid compositions (mol/100 ml) of purified HMG-like proteins from the fruit fly *Ceratitis capitata* [126], *Drosophila melanogaster* [199], wheat [183] and yeast [38, 183]

| Amino acid | Ceratitis capitata HMG | D. melanogaster DI | Wheat HMG | Yeast HMG proteins | |
|---|---|---|---|---|---|
| Asp | 17.2 | 16.0 | 11.7 | 8.5 | 10.0 |
| Thr | 4.8 | 3.1 | 1.4 | 3.4 | 2.9 |
| Ser | 7.7 | 10.6 | 8.3 | 7.5 | 7.1 |
| Glu | 10.2 | 10.7 | 12.7 | 15.6 | 15.5 |
| Pro | 7.2 | 8.0 | 4.7 | 5.9 | 5.9 |
| Gly | 11.4 | 13.2 | 13.7 | 3.6 | 7.7 |
| Ala | 8.3 | 9.6 | 15.9 | 8.8 | 7.4 |
| Val | 2.3 | 4.8 | 3.6 | 2.3 | 3.2 |
| ½-Cys | — | — | — | — | — |
| Met | 0.4 | 0.2 | — | — | < 1 |
| Ile | 1.8 | 1.7 | 1.5 | 6.5 | 5.0 |
| Leu | 1.5 | 1.5 | 2.2 | 7.5 | 7.3 |
| Tyr | 0.8 | 0.7 | 0.8 | 4.4 | — |
| Phe | Trace | 0.1 | 2.7 | 2.8 | 4.2 |
| Lys | 17.6 | 11.5 | 17.0 | 15.9 | 14.7 |
| His | 0.6 | 1.3 | 1.3 | 1.3 | 1.3 |
| Arg | 8.2 | 7.4 | 2.5 | 5.5 | 4.7 |

— indicates values close to zero.

appear to be like HMG1 and 2 in that they are readily washed out of the nucleus. The amino acid compositions of these two proteins A and B are typically HMG-like in having high basic and acidic amino acids (Table 7). It is not known at the moment whether these are true chromosomal proteins or simply nucleoplasmic proteins, but they are so similar in composition to the HMG proteins that an involvement with the chromatin seems highly likely.

### 3.5. Primary structures

*3.5.1. Calf thymus HMG14.* The primary structure of calf thymus HMG14 has been determined [206] and is shown in Fig. 12. The protein is a single polypeptide chain of 100 amino acids. 85% of

*References p. 566*

TABLE VII

Amino acid analysis (mol/100 ml) of two low molecular weight HMG-like proteins A and B from calf thymus nuclei [184].

| Amino acid | A | B |
|---|---|---|
| Asp | 4.3 | 9.4 |
| Thr | 2.2 | 2.5 |
| Ser | 3.8 | 6.0 |
| Glu | 18.0 | 18.1 |
| Pro | 5.4 | 4.2 |
| Gly | 10.5 | 7.2 |
| Ala | 10.6 | 10.2 |
| ½-Cys | — | — |
| Val | 1.1 | 2.0 |
| Met | 1.3 | 1.8 |
| Ile | 1.3 | 1.5 |
| Leu | 5.1 | 3.9 |
| Tyr | 0.8 | 0.9 |
| Phe | 2.0 | 1.7 |
| Lys | 29.8 | 22.4 |
| His | 0.4 | 0.6 |
| Arg | 3.4 | 7.8 |

— indicates values close to zero.

the basic amino acids (5 arginine and 17 lysine residues) are concentrated in the first 61 residues of the molecules. This situation is similar to that found with the histones where the N-terminal regions are rich in basic amino acids. However, in contrast to the histones, the C-terminal region of the molecule (residues 62—100) contains a high proportion of acidic amino acids (14 glutamic and aspartic residues, but only 4 lysine residues). This overall negative charge in the C-terminal region of the molecule is a feature not found with any of the histones. No hydrophobic regions exist in the molecule. Sequence comparisons with other non-histone proteins are discussed in Section 3.5.6. below. Protein HMG14 has also been isolated from chicken erythrocyte chromatin [160]. Comparison of the amino acid analyses of the calf and chicken proteins (Tables III and IV) shows a number of differences between the two proteins. The most striking difference relates to the hydrophobic amino acids where there is an absence of the 4 valine

```
                                                              10
Pro — Lys — Arg — Lys — Val — Ser — Ser — Ala — Glu — Gly —
                                                              20
Ala — Ala — Lys — Glu — Glu — Pro — Lys — Arg — Arg — Ser —
                                                              30
Ala — Arg — Leu — Ser — Ala — Lys — Pro — Ala — Pro — Ala —
                                                              40
Lys — Val — Glu — Thr — Lys — Pro — Lys — Lys — Ala — Ala —
                                                              50
Gly — Lys — Asp — Lys — Ser — Ser — Asp — Lys — Lys — Val —
                                                              60
Gln — Thr — Lys — Gly — Lys — Arg — Gly — Ala — Lys — Gly —
                                                              70
Lys — Gln — Ala — Glu — Val — Ala — Asn — Gln — Gln — Thr —
                                                              80
Lys — Glu — Asp — Leu — Pro — Ala — Glu — Asn — Gly — Glu —
                                                              90
Thr — Lys — Asn — Glu — Glu — Ser — Pro — Ala — Ser — Asp —
                                                              100
Glu — Ala — Glu — Glu — Lys — Glu — Ala — Lys — Ser — Asp
```

Fig. 12. The amino acid sequence of calf thymus HMG14.

residues and one of the 2 leucine residues in the chicken protein. Comparisons of the two analyses suggests at least 11 amino acid differences between the two proteins. This sequence variation of 11% between the calf and chicken proteins is greater than that observed for the nucleosome core histones. The sequence of histone H3 differs in only 1 position in 135 residues between the chicken and calf [21], and the sequence of calf H2B differs from that of chicken H2B at 3 positions in 115 residues [196]. HMG14 does not, therefore, exhibit the degree of sequence conservation that is known to exist for the four nucleosome core histones.

The N-terminal sequence of the first 30 residues of chicken erythrocyte HMG14 is shown in Fig. 13. Comparison with the calf thymus sequence shows 3 of the expected amino acid changes, and it is also necessary to introduce a deletion in the chicken sequence in order to maintain the sequence homology between the two proteins.

*3.5.2. Calf thymus HMG17.* The complete amino acid sequence of calf thymus HMG17 has been determined [203] and is shown in

*References p. 566*

|            | 1   |     |     | 4   |     |     |     |     |     |
|------------|-----|-----|-----|-----|-----|-----|-----|-----|-----|
| CT HMG17   | Pro — | Lys — | Arg — | Lys — |     |     |     | — Ala — | Glu — |
| CE HMG17   | Pro — | Lys — | Arg — | Lys — |     |     |     | — Ala — | Glu — |
|            | 1   |     |     | 4   |     |     |     |     |     |
| CT HMG14   | Pro — | Lys — | Arg — | Lys — | Val — | Ser — | Ser — | Ala — | Glu — |
| CE HMG14   | Pro — | Lys — | Arg — | Lys — | Ala — | Pro — | O — | Ala — | Glu — |
|            | 1   |     |     | 4   |     |     |     |     |     |
| Trout H6   | Pro — | Lys — | Arg — | Lys — | Ser — | Ala — | Thr — |     |     |

|            |     |     | 10  |     |     |     |     |     |
|------------|-----|-----|-----|-----|-----|-----|-----|-----|
| CT HMG17   | Gly — | Asp — | Ala — | Lys — | Gly — | Asp — | Lys — | Ala — | Lys — |
| CE HMG17   | Gly — | Asp — | Thr — | Lys — | Gly — | Asp — | Lys — | Ala — | Lys — |
|            | 10  |     |     |     |     |     |     |     |
| CT MHG14   | Gly — | Ala — | Ala — |
| CE HMG14   | Gly — | Glu — | Ala — |

| Trout H6   |     |     |     | — Lys — | Gly — |

|            |     |     |     | 20  |     |
|------------|-----|-----|-----|-----|-----|
| CT HMG17   | Val — | Lys — | Asp — | Glu — | Pro — |
| CE HMG17   | Val — | Lys — | Asp — | Glu — | Pro — |
|            |     |     | 16  |     |     |
| CT HMG14   | — Lys — | Glu — | Glu — | Pro — |
| CE HMG14   | — Lys — | Glu — | Glu — | Pro — |
|            | 10  |     | 12  |     |
| Trout H6   | — Asp — | Glu — | Pro — |

|            |     |     |     | 24  |     |     |     |     |     |
|------------|-----|-----|-----|-----|-----|-----|-----|-----|-----|
| CT HMG17   | Gln — | Arg — | Arg — | Ser — | Ala — | Arg — | Leu — | Ser — | Ala — |
| CE HMG17   | Gln — | Arg — | Arg — | Ser — | Ala — | Arg — | Leu — | Ser — | Ala — |
|            |     |     |     | 20  |     |     |     |     |     |
| CT HMG14   | Lys — | Arg — | Arg — | Ser — | Ala — | Arg — | Leu — | Ser — | Ala — |
| CE HMG14   | Lys — | Arg — | Arg — | Ser — | Ala — | Arg — | Leu — | Ser — | Ala — |
|            |     |     |     | 16  |     |     |     | 20  |     |
| Trout H6   | Ala — | Arg — | Arg — | Ser — | Ala — | Arg — | Leu — | Ser — | Gly — |

|            | 30  |     |     |     |     |     | 37  |
|------------|-----|-----|-----|-----|-----|-----|-----|
| CT HMG17   | Lys — | Pro — | Ala — | Pro — | Pro — | Lys — | Pro — | Glu — |
| CE HMG17   | Lys — | Pro — | Ala — | Pro — | Pro — | Lys — | Pro — | Glu — |
|            |     |     |     | 30  |     |     |     |
| CT HMG14   | Lys — | Pro — | Ala — | Pro — | Ala — | Lys — | Val — | Glu — |
| CE HMG14   | Lys — | Pro — | Ala — | Pro — | Pro — | Lys — | Pro — | Glu — |
|            |     |     |     |     |     |     | 28  |
| Trout H6   | Arg — | Pro — | Val — | Pro — | O — | Lys — | Pro — | Ala — |

Fig. 13. Comparison of the amino terminal sequences of calf thymus and chicken erythrocyte proteins HMG14 and 17, and the trout testis portein H6. CT = calf thymus; CE = chicken erythrocyte.

```
                                                     10
Pro — Lys — Arg — Lys — Ala — Glu — Gly — Asp — Ala — Lys —
                                                     20
Gly — Asp — Lys — Ala — Lys — Val — Lys — Asp — Glu — Pro —
                                                     30
Gln — Arg — Arg — Ser — Ala — Arg — Leu — Ser — Ala — Lys —
                                                     40
Pro — Ala — Pro — Pro — Lys — Pro — Glu — Pro — Lys — Pro —
                                                     50
Lys — Lys — Ala — Pro — Ala — Lys — Lys — Gly — Glu — Lys —
                                                     60
Val — Pro — Lys — Gly — Lys — Lys — Gly — Lys — Ala — Asp —
                                                     70
Ala — Gly — Lys — Asx — Gly — Asx — Asx — Pro — Ala — Glx —
                                                     80
Asx — Gly — Asx — Ala — Lys — Thr — Asx — Glx — Ala — Glx —

Lys — Ala — Glu — Gly — Ala — Gly — Asp — Ala — Lys
```

Fig. 14. The amino acid sequence of calf thymus HMG17.

Fig. 14. HMG17 is a single polypeptide chain of 89 amino acid residues with an $M_r$ of ~ 9 250. Charges have yet to be assigned to 6 Asx and Glx residues. However, it is known from peptide mobility measurements that of the two Glx residues and one Asx residue between residues 77 and 80, two are amidated and one is an acid, and of the 5 Asx and 1 Glx residues between residues 64 and 73, three are amidated and three are acids. The sequence shows that the N-terminal two-thirds of the molecule is strongly basic (22 basic amino acids and 7 acidic amino acids in the first 58 residues). In contrast, the C-terminal one-third of the molecule (residues 59—89) has an overall negative charge (4 basic amino acids and 7 acidic amino acids). The overall architecture of the HMG17 molecule is, therefore, very similar to that of HMG14. A remarkable structural feature of this molecule is a region of high density of proline residues, there being 6 proline residues between residues 31 and 40. None of the histone shows this feature. Sequence homologies with other non-histone proteins are discussed in Section 3.5.6. below.

Protein HMG17 has also been isolated from chicken erythrocyte chromatin [160]. Comparison of the amino acid analyses of the calf and chicken proteins (Tables 3 and 4) suggests only about 5 differences between the two proteins. The N-terminal sequence of chicken HMG17 (Fig. 13) shows only one substitution at position

9 when compared with calf HMG17. HMG17 would, therefore, appear to be more evolutionarily stable than HMG14.

*3.5.3. Trout testis protein H6.* The complete amino acid sequence of protein H6 has been determined [210] and is shown in Fig. 15. The protein is a single polypeptide chain of 69 amino acid residues with an $M_r$ of 7 200 which makes it the smallest chromosomal protein so far described, with the exception of the sperm protamines. Like HMG14 and 17 the distribution of charged residues is not symmetrical. The first 45 residues contain 17 basic residues and only two acidic residues. In contrast, the C-terminal third of the molecule (residues 46—69) contains only 4 acidic amino acids and 3 basic amino acids. There are no extensive regions of hydrophobic amino acids.

```
                                                        10
Pro — Lys — Arg — Lys — Ser — Ala — Thr — Lys — Gly — Asp —
                                                        20
Glu — Pro — Ala — Arg — Arg — Ser — Ala — Arg — Leu — Ser —
                                                        30
Ala — Arg — Pro — Val — Pro — Lys — Pro — Ala — Ala — Lys —
                                                        40
Pro — Lys — Lys — Ala — Ala — Ala — Pro — Lys — Lys — Ala —
                                                        50
Val — Lys — Gly — Lys — Lys — Ala — Ala — Glu — Asn — Gly —
                                                        60
Asp — Ala — Lys — Ala — Glu — Ala — Lys — Val — Gln — Ala —
Ala — Gly — Asp — Gly — Ala — Gly — Asn — Ala — Lys
```

Fig. 15. The amino acid sequence of trout testis protein H6.

*3.5.4. Trout testis HMG-T.* The amino-terminal sequence of HMG-T has been determined [208] and is shown in Fig. 16. This sequence shows considerable similarity with that of HMG1 and HMG2. Sixteen of the first 26 residues are common to all three proteins, which clearly shows that protein HMG-T belongs to the high mobility class of non-histone chromosomal proteins. The amino acid analysis of HMG-T is shown in Table V.

# PRIMARY STRUCTURES

|      |                  |                  |                  |                  |                  |     |
|------|------------------|------------------|------------------|------------------|------------------|-----|
| HMG1 | Gly — Lys — Gly — | Asp — Pro — | — Lys — | Lys — Pro — | — Arg — | Gly — Lys — |
| HMG2 | Gly — Lys — Gly — | Asp — Pro — | — Asn — | Lys — Pro — | — Arg — | Gly — Lys — — Met — |
| HMGT | Pro — Gly — Lys — | Asp — Pro — | — Asn — | Lys — Pro — | — Lys — | Gly — Lys — — Met — |
|      |                  |                  |                  |                  |                  | — Thr — |

(position 10)

|      |                 |          |                   |                     |
|------|-----------------|----------|-------------------|---------------------|
| HMG1 | Ser — Ser —     | — Tyr —  | Ala — Phe — Phe — Val — | — Gln — Thr — Ser — |
| HMG2 | Ser — Ser —     | — Lys —  | Ala — Phe — Phe — Val — | — Gln — Thr — Ser — |
| HMGT | Ser — Ser —     | — Ser —  | Ala — Phe — Phe — Val — | — Ala — Gln — Arg — |

(position 20)

|      |                         |                              |
|------|-------------------------|------------------------------|
| HMG1 | Arg — Glu — Glu — His — | — Lys — Lys — Lys — His       |
| HMG2 | Arg — Glu — Glu — His — | — Lys — Lys — Lys — His       |
| HMGT | Arg — Glx — Glx — His — |                               |

(position 30)

Fig. 16. Comparison of the amino terminal sequences of calf thymus HMG1 and 2, and trout testis HMG-T.

*References p. 566*

*3.5.5. Calf thymus HMG1 and 2.* Both proteins HMG1 and 2 are approximately 250 amino acid residues long. One of the interesting features of HMG1 and 2 is the fact that over half of their amino acids are either charged or potentially charged and sequence studies have revealed an asymmetrical distribution of these charged groups within the molecule (see below). Initial studies on HMG1 and 2 suggested that they had closely related primary structures [202]. Recent sequence data on HMG1 and 2 have confirmed the similarities between the two proteins [207]. Figure 17 shows a comparison of the sequence data for two major peptides from HMG1 and 2. Of 130 residues shown in this figure, only 29 residues differ between the two proteins. The most intriguing feature of the structures of both HMG1 and 2 is the presence in the C-terminal region of a continuous sequence of about 40 aspartic and glutamic acid residues [205]. In contrast to this highly acidic region, the basic amino acids are fairly evenly distributed, although small clusters of basic amino acids do occur. For example, the sequences Arg-Glu-Glu-His-Lys-Lys-Lys-His- and Lys-Lys-Gly-Lys-Lys-Lys- occur in the N-terminal half of the molecule. Clusters of hydrophobic amino acids such as Ala-Phe-Phe-Leu-Phe-Ala- and Tyr-Ala-Phe-Phe-Val-Gln- also occur in the N-terminal half of the molecule. The overall picture of the HMG1 and 2 molecule is, therefore, one of a basic N-terminal half containing clusters of basic and hydrophobic amino acids but few acidic amino acids, whereas the C-terminal region contains the majority of the acidic residues in a continuous sequence. The function of this highly acidic region of the molecule has yet to be determined. The only side chain modification to have been positively identified in the HMG proteins to date is the acetylation of 4 lysines. Recent results have shown that 2—3% of the lysine residues in HMG1 and 2 are acetylated [190].

*3.5.6. Sequence homologies within the HMG proteins.* Comparison of the amino acid analyses of the calf thymus proteins HMG14, HMG17 and the trout protein H6 (Tables III and V), shows an absence in all three proteins of the amino acids cysteine, methionine, isoleucine, tyrosine, phenylalanine, tryptophan and histidine. Comparison of the amino terminal sequences of calf thymus HMG14 and HMG17 and trout testis H6 shows lengthy regions of sequence homology between these three proteins. A comparison of the amino terminal sequences of calf and chicken HMG14 and

| | | |
|---|---|---|
| PEPTIDE 1 | HMG1 | Trp — Asn — Asn — Thr — Ala — Ala — Asp — Asp — Lys — Gln — Pro — Tyr — Glu — Lys — Lys — Ala — Ala — Lys — Leu — |
| | HMG2 | Trp — Ser — Gln — Glu — Ser — Ala — Lys — Asp — Lys — Gln — Pro — Tyr — Glu — Gln — Lys — Ala — Ser — Lys — Leu — |
| | HMG1 | Lys — Glu — Lys — Tyr — Glu — Lys — Asp — O — Ala — Ala — Tyr — Arg — Ala — Lys — Gly — Lys — Pro — Asp — Ala — |
| | HMG2 | Lys — Glu — Lys — Tyr — Glu — Lys — Asp — ( ) — Ala — Ala — Tyr — Arg — Ala — Lys — Gly — Lys — Lys — Glx — Ala — |
| | HMG1 | Ala — Lys — Lys — Gly — Val — Val — Lys — Ala — Glu — Lys — Ser — Lys — Lys — Lys — Lys — Glu — Glu — Glu — Glu — |
| | HMG2 | Gly — Lys — Lys — Gly — Pro — Gly — Arg — Pro — Thr — Gly — Ser — O — Lys — Lys — Lys — Asn — Glu — Pro — Glu — |
| | HMG1 | Asp — ($Glu_{25}$, $Asp_{11}$, $Lys_{1}$) |
| | HMG2 | Asp — ($Gly_{27}$, $Asp_{5}$, $Pro_{1}$, $Lys_{1}$) |
| PEPTIDE 2 | HMG1 | Ser — Ala — Lys — Glx — Asx — Gly — Lys — Phe — Glx — Lys — Glx — Met — Ala — Lys — Ala — Asx — Lys — Ala — Arg — |
| | HMG2 | Ser — Ala — Lys — Glx — Asx — Ser — Lys — Phe — Glx — Lys — Glx — Met — Ala — Lys — Ser — Asx — Lys — Ala — Arg — |
| | HMG1 | Tyr — Glx — Arg — Glx — Met — Lys — Thr — Tyr — Ile — Pro — Pro — Lys — Gly — Glu — Thr — Lys — Lys — Lys — Phe — |
| | HMG2 | Tyr — Asx — Arg — Glx — Met — Lys — Asn — Tyr — Val — Pro — Pro — Lys — Gly — Asp — Lys — Lys — Gly — Lys — Lys — |
| | HMG1 | Lys — Asp — Pro — Asn — Ala — Pro — Lys — Arg — Pro — Pro — Ser — Ala — Phe — Phe — Leu — Phe — Ala — Ser — Glu — |
| | HMG2 | Lys — Asp — Pro — Asn — Ala — Pro — Lys — Arg — Pro — Pro — Ser — Asp — Phe — Phe — Leu — Phe — Ala — Ser — Glx — |
| | HMG1 | Tyr — Arg — Pro — Lys — Lys — Lys — Gly — Glu — His — Pro — Gly — Leu — Ser — Ile — Gly — Asx — |
| | HMG2 | His — Arg — Pro — Lys — Ile — Lys — Ala — Glx — His — Pro — Gly — Leu — Ser — Ile — Gly — Asx — |

Fig. 17. A comparison of two homologous peptides from calf thymus proteins HMG1 and HMG2. Parentheses indicate amino acid residues not identified.

17 and trout testis H6 is shown in Fig. 13. It has previously been shown that the principal DNA-binding region of HMG17 is between about residues 15 and 40 [168]. Since it is this region that shows greatest sequence homology between these proteins (Fig. 13), it seems likely that this region of common sequence represents a common DNA-binding site in all three proteins. In contrast, comparison of the acidic C-terminal sequences of these three proteins shows only limited sequence homology between the three proteins.

*3.5.7. Other HMG proteins.* Two other HMG-like proteins (proteins A and B, Table 7) have been mentioned in Section 3.2. Their amino acid analysis and gel electrophoretic mobilities suggests that both proteins are approximately 75 residues long. Protein A has a blocked N-terminal amino acid, but protein B has a free amino terminus and its N-terminal amino acid sequence has been determined [201] as Thr-Arg-Gly-Asn-Gln-Arg-Glu-Leu-Ala-Gly-Gln-Lys-Asn-Met-Lys-Gln.

## 3.6. Chromosomal and cellular localisation

When thymus chromatin is digested with micrococcal nuclease and the resulting nucleosomal particles are separated by gel filtration or gradient centrifugation, the 4 HMG proteins are found to be associated with the isolated monomer nucleosomes [67, 126]. However, because the HMG proteins comprise only a few percent ($\sim$ 3%) of the nucleosomal protein, it follows that only a small proportion of the total population of nucleosomes will have HMG proteins bound to them. Further micrococcal nuclease digestion studies on these HMG-containing nucleosomes suggest that HMG14 and HMG17 in mammals and H6 in trout are associated with nucleosome core particles, i.e., the relatively nuclease-resistant DNA bound to the 4 histones, H2A, H2B, H3 and H4 [113, 126]. However, the size of the DNA of the HMG14 and 17 core particles (160 base pairs) [71] is somewhat larger than the H6 core particle (140 b.p.) [113] and the bulk of the chromosomal core particle (145 b.p.). Quantitative analysis of the H6 in the testis H6-containing core particles, suggest that there are one or two H6 molecules per core particle [113].

Digestion of nuclei with micrococcal nuclease results in the rapid release of some of the HMG1 and 2 (and HMG-T in testis) from the nuclei [112, 113, 126]. This has been interpreted as indicating that HMG1 and 2 (HMG-T) are associated with the linker region of the nucleosome [112]. This may not be so since in thymus tissue there are two populations of HMG1 and 2 molecules [126]. One population, comprising about half the HMG1 and 2, is very loosely bound within the nucleus and can readily be washed out of the nucleus with low ionic strength buffers or by briefly digesting the chromatin with nuclease, thereby breaking down the chromosomal structure. This population may not actually be bound to the DNA but simply occluded within the nucleus. The second population of HMG1 and 2 proteins are found bound to the nucleosomes and are associated with a group of nucleosomes which are rapidly excised out of the chromatin. In some cells, such as liver, there is a considerable amount of HMG1 and 2 protein in the cytoplasm [24]. Thus, there is probably a dynamic equilibrium of HMG1 and 2 between the cytoplasm, the nucleoplasm and nucleosomes. This equilibrium probably varies from cell type to cell type and possibly during the cell cycle, for there is some evidence that nuclear HMG1 and/or HMG2 change qualitatively (or possibly quantitatively) in a manner related to the cell cycle — they are detected by immunoperoxidase methods only during later S, G2, mitosis and early G1 phases [198]. This distribution of protein in the cytoplasm, nucleoplasm and chromatin is also observed with the *Drosphila melanogaster* protein D1 (see Table VI) and hence this protein may be the *Drosophila* counterpart of the vertebrate HMG1 and 2. D1 is also found to be located in specific regions of the salivary gland chromosomes, probably AT-rich DNA regions including the AT-rich satellite regions [4].

### 3.7. Functions

Of the total population of nucleosomes in the cell nucleus only a small proportion of them will have HMG proteins attached. Are the HMG-containing nucleosomes evenly distributed throughout the chromatin, the HMG proteins playing some general structural role like the histones, or are they clustered in specific regions of

the genome playing some regulatory role(s)? It was suggested, for example, that the HMG-containing nucleosomes may be associated with transcribed sequences [67], and there is now indeed some evidence that at least the smaller HMG proteins (HMG14, 17, H6) may be thus bound to such sequences. The evidence for this is:
  (1) Digestion of trout testis nuclei with DNase I, a nuclease which is known to preferentially degrade transcribed sequences, releases H6 from the chromatin [112].
  (2) Core particles enriched in H6 can be prepared from trout testis by a salt fractionation procedure. These core particles are also found to be enriched in transcribed sequences [113].
  (3) Digestion of embryonic chicken erythrocyte nuclei with DNase I releases HMG14 and 17 [214] (oddly enough these proteins are not released from thymus or liver nuclei) [69].
  (4) If embryonic erythrocyte nuclei are extracted with 0.35 M NaCl (removing HMG proteins) prior to DNase I digestion, the preferential susceptibility of the globin genes to DNase I digestion is destroyed; when HMG14 and 17 are added back to the nuclei this DNase I susceptibility is restored [214].
  (5) Nucleosomes enriched in ovalbumin gene sequences are preferentially released from oviduct nuclei as a result of micrococcal nuclease digestion. These nucleosomes are enriched in HMG14 and slightly in HMG17 [71].
  (6) A sub-nucleosomal fragment, consisting of a small piece of DNA with an HMG protein attached (probably HMG14 or 17), can be isolated from extensive micrococcal nuclease digests. The DNA of this sub-nucleosome is enriched in transcribed sequences [10].

Thus, the picture that is emerging of transcriptionally active chromatin is one in which this portion of the genome, about 5—20% of the total, has highly acetylated histones which are turning over rapidly, and HMG14 and HMG17 (or H6 in trout).

It is not known for certain whether or not HMG1 and 2 (and HMG-T) are also associated with transcribed sequences. It is striking, however, that the behaviour of HMG1 and 2 in micrococcal nuclease digestions (see previous section) is very similar to that of the estradiol-receptor complex in the uterus, suggesting similar roles in transcription [174]. It is also noteworthy that yeast, which has a smaller genome than higher eukaryotes and, therefore, probably expresses a higher proportion of its genome than a typical mammalian cell, has a large amount of the protein similar to HMG1 and 2 [181, 183]. It is, thus, possible that HMG1 and HMG2 are required to open up the chromatin structure to allow transcription to take place. The fact that the Drosophila protein D1 (which is similar to HMG1 and 2) is located in specific regions of the chromatin [4] does suggest some important regulatory function.

On the other hand, the recent finding that HMG1 and 2 are DNA unwinding proteins, that they bind preferentially to single-stranded DNA, and the cell-cycle changes of HMG1 and 2 at S-phase and mitosis, do suggest that these two proteins are associated in some ways with replication. One fact militates against such a hypothesis and that is the presence of HMG1 and 2 in the non-dividing chicken erythrocyte and in $G_O$ phase mouse salivary gland cells [179]. It, therefore, remains to be seen whether HMG1 and 2 are associated with transcription or replication.

## 4. Other non-histone proteins

### 4.1. Protein A24

Protein A24 was initially identified by two-dimensional gel electrophoresis of nucleolar proteins [148]. It can be extracted from rat liver with 0.4 N $H_2SO_4$ [148]. The amino acid analysis of this protein [57] is given in Table 8. The level of protein A24 in rat liver nucleoli has been shown to be markedly reduced during nucleolar hypertrophy induced by thioacetamide administration [12]. Chemical analysis of protein A24 has shown the protein to have a branched structure. One arm and the stem of this structure is composed of histone H2A and the other arm contains the protein ubiquitin [55, 57, 85]. The carboxy terminus of ubiquitin is attached to a Gly-Gly peptide which is in an isopeptide linkage to the $\epsilon$-$NH_2$ group of lysine residue 119 of histone H2A. The overall structure of protein A24 is therefore:

Ubiquitin-(X)-Gly-Gly ─────┐

Histone H2A      R-Lys-Lys-Thr-Glu-Ser-His-His-Lys-Ala-Lys-Gly-Lys-COOH
                    119                                              129

TABLE VIII

Amino acid analyses of some non-histone chromosomal proteins

| Amino acid | C14 | BA | A24 |
|---|---|---|---|
| Asp | 11.5 | 8.3 | 7.3 |
| Thr | 6.8 | 3.8 | 6.5 |
| Ser | 7.2 | 4.8 | 4.5 |
| Glu | 13.5 | 11.0 | 12.3 |
| Pro | 4.1 | 6.5 | 5.6 |
| Gly | 8.7 | 7.3 | 9.2 |
| Ala | 8.0 | 5.2 | 9.6 |
| Cys | 1.5 | 0 | — |
| Val | 5.1 | 4.1 | 4.9 |
| Met | 2.1 | 2.5 | 0.3 |
| Ile | 4.7 | 3.9 | 5.8 |
| Leu | 7.4 | 9.1 | 10.9 |
| Tyr | 2.2 | 4.9 | 1.3 |
| Phe | 3.3 | 5.9 | 0.9 |
| His | 1.3 | 4.7 | 2.4 |
| Lys | 7.9 | 9.7 | 11.3 |
| Arg | 4.6 | 8.3 | 7.4 |
| Trp | 0 | 0.5 | 0 |
| N-terminal amino acid | Lysine | Blocked | Methionine |

— indicates values close to zero.

The complete amino acid sequences of ubiquitin and histone H2A are shown in Figs. 18 and 4, respectively. Protein A24 has recently been shown to be present in isolated nucleosomes [56], but not in sufficient quantity for there to be one per nucleosome. Protein A24 is found in approximately one-tenth of the amount of histone H2A. Like the HMG proteins, therefore protein A24 might be associated with a specific group of nucleosomes. ADP-ribosylation of protein A24 has recently been shown to occur in isolated rat liver nuclei [144].

### 4.2. Nucleolar protein C23

A nucleolar protein, C23, has been isolated from Novikoff hepatoma [120]. The protein has an $M_r$ of ~ 114 000 and the N-terminal amino acid sequence has been determined as Val-Lys-Leu-Ala-

Lys-Ala-Gly-Lys-Thr- [147]. Large, highly negatively charged, phosphopeptide sequences have been found in this protein. At least three highly acidic phosphorylated regions exist in protein C23. One of these regions (peptide C23-Ca) has been partially sequenced [120] and has the following structure:

```
                                                          10
Ala — Ala — Pro — Ala — Ala — Pro — Ala — Ser — Glu — Asp —
                                                          20
Glu — Asp — Glu — Glu — Asp — Asp — Asp — Asp — Glu — Asp —
                                                          30
Asp — Asp — Asp — Asp — Ser — Gln — Glu — Ser — Glu — Glu —
                                                          40
Glu — Asp — Glu — Glu — Val — Met — Glu — Ile — Thr — Pro —

Ala — Lys — ..............
```

The high concentration of acidic residues is similar to the HGA region of proteins HMG1 and 2 (Section 3.5.5). The above peptide has been found in three phosphorylated forms with 1—3 phosphorylated serine residues at positions 8, 25 and probably 28. The highly acidic sequences adjacent to the phosphorylation sites represent a unique class of phosphorylation sites different from those in histones or substrates for cytoplasmic cAMP-dependent kinases. These sequence data do, of course, pose the question as to what role the addition of phosphate groups to an already highly negatively-charged region of the protein plays in the function of the protein.

## 4.3. Ubiquitin

Ubiquitin was initially isolated from bovine thymus, and the amino acid sequence of ubiquitin from this source has been determined [171] (Fig. 18). Initial interest in ubiquitin was due to its ability to induce differentiation antigens in thymocytes and bursocytes [59]. Radioimmunoassay has demonstrated ubiquitin in a variety of mammalian tissues and in many organisms [60] (including bacteria, yeast, higher green plants, invertebrates and vertebrates) and it is thought to be present in all living cells. The wide distribution of ubiquitin suggests that it has an ancient, important function

```
Met — Gln — Ile   — Phe — Val — Lys — Thr — Leu — Thr — Gly —
                                                         10
                                                         20
Lys — Thr — Ile   — Thr — Leu — Glu — Val — Glu — Pro — Ser —
                                                         30
Asp — Thr — Ile   — Glu — Asn — Val — Lys — Ala — Lys — Ile —
                                                         40
Gln — Asp — Lys — Glu — Gly — Ile  — Pro — Pro — Asp — Gln —
                                                         50
Gln — Arg — Leu — Ile   — Phe — Ala — Gly — Lys — Gln — Leu —
                                                         60
Glu — Asp — Gly — Arg — Thr — Leu — Ser — Asp — Tyr — Asn —
                                                         70
Ile  — Gln — Lys — Glu — Ser — Thr — Leu — His — Leu — Val —

Leu — Arg — Leu — Arg
```

Fig. 18. The amino acid sequence of calf thymus ubiquitin.

in all cells, prokaryotic and eukaryotic. However, this function is still unknown.

The first evidence for the involvement of ubiquitin in chromatin structure came when it was realised that part of the structure of protein A24 comprised the ubiquitin molecule (see Section 3.1, above). This discovery was soon followed by the isolation of free ubiquitin from both calf thymus chromatin [204] and trout testis chromatin [209]. The amino acid sequence of ubiquitin from trout testis chromatin has been determined and was shown to be identical with the sequence of the calf thymus ubiquitin in *at least* 71 of the 74 positions [209]. This considerable sequence conservation suggests that the structure of ubiquitin is probably as highly conserved during evolution as that of the arginine-rich histones H3 and H4. Ubiquitin is present in chromatin at ~ 2.5% of the concentration of the individual nucleosomal core histones [209] and can, therefore, only be present in a small subset of the total nucleosomes. Recently, a reciprocal relationship between "conjugated" ubiquitin (as protein A24) and free ubiquitin, has been observed [58]. Analysis of active and inactive chromatin showed that conjugated ubiquitin (protein A24) was present in gene-inactive fractions but markedly diminished in content in the gene-active fraction. Conversely, free ubiquitin was found in the gene-active fraction, but was absent in the inactive fractions. It has been suggested that this reciprocal distribution between ubiquitin and its conjugate may reflect a switch in chromatin structure due to gene activation or repression.

### 4.4. Protein BA

Protein BA remains bound to chromatin after extraction of rat liver chromatin with 0.4 N $H_2SO_4$. Purification of this protein from rat liver has been described [29]. The protein has an $M_r$ of ~ 31 000 and the N-terminal amino acid is blocked. The amino acid analysis of BA is shown in Table 8. Protein BA is present in the chromatin of non-growing tissues, but is absent or greatly reduced in Novikoff and Morris hepatomas and also in Walker 256 carcinosarcoma. A marked reduction is also found in the chromatin of 18 h regenerating rat liver and PHA-stimulated human lymphocytes.

### 4.5. Protein C14

Protein C14 has been purified from Novikoff hepatoma ascites cell nuclei [88]. The protein has an $M_r$ of ~ 70 000 and the N-terminal amino acid is lysine. The amino acid analysis of C14 is given in Table 8. This protein has been shown to decrease relatively in amount during liver hypertrophy produced by administration of thioacetamide to rats or following partial hepatectomy, and has been shown to stimulate RNA synthesis [88].

### 4.6. Nucleosome assembly factor

A protein factor that will induce nucleosome formation between DNA and histones at physiological ionic strength has been purified from eggs of *Xenopus laevis* [111]. SDS gel electrophoresis of the factor suggests an $M_r$ ~ 29 000. However, sedimentation and gel filtration properties suggest that the native protein has an $M_r >$ 100 000. Electrophoresis over long distances resolves the factor into two bands, suggesting the presence of more than one polypeptide chain. Its binding characteristics to ion exchange resins suggest that the protein is highly acidic and this has been confirmed by isoelectric focusing where the protein has been shown to have a pI of 5. Although perhaps not strictly a chromosomal protein, the assembly factor is worthy of mention here, since it is obviously instrumental in bringing about the correct binding of certain chromosomal proteins.

### 4.7. Component 10

Component 10 is a chromosomal protein with $M_r \sim 10\,000$ which has been isolated from mouse liver nuclei [118]. Isolation of this protein from nuclei which have been labelled with [$\gamma$-$^{32}$P] ATP has shown Component 10 to be a phosphoprotein with phosphorylation occurring at both serine and threonine residues. The function of Component 10 is not known.

### 4.8. Contractile proteins

Several laboratories have reported that actin is a normal constituent of interphase eukaryotic nuclei. In one study, both actin ($M_r = 200\,000$) and myosin ($M_r = 45\,000$) were isolated from rat liver chromatin, and three other contractile proteins, tubulin and heavy and light tropomyosin, were identified by their electrophoretic mobilities [42]. In a separate study both $\alpha$ and $\beta$ tubulin have been identified in rat liver chromatin extracts by their mobilities in SDS polyacrylamide gel electrophoresis [6]. It has been suggested that the contractile proteins may be involved in the condensation of chromatin into heterochromatin, the condensation of mitotic chromosomes, and/or chromosome movement during mitosis.

### 4.9. DNA and RNA polymerases

*4.9.1. DNA polymerases.* Higher eukaryotic cells contain at least three distinct DNA polymerases, which have been named DNA polymerases $\alpha$, $\beta$ and $\gamma$. Although obviously sharing the same function, namely that of producing a DNA copy of a DNA template, the polymerases can be distinguished from one another by their molecular weights, sensitivity to salts and ability to copy various DNA templates. DNA polymerase α seems to be ubiquitous in growing cells and has been found in calf thymus, human, murine and hamster cells, and in avian tissues, sea urchins and yeast. Some uncertainty exists over the molecular weight of this protein since values in the range 120 000—300 000 have been reported from various sources. However, the DNA polymerase α of calf thymus is a single polypeptide of $M_r$ 155 000—170 000 and seems to contain an additional subunit of 50 000—70 000. DNA polymerase β is a low molecular weight polymerase. The purified enzyme from calf thymus has an $M_r \sim$ 45 000. DNA polymerase γ has an $M_r$ in the range 150 000—300 000. This complex subject has recently been excellently reviewed by Weissbach [215].

*4.9.2. RNA polymerases.* Eukaryotic cells contain multiple forms of nuclear RNA polymerases. These enzymes fall into three distinct classes, designated I, II and III, which are distinguished on the basis of their enzymatic properties and subunit structure. In mammalian cells two major class I enzymes, three major class I enzymes and two major class III enzymes have been identified. Each of these nuclear enzymes comprises two polypeptide chains with molecular weight greater than 100 000, together with a number of smaller polypeptide chains giving overall molecular weights for the enzyme complexes of between 400 000 and 650 000. The number of smaller polypeptide chains varies between 3 and 8 in the various enzyme complexes. RNA polymerase stimulatory and inhibitory factors have also been described [6, 26, 27]. The subject of animal RNA polymerases has recently been the subject of an excellent review by Roeder et al. [163].

## REFERENCES

1. B. D. Abercrombie, G. G. Kneale, C. Crane-Robinson, E. M. Bradbury, G. H. Goodwin, J. M. Walker and E. W. Johns, Eur. J. Biochem., 84 (1978) 173.
2. P. Adamietz, R. Bredehorst and H. Hilz, Eur. J. Biochem., 91 (1978) 317.
3. E. D. Adamson and H. R. Woodland, J. Mol. Biol., 88 (1974) 263.
4. C. R. Alfagame, G. T. Rudkin and L. H. Cohen, Cell, (1979) submitted for publication.
5. V. G. Allfrey in H. J. Li and R. Eckhordt (Eds.), Chromatin and Chromosome Structures, (1977) p. 167.
6. B. Anachkova and G. Russev, FEBS Lett. 81 (1977) 37.
7. R. Appels, L. Bolund and N. R. Ringertz, J. Mol. Biol., 87 (1974) 339.
8. G. S. Bailey and G. H. Dixon, J. biol. Chem., 248 (1973) 5463.
9. T. G. Bakayeva and V. V. Bakayev, Mol. Biol. Rept., 4 (1978) 185.
10. V. Bakayev, T. Bakayeva, V. V. Schmatchenko and G. P. Georgiev, Eur. J. Biochem., 91 (1978) 291.
11. C. Baker, I. Isenberg, G. H. Goodwin and E. W. Johns, Biochemistry, 15 (1976) 1645.
12. N. R. Ballal and H. Busch, Cancer Res., 33 (1973) 2737.
13. P. Beard, Cell, 15 (1978) 955.
14. D. L. Bidney and G. R. Reeck, Biochem. Biophys. Res. Commun., 85 (1978) 1211.
15. E. M. Bradbury in D. W. Fitzsimmons and G. E. W. Wolstenholme (Eds.), The Structure and Function of Chromatin, Vol. 28, Associated Scientific Publishers, Amsterdam, (1975) p. 1.
16. E. M. Bradbury in The Organisation and Expression of the Eukaryotic Genome, Academic Press, (1977) p. 99.
17. E. M. Bradbury, B. G. Carpenter and H. W. E. Rattle, Nature (Lond.), 241 (1973) 123.
18. E. M. Bradbury, P. D. Cary, C. Crane-Robinson, Ann. N.Y. Acad. Sci., 222 (1973) 266.
19. E. M. Bradbury, R. J. Inglis and H. R. Matthews, Nature (Lond.), 247 (1974) 257.
20. E. M. Bradbury, R. J. Inglis, H. R. Matthews and T. A. Langan, Nature (Lond.), 249 (1974) 553.
21. W. F. Brandt and C. Von Holt, Eur. J. Biochem., 46 (1974) 419.
22. W. F. Brandt, W. N. Strickland, M. Strickland, L. Carlisle, D. Woods and C. Von Holt, Eur. J. Biochem., 94 (1979) 1.
23. R. Bredehorst, K. Wielckens, A. Garternann, H. Lengyel, K. Klapproth and H. Hilz, Eur. J. Biochem., 92 (1978) 129.
24. M. Bustin and N. K. Neihart, Cell, 16 (1970) 181.
25. J. A. V. Butler, E. W. Johns and D. M. P. Phillips, Progr. Biophys. Mol. Biol., 18 (1968) 209.
26. H. G. Callan, Nature (Lond.), 152 (1943) 503.
27. A. Caplan, M. G. Ord and L. A. Stocken, Biochem. J., 174 (1978) 475.

## REFERENCES

28  P. S. Cary, C. Crane-Robinson, E. M. Bradbury, K. Javaherian, G. H. Goodwin and E. W. Johns, Eur. J. Biochem., 62 (1976) 583.
29  J. J. Catino, L. C. Yeoman, M. Mandel and H. Busch, Biochemistry, 17 (1978) 983.
30  R. D. Cole in P. Ts'o (Ed.), The Molecular Biology of the Mammalian Genetic Apparatus, Elsevier/North Holland Biomedical Press, (1977) p. 93.
31  R. D. Cole, M. W. Hsiang, G. M. Lawson, R. O'Neal and S. L. Welch, ICN-UCLA Symp. Mol. Cell Biol., 1 (1977) 179.
32  J. L. Compton, M. Bellard and P. Chambon, Proc. natl. Acad. Sci. USA, 73 (1978) 4382.
33  J. R. Davie and E. P. M. Candido, Proc. natl. Acad. Sci. USA, 75 (1978) 3574.
34  R. J. DeLange, D. M. Farnbrough, E. L. Smith and J. Bonner, J. biol. Chem., 244 (1969) 319.
35  R. J. DeLange, D. M. Farnbrough, E. L. Smith and J. Bonner, J. biol. Chem., 244 (1969) 5669.
36  R. J. DeLange, J. A. Hooper and E. L. Smith, Proc. natl. Acad. Sci. USA, 69 (1972) 882.
37  R. J. DeLange, J. A. Hooper and E. L. Smith, J. biol. Chem., 248 (1973) 3261.
38  R. J. DeLange and E. L. Smith in The Structure and Function of Chromatin, Ciba Foundation Symposium, Vol. 28 (1975) p. 60.
39  G. H. Dixon, E. P. M. Candido, B. M. Honda, A. J. Louie, A. R. Macleod, and M. T. Sung in The Structure and Function of Chromatin, Ciba Foundation Symposium, Vol. 28 (1975) 223.
40  G. H. Dixon, E. P. M. Candido, B. M. Honda, A. J. Louie, A. R. Macleod and M. T. Sung in The Structure and Function of Chromatin, Ciba Foundation Symposium, Vol. 28 (1975) p. 229.
41  G. H. Dixon, N. Wong and G. G. Poirier, Fed. Proc., 35 (1976) 1623.
42  A. S. Douvas, C. A. Harrington and J. Bonner, Proc. nat. Acad. Sci. USA, 72 (1975) 3902.
43  S. C. R. Elgin and H. Weintraub, Ann. Rev. Biochm, 44 (1975) 725.
44  M. Ermini and C. C. Kuenzle, FEBS Lett., 90 (1978) 167.
45  M. Fazal and R. D. Cole, J. biol. Chem., 252 (1977) 4068.
46  G. Felsenfeld, Nature (Lond.), 271 (1978) 115.
47  A. Fonagy, H. G. Ord and L. A. Stocken, Biochem. J., 162 (1977) 171.
48  L. Franco, F. Montero and J. J. Rodriguez-Molina, FEBS Lett., 78 (1977) 317.
49  A. Garel and R. Axel, Proc. nat. Acad. Sci. USA., 73 (1976) 3966.
50  G. P. Georgiev, S. A. Nedosposov and V. V. Bakayev in H. Busch (Ed.), The Cell Nucleus, Vol.4 part C, Academic Press, New York, San Francisco, London (1978) p. 3.
51  R. S. Gilmour and J. Paul, FEBS Lett., 9 (1970) 242.
52  P. Giri, M. H. P. West and M. Smulson, Biochemistry, 17 (1978) 3495.
53  C. P. Giri, M. H. P. West, M. L. Ramirez and M. Smulson, Biochemistry, 17 (1978) 3501.

54  C. V. C. Glover and M. A. Gorovsky, Proc. nat. Acad. Sci. USA, 76 (1979) 585.
55  I. L. Goldknopf and H. Busch, Proc. nat. Acad. Sci. USA, 74 (1977) 864.
56  I. L. Goldknopf, M. F. French, R. Musso and H. Busch, Proc. nat. Acad. Sci. USA, 74 (1977) 5492.
57  I. Goldknopf and H. Busch in H. Busch (Ed.), The Cell Nucleus, Vol. 4 Part C, Academic Press, New York, San Francisco, London, (1978) p. 149.
58  I. L. Goldknopf, M. F. French, Y. Daskal and H. Busch, Biochem. Biophys. Res. Commun., 84 (1978) 786.
59  G. Goldstein, M. Scheid, E. A. Boyse, A. Brand and D. G. Gilmour, Cold Spring Harbor Symp. Quant. Biol., XLI, 5 (1977).
60  G. Goldstein, M. S. Scheid, V. Hammerling, E. A. Boyse, D. H. Schlesinger and H. D. Niall, Proc. nat. Acad. Sci. USA, 72 (1975) 11.
61  G. H. Goodwin, C. Sanders and E. W. Johns, Eur. J. Biochem., 38 (1973) 14.
62  G. H. Goodwin and E. W. Johns, Eur. J. Biochem., 40 (1973) 215.
63  G. H. Goodwin, R. H. Nicolas and E. W. Johns, Biochem. Biophys. Acta, 405 (1975a) 280.
64  G. H. Goodwin, K. V. Shooter and E. W. Johns, Eur. J. Biochem., 54 (1975) 427.
65  G. H. Goodwin, R. H. Nicolas and E. W. Johns, FEBS Lett., 64 (1976) 412.
66  G. H. Goodwin and E. W. Johns, Methods Cell Biol., 6 (1977) 257.
67  G. H. Goodwin, L. Woodhead and E. W. Johns, FEBS Lett., 73 (1977) 85.
68  G. H. Goodwin, J. M. Walker and E. W. Johns, Biochem Biophys. Acta, 519 (1978) 233.
69  G. H. Goodwin and E. W. Johns, Biochem. Biophys. Acta, 519 (1978) 279.
70  G. H. Goodwin, J. M. Walker and E. W. Johns in H. Busch (Ed.), The Cell Nucleus, Vol. VI (1978) 181.
71  G. H. Goodwin and E. W. Johns, (1979) in preparation.
72  G. H. Goodwin, J. M. Walker and E. W. Johns. Unpublished results.
73  M. A. Gorovsky and J. B. Keevert, Proc. nat. Acad. Sci. USA, 72, (1975) 2672.
74  M. A. Gorovsky and J. B. Keevert, Proc. nat. Acad. Sci. USA, 72 (1975) 3536.
75  J. M. Gottesfeld and P. J. G. Butler, Nucleic Acid Res., 4 (1977) 3155.
76  J. M. Gottesfeld and D. A. Melton, Nature (Lond.), 273 (1978) 317.
77  P. J. Greenaway and K. Murray, Nature, New Biol., 229 (1971) 233.
78  L. R. Gurley, R. A. Walters and R. A. Tobey, J. biol. Chem., 250 (1975) 3936.
79  L. R. Gurley, J. A. D'Anna, S. S. Barham, L. L. Deaven and R. A. Tobey, Eur. J. Biochem., 84 (1978) 1.
80  D. G. Hardie, H. R. Matthews and E. M. Bradbury, Eur. J. Biochem., 66 (1976) 37.

## REFERENCES

81  D. R. Hewish and L. A. Burgoyne, Biochem. Biophys. Res. Commun., 52 (1973) 504.
82  L. S. Hnilica, Progr. Nucleic Acid Res. Mol. Biol., 7 (1967) 25.
83  P. Hohmann and R. D. Cole, Nature (Lond.), 223 (1969) 1064.
84  J. A. Hooper, E. L. Smith, K. R. Sommer and R. Chalkley, J. biol. Chem., 248 (1973) 3275.
85  L. T. Hunt and M. O. Dayhoff, Biochem. Biophys. Res. Commun., 74 (1977) 650.
86  G. H. Huntley and G. H. Dixon, J. biol. Chem., 247 (1972) 4916.
87  K. Iwai, K. Ishikawa and H. Hayashi, Nature (Lond.), 226 (1970) 1056.
88  G. T. James, L. C. Yeoman, S. Matsui, A. H. Goldberg and H. Busch, Biochemistry, 16 (1977) 2384.
89  K. Javaherian in E. M. Bradbury and K. Javaherian (Eds.), The Organisation and Expression of the Eukaryotic Genome, Academic Press, New York, (1977) p. 51.
90  K. Javaherian and S. Amini, Biochem. Biophys. Acta, 478 (1977) 295.
91  K. Javaherian, L. F. Liu and J. C. Wang, Science, 199 (1978) 1345.
92  K. Javaherian and S. Amini, Biochem. Biophys. Res. Commun., 85 (1978) 1385.
93  E. W. Johns,Biochem. J., 92 (1964) 55.
94  E. W. Johns in J. Bonner and P. O. P. Ts'o (Eds.), The Nucleohistones, Holden-Day, San Francisco, CA, (1964) p. 52.
95  E. W. Johns, Ph. D. Thesis, London University (1965).
96  E. W. Johns, Biochem. J., 105 (1967) 611.
97  E. W. Johns in D. M. P. Phillips (Ed.), Histones and Nucleohistones, Plenum, New York and London, (1971) p. 1.
98  E. W. Johns in G. D. Birnie (Ed.), Subnuclear components, Butterworths, London, Boston, (1976) p. 187.
99  E. W. Johns in G. Stein, J. Stein and L. J. Kleinsmith (Eds.), Methods in Cell Biology, Vol. 16 (1977) p. 183.
100  E. W. Johns and S. Forrester, Eur. J. Biochem., 8 (1969) 547.
101  A. C. Kay and M. F. Singer, Nucleic Acid Res., 4 (1977) 3371.
102  J. M. Kinkade, J. biol. Chem., 244 (1969) 3375.
103  L. Klevan and D. M. Crothers, Nucleic Acid Res., 4 (1977) 4077.
104  A. Koostra and G. S. Bailey, Biochemistry, 17 (1978) 2502.
105  R. D. Kornberg, Science, 184 (1974) 868.
106  R. D. Kornberg, Annu. Rev. Biochem., 46 (1977) 931.
107  A. Kossel, Z. Physiol. Chem., 8 (1884) 511.
108  B. Laine, P. Sautiere and G. Biserte, Biochemistry, 15 (1976) 1640.
109  T. A. Langan, Ann. N.Y. Acad. Sci., 185 (1971) 166.
110  T. A. Langan and P. Hohmann in Chromosomal Proteins and their Role in the Regulation of Gene Expression, Academic Press, New York, (1975) p. 113.
111  R. A. Laskey, B. M. Honda, A. D. Mills and J. T. Finch, Nature (Lond.), 275 (1978) 416.
112  W. B. Levy, N. C. W. Wong and G. H. Dixon, Proc. nat. Acad. Sci. USA, 74 (1977) 2810.

113  W. B. Levy, W. Connor and G. H. Dixon, J. Biol. Chem., 254 (1979) 609.
114  D. M. J. Lilley, Cell Biol. Int. Rep., 2 (1978) 1.
115  J. M. Luck in J. Bonner and P. O. P. Ts'o (Eds.), The Nucleohistones, Holden-Day Inc., San Francisco, London, Amsterdam, (1964) p. 3.
116  J. M. Luck, H. A. Cook, N. T. Eldredge, M. I. Haley, D. W. Kupke and P. S. Rasmussen, Arch. Biochem. Biophys., 65 (1956) 449.
117  N. Lukas and H. Stein, FEBS Lett., 69 (1976) 295.
118  A. J. MacGillivray, C. Johnston, R. MacFarlane and D. Rickwood, Biochem. J., 175 (1978) 35.
119  S. Macleod, N. C. W. Wong and G. H. Dixon, Eur. J. Biochem., 78 (1977) 281.
120  M. D. Mamrack, M. O. J. Olson and H. Busch, Fed. Proc., 37 (1978) 1786.
121  P. Mandel, H. Okazaki and C. Niedergang, FEBS Lett., 84 (1977) 331.
122  W. H. Marsh and P. J. Fitzgerald, Fed. Proc., 32 (1973) 2119.
123  D. Z. Martin, R. D. Todd, D. Lang, P. N. Pei and W. T. Garrard, J. Biol. Chem., 252 (1977) 8269.
124  H. G. Martinson, R. True, J. B. E. Burch and G. Kunkel, Proc. Natl. Acad. Sci. USA, 76 (1979) 1030.
125  K. Marushige and G. H. Dixon, J. Biol. Chem., 246 (1971) 5799.
126  C. G. P. Mathew, G. H. Goodwin and E. W. Johns, Nucleic Acid Res., 6 (1979) 167.
127  C. G. P. Mathew, G. H. Goodwin, K. Gooderham, J. M. Walker and E. W. Johns, Biochem. Biophys. Res. Commun., 87 (1979) 1243.
128  D. T. Mayer and A. Gullick, J. biol. Chem., 146 (1942) 433.
129  Z. A. Medvedev, M. N. Medvedeva and L. I. Huschtscha, Gerontology, 23 (1977) 334.
130  M. L. Meistrich, W. A. Brock, S. R. Grimes, R. D. Platry and L. S. Hnilica, Fed. Proc., 37 (1978) 2522.
131  F. Miescher, Verhandl. d. Naturforsch. Ges. in Basel, 6 (1874) 138.
132  N. R. Morris, Cell, 8 (1976) 357.
133  N. R. Morris, Cell, 9 (1976) 627.
134  D. W. Mullins Jr., C. P. Giri and M. Smulson, Biochemistry, 16 (1977) 506.
135  R. F. Murphy, R. B. Wallace and J. Bonner, Proc. nat. Acad. Sci. USA, 75 (1978) 5903.
136  K. Murray, Ann. Rev. Biochem., 34 (1965) 209.
137  S. A. Nedospasov, V. V. Bakayev and G. P. Georgiev, Nucleic Acid Res., 65 (1978) 2847.
138  J. M. Neelin and G. E. Connell, Biochim. Biophys. Acta, 31 (1959) 539.
139  D. A. Nelson, M. Perry, L. Sealy and R. Chalkley, Biochem. Biophys. Res. Commun., 82 (1978) 1346.
140  M. Noll, Nature (Lond.), 251 (1974) 249.
141  M. Noll, Cell, 8 (1976) 349.
142  Y. Ogawa, G. Quagliarotti, J. Jordan, C. W. Taylor, W. C. Starbuck and H. Busch, J. biol. Chem., 244 (1969) 4387.

143 H. Okayama, C. M. Fukushima, K. Ueda and O. Hayaishi, J. biol. Chem., 252 (1977) 7000.
144 H. Okayama and O. Hayaishi, Biochem. Biophys. Res. Commun., 84 (1978) 755.
145 A. L. Olins and D. E. Olins, Science, 183 (1974) 330.
146 A. L. Olins, R. D. Carlson, E. B. Wright and D. E. Olins, Nucl. Acids Res., 3 (1976) 3271.
147 M. O. J. Olson, personal communication.
148 L. R. Orrick, M. O. J. Olson and H. Busch, Proc. Natl. Acad. Sci. USA, 70 (1973) 1316.
149 P. Oudet, M. Cross-Bellard and P. Chambon, Cell, 4 (1975) 281.
150 S. Panyim and R. Chalkley, Biochem. Biophys. Res. Commun., 37 (1969) 1042.
151 S. Panyim and R. Chalkley, Biochemistry, 8 (1969) 3972.
152 A. Pastink, T. A. Berkhaut, W. H. Mager and R. J. Planta, Biochem. J., 177 (1979) 917.
153 L. Patthy, E. L. Smith and J. Johnson, J. Biol. Chem., 248 (1973) 6834.
154 J. Paul and R. S. Gilmour, J. Mol. Biol., 34 (1968) 305.
155 T. Pederson, Int. Rev. Cytol., 55 (1978) 1.
156 D. M. P. Phillips, Biochem. J., 68 (1958) 35.
157 V. A. Pospelov, S. B. Svetlikova and V. I. Vorob'ev, FEBS Lett., 99, (1979) 123.
158 A. Rabbani, Ph.D. Thesis, (178) London University.
159 A. Rabbani, G. H. Goodwin and E. W. Johns, Biochem. J., 173 (1978) 497.
160 A. Rabbani, G. H. Goodwin and E. W. Johns, Biochem. Biophys. Res. Commun., 81 (1978) 351.
161 P. S. Rasmussen, K. Murray and J. M. Luck, Biochemistry, 1 (1962) 79.
162 J. B. Rattner and B. A. Hamkalo, J. Cell Biol., 81 (1979) 453.
163 R. G. Roeder, C. S. Parker, J. A. Jaehning, Ng, S-Y., and SKLAR, V.E.F. in G. F. Saunders (Ed.), Cell Differentiation and Neoplasia, Raven Press, New York, (1978) p. 305.
164 J. V. Ruderman, C. Baglioni and P. R. Gross, Nature (Lond.), 247 (1974) 36.
165 A. Ruiz-Carrillo, L. J. Wangh and V. G. Allfrey, Science, 190 (1975) 117.
166 C. Sanders and E. W. Johns, Biochem. Soc. Trans., 2 (1974) 547.
167 C. Sanders, Ph.D. Thesis, 1975, London University.
168 C. Sanders, Biochem. Biophys. Res. Commun., 78 (1977) 1034.
169 P. Sautiere, G. Briand, D. Kmiecik, O. Loy and G. Biserte, FEBS Lett., 63 (1976) 164.
170 J. Schlepper and R. Knippers, Eur. J. Biochem., 60 (1975) 209.
171 D. H. Schlesinger, G. Goldstein and H. D. Niall, Biochemistry, 14 (1975) 2214.
172 R. L. Seale, in H. Busch (Ed.), The Cell Nucleus, Vol. 4 part A, Academic Press, New York, San Francisco, London, 1978, p. 155.

173  R. L. Seale, Proc. nat. Acad. Sci. USA, 75 (1978) 2717.
174  M. B. Senior and F. R. Frankel, Cell, 14 (1978) 857.
175  K. V. Shooter, G. H. Goodwin and E. W. Johns, Eur. J. Biochem., 47 (1974) 263.
176  G. Siebert, M. Ord and L. Stocken, Biochem. J., 122 (1971) 721.
177  Smerdon, M. J. and I. Isenberg, Biochemistry, 15 (1976) 4233.
178  B. J. Smith, D. Robertson, M. S. C. Birbeck, G. H. Goodwin and E. W. Johns, Expt. Cell Res., 115 (1978) 420.
179  B. J. Smith, unpublished results.
180  J. A. Smith and L. A. Stocken, Biochem. Biophys. Res. Commun., 54 (1973) 297.
181  A. Sommer, Molec. Gen. Genet., 161 (1978) 323.
182  S. Spiker, J. L. Key and B. Wakim, Arch. Biochem. Biophys., 176 (1976) 510.
183  S. Spiker, J. K. W. Mardian and I. Isenberg, Biochem. Biophys. Res. Commun., 82 (1978) 129.
184  E. Stedman and E. Stedman, Nature (Lond.), 152 (1943) 267.
185  E. Stedman and E. Stedman, Nature (Lond.), 780 (1950) 166.
186  E. Stedman and E. Stedman, Phil. Trans. B. roy. Soc., 235 (1951) 565.
187  G. S. Stein, J. S. Stein and L. J. Kleinsmith, Scientific American, 232 (1975) 47.
188  M. Steinmetz, R. E. Streeck and H. G. Zachau in H. Busch (Ed.), The Cell Nucleus, Vol. 4 part B, Academic Press, New York, San Francisco, London, (1978) pp. 167.
189  R. Sterner, L. C. Boffa and G. Vidali, J. biol. Chem., 253 (1978) 3830.
190  R. Sterner, G. Vidali, R. L. Henrikson and V. G. Allfrey, J. biol. Chem., 253 (1978) 7601.
191  P. R. Stone, W. S. Lorimer III and W. R. Kidwell, Eur. J. Biochem., 81 (1977) 9.
192  M. T. Sung and E. F. Freedlender, Biochemistry, 17 (1978) 1884.
193  C. S. Teng, K. Gallagher and C. T. Teng, Biochem. J., 176 (1978) 1003.
194  J. O. Thomas and R. D. Kornberg, Proc. nat. Acad. Sci. USA, 72 (1975) 2626.
195  K. Ueno, K. Sekimizu, D. Mizuno and S. Natori, Nature (Lond.), 277 (1979) 145.
196  P. Van Helden, W. N. Strickland, W. F. Brandt and C. Von Holt, Biochim. Biophys. Acta, 533 (1978) 278.
197  K. E. Van Holde and W. O. Weischert, in H. Busch (Ed.), The Cell Nucleus, Vol. 4 part A, Academic Press, New York, San Francisco, London, 1978, p. 75.
198  F. Varrichio, Arch. Biochem. Biophys., 179 (1977) 715.
199  G. Vidali, L. C. Boffa, E. M. Bradbury and V. G. Allfrey, Biochemistry, 75 (1978) 2239.
200  C. Von Holt, W. N. Strickland, W. F. Brandt and M. S. Strickland, FEBS Lett., 100 (1979) 201.
201  J. M. Walker, unpublished results.

## REFERENCES

202 J. M. Walker, G. H. Goodwin and E. W. Johns, Eur. J. Biochem, 62 (1976) 461.
203 J. M. Walker, J. R. B. Hastings and E. W. Johns, Eur. J. Biochem., 76 (1977) 461.
204 J. M. Walker, G. H. Goodwin and E. W. Johns, FEBS Lett., 90 (1978) 327.
205 J. M. Walker, J. R. B. Hastings and E. W. Johns, Nature (Lond.), 271 (1978) 281.
206 J. M. Walker, G. H. Goodwin and E. W. Johns, FEBS Lett., 100 (1979) 394.
207 J. M. Walker, K. Gooderham and E. W. Johns, Biochem. J. 181 (1979) 659.
208 D. C. Watson, E. H. Peters and G. H. Dixon, Eur. J. Biochem., 74 (1977) 53.
209 D. Watson, W. B. Levy and G. H. Dixon, Nature (Lond.), 276 (1978) 196.
210 D. C. Watson, N. C. W. Wong and G. H. Dixon, Eur. J. Biochem., 95 (1979) 193.
211 H. Weintraub and F. Van Lente, Proc. nat. Acad. Sci. USA, 71 (1974) 4249.
212 H. Weintraub, Nucleic Acid Res., 5 (1978) 1179.
213 H. Weintraub and M. Groudine, Science, 193 (1976) 848.
214 S. Weisbrod and H. Weintraub, Proc. nat. Acad. Sci. USA, 76 (1979) 630.
215 A. Weissbach, Ann. Rev. Biochem., 46 (1977) 25.
216 S. L. Welch and R. D. Cole, J. biol. Chem., 254 (1979) 662.
217 D. T. Wigle and G. H. Dixon, J. biol. Chem., 246 (1971) 5636.
218 R. E. Williams, Science, 192 (1976) 473.
219 N. C. Wong, G. G. Poivier and G. H. Dixon, Eur, J. Biochem., 77 (1977) 11.
220 T. Y. Wong and N. C. Kostraba in Methods in Cell Biology, Vol. XVI (1977) p. 317.
221 C. L. F. Woodcock, J. Cell Biol., 59 (1973) 368.
222 H. R. Woodland and E. D. Adamson, Develop. Biol., 57 (1977) 118.
223 A. Worcel, Cold Spr. Harb. Symp. Quant. Biol., 42 (1977) 313.
224 C. Wu, P. M. Bingham, K. J. Livak, R. Holmgren and S. C. R. Elgin, Cell, 16 (1979) 797.
225 C. Wu, Y-C. Wong and S. C. R. Elgin, Cell, 16 (1979) 807.
226 L. C. Yeoman, M. O. J. Olson, J. Sugano, J. J. Jordan, C. W. Taylor, W. C. Starbuck and H. Busch, J. biol. Chem., 247 (1972) 6018.
227 M. Yukioka, Y. Okai, T. Hasuma and A. Inoue, FEBS Lett., 86 (1978) 85.
228 V. Zongza and A. P. Mathias, Biochem. J., 179 (1979) 291.

# Subject Index

Abetalipoproteinaemia, 475, 482
Acetal groups in glycans, 76
2-Acetamido-1-(L-β-aspartamido)-1,2-dideoxy-β-D-glycopyranose, 43—44
Acetylation in histones, 527—528
— — —, reversible, 527
α-N-Acetylgalactosamine linkage in glycoproteins, 41, 44—45, 58
β-N-Acetylglucosamine linkage in glycoproteins, 41, 43—44, 58
Acetylglucosamine phosphodiesterase, 128
N-Acetylglucosamine, role in oligosaccharide degradation, 128
— transferases I, II and III, mode of action, 107—109
N-Acetyl-α-glucosaminidase, 132
N-Acetylglucosaminyl-asparagine, 43—44
N-Acetylhexosamine sulfate sulfatases, 124, 131
N-Acetyl-lactosaminic glycans, role, 152
$\alpha_1$-Acid glycoprotein, asialo-glycans, 68
— —, early work, 17—18
— —, heterogeneity, 78—79
— —, NMR, 38—40
— —, secondary structure and glycosylation, 105
— —, structure, 87, 88
Acid hydrolysis in sugar analysis, 31
Actin in chromosomes, 563

Actinomycin, effect on tryptophan pyrrolase, 331
Acyl co-enzyme A: cholesterol acyl transferase, 476, 481
ADP-ribose-protein linkages, 530
ADP-ribose units attached to en- chromatin, 529
— — linked to histones, 529
— —, location in linker region, 530
ADP-ribosylation and chromosome condensation, 530
— and histones H1 and H2B, 530
Affinity labelling in immunological studies, 217—221
Albumin-IgM complexes, 153
Allelic exclusion, 268
Allo-reactive T cells, prevalence, 305—306
Allotypes and amino acid substitution, 209—212
— in IgA2, 210
— in rabbit IgG, 210, 213
— of immunoglobulins, 209—212
Allotypic markers, 210—213
Altered self hypothesis, 303
Amino acid sequences of α and β chains, 352—356
Antennae, definition, 51
Anti-allotype sera, 209
Antibody-antigen complex clearance, 152—153
Antibody-binding site, 192
Antidextran serum, 215
Antifreeze glycoprotein, 145—146

Antigen-binding site, 214—221, 243—244
— —, location, 215
— —, size and shape, 215—217
— —, stereochemistry, 235—236
Anti-immunoglobulins, 267
Apolipoprotein, activation of LCAT, 429
—, α-helix and amphipathic properties, 428
—, detection in immunoassays, 429—430
—, effects on structure, uptake and metabolism of lipoproteins, 428
—, molecular weight and amino acid composition, 429
—, movement between lipoproteins, 441
ApoA abnormal metabolism, 473
ApoA-1, association with HDL, 429
—, plasma levels, 471
—, production by the intestine, 448—449
—, production by the liver, 453
— and II, kinetics of metabolism, 471—472
ApoA-II, amino acid composition and molecular weight, 430
—, detection, 430
—, plasma levels, 471
—, presence in HDL, 430
—, production by the intestine, 449
ApoA-IV, production by the liver, 453
ApoB, binding to receptor, 477
—, derivation from VLDL-B in LDL, 466
—, formation and secretion by the intestine, 449—450
—, kinetics of conversion or degradation in normals and hypertriglyceridaemia, 466—467
—, plasma levels, 466
—, production by the liver, 453—454
—, requirement for secretion of VLDL and LDL, 430
—, structure, 431

ApoC-I, activation of LCAT, 431
—, amino acid composition and mol. wt., 431
—, and remnants, 459—460
—, an inhibitor of remnant uptake, 461
—, binding of phospholipids, 431
—, formation by the intestine, 450—451
—, occurrence in lipoprotein, 431
—, production in the liver, 454
—, transfer from HDL to VLDL, 456, 457
ApoC-II, 431
— as promoter of triglyceride hydrolysis, 461
—, deficiency, 434—435
— peptides, activation of protein lipase, 434
— and III, plasma levels, 473
ApoC-III, amino acid composition and mol. wt., 431
—, carbohydrate composition, 431
ApoD, 432
—, a cholesteryl ester exchange protein, 442
ApoE and remnants, 475
—, a promoter of remnant uptake, 461
—, binding to receptor, 478
—, composition, 432
—, formation by the intestine, 451
—, in hyperlipidaemia, 475
—, kinetics of metabolism, 475
—, LCAT deficiency, 474
—, occurrence in lipoproteins, 432
—, phospholipid complex, binding to receptor, 478
—, plasma levels, 475
—, production in the liver, 454
ApoC lipoproteins, catabolism and production, 473—474
— — in hyperlipidaemia, 473—474
— —, recycling, 473—474
— —, transfer from HDL to VLDL, 474
β-L-Arabinofuranose in glycoprotein linkage, 41, 46, 47

# SUBJECT INDEX

Aromatic residues in antigen-binding site, 237—238
Arginine-rich histones, 516
Arrangement of non-α globin genes, 385
Asialo-agalacto-glycoproteins, catabolism, 125
Asialo-glycoproteins, catabolism, 125
Asparaginylglucosaminuria, 130, 133, 140
Aspartylglycosylamine amidohydrolase, 33—34, 121, 123
Atherosclerosis, 482
— and lipoproteins, 420
Azulene in affinity labelling, 221

Balanced polymorphism, 285
Basophils, 248
Bence-Jones protein, 199
— —, X-ray structure, 230—231
BHK cell surface glycoproteins, 151
Biantennary glycans, 68
— —, 'bird-conformation', 82—84
— —, molecular model, 81—83
— —, T-conformation, 81—82
Biosynthesis of glycans, control by glycosidases, 99, 157
— — —, — — phosphodiesterase, 102
— — glycoproteins, control by substrate availability, 101
Bird conformation, 82—84
Blood group substances, early history, 21—22
— — —, megaglycan, 57
Blue whale tail oxygenation, 348
Body pools of immunoglobulins, 277
Bohr effect, 349—350
— —, acid, 373, 410
— —, alkaline, 372—373
Bromelain glycans, 76
Bromolacetyl derivatives in affinity labelling, 220
Bronchial mucus, 54—56
Butyration and histone acetylation, 527

Cancer, changes in glycan moieties, 148
Cap formation on lymphocytes, 269
Capping on lymphocytes, 268—269
Carbamino compounds, 374
Carbohydrate addition to antibodies, 254
— in bound and secreted immunoglobulins, 272
— incorporation into membrane-bound immunoglobulins, 273—274
Carbohydrate-peptide linkages in glycoprotein, 41
— —, evolution , 48—49
Carbon dioxide binding of haemoglobin, 374
Carboxypeptidase Y
κ-Casein, 55, 144
Catabolic rates of immunoglobulin classes, 276—277
Catabolism of glycans of the $N$-acetyl-lactosaminic type, 141
Cathepsins, role in glycoprotein catabolism, 120
Cell adhesiveness, 148—150
Cell cycle and phosphorylation of histones, 529, 533
Cell cycle and synthesis of immunoglobulins, 260—261
Cell-free synthesis, 253—254
Cell surface antigens, 261
Chain formation balance in Hb synthesis, 401—402
α- and β-Chain formation and control, 401
β-Chain, inefficient synthesis, 407
ε-Chain, 397—399
ʃ-Chains, 397, 399
γ-Chains, 397, 399
δ-Chains, 397, 399
μ-Chain differences, 271—272
Chicken oviduct, specific HMG protein, 544
Chinese hamster ovary cell glycoproteins, 61, 62, 71, 73
Chitosamine, 4

Cholesterol analogues, effects, 480
— feeding and mRNA for apoE, 455
Cholesteryl ester exchange protein, 432, 442
Cholestyramine, 468, 471, 472
Chondroitin, 8
Chondroitin sulfates, biosynthesis, 116—118
— —, early formula, 9
Chondroitin-4-sulfate, 58—59
Chondroproteins, 9
Chondrosamine, 4—5
Chondrosin, 8
Chorionic gonadotropin, 55
Chromatin-associated proteins, 509, 511
Chromatin structure, 508—510
Chromosomal proteins, definition, 507
Chromosome 6, presence of gene coding for antigen peptide, 296
Chromosomin, 508
Chylomicrons, effect of unsaturated fatty acids on site, 423
— as carriers of triglycerides, 422
—, physical properties and composition, 421
—, site, 423
Circular polarization in antigen-binding, 238
Clofibrate, 469
Clonal selection hypothesis, 251
CMP-NeuAc, location in nucleus, 94
Cold insoluble globulin, 151
— — — glycans, 66, 75
Collagens, 54, 144—145
Collagen, insertion of galactose and glucose, 118—119
Colostrum IgG, 65
—, transfer of antibodies, 246
Complement binding mechanism, 250—251
Complement fixation, 248—251
— — of immunoglobulins, 243
Complement CIg, 65, 66
Complex alleles, 285

Complex type structure, 51
Concanavalin A, reaction with glycogen as lectin, 29
Conformational changes on antigen-binding, 238
Conformation of glycans, 80—85
CIg binding to IgM, 250
CIg of complement, sugar-peptide linkage, 41, 46
Constant region genes, 262
Contact inhibition, 148
Control gene model, 285—287
Cores A, B and C, 50
Core glycan, block transfer, 111—112
Core particle, 510
Corticotropin, 145
Covalent assembly of antibodies, 254
Crossover of globin genes, 384—385
Cyanosis due to HbM, 390
Cysteine in histone H3, 516
Cytoskeleton, involvement in secretion of lipoproteins, 447—448

Delayed hypersensitivity to genes, 307
Deoxyhaemoglobin, 350
Dermatan sulfate, 59—60, 63
— —, biosynthesis, 116—118
Dextrans as antigens, 216
Diabetes and $HbA_{1c}$, 403
Diazonium compounds in affinity labelling, 217—220
Di-$N$-acetylchitobiose in inner core, 50
2,3-Diphosphoglycerate, 364
—, binding to haemoglobin, 374—376
Discoidin, 150
Disialyl groups, 70
Disorders of glycoprotein catabolism, symptoms, 128—129
Disulphide bridge formation in antibodies, 256
— — — in IgM, 259
DNA folding, 531

## SUBJECT INDEX

DNA for apoVLDL-II, 455
DNA interaction with HMG proteins, 542, 555
DNA polymerases, 563—564
Dolichol mono- and di-phosphate sugars, 95
— monophosphate-$\beta$-Glc, 95
— monophosphate-$\beta$-Man, 95
— pathway, 95—97
— pyrophosphate-$\alpha$-GlcNAc, 95
— phosphate synthesis, rate, 102
— diphosphate-$N$-acetylglucosamine, 141—142
Dol-P-P-oligosaccharide degradation, 122
Domain theory of immunoglobulins, 231
DNP-binding of IgA, 244
DNP-diazoketones in affinity labelling, 221
*Drosophila melanogaster* protein D1, 557
Dual recognition hypothesis, 303, 305
Duplication-deletion model, 285—287

Ectoglycosyltransferases, 150
Electron microscopy of antigen antibody complexes, 221—225
Electron spin resonance in immunoglobulin structural studies, 237—238
Endo-$N$-acetylgalactosaminidase, 33
Endo-galactosidase, 33—35
Endo-$N$-acetylglucosaminidases, 33—35, 122—123
Endo-$N$-acetyl-$\beta$-glucosaminidase, 131, 141
Endoglycosidases, 32—35
— in catabolism, 121—122
Endoplasmic reticulum in formation of lipoproteins, 445—447
$\beta$-Endorphin, 145, 155
Epiglycanin, 53, 54
Epitopes, 192
Erythrocytes, aging and desialation, 149

—, interaction with viruses, 149
Erythrocyte membrane glycans, 71, 73
Ester linkages (assumed) in glycoproteins, 40—42
Estrogen and lipoprotein mRNA, 458
Ether bonds (assumed) in glycoproteins-peptide, 42
Excision enzymes, 266—267
'Exit passport' hypothesis, 146—147
Exoglycosidases, 32
Extensin, 47, 53
—, biosynthesis, 119

Fab, 193—194
Fab and F(ab')$_2$ half-lives, 278—279
Fc, 193—194
Fc and Fc' catabolism, 278—279
Fc in binding of immunoglobulins to lymphocyte membrane, 270—271
Fc receptors, 268
Fc receptors and antibody binding, 246
Ferritin DNP derivative, 221
Fetuin, 55, 69
Fibroblast, lipoprotein physiology, 476—486
Fibronectin, 151
Flagella of *Salmonella* as antigen, 221, 229, 230
Flexibility of glycan structures, 84
Foetal Hb, hereditary persistence, 406
Foetal Hb (HbF($\alpha_2\gamma_2$)), 397, 399, 405
Foetal switch, 399
Fucose in antibody synthesis, 254
$\alpha$-L-Fucose linkage to peptide, 47
Fucosidosis, 133, 136, 137
Fucosylation and sialylation, 109
Fucosyltransferase, mode of action, 107—110
Fucose as recognition signal, 153

Galactoprotein receptor, 125

$O$-$\beta$-D-Galactopyranosyl(1 → 4)-hydroxy-L-proline, 47
Galactose addition in proteoglycan synthesis, 116—117
— attachment to antibodies, 254
$\beta$-Galactose-hydroxylysine linkage in collagen, 41, 46
$\alpha$-D-Galactose-L-serine linkage, 41, 45
Galactose transfer in mucin biosynthesis, 115
Galactosylation and sialylation in $O$-glycosylproteins, 115
Galactosyltransferases, mode of action, 109
Gamma ($\gamma$) chains, allotypes, 210, 211
Gastric mucin, 55, 56
Gene duplication with haemoglobin, 356
$\psi$ Genes, 409
Genes of the Hb chains, 396—397
Genetic control of immune systems, 280—281
Genetic variants of haemoglobin detection, 379
Globin ($\alpha$) genes, location on chromosome, 409
— effect on oxygen affinity, 360
— genome of *Xenopus laevis*, 409—410
— (non-$\alpha$) chains sequencing, 409
— synthesis control at transcription, 400—401
— synthesis control at translation, 400—402
Glucuronic acid in chondroitin sulfate, 4
Glucosamine, attachment to antibodies, 254
Glucose in collagen, 46
— removal from nascent glycopeptide, 111—112
$\beta$-Glucuronosidase, 128, 129, 132
Glucuronyltransferase in proteoglycan synthesis, 116—117
Glucuronosyl C-5 epimerase, 118
Glycans, history of classification, 23—26

—, linear, 49
—, maturation, 109—113
Glycan moiety, effect on membrane transport, 146
— —, — — protein folding, 143—145
— —, — — proteolysis, 145
— structures, foreseen but not yet discovered, 154—155
Glycans, sugar composition, 31
$n$-Glycans, 49
Glycine, replacement by another residue in Hb, 389
Glycoasparagines, 140
Glycophorin, 55
—, amino acid sequence, 90—92
—, glycan structure, 92
—, integration into cellular membrane, 91—92
Glycophorin, sialyl residues, 149
Glycoproteins and cancer, 20—21, 22
— as precursor glycogen, 12
— as recognition signals, 21, 22
—, birth of, 7
—, catabolism of, 22
—, definition, 1
—, degradation by hepatocytes, 125
—, distribution, 1
—, early work on biosynthesis, 18
—, ester linkages (assumed), 40—42
—, ether bonds (assumed), 42
—, functions, 2—3
—, isolation, 27—30
—, Levene's concepts, 9—14
—, tentative classification, 26—27
—, uptake by cell, 125
—, use of lectins, 28—29
Glycosaminoglycans, 58—60
Glycosidases, 32
— in lysosomes, 121
$N$-Glycosidic bond in glycoproteins, 42—43
$O$-Glycosidic linkages in glycoproteins, 41, 44—47
Glycosylation and translation, 105—106
Glycosyltransferases, acceptors, 100
— as receptors, 150

# SUBJECT INDEX

—, cation requirements, 97
—, cellular location, 98—99
—, general reaction, 92, 93
—, occurrence and properties, 97
— on the cell surface, 99—100
—, specificity, 98
—, topographical arrangement, 99, 157
Gly-gly peptide, part of protein A24, 558
GM 1-gangliosidosis, 133, 138, 130
GM 3-gangliosidosis, 133
Golgi apparatus and mucin synthesis, 115
— —, protein synthesis of lipoproteins, 447
Golgi in antibody synthesis, 254—255
Graft rejection, 279

H2 complex, H and D regions, 284
H2 polymorphism, 281—282
H2 systems, 280
Haem and globin synthesis co-ordination, 402
Haem—haem interaction, 349—350
Haemichrome I and II, 377—378
Haem inhibition of aminolaevulinate synthesis, 402
Haem interactions, 356—358
Haem ligands, 352
Haemoglobins, 390
Haemoglobin $\beta$-unit structure, 357
Haemoglobins, distribution over the globe, 386—387
Haemoglobin, evolution, 348, 407, 408
— F and DPG binding, 374
— glycosylation, 403
—, quaternary structure, 359—367
—, secondary structure, 354—356
—, spectra, 351—352
—, tertiary structure, 356—357
— tetrameric, 349
Haemolytic anaemia due to unstable Hb, 389
— disease of the newborn, 246
Haem pocket substitution, 388

— structure, 351
Hb Agenogi, 392
Hb anti-Kenya, 384—385
Hb anti-Lepore, 384—385
Hb Bart's ($\gamma_4$), 404—405, 410
Hb, $\beta$-chain, 384—385
Hb Bethesda, 393
Hb Bristol, 380
Hb Chesapeake, 393
Hb code, deletions, 383
— —, insertions, 383
Hb Constant Spring, 381, 405
Hb Coventry, 385
Hb Cranston, 382
Hb, $\delta$-chain, 384—385
Hb, fusion variants, 384—385
Hb Gower 1 and 2, 396—397
Hb Grady, 383
Hb Hammersmith, 388
Hb Hasharon, 387
Hb Heathrow, 393
Hb Hiroshima, 393
Hb Icaria, 381
Hb Kansas, 391—392
Hb Kenya, 384—385
Hb Koellickey, 386
Hb Korle-Bu, 387, 395—396
Hb Koya Dora, 381
Hb Lepore, 384—385, 406
Hb McKees Rock, 383—384
Hb Mosera, 380
Hb, non-genetic variations, 386
Hb, nonsense mutants, 383—384
Hb Philly, 390
Hb Portland, 373, 397, 399
Hb Rahere, 393—394
Hb Riverdale-Bronx, 380
Hb Savannah, 380
Hb Seal Rock, 381
Hb Sydney, 380
Hb synthesis, frameshift variants, 381—383
— —, polymorphism, 380
— —, termination errors, 381
Hb Tak, 382, 407
Hb Titusville, 391—392
Hb, tyrosine mutation, 383
Hb Volga, 389

Hb Wayne, 382
HbA$_{Ic}$ and HbA$_{1a}$, 403
HbA$_2$ ($\alpha_2\gamma_2$), 397, 399
HbC, occurrence, 386
HbD, occurrence, 386
HbE, occurrence, 386
HbE Saskatoon occurrence, 387
HbG Galveston occurrence, 387
HbG Philadelphia occurrence, 387
HbH (B$_4$), 404—405
HbK Woolwich, 407
HbM Boston, 391
HbM Hyde Park, 391
HbM Iwate, 391
HbM-Milwaukee, 380
HbM Saskatoon, 391
HbO-Arab occurrence, 387
HbS and malaria, 386—387
HbS, occurrence, 386—387
H-chain in RNA, 262
H-chain synthesis, 253
HDL (high density lipoprotein) and heart disease, 468
HDL, apoE, 424
HDL catabolism, location, 470
HDL cellular uptake, 484
HDL, conversion from discoid to spherical shape, 452, 456
HDL, discoid particles in LCAT deficiency, 441
HDL, disc particles, 452—453
HDL, effects of cholesterol pools, 485
HDL, effect of LPL, 469
HDL, gain of cholesteryl esters, 456
HDL, general scheme of metabolism, 456
HDL, heterogeneity, 469
HDL, physical properties and composition, 421, 424
HDL, protein composition, 424
HDL, receptor, 483
HDL-cholesterol, effect of carbohydrate feeding, 469
—, factors causing an increase, 469
HDL$_2$ and HDL$_3$, different effects, 484
HDL$_2$, specific effects, 456, 469

Heinz bodies, 389
Helical regions in $\alpha$ and $\beta$ chains, 354—356
Helper T cells, 310—312
Helper T lymphocytes, 251
Heparan sulfate, biosynthesis, 116—118
—, catabolism, 123—124
Heparin, 59—60, 63
— biosynthesis, 116—118
—, effect on release of lipase, 433, 435
— sulfate, 59—60, 63
Hepatic triglyceride lipase (HTL), physiological role, 436—437
Hepatoma, tryptophan pyrrolase, 340—342
Heterogeneity of glycans, causes, 80
Heterogeneous nuclear RNA (mRNA) for globin chains, 398
Heterotropic interaction, 375
High mannose glycans, 51
High mobility group (HMG) proteins, 511
— — — as proteolytic products, 535—536
— — — isolation, 536—537
— — — properties, 535
Hill equation, 370—371
Hinge region, carbohydrate linkage, 208, 209
Histamine release, 248
Histidine in antigen-binding site, 238
— and haemichrome formation, 378
Histidines, C-terminal in Bohr effect, 373, 393
Histidine in haemoglobin, 356—357, 368
Histocompatibility antigens, evolution, 299
— —, homology with immunoglobulins, 299—300
Histocompatibility gene maps in man, 283
Histocompatibility K or D-antigens, 303

# SUBJECT INDEX

Histone, variants in differentiation, 526—527, 532
Histones, amino acid composition, 516
— and DNA, 531—532
— and polyoma viruses, 514
— and SV40, 514
—, arrangement on nucleosome, 516—517
—, basic $N$-terminal regions, 517
—, blocking of terminal amino group by acetyl, 516
—, central and C-terminal apolar and helical regions, 517
—, classification, 508
—, conservation of structure, 512—514
—, free terminal amino groups, 515
—, general characteristics, 512
—, H1, H2A, H2B, H3, H4, 512
—, history of definition, 507—508
—, in unfertilized frog's egg, 541
—, microheterogeneity, 512—514
— of calf thymus, 512—514
— of *Tetrahymena*, 512—514
—, preparation, 515
—, stoichiometry in nucleosome, 510, 514
Histone octamer, 510
Histone-H1°, 513—514
—, and suppression of DNA synthesis, 533—534
Histone-H1 C-terminal basic region, 523—524
— heterogeneity, 523
— of calf thymus, 523—524
— of rabbit thymus, 523
— of trout, 524
—, variability of structure, 525
Histone-H2A, amino acid sequence, 521
—, heterogeneity, 522
—, molecular weight, 521
—, resemblance to H4, 521
Histone-H2B, amino acid sequence, 523
—, molecular weight, 523
Histone-H3, amino acid sequence, 519—521

—, carp, pea and thymus, 520
—, comparison of sequence in chicken and calf, 548
—, molecular weight, 519—521
—, cysteine(s), 519—520
Histone-H4, 514
—, amino acid sequence, 518—519
—, molecular weight, 518—519
—, strict conservation of sequence, 519
Histone-H5 as repressor of transcription, 533—534
—, relationship to H1, 525
—, structure and variation, 526
Histone-H6, 514
—, amino acid sequence of trout testis, 551
—, association with core particles of nucleosomes, 555
—, molecular weight, trout testis, 551,
—, N-terminal sequence of trout testis, 549
HLA-A(LA) and HLA-B genes, 280
HLA and HLB regions, 284
HLA antigens, association with $\beta_2$-microglobulin, 288, 292
— —, carbohydrate content, 288, 292
— —, hydrophobic domain, 288—290
— — in man, 294—295
— —, light chains, 292
— —, molecular weights, 288, 291, 292
— —, structures, 288—292
— —, transmembrane orientation, 291
HMG proteins, amino acid analysis, 538—539
— —, $\alpha$-helix content, 542
— — and transcription, 557—558
— — as DNA unwinders, 542, 558
— — -DNA complex, 557
— —, function in replication of DNA, 558
— — in different species, 543
— —, microheterogeneity and subfractions, 539—540

— —, molecular weights, 538
— —, N-terminal groups, 541
— —, possible functions, 556—558
— —, post-synthetic modifications, 541
— —, sequence homologies, 553
— —, tissue specificities, 543—544
— — of calf HMG1, HMG2, HMG14, HMG17, 538
HMG-like proteins of calf thymus, 546—547, 555
HMG-like proteins in invertebrates, plants and yeast, 545—546
HMG1 and HMG2, acetylation of lysine residues, 558
— — —, acidic C-terminal region, 553
— — —, changes of distribution with cell cycle, 556
— — —, fraction in cytoplasm, 557
— — —, fraction loosely bound to nucleosomes, 557
— — —, fraction tightly bound to nucleosomes, 556
— — — of calf thymus, N-terminal sequence, 552, 554
HMG-T and H6 in trout liver and testis, 544—545, 551—552
HMG14, association with core particles of nucleosomes, 555
—, from chicken erythrocytes, 547—548
— — —, N-terminal sequence, 548—549
HMG14 of calf thymus, N-terminal sequence, 548—549
— — — —, primary structure, 546—548
HMG17, association with core particles of nucleosomes, 555
HMG17-DNA complexes, 542
HMG17 of calf thymus, amino acid sequence, 548—550
HMG17 of chicken, N-terminal sequence, 549—551
HMG17, molecular weight and approximate sequence, 550
HMG17 region of high density of proline residues, 550

Homotropic interaction, 375
Hurler's disease, 128—129, 132
HTL, liver origin, 437
'Hyaloidins', 13—14
Hyaluronic acid, 116
— —, discovery, 14
Hybrid type structures of glycans, 52, 73—74
Hydantoin-like bond (assumed) in glycoproteins, 42
Hydrogen bonding in antigen-binding, 237
Hybridomas, 261
Hydroxy-L-proline in peptide carbohydrate linkage, 46—47, 63
3-Hydroxy-3-methylglutaryl CoA reductase, 476, 480
5-Hydroxy-L-lysine in peptide-carbohydrate linkage, 46
Hyperbolic oxygen saturation curves, 350
Hypercholesterolaemia, abnormality in apo-B metabolism, 467—468
—, apo-B plasma levels, 466
— (familial), mechanism, 476, 479, 481, 482
Hyperlipidaemia, 471, 472
Hypertriglyceridaemia, 486

Ia antigen functions, 310—311
Ia antigens in guinea pigs, 293
— — in mice, 292—293, 309
— —, structures, 296
Ice fish, haemoglobin lack, 348
I-cell disease, 126
Idiotypes, 212—214
Idiotypic specificity, 214
L-Iduronic acid formation in proteoglycans, 118
α-L-Iduronidase, 132
IgA (human), 67
— —, electron microscopy, 221—224
— —, general structure, 193—194
— —, J chain, 194, 260
— —, local synthesis, 245
— —, occurrence in secretions, 191
— —, of milk, hinge region, 57

# SUBJECT INDEX

— —, physicochemical properties, 190
— —, polymerisation, 260
— —, secretory piece, 195, 260
— —, — —, carbohydrate content, 208
— —, — —, synthesis, 195—197, 260
— —, transfer to the young, 245—246
IgA 1 and IgA 2 catabolism, 277—278
IgA 7S, 260
IgA 11S (dimer), 260
IgD on β-lymphocytes, 268, 270
IgD catabolism, 277—278
IgD function and occurrence, 191
IgD general structure, 199
IgD physicochemical properties, 190
IgD possible functions, 245
IgD synthesis in lymphocytes, 270
IgE allergic reactions, 191—192
IgE and immediate hypersensitivity, 245
IgE antibody and helminthic infection, 248
IgE catabolism, 277—278
IgE general structure, 199
IgE physicochemical properties, 190
IgG (human), 65, 66, 67
— —, digestion with papain, 193, 195
— —, — — pepsin, 193, 195
— —, disulphide bonds, 193—194
— —, electron microscopy, 221—229
— —, general structure, 193—194
— —, heavy chains, 193—194
— —, light chains, 193—194
— —, NMR, 38
— —, physicochemical properties, 190
— —, reduction with mercaptoethanol, 193
— —, subclasses, structures, 196
IgG1 and IgG3 as activators of complement, 249
IgG1 $C_H1$ and $C_H2$ domains, 143

IgG1 conformations Y and T, 143
IgG1 Fc fragment, 143
IgG1, stereochemistry of backbone and carbohydrate units, 234
IgM, position of carbohydrate, 198
—, the primary response, 245
—, structures, 197—199
IgM and flagella antigen, 229, 230
IgM antigen receptor, 257
IgM as activator of complement, 249
IgM catabolism, 277—278
IgM complex with antibody, 153
IgM electron microscopy, 225—228
IgM heavy chains, 198
IgM J-chain, 198, 258—259
IgM models, 240—241
IgM in mouse, 198
IgM myeloma glycopeptides, 74
IgM on β-lymphocytes, 268—270
IgM 19S pentamer, 257
IgM physicochemical properties, 190, 198
IgM polymerisation, 257—258
IgM-like protein, from the carp, 225, 228
— —, from the toad, 227
IgT, 273
Immune response (Ir) genes, 281, 307—310
Immunoglobulins, binding to cell membranes, 243
—, carbohydrate content, 190
—, — linkages, 208
—, catabolism, 274—279
—, — control, 278—279
—, constant regions, 199—206
—, $C_\gamma 1$ domain model, 238—239
—, $C_L C_\gamma 1$ domain pair model, 238—239
—, distribution ratio, 274—276
—, duplication of genes, 207
—, Fab electron microscopy, 221—223
—, Fab (mouse), structure, 233—234
—, Fc region electron microscopy, 221, 226
—, Fc stereochemistry, 234
—, Fc (human), structure, 233—234

—, fixation to complement, 191–192
—, fold, 231
—, framework regions, 206
—, glycosylation, 207–209
—, N-Glycosidic linkages, 208–209
—, O-Glycosidic linkages, 208–209
—, heavy chain classes and subclasses, 207
—, heavy chain classification $\alpha$, $\gamma$, $\mu$ and $\delta$, 207
—, hinge region, 152
—, hypervariable regions, 201–206
—, isotypes, 206
—, light chain types, differences between species, 206
—, model-building, 235–241
—, model of polypeptide chain, 232, 233
—, monoclonal, 199
—, occurrence in animals, 191
—, physical properties, 190
—, plasma half-life, 274–276
—, plasma pool, 274–276
—, serum concentration, 190
—, structures (see also under individual proteins, such as IgG, etc.), 192–215
—, sub-classes, 190
—, sub-groups of variable regions, 200–201
—, transmission to, 191
—, variability, genetic origin, 201
—, variable regions, 199–206
Influenza virus haem, 21–22
Inhibitor of globin synthesis, 402
Inner core, 49–50
*Inv* fraction of glycans, 50
$\alpha_1\beta_2$ Interface mutations, 393
$\beta$-$\beta$ Interface mutations, 393
Interferon secretion, 147
Intermediate density lipoproteins (IDL), absence from fasting plasma, 423
— — —, increase after alimentary lipaemia and heparin, 423
— — —, physical properties and composition, 421

— — —, presence in type III hyperlipoproteinaemia, 423
Intrinsic factor, 145
Introns in globin DNA, 398, 310
—, in V and C genes, 263
Invariant residues in globin, 408
Invertase, 143
I region, 281
Ir function, 309
Ischaemic heart disease, 471, 472, 485
— — —, and pre-beta-1 lipoproteins, 425
Isoglycans, 49
Isomaltohexase as antigen, 215
Iso-N-acetyl-lactosamine structures, 74
Isopeptide linkage in protein A24, 558
Isoprenols, 95–97

'J' segment, 265
'J' sequences, 266

Kappa ($\kappa$) chains, allotypes, 210, 212
Keratan sulphate, 59–60, 63
— —, biosynthesis, 116, 118
Kinases A, B and G in histone H1 phosphorylation, 528–529
Km antigenic determinants, 211

Lactosaminic structures, 51, 64–73
Lactotransferrin, 61, 65, 67, 71, 73
Lamprey haemoglobin, 350
Leader of precursor light chains, 263–264
LCAT, action on HDL, 452, 456
—, activation by A-I and apoC-I, 438
—, deficiency and lipoprotein-X, 427
—, —, biochemical findings, 439
—, —, clinical manifestations, 439
—, effect of apoA-II, 440
—, effect on metabolism of triglyceride-rich lipoproteins, 438–441

—, physiological role, 438—440
—, source, 437
—, specificity and preferred substrates, 438
Lectins, 20
—, recognition of polysaccharide sequences, 29—30
—, sugar specificity, 29—30
Leghaemoglobin, 349
*Lens escularis* lectin, 29
LETS, 150—151
Leukocyte agglutinins, 280
Ligase enzymes, 266, 267
Light (L) chain excretion, 255
L chain synthesis, 253
L chain in RNA, 262
Lima bean lectin glycan, 76
Linkage disequilibrium, 286
Linker region, 510
— — and cell function, 537
Lipid intermediates, discovery, 19
— —, in glycoprotein biosynthesis, 94—97
Lipids, possible intermediates in proteoglycan biosynthesis, 118
Lipid transfer between lipoproteins, 438—440
Lipoprotein lipase, action on VLDL, 456, 465
— —, activation by apoC-II, 434
— — activity, influence of diet and hormones, 436
— —, different forms, 433—434
— —, inhibition by protamine and salt, 434
— —, specificity, 434
Lipoprotein lipids as membrane constituents, 420
Lipoprotein-X in cholestatic liver disease, 426
— phospholipid, 426
— protein composition, 427
— unesterified cholesterol, 426
Lipoproteins and cholesterol esterification, 427—428
Lipoproteins, electrophoretic migration, 426
—, — mobility, 421

—, structure, 429
β-Lipotropin, 145
Liver, endoplasmic reticulum and synthesis of lipids, 451
—, production of chylomicrons VLDL and HDL, 444, 451, 455
—, RER, location of apoprotein synthesis, 452
—, SER, location of lipid synthesis, 452
—, synthesis of phospholipids, 451
Location of the Hb genes on chromosomes, 396—397
Lotus lectin, 29
LDL (low density lipoproteins), atherosclerosis, 485
—, binding, effects of hormones, 479
—, binding to plasma membrane, 476—478
—, general scheme of metabolism, 456
—, incorporation into lysosomes, 477, 486
—, internalisation, 479—480
—, physical properties and composition, 421
—, apoprotein B, 423
—, receptor binding, charge effects, 478—479
LDL catabolism, location, 463—465
LDL concentration in tissues, 482—483
—, secondary produkt of VLDL metabolism, 444, 448
—, secretion and formation, 462—463
—, secretion by liver, 452
LDL receptor, general properties, 476—479
— —, function in vivo, 482—483
— —, in coated pits, 479
— —, modification of lysine groups, 479
LDL uptake, effect on cholesterol synthesis, 480
— —, effect on LDL binding, 481

— —, (non-specific), 483—484
Lp(a) heritability, 425—426
Lp(a), carbohydrate content, 425
—, protein composition, 424—425
L-type structures, 52
Lymphocytes and L-fucose, 149
Lymphocyte antigen receptor, 267—274
— immunoglobulin binding to membrane, 270
Lymphocytes (B) and synthesis of immunoglobulins, 247, 251
— (B), receptors for Fc, 247
— (T) function, 247, 251
Lysine in antigen-binding site, 220
Lysine-rich histones, 516
Lysines and Bohr effect, 373
Lysosomal enzyme, intra-cellular transport, 126—128
— —, phosphate, 126
— —, uptake by fibroblasts, 126—127
Lysosomes, role in glycoprotein catabolism, 120—121, 123—125
Lysozyme antibodies, 238

Macrophage, 312—313
— and phagocytes, 247—248
Magnetic resonance and immunoglobulin structural studies, 237—238
Major histocompatibility complex, 279
Mannoproteins of *S. cerivisiae*, 64
Mannose in antibody synthesis, 255
— removal from nascent glycopeptide, 111—112
— -serine linkage, 41, 45
— -6-phosphate as recognition signal, 126—128
— -threonine linkage, 41, 45
Mannosidosis, 130—131, 133
—, unusual glycan, 77
Mannosyl-serine in brain glycopeptide, 63
Mannotriose in inner core, 50
Mast cells, 248
Megaloglycans, 73

Membrane-bound immunoglobulin biosynthesis, 273—276
Membrane lectins, 148, 150
Memory cells, 251—252
Mesenteric lymph and lipoprotein synthesis, 444—445
Messenger ribonucleic acid (mRNA) and antibody synthesis, 253
— for apoproteins, 455
— of $\alpha$- and $\beta$-globin chains, 398
Methaemoglobin, 377, 390
— reductase, 377
Methaemoglobinaemia, 377
Methylation of histones, 528
MHC, I region, 308
MHC of mouse, genetic maps, 281—283
MHC products, amino-terminal sequence data, 298
— —, allelic products homology, 299
— — and cytotoxic T-cells, 301—302
— — and graft rejections, 302
— —, experimental design, 297—298
— —, function, 301—302
— —, polymorphism, 299
— —, sequence analysis, 296—301
MHC restriction, 303
$\alpha_1$-Microglobulin, 66
$\beta_2$-Microglobulin and HLA antigens, 288, 292
$\beta_2$-Microglobulin sequence, comparison with that of immunoglobulins, 299—300
Microheterogeneity of glycans, 77—78
Milk fat globules, 54, 65
Milk oligosaccharide, early work, 16
Milk, transfer of antibodies, 246
Mini-chromosome, 514
Mitosis and H1 phosphorylation, 533
Mixed lymphocyte reaction, 284, 301
Mixed type structures, 52, 73—74
ML-type structures, 52, 73—74
Molisch reaction, 4
Monoclonal antibodies, 261

— antibody and antigen-binding, 244
Monocytes and phagocytes, 247—248
2-Monoglycerides, synthesis of triglycerides, 445
Monomethyl and dimethyl lysine in histones, 528
Monosaccharide structure of glycopeptides, 53
α-MSH, 145
M-type structures, 52
Mucin and mucin-like glycoproteins, biosynthesis, 113—116
Mucins and mucoids, 6—7, 9
—, rod-like conformation, 142—143
Mucopolysaccharidosis I—VII, 132
Multiglycosyltransferase (MGT) system, 19, 99, 157
Multiple binding sites, effects, 214—215
Multiple myeloma, depression of normal IgG, 278
Mutants of haemoglobin, single point, 379—380
Mutations of Hb affecting $O_2$ affinity, 390—391
— on the 2,3-DPG, binding site of Hb, 393—394
Myeloma glycopeptide, 72
— IgM, 61, 62
— proteins, 192, 199
— subtypes, mouse, 207
Myoglobin, 350
Myoglobins, antiserum, 216—217
Myosin in chromosomes, 563

Nephrotic syndrome, increase in HDL catabolism, 472
Neuraminidases in plasma membranes, 121
Neutrophils and antibodies, 248
Nicotinic acid, mode of action, 468, 471, 472
2-Nitro-azidophenyl (NAP) lysine, 220—221
NMR spectroscopy, use in glycan structured studies, 35

Non-histone chromatin proteins, 401
Non-histone chromosomal proteins, 511
Nucleolar protein component 10, 563
Nucleolar protein A24, amino acid analysis, 559
— — — and ADP-ribosylation, 559
— — —, histone H2A content, 558
— — —, structure, 558—560
Nucleolar protein BA, amino acid analysis, 559
— — —, molecular weight and occurrence, 562
Nucleolar protein C14, amino acid analysis, 559
— — —, molecular weight, 562
— — —, possible function, 562
Nucleolar protein C23, amino acid analysis, 560
— — —, molecular weight, 559
— — —, phosphorylation, 560
Nucleosome assembly factor from *Xenopus laevis* eggs, 562—563
Nucleotide sugars, degradation, 102
— —, location, 93
— —, stereochemistry, 94, 97

Oligomannosidic glycans, role, 152
— —, structures, 61—64
Oligomannosidic structures, 51, 63—64
Orotic acid blocks secretion of apo-B containing lipoproteins by liver, 454
Ovalbumin, 70
—, early work, 15
—, glycopeptides, 61, 72—73
—, glycosylation, 106
—, heterogeneity, 79—80
—, secretion, 147
Ovomucin, 55
Ovomucoid, early structures, 15—16
— (turtle-dove), 69
Ovotransferrin glycan, 70

Oxygen reaction with haemoglobin, 368—371
Oxyhaemoglobin, 350

(kappa)-Paracasein, 144
Paradoxical breathing of haemoglobin, 359
Paramecium, occurrence of haemoglobin, 348
Paramyxovirus SV-5 glycan, 76
— — —, unusual features, 76
Penta-antennary glycans, 68
Pepsinogen, 143—144
Peripheral heterogeneity, 77
$P_{50}$ value, 366
Phenylalanine and haem, 357
Phosphatases, 143
Phospholipid exchange protein, 443—444
Phosphorylation in histones, 528—529
— of histone H1, 528—529
Phosphorylcholine as antigen, 235
Phosphoserine in histones, 528
Phosphothreonine in histones, 528
Photoactivation in affinity labelling, 221
Phylogeny and globin structure, 408
Phytohaemagglutinin (PHA), 29
Pinocytosis in glycoprotein degradation, 125
— (reverse) and secretion of lipoproteins, 452
Placental transfer of immunoglobulins, 243
— transport of IgG, 246
Plasma cells, antibody production, 251—252
Plasmacytoma cells, antibody production, 252
Plasmacytomas, 192
Plasminogen, 66
Poly(ADP-ribose)-polymerase, 529
Poly(ADP-ribose) units attached to enchromatin, 529
— —, location in linker region, 530
Poly(A) tail in Ig mRNA, 264
Poly(A) tail of globin mRNA, 398

Polycythaemia due to high affinity Hb, 394
Polyglycosyl peptides, 70
Poly-lysine, 307
Poly-L-alanine as antigen, 215
Poly-L-lysine as antigen, 215
Polymorphism of genes, 280, 284
Poly-$N$-acetyl-lactosaminyl sequences, 70
Polyribosomes in antibody synthesis, 252—253
Postnatal development of tryptophan pyrrolase, in the rat, 338—340
— — — — —, and mRNA, 338—341
Prebeta-1 lipoproteins and glucose tolerance, 425
Precursor sequence, 253—254
Procollagen secretion, 147
Proline, replacement of another amino acid in Hb, 389
—, role in $O$-glycosylation, 103—104
Protamines, 507, 511
Proteases, role in glycoprotein catabolism, 120
Protein secretion, 146
Proteoglycans, 144
—, core protein, 89
—, link protein, 90
—, structure, 89—90
Proteoglycans and hyaluronic acid, 90
Proteolysis, effect of glycan conformation, 84
Prothrombin, 66
— glycans, 76

Radioactive derivatives in affinity labelling, 217
Rearrangement of V and C genes, 263
Recombination of genes, 266
Rejection of skin grafts, 280
Relaxed (R) state, 376—377, 392—393
Remnant association with LPL, 460

Remnant particles, composition, 459
— uptake by extrahepatic tissues, 462
— uptake by the liver, 460—461
Remnants as substrates for LPL, 460
Retina aggregation factor, 150
Retinol role in sugar incorporation, 97
Reverse cholesterol transport, 439
*Rhizobium*, 349
*Rhodopsin* glycopeptide, 72, 74
Ribonuclease antibodies, 238
Ribonucleases A and B, 144
Ribosomes, membrane-bound and antibody synthesis, 254
*Ricinus communis* lectin, 29
Rigidity of glycan structures, 84
RNA polymerase, 564

Salt bridges in haemoglobin, 365—367
Sanfilippo diseases A, B, and C, 132
Sandhoff's disease, 130, 133, 134
Schiff reaction, 4
Secretory IgA, 260
Secretory piece, carbohydrate content, 208
Self-recognition, 304—305
Sequon, 103—104
L-Serine in peptide-carbohydrate linkage, 41, 44—45
Serotonin release, 248
Serotransferrin, 66
Serotransferrin and ovotransferrin, 87
—, comparison with lactotransferrin, 86
—, differences in structure, 86
—, function, 86
— glycan structures by NMR spectroscopy, 35—36, 38
— glycans, 86
—, molecular weight and structure, 86—87
Serum albumin secretion, 147
Serum IgA and IgE, 65

Sialic acid and plasma protein clearance, 22
— — derivatives, 153
— — residue, role in degradation, 125
— — residue of antifreeze protein, 146
— — structure, 5—6
— — $\alpha$-(2→3) and $\alpha$-(2→6) linkages and NMR, 37, 39
Sialic acids, discovery, 5
— —, effect of point of substitution on conformation, 84
— —, electrostatic repulsion, 142—143
— — in antibody synthesis, 254
Sialidosis, 130, 131—134
Sialylation and fucosylation, 109
— and galactosylation in O-glycosylproteins, 115
Sialyl groups as recognition signals, 153
— transferases, specificity, 94
Sickle-cell haemoglobin, 394—396
Sickling process, 394—396
Sigmoid oxygen equilibrium, 349, 370, 372
Signal sequences in apoproteins, 455
Sindbis virus glycoproteins, 62, 66
Small intestine, production of chylomicrons, VLDL and HDL, 444—448
Somatic rearrangement of V and C genes, 262—263
Soybean lectin, 64, 157
Sponge-cell aggregation, 150
Stereochemistry of sugar transfer and incorporation, 97
Sublingual glycoprotein, 57
Submaxillary mucin, 53, 54, 56
— —, biosynthesis, 114
Substitution rules in glycoprotein glycans, 52—53
Subtypes of human $\lambda$ (lambda) chains, 207
Subunit interaction in haemoglobin, 361—365
Sulfamidase, 132

Sulfatases in lysosomes, 121
Sulfation in proteoglycan biosynthesis, 117—118
Superinduction, 333—336
Superoxide dismutase, 378—379

Tail peptide in membrane IgM, 272
Taka-amylase glycopeptide, 62
Tangui disease, 473
Target cell hypothesis, 147
Tay—Sachs disease, 132
T-cell antigen receptor, 272—273
T-cell macrophage interaction, 313
T-conformation, 81—83
Tetramer of Hb assembly control, 402
Tetra-antennary glycans, 68
Thalassaemias, 403—406, 410
α-Thalassaemia 1 and 2, 403—405, 410
β-Thalassaemia due to mutation of a lysine codon, 409
β-Thalassaemia ($\beta^+,\beta^\circ$), 405—406, 409
Thioglycosidic linkage, 47
Thiol groups on IgM, 258
L-Threonine in peptide-carbohydrate linkage, 41, 44—45
Thymus in self-recognition, 305
Thyroglobulin, structure, 87
— unit A, 61
— units A and B, 87
— unit B glycopeptide, 69
Thyroglobulin, unit C, 89
Transcription and histones, 533
— and histone acetylation, 527—528
— control of tryptophan pyrrolase genes, 342—343
Transport of membrane-bound immunoglobulins, 275
— of secreted immunoglobulins, 275
Triantennary glycans, 68
Triglyceride exchange protein, 443
— hydrolysis, 459
Trimming of nascent glycopeptide, 111—112

Tri-$N$-acetylchitobiose structures, 77
Tropomyosin in chromosomes, 563
Trout testis HMG-T, N-terminal sequence, 551—552
Tryptophan in antigen-binding site, 237
Tryptophan 2,3-dioxygenase mRNA, 331—333
— — —, increase after hydrocortisone, 333—334
— — —, induction by glucocorticoid hormones, 331—332
— — —, induction by tryptophan, 331, 336—338
— — —, induction in the developing rat, 331—332
Tryptophan pyrrolase, absence in hepatoma, 331—332
— —, induction, 331
T-state of haemoglobin, 365, 376—377, 392—393
Tubulin in chromosomes, 563
Tunicamycin, 137, 140—142
Turnover rate of membrane-bound immunoglobulins, 273
β-Turns and glycosylation, 105
Tyrosine in affinity labelling, 220
— aminotransferase, 335, 338
— in antigen-binding site, 237
— in haemoglobin, 369
— replacement of another residue, 390—391

Ubiquitin, and gene activation, 561—562
—, conjugated and free, 561—562
—, highly conserved structure, 561
— induces differentiation antigens, 560
—, part of protein A24, 60—61, 558—559
—, present in all cells, 560
UDP-D-glucose, Leloir's work, 18
UDP-$N$-acetylglucosamine, 141—142
Ulex lectin, 29
Untranslated sequences of globin mRNA, 398

SUBJECT INDEX

Valines, N-terminal in haemoglobin, 373
Var-fraction, 50—51
Variable (V) region genes, 262
V—C joining, 262
Very low density lipoproteins (VLDL), clearance in the rat, 456
—, effects in cell culture, 485—486
—, enzyme repression, 486
—, general scheme of metabolism, 451
—, physical properties and composition, 421
—, size, 423
Vesicular stomatitis virus, glycoprotein, 69
— — —, envelope, 144
— — —, formation, 111
*Vicia faba* lectin, 29

Vitamin K as antigen, 235, 236
V region synthesis, 262

Waldenström IgM J-chain, 66
Wax bean lectin, 29
Wolman syndrome, 480
X-ray crystallography, of haemoglobin, 350
— —, of immunoglobulin fragments and derivatives, 228—235
$\beta$-D-Xylose linkage in proteoglycans, 45
$\beta$-D-Xylosyl-L-serine, 45
$\beta$-D-Xylosyl-L-threonine, 45
Xylosylation in proteoglycan synthesis, 116—117

Yeast, occurrence of haemoglobin, 348